ESTATES GAZETTE LAW REPORTS

Estates Gazette Law Reports

1979

Volume 1

Series edited by
Barry Denyer-Green LLM PhD FRICS
of the Middle Temple, barrister

Cases edited by
J Muir Watt OBE MA
of the Inner Temple, barrister

Audrey Boyle BA
Publisher's editor

Estates Gazette

Routledge
Taylor & Francis Group
LONDON AND NEW YORK

EG Books is an imprint of Routledge
2 Park Square, Milton Park, Abingdon, Oxon OX14 4RN
711 Third Avenue, New York, NY 10017, USA
Routledge is an imprint of the Taylor & Francis Group, an informa business

ISBN for complete set of 2 volumes: 0 7282 0227 1
ISBN for this volume 0 7282 0228 X
ISSN 0951-9289

For information on all EG Books publications
visit our website www.routledge.com

CONTENTS

TABLE OF CASES

Titles of cases shown in bold type
Names of parties reversed shown in ordinary type

INDEX OF SUBJECT-MATTER

AGRICULTURE

COURT OF APPEAL

November 17 1978

(Before Lord Justice MEGAW, Lord Justice BROWNE and Lord Justice WALLER)

LUTON AND ANOTHER v TINSEY

Estates Gazette January 20 1979

(1978) 249 EG 239

Agricultural Holdings Act 1948, section 2—Appeal from decision of county court judge that the proviso to section 2(1) applied to agreement—Whether a "grass keep" agreement within proviso or a deemed tenancy from year to year within main part of subsection—Original agreement for grass keep clearly for less than a year—Transfer of benefit of agreement to new occupier—Evidence of arrangements in subsequent years—No change in nature of arrangement despite transfer and despite animals being kept in fields for full 12 months—Proviso to section 2(1) held to apply and order for possession upheld

In this case Donald John Tinsey, the unsuccessful defendant in an action for possession of 16 acres of land at Henroost Farm, Winterbourne, in the County of Avon, appealed against an order for possession of the land granted by Judge Russell at Bristol County Court in favour of the plaintiffs, Kenneth Luton and his wife, Mary Luton.

David Spens (instructed by Field Fisher & Martineau, agents for A E Smith & Son, Nailsworth, Gloucestershire) appeared on behalf of the appellant; C L Boothman (instructed by L Montague Harris & Co, Chipping Sodbury, Bristol) represented the respondents.

Giving judgment, MEGAW LJ said: This appeal from the reserved judgment of Judge Russell in the Bristol County Court delivered on May 16 1978 raises the question whether the defendant, Donald John Tinsey, does or does not have an agricultural tenancy of 16 acres of land at Henroost Farm, Winterbourne, in the County of Avon. The plaintiffs, Kenneth Luton and his wife, Mary Luton, who are the owners of the land in question as the result of a devise of it to them under the will of Mrs Mary Victoria Taylor, who died on May 20 1974, claimed in the county court proceedings that the defendant, Mr Tinsey, wrongfully refused to give up possession of the land to them. The defendant's answer was that he was entitled to possession under the provisions of section 2 (1) of the Agricultural Holdings Act 1948. Judge Russell, having heard a good deal of evidence on April 24 and 26 1978, decided the issue in favour of the plaintiffs. The defendant appeals.

Section 2 (1) of the Agricultural Holdings Act 1948 reads:

(1) Subject to the provisions of this section, where under an agreement made on or after the first day of March, 1948, any land is let to a person for use as agricultural land for an interest less than a tenancy from year to year, or a person is granted a licence to occupy land for use as agricultural land, and circumstances are such that if his interest were a tenancy from year to year he would in respect of that land be the tenant of an agricultural holding, then, unless the letting or grant was approved by the Minister before the agreement was entered into, the agreement shall take effect, with the necessary modifications, as if it were an agreement for the letting of the land for a tenancy from year to year: Provided that this sub-section shall not have effect in relation to an agreement for the letting of land, or the granting of a licence to occupy land (whether or not the agreement expressly so provides) in contemplation of the use of the land only for grazing or mowing during some specified period of the year. . . .

I need not read the rest of the proviso.

The question in the present case was whether an agreement which the defendant had with Mrs Taylor, the owner of the land up to the date of her death, was or was not within the proviso. It was "a licence to occupy land." But was it made "(whether or not the agreement expressly so provides) in contemplation of the use of the land only for grazing or mowing during some specified period of the year"? If it was not within the proviso, the defendant was entitled to succeed. If it was within the proviso, the plaintiffs were entitled to succeed.

The facts are all clearly set out in the judgment of Judge Russell. I do not think that his findings of primary fact are challenged, though the inferences or conclusions that he drew from them are challenged. If there is any challenge to his findings of primary fact, in my view those findings were all amply justified on the evidence. In so far as there was conflict between the evidence of the plaintiff, Mr Luton, and the defendant, Mr Tinsey, the judge preferred the evidence of Mr Luton. Mr Spens, who appears for the appellant defendant in this court, does not seek to persuade this court to take a contrary view in relation to the relative credibility of those witnesses.

Mrs Taylor had had an agreement with Mr Luton, year by year, for a number of years, starting in 1964, whereby Mr Luton had what is called a "grass keep" agreement in respect of the 16 acres. There is no doubt what "grass keep" means, at any rate in that farming community. There is no doubt as to the meaning to be attached to the phrase as it was used in evidence and documents in this case. There is no doubt that its meaning was well understood by Mrs Taylor, who had farming experience, by Mr Luton the plaintiff, by Mr Tinsey the defendant, and by Mr Heaven, a solicitor who acted for Mrs Taylor before her death. It was not a phrase which was known to, or understood by, Mrs Pearce, a lady who, when she gave evidence in the action in the county court—evidence given in fact in her own home—was 89 years of age. Her sight was bad, but the judge was satisfied that mentally she was alert. She was not familiar with agricultural matters and terms, as the others were. A "grass keep" agreement was an agreement under which the landlord gave permission—a licence, to put it in legal jargon—to the licensee to pasture his beasts on the land, and to mow the grass if he wished, and to take away the mown grass or hay. Apart from that, the licensee has no other rights in respect of the land by virtue of the licence. The other feature of a "grass keep agreement" which is of vital importance is that it is for less than one year. That this was the meaning in this community, and the relevant meaning for the purposes of this case, is shown by an answer that was given in cross-examina-

A tion by the defendant, Mr Tinsey, which answer is clear and unambiguous and final. The answer is: "I know grass keep agreement is for less than one year."

Now the reason why a "grass keep agreement" is for less than one year is quite simple. It is for the purpose of trying to ensure that the legal position of the licensee will be within the proviso in section 2(1) of the Agricultural Holdings Act 1948 and not within the body of that subsection. It is a perfectly proper way of avoiding the creation of a deemed agricultural tenancy under the body of section 2(1): but, to succeed, the requirements of the proviso must be shown to be satisfied. Nothing in this case turns on the burden of proof.

B The agreement which Mr Luton had had, year by year, for seven or eight years up to 1972, with Mrs Taylor, then the owner of the 16 acres, was a grass keep agreement under which, each year, the permission to Mr Luton to graze his beasts on the fields began on January 1 and ended on December 24. So it was for a specified period less than a year. Whether those annual agreements between Mrs Taylor and Mr Luton were in writing, I very much doubt. Whether, if they were oral, they were expressed solemnly and verbally each year with the date orally stated in the form of an express offer and an express acceptance, I very much doubt. But, written or verbal, express or implied, they were grass C keep agreements, with the last week of the year not included: they were so known and understood. This is not challenged. They were licences, and made "in contemplation of" the use of the land as set out in the proviso.

In the spring of 1972 Mr Luton had an illness and wished to give up or reduce his farming activities. He made an arrangement with Mr Tinsey, the defendant, which one hopes was contemporaneously approved by or on behalf of Mrs Taylor (but for present purposes it does not matter whether it was or was not), by which the unexpired portion of Mr Luton's grass keep agreement for the year 1972 was trans-D ferred to Mr Tinsey. Mr Tinsey thus became entitled to pasture his beasts on the 16 acres up to and including December 24 1972. Mr Tinsey in his evidence denied that he knew that Mr Luton's right in respect of the 16 acres was thus limited; but we need not trouble about that because the judge disbelieved him. The judge accepted, indeed, that there had been a written agreement between Mr Luton and Mr Tinsey purporting to effect the transfer of Mr Luton's licence from Mrs Taylor. In that written agreement the relevant grass keep term had been actually set out. That document has been lost.

Mrs Taylor, unfortunately, was seriously ill, suffering E from multiple sclerosis, and she was in and out of hospital for a considerable time before her death in May 1974. To anticipate in time, when she did die, in May 1974, by her will she left Henroost Farm to the plaintiffs, Mr and Mrs Luton. Mr Tinsey thereafter claimed to be entitled to remain in possession by reason of having become what I may call a deemed tenant from year to year under section 2(1) of the Act. Mr Luton disputed that claim. Hence this action.

I return now to December 1972, when Mrs Taylor had still some 16 months to live and when Mr Tinsey's grass keep agreement, taken over from Mr Luton in the spring of 1972, was coming towards its end on December 24 of that year. F Some time about December 1972, probably about December 7, Mr Tinsey, wishing to make further arrangements for the use of the 16 acres after the expiration of the current agreement taken over from Mr Luton, approached Mrs Pearce, the very old lady whom I have already mentioned. Mrs Taylor at the time was in hospital. Mrs Pearce had been a friend of Mr Taylor's mother, and was now a close friend of Mrs Taylor herself. She was very much older than Mrs Taylor. Mrs Pearce saw Mrs Taylor in hospital, and thereafter she conveyed an answer to Mr Tinsey. It is common ground that Mrs Pearce was authorised by Mrs Taylor to act as her agent in this matter, though no doubt there was no formal, verbal appointment as such. We do not have any evidence from Mrs Pearce as to what she said to Mr Tinsey, G but we have her evidence that Mrs Taylor said to her: "Tell Mr Tinsey that he can have the fields." Mr Tinsey gave Mrs Pearce a cheque for £180. That was a cheque made out to Mrs Taylor. Mrs Pearce sent that cheque on to Mr Heaven, Mrs Taylor's solicitor.

At the end of 1973—not surprisingly, Mrs Pearce has no clear recollection of dates with any precision—Mr Tinsey came to see her again—as she put it, "about the fields." He gave her another cheque, this time for £210. That cheque is dated November 8 1973 and it may therefore be some guidance, at any rate, as to the approximate date of this H meeting. According to Mrs Pearce, when she saw Mrs Taylor on that occasion about Mr Tinsey's request, Mrs Taylor said to Mrs Pearce: "Yes: he can have the fields. He is a very hard-working man and a very decent man." Mrs Taylor was in hospital at the time. Mrs Pearce also said in her evidence, as regards the 1972 meeting as well as the 1973 meeting, "Tinsey had the fields for 12 months," or, "Tinsey had the land for 12 months." I do not know what that means. It cannot be evidence of relevance, or, if relevant, of weight. It does not purport to describe something that was part of an agreement, written or verbal. Later in her evidence Mrs Pearce said this: "She" (that is Mrs Taylor) "said to me J 'You tell Mr Tinsey not to take anything out of the fields'." Then Mrs Pearce said "(or put anything in, was it?)—I'm sure it was 'take anything out'"; and then, going on with what she said Mrs Taylor said to her, "'and then he'll have them'." Then Mrs Pearce's evidence went on, "... and she" —that is, Mrs Taylor—"put her hands out. I don't understand farming matters. I didn't have a farming background."

Mr Spens for the defendant seeks to place very great weight on that passage in Mrs Pearce's evidence. He submits that the only meaning that makes sense of it, or at any rate the meaning that ought on balance to be given to it, is that Mrs Taylor was telling Mrs Pearce to tell Mr Tinsey not to K take his beasts out of the fields. The word "anything," in the phrase "not to take anything out of the fields," was intended by Mrs Taylor to refer to Mr Tinsey's animals. Mrs Pearce was, Mr Spens submits, being told by Mrs Taylor to advise Mr Tinsey to keep his animals in the fields that year after December 24. So, "he'll have them"—"them" being the fields. He will escape from the proviso to section 2 (1) of the Act and will obtain a deemed agricultural tenancy. The reason, says Mr Spens, why Mrs Taylor was thus inspiring Mrs Pearce to give advice or instructions to Mr Tinsey as to what he should do was because Mrs Taylor had at that L time made up her mind to leave the fields in question to Mr Tinsey's daughter, or to Mr Tinsey on trust for his daughter. There is a manuscript note in Mrs Taylor's writing which indicates that at some time, whenever it was, that was something which was Mrs Taylor's wish. Mr Spens submits that it was at about this time, and that Mrs Taylor's advice or instruction to Mrs Pearce as to what she was to advise or instruct Mr Tinsey to do was by reference to that intention that she then had of leaving in her will these 16 acres to Mr Tinsey's daughter, or to Mr Tinsey on trust for his daughter.

I am afraid that I am unable to accept this, any more than M the judge, who saw and heard the witnesses, was able to accept it, as a probable interpretation of Mrs Pearce's account of what Mrs Taylor said to her. But in any event, as Mr Boothman for the plaintiffs pertinently points out, what we are concerned with is not what Mrs Taylor said to Mrs Pearce, her agent, but what Mrs Pearce, the agent, said to Mr Tinsey. Now, of course, an inference might be drawn that Mrs Pearce, in recounting in evidence what Mrs Taylor told her to tell Mr Tinsey, was really seeking to give evidence of what she, Mrs Pearce, thereafter did say to Mr Tinsey and that it is, as it were, accidental that she had not expressed it in that form in giving her evidence. In some circumstances (and Mr Spens in his submissions this morning invited us to say such circumstances applied here) it would be legalistic, in the bad sense of that word, to refuse to draw such an inference. But in this case it seems to me that that will not do; and I suspect that it may have been for good and sufficient reasons that Mrs Pearce was not pressed in cross-examination with the obvious question whether or not she did say to Mr Tinsey what she said in evidence Mrs Taylor had told her to say. For when Mr Tinsey himself came to give evidence of this meeting with Mrs Pearce, Mrs Pearce having already given her account of it in evidence, Mr Tinsey's evidence was: "Second visit"—that is the 1973 meeting—"she" (that is Mrs Pearce) "said 'You continue on land'." There would then perhaps have been difficulties in the way of cross-examining Mrs Pearce so as to suggest to her that she had said something to Mr Tinsey which did not accord with the evidence which he himself was going to give as to his recollection of what was said at that meeting.

Mr Boothman, not unnaturally, lays stress on Mr Tinsey's version of the December 1972 and December 1973 conversations with Mrs Pearce. Though the judge did not accept Mr Tinsey's evidence when it conflicted with Mr Luton's, that does not prevent the plaintiffs from relying on those parts of Mr Tinsey's evidence which are not in conflict with Mr Luton's evidence and which tell in the plaintiffs' favour. Mr Tinsey's versions of the two conversations at the end of 1972 and 1973 are, that Mrs Pearce in 1972 simply said "Yes, all right; you stay on land," and in 1973, in the answer to which I have already referred, "You continue on land." The obvious inference is, Mr Boothman submits, that Mr Tinsey is being told by Mrs Pearce that he is to "stay," or "continue," on the same basis as previously: that is, the 1972 agreement, taken over from Mr Luton; which, as Mr Tinsey well knew, was a grass keep agreement, starting on January 1 and terminating on December 24.

I agree entirely with the reasons given by Judge Russell in his admirably clear and reasoned judgment. The judgment is criticised by Mr Spens, in what, if I may say so, was an admirably forceful and tenacious argument. The criticism is on the ground that the judge failed to find that fact which he would have had to find before he could properly hold that the proviso to section 2 (1) of the 1948 Act is applicable in this case. The judge says, in the concluding sentence of his judgment, "In my view the circumstances are such that the interest of the defendant did not amount to a tenancy from year to year, and the agreement in respect of the fields at Henroost Farm was never for any interest greater than the sale of the grass keep." That, says Mr Spens, is not enough. It does not matter that there is no tenancy. A licence will do. It is only if the licence is in contemplation of the use of the land "during some specified period of the year" that the proviso can operate. This licence, says Mr Spens, was not found by the judge to have been for a specified period of the year, and he asks, rhetorically, the question: if the judge did find that it was for a specified period of the year, for what specified period of the year was it? He submits that, on balance of probability, and particularly having regard to the evidence of Mrs Pearce as to what she was told by Mrs Taylor in December 1973, the licence was for a full 12 months, with cheques paid in advance in each case for 12 months' rent or licence fee, and with the intention or understanding or contemplation that Mr Tinsey would keep his beasts in the fields in accordance with the licence for the full 12 months, so that there would be no gap between one year and another and the licence would not be for a specified period of the year. The licence, Mr Spens submits, the judge ought to have held, both in 1973 and 1974, ran continuously, with no contractual provision for cessation for any day or days, and with no break between the annual licences.

It was not necessary, in my opinion, for the judge to express in words what quite plainly, by inference, was his finding of fact, and a finding of fact which had, at the least, ample, if not overwhelming, evidence to support it. That is, that what was agreed between Mrs Pearce, on behalf of Mrs Taylor, and Mr Tinsey was, on each of the two occasions, by implication, as was well understood by Mr Tinsey, a new agreement on the same terms as the agreement with Mr Tinsey which he had taken over from Mr Luton, the terms of which he well knew and which were in existence when he had his discussion with Mrs Pearce in early December 1972. That was an agreement with the specific period of the year, as it had been in each of the preceding agreements between Mrs Taylor and Mr Luton—that is to say, the specified period of the year being a period starting on January 1 and ending at midnight on December 24; so that it did not cover the period from December 25 to December 31.

The evidence in support of that conclusion is, in my view, irresistible. I referred earlier to Mr Tinsey's admission "I know grass keep agreement is for less than one year." In a letter which was written by Mr Tinsey to Mr Heaven, Mrs Taylor's solicitor, on January 9 1973—that is, after the 1972 meeting between Mrs Pearce and Mr Tinsey—Mr Tinsey wrote: "PS I would be pleased to receive the agreement to sign for the grass keep at Henroost Farm." Put together the answer in cross-examination and what Mr Tinsey says in that letter, and the result is as unchallengeable as that 2 and 2 make 4. Mr Tinsey, when he says that he had got an agreement for a tenancy, was asking for a written agreement which he knew was for less than one year and which he knew was for the period of Mr Luton's agreement. That was the agreement in writing for the grass keep at Henroost Farm.

I add only one other reference to the evidence. That is the evidence given by Mr Heaven, Mrs Taylor's solicitor, in examination in chief. Mr Heaven was referring to August 1974. He said: "Defendant spoke to me on phone. This would appear to have been in August 1974. Defendant said he had paid £210 for grass keep to Christmas 1974, and £200 for grass keep to Christmas 1973. Mrs Taylor had paid him £5 for cutting hedges. I am clear he was talking of grass keep: otherwise £5 for cutting hedges would not have arisen." That is what Mr Tinsey himself said to Mr Heaven in August 1974.

Mr Spens in his submissions in reply this morning amplified something that he had put earlier in his submissions. He suggested that, as the evidence was uncontradicted that in fact Mr Tinsey left his beasts on the fields in question between December 25 1972 and January 1 1973, and between December 25 1973 and January 1 1974, when you take that along with what Mrs Pearce said Mrs Taylor told her to tell Mr Tinsey, the conclusion must be that Mrs Taylor could not have succeeded in a claim against Mr Tinsey for trespass in respect of his keeping of the cattle on the land at any rate during the period at the end of 1973, and that, therefore, the conditions of the proviso were not satisfied. I do not accept that submission, for various reasons, of which one is, as I have already said, that I find myself quite unable to accept the interpretation which Mr Spens seeks to put upon Mrs Pearce's evidence as to what Mrs Taylor told her in December 1973.

Accordingly, I would dismiss the appeal.

Agreeing, BROWNE LJ said: In opening this appeal Mr Spens, for the appellant, made, as I understood him, two criticisms of the judge. First, he said that the judge wrongly thought that the question he had to consider was whether the defendant was occupying these fields under a tenancy or a licence, whereas the body of section 2 (1) of the 1948 Act applies both to tenancies and licences, subject only to the proviso. The second criticism was that to bring a licence within the proviso it must be for a specified period of less than a year, and the judge made no finding that the defendant's agreement was for such a period. Mr Spens said that the judge wrongly thought that a grass keep agreement was necessarily an agreement for a specified period of less than a year, which it is not, and that a grass keep agreement for a year or more would be within the body of section 2 (1) and not within the proviso.

I think that the answer to both of these criticisms becomes clear when one sees what the issue really was at the trial. The defendant's case as pleaded in his defence and as stated in his evidence was that Mr Luton had been a tenant and that he took over that tenancy. He said in his evidence " Luton told me that he was tenant. He said 'I have the tenancy of the land. I don't want it. I'm packing up.' " Mr Luton, on the other hand, in his evidence said: " I told the defendant it was grass keep; that I had no tenancy." On that point the judge specifically accepted the evidence of Mr Luton and rejected the evidence of Mr Tinsey.

It is now accepted, as I understand it, by Mr Spens that between 1964 and 1972 Mr Luton had a grass keep agreement falling within the proviso. That is, an agreement for a specified period of less than a year. He also accepts, as I understand it, that the defendant took over this agreement on the same terms from May to December 1972. The defendant's case now is that in late 1972 or, alternatively, in late 1973, there was a new agreement changing the previous arrangement with Mr Luton into a tenancy or licence for a year. The question being whether the defendant's position after 1972 was a continuation of the arrangement which there had been with Mr Luton or whether it was a new agreement on changed and different terms, it is not surprising that the judge did not specifically deal with the questions whether the new agreement was a licence or a tenancy, and whether it was for a period of less than a year. No written agreement between Mr Tinsey and Mrs Taylor has been produced, and the question what was the arrangement made between the defendant and Mrs Taylor through Mrs Pearce, her agent, in 1972 and 1973 was a question of fact for the judge. The judge found that from 1964 to 1972 Mr Luton had a grass keep arrangement falling within the proviso. It appears from the evidence that Mr Luton asked Mrs Taylor on various occasions to sell him the land or grant him a tenancy but that she always refused. In May 1972 the agreement between Mr Luton and Mrs Taylor was, with the consent of Mrs Taylor, transferred to Mr Tinsey. Quite plainly, at that stage the agreement with Mr Tinsey was on the same terms as the agreement with Mr Luton had been. As I have said, the judge accepted Mr Luton's evidence that he actually told Mr Tinsey that the agreement was for the purpose of grass keep and he did not have a tenancy. From May until December 1972 Mr Tinsey's agreement with Mrs Taylor was plainly within the proviso.

My Lord has already dealt with the evidence as to the arrangement made in late 1972 between Mr Tinsey and Mrs Pearce on behalf of Mrs Taylor. I add a few words about that, at the risk of some repetition. The story begins with a letter from Mr Heaven to Mr Tinsey, in which it is said: " We are acting for Mrs Taylor and we understand you are now in possession of her fields forming part of Henroost Farm. Mrs Taylor instructed us to prepare the necessary agreement. We understand you purchased the grazing for a period of less than 12 months. We understand the payment is £200 and that you will keep the hedges in order. If you will kindly confirm this is so and let us have a note of your full name we will prepare the necessary agreement." Mr Tinsey said that he never received that letter. But, as my Lord has said, on January 9 1973 after the arrangement, whatever it was, which he had made with Mrs Pearce, he wrote the letter to which my Lord has already referred, in which he said: " I would be pleased to receive the agreement to sign for the grass keep at Henroost Farm." Mr Tinsey said in evidence that in that postscript he was referring to something that had been said on the telephone. I do not think that was put to Mr Heaven in cross-examination; I may be wrong. But I am quite prepared to assume that if there was a telephone conversation between Mr Heaven and Mr Tinsey before Mr Tinsey wrote that letter, Mr Heaven would have repeated what he said in his letter.

Be that as it may, Mr Tinsey's case at the trial was that by the time he wrote that letter of January 9 1973 he was already a tenant of the fields. He said: " I realised at end of November 1972 and early December that I was tenant." If that is true it is really incredible that he should have written that letter of January 9. He also said, as my Lord has already said, " I knew grass keep agreement is for less than one year." Mrs Pearce's evidence about what she said to Mr Tinsey in November or December 1972 is really completely ambiguous. She said: " I told Mr Tinsey Mrs Taylor said 'Tell Mr Tinsey that he can have the fields'." The defendant's own evidence (to which also my Lord has already referred) was: " I asked Mrs P."—that is Mrs Pearce—" if I could continue to have land. She said 'I'll see Mary'. I went and saw Mrs Pearce again, perhaps a week later. She said she had seen Mary: she says 'Yes all right, you stay on land'."

So far as the 1973 arrangement is concerned there was no evidence, I think, from Mrs Pearce of what she had said to Mr Tinsey, although there was the evidence, to which my Lord has already referred, of what Mrs Taylor had said to Mrs Pearce. As my Lord has pointed out, there is no evidence that that was passed on to Mr Tinsey. The defendant's own evidence about the 1973 arrangement was that Mrs Pearce simply said " You continue on land." Finally, there is the evidence of Mr Heaven about the telephone conversation in August 1974, to which my Lord has already referred.

As I have said, the question was, what was the arrangement made in late 1972 and late 1973 between the defendant and Mrs Pearce on behalf of Mrs Taylor, and that was a question of fact. Was the arrangement on the same terms as the arrangement with Mr Luton, or was there a change in the arrangement so that the new arrangement was on different terms? I agree with my Lord that there was ample, and indeed overwhelming, evidence on which the judge could find, as he did, that the arrangement was on the same terms as the arrangement with Mr Luton, that it did not amount to a tenancy from year to year, and was within the proviso. For these reasons, which are I think really the same as the reasons which my Lord has already given, I agree that the appeal should be dismissed.

Also agreeing, Waller LJ said: I will just summarise the points which seem to me to point irresistibly to the conclusions at which the judge arrived—points which arise either on the evidence of Mr Tinsey or on evidence that was given as to something he is said to have said.

At the trial, Mr Tinsey was suggesting that there was a tenancy, and that there had been a tenancy even before he got it. The learned judge said that when the transfer came from Mr Luton to Mr Tinsey " the defendant said that Mr Luton did not tell him at that time of the nature of his interest, but I accept Mr Luton's evidence that defendant was told by Mr Luton that the agreement was for the purchase of grass keep and that Mr Luton did not have a tenancy."

My Lords have already drawn attention to the evidence that Mr Tinsey knew that " grass keep " meant something less than a year; and my Lord, Megaw LJ, has already drawn attention also to the statement made by Mr Tinsey to Mr Heaven, partially supported by the documents, that for both 1973 and 1974 he had been paying for grass keep. There is one other piece of evidence which it may be throws light on exactly what was happening in Mr Tinsey's mind, namely, the evidence of Mrs Luton. Mrs Luton's evidence, I should say, was accepted by the learned judge if it conflicted with Mr Tinsey's. She said this: " The defendant came to the door one day and we discussed Mrs Taylor's death." At that stage Mrs Luton did not know that they were going to inherit the

A farm. "He said 'I am going to try to claim an agricultural tenancy on the land, whoever it's left to'." Mrs Luton frankly said that she did not know who it had been left to, and said "I am afraid I said 'I should'; but I did not know it had been left to us at all." It seems to me that that clearly indicates a situation in which Mr Tinsey knew perfectly well that he had got grass keep only, and that he was going to try to establish that it was not merely grass keep for a limited period but was for a full period of the year. In my judgment the grass keep agreement was the same in 1973 and in 1974 as it had been every year since 1964, namely, from January 1 to December 24.

B The appeal was dismissed and the appellant ordered to give up possession on December 24 1978 and to pay £800 by way of mesne profits. Provisional order made for payment of costs out of Legal Aid Fund.

COURT OF APPEAL

November 17 1978

(Before Lord Justice STEPHENSON, Lord Justice GEOFFREY LANE and Lord Justice BRANDON)

SHORT BROS (PLANT) LTD v EDWARDS

C Estates Gazette February 10 1979

(1978) 249 EG 539

Agricultural Holdings Act 1948, section 2—Whether a series of transactions over a period of years resulted in a licence for grazing for a specified period of the year or a tenancy from year to year—Succession of written agreements expressed to be for grazing for six months only in each case—Oral agreement before signature of first written agreement that, despite its wording, the licensee could remain in occupation for several years—In fact he used land for mixed farming for some 13 years and latterly D integrated the farming with his own neighbouring farm—Occupier given notice to vacate the land on the expiry of the last written agreement—Deputy circuit judge's view that final written agreement resulted in surrender of yearly tenancy and restoration of grazing licence rejected by Court of Appeal—Occupier remained, by virtue of section 2(1), a tenant from year to year

This was an appeal by Mr A A Edwards, the occupier of land forming part of Tredomen Farm, Ystrad Mynach, Nelson, Mid-Glamorgan, from a decision by Deputy Circuit Judge J Q Evans, sitting at Caerphilly County Court, in an action in which Short Bros (Plant) Ltd, the plaintiffs, E succeeded in obtaining an order for possession of the land against Mr Edwards as defendant. In addition to occupying the land which was the subject of the action Mr Edwards owned a neighbouring farm known as Wern Isaf which lay south of the road between Nelson and Ystrad Mynach. The facts are fully set out in the judgment of Brandon LJ.

M T Pill QC, V Pugh and N Cook (instructed by O H Parsons, agents for Rees, Currie-Jones, Davies & Evans, of Cardiff) appeared on behalf of the appellant; Derek Wood QC and C Priday (instructed by Robbins, Olivey & Lake, agents for Burges, Salmon & Co, of Bristol) represented the F respondents.

Giving the first judgment at the invitation of Stephenson LJ, BRANDON LJ said: This is an appeal from an order made by J Q Evans, sitting as deputy circuit judge in the Caerphilly County Court on May 26 1978. By that order he adjudged that the plaintiff company was entitled to possession of a farm known as Tredomen Farm, Ystrad Mynach, Nelson, Mid-Glamorgan, and in addition to damages at the rate of £1,400 per annum from December 1 1977 until the date on which possession should be given.

The parties in the action were these. First, the appellant, who was the defendant in the action, A A Edwards. He owns a farm called Wern Isaf, which lies south of the road G between Nelson and Ystrad Mynach in Mid-Glamorgan. He is about 55 years of age and has been farming all his life. I shall refer to him as the farmer. Second, the respondents, who were the plaintiffs in the action, Short Bros (Plant) Ltd, a company having its head office at Taffs Well, near Cardiff, and formerly at Hirwaun, Aberdare, in Mid-Glamorgan. They are plant hire and haulage contractors and have two principal shareholders and directors, J G Short and V Short, who are brothers. I shall refer to the respondents as the owners, and to J G Short and V Short together as the Short brothers.

Tredomen Farm, the land to which the action related and H the possession of which was the subject matter of it, consists of a farmhouse, certain other buildings and some 93½ acres of land. Its western edge adjoins the farmer's own farm, to which I have referred, Wern Isaf Farm.

The court in this case has the advantage of a very full and careful judgment given by the deputy circuit judge, in which he has set out, with great clarity, both his findings of fact and the conclusions of law reached by him on those findings. The quality of the judgment below in these respects has given the court much assistance in dealing with what is by no means an easy case.

The material facts, as found by the deputy circuit judge, J are these. On November 25 1963 Tredomen Farm was conveyed to the Short brothers jointly. They had bought it as a speculation in the hope that, not immediately but after some years, they would be able to obtain planning permission for the building of dwelling-houses on the land. At the time of the conveyance, the farmer, by arrangement with the previous owners, was grazing some of his cattle on part of the land. After the conveyance he wished to make a fresh arrangement with the new owners for the use of the land by him. He arranged by telephone a meeting with Mr J G Short and at that meeting Mr J G Short made three statements. K First, that the farmer would only be allowed to use the land at Tredomen Farm for grazing purposes. Second, that there would be a succession of written agreements in respect of such use, each of which would be for a period of six months only. Third, that there was no reason why such an agreement should not go on for several years. As a result of these statements, both J G Short and the farmer contemplated that the land would be used for grazing only and that the farmer would remain in occupation of the land for several years.

Following that meeting, a form of written agreement was prepared by solicitors for the Short brothers and sent to the farmer for signature by him. That was dated December 1 L 1963. Its main provisions were these. First, that the farmer should have grazing rights over Tredomen Farm from December 1 1963 to May 31 1964. Second, that the farmer should pay a sum of £162 15s for such rights. Third, that the farmer should not use the land for any purpose other than grazing. The written agreement contained certain other provisions, but I do not think it is necessary that I should refer to them. The farmer signed this form of agreement in reliance on the assurance given to him by J G Short at the previous meeting that he would be able to occupy the land for several years. Subsequently, from 1964 to 1977, the farmer occupied Tredomen Farm. He did not use the land M for grazing only, but for mixed farming purposes, including the growing of various crops. From 1968 onwards at least he farmed Tredomen Farm with his own farm as a single agricultural unit. The Short brothers took no interest in the use to which the land was being put, and did not know that it was being used for mixed farming purposes, rather than for grazing only, until August 25 1977.

During this period of over 13 years a succession of further written agreements in the same form as the original agreement dated December 1 1963 were sent to the farmer and signed by him. Each agreement related to a six months' period, so that, if there had been one in respect of every such period there would have been some 28 in all. In fact, due to oversights of one kind or another by one side or the other, there were only 23 such written agreements. In particular, no written agreement was signed for the period from June 1 to November 30 1976. The last agreement to be signed was one dated June 1 1977, relating to the period from June 1 1977 to November 30 1977. During this period of 13 years the farmer paid a regular sum to the Short brothers every six months, both when there was a signed agreement relating to the period concerned and also when there was not. The amount of this sum was increased from time to time from the original sum of £162 15s. It went up to £170.89 in June 1973 and to £187.96 in June 1974. Subsequently it remained at the latter figure.

In July 1977, after the last agreement to which I have referred had been signed, the Short brothers decided to sell Tredomen Farm. They instructed agents to effect the sale for them by auction. On September 15 1977 the agents wrote a letter to the farmer in which they asked him, in effect, to vacate the land at the end of the six-month period covered by the last written agreement, that is to say to vacate it after November 30 1977. They also referred to the possibility of the farmer himself buying the land, but that suggestion did not come to anything. They enclosed with the letter a form which they hoped that the farmer would sign, and that form read in this way: "Dear Sirs, Land at Tredomen. I write to confirm that, in accordance with the terms of the Grazing Licence, entered into by me with you on June 1 1977, in respect of the Tredomen Land, I agree that by December 1 1977 I will remove all my stock implements and chattels from the land, and give you vacant possession of the land." The farmer consulted solicitors about his position and on September 29 1977 they wrote a letter to the Short brothers' agents in which they said, in effect, that he had been on the land for 14 years, that he was there as the holder of a normal agricultural holding, and that he had the right to remain there by virtue of the Agricultural Holdings Act 1948. On the same day as that letter was written, September 29 1977, the Short brothers conveyed the land to the owners. This change of title is not significant and was presumably effected by some fiscal or similar reasons. On January 12 1978 the owners began the action for possession against the farmer in the Caerphilly County Court.

The Agricultural Holdings Act 1948, provides by section 2(1) as follows:

> Subject to the provisions of this section, where under an agreement made on or after the first day of March, nineteen hundred and forty eight, any land is let to a person for use as agricultural land for an interest less than a tenancy from year to year, or a person is granted a licence to occupy land for use as agricultural land, and the circumstances are such that if his interest were a tenancy from year to year he would in respect of that land be the tenant of an agricultural holding, then, unless the letting or grant was approved by the Minister before the agreement was entered into, the agreement shall take effect, with the necessary modifications, as if it were an agreement for the letting of the land for a tenancy from year to year: Provided that this subsection shall not have effect in relation to an agreement for the letting of land, or the granting of a licence to occupy land, made (whether or not the agreement expressly so provides) in contemplation of the use of the land only for grazing or mowing during some specified period of the year, or to an agreement for the letting of land, or the granting of a licence to occupy land, by a person whose interest in the land is less than a tenancy from year to year and has not by virtue of this section taken effect as such a tenancy.

On the primary findings of fact set out above, there were as it seems to me, three questions upon which the deputy circuit judge had to reach conclusions of law. The first question was: what was the effect of the original transaction or set of transactions between the farmer and the Short brothers, which took place either at the end of November or the beginning of December 1963, apart from the application of the Agricultural Holdings Act 1948? The second question was: how was the effect of that transaction or set of transactions altered by the application of that Act? The third question was: what was the effect, if any, of subsequent transactions between the parties?

With regard to the first and second of these questions the judge reached the following conclusions. First, that there was a collateral oral agreement made at the meeting between the farmer and J G Short before the first written agreement dated December 1 1963 was signed.. Second, that the agreement of December 1 1963 was signed by the farmer in reliance on that collateral agreement. Third, that the agreement of December 1 1963 was not a genuine agreement so far as the period for which grazing rights were being granted, namely the period of six months, was concerned, because the oral agreement between the parties was for a period of several years. Fourth, that, on the footing that grazing rights were granted for a period exceeding one year, the effect of section 2 of the Agricultural Holdings Act 1948 was to create a letting from year to year between the Short brothers as landlords and the farmer as tenant.

With regard to the third question, the judge held this. First, that a time came, certainly by June 1 1977, and probably earlier, when the collateral agreement, under which the period for which grazing rights were being granted was for several years, was discharged by performance. Second, that in these circumstances the signing by the farmer of a further written agreement dated June 1 1977 constituted, by operation of law, the surrender by him of the letting from year to year created by the Agricultural Holdings Act. Third, that, as a result of that surrender, the farmer's rights were limited to those granted by the written agreement dated June 1 1977, that is to say were limited to grazing rights for six months until November 30 1977, after which date, if he remained on the land at all, he did so only as a trespasser.

Having reached those conclusions of law on the questions which I have set out, he adjudged that the owners were entitled to possession and also to damages for continued occupation after November 30 1977.

I propose to examine the three questions which I have mentioned again and see to what extent I feel able to agree with the view which the deputy circuit judge formed upon them. The most crucial of these questions seems to me to be the first: what was the effect of the original transaction or set of transactions between the farmer and the Short brothers apart from the application of the Agricultural Holdings Act 1948? The learned judge, as I have said, found that it was contemplated by both parties that the farmer would occupy the land for several years, and although he was not prepared to put a figure on the term of years concerned, he concluded that what was contemplated was a period substantially less than the 13 years which had elapsed by 1977. The question is whether that conclusion was the correct one on the primary facts which he found. When the Act of 1948 talks about what is in contemplation of the parties, it seems to me that it must refer to something which is sufficiently certain to be capable of being given some effect to in law. I do not think that it is possible to grant a lease or to grant a licence which, by its terms, is for an indefinite period of years. It seems to me that is something which cannot be done as a matter of law. And, since that cannot be done, I do not think that, for the purposes of section 2(1) of the Act of 1948, the parties can be held to have contemplated that it will be done.

It seems to me that the right conclusion of law on the primary facts found by the deputy circuit judge with regard to the original set of transactions is this. The farmer was going to get a series of six months' periods on the land which in total would amount to at least three years, because I do not think that any lower figure could be described as

A "several" years, and that thereafter he would continue to have further six months' periods in succession until he was given reasonable notice to the contrary. It would be possible for the owner of land to grant a licence in such terms, and, if it would be possible for him to grant a licence in such terms, it would be possible for the parties to contemplate, within the meaning of section 2(1) of the Act of 1948, that that was what was going to happen. It seems to me that, on a true view of the two transactions—that is the oral meeting before the written agreement was signed and the written agreement itself—that was what was in contemplation I would therefore, to that extent, disagree with the view B of the deputy circuit judge that what was in contemplation was a term of years substantially less than thirteen which ended automatically when, whatever the number of years concerned was, they had elapsed. As to the second question, there is no dispute about the effect of the Act, and I agree that, since what was contemplated involved occupation of the land for more than a year, the transaction or set of transactions took effect as a letting from year to year under the Act.

Then I come to the third question: what was the effect, if any, of subsequent transactions between the parties? The deputy circuit judge came to the conclusion that there had been a surrender by operation of law because, by the time C the farmer came to sign the last agreement, or possibly earlier, the contemplated period of years had elapsed and therefore the effect of the oral agreement which was collateral to the first written agreement had been spent. He considered that, by the time the farmer came to sign that last agreement, he must be taken to have been aware that what was originally contemplated no longer applied, and that, by signing that agreement, he impliedly surrendered the Agricultural Holdings Act tenancy which he had.

I do not think that there is any dispute or question in this case as to the doctrine of surrender by operation of law. The only question is whether it applies on the facts of this D case. It seems to me that these written agreements were all subject to the same vice, that is to say, in so far as they provided that the grazing rights were to last only for a period of six months, they were not genuine. It seems to me that that analysis of them is no less applicable to the last agreement signed on June 1 1977 than to the first agreement signed on December 1 1963. The real arrangement between the parties, apart from the Act, was that the farmer would occupy the land for at least three years and thereafter until his occupation was terminated by reasonable notice. No such notice was ever given by the Short brothers to the farmer; E indeed, the evidence is that the instructions to sell the farm were not given until July 1977, and it seems to me, therefore, that what had been contemplated originally at the beginning of December 1963 was still being contemplated on June 1 1977.

Now it was conceded by Mr Wood, and very properly in my view, for the owners that, if the last agreement, that is to say the agreement of June 1 1977, did suffer from the same vice as the first agreement, in that it was not genuine so far as it provided for occupation and grazing rights for a period of six months only, then the signing of it could not operate as a surrender by operation of law. The principle involved F in surrender by operation of law is that a tenant who has one grant from an owner of land and who then accepts a second and inconsistent grant must be taken to have surrendered the first grant because, by taking the second grant, he is accepting that the owner can make it and the owner can only make it if the first grant had been surrendered. But the principle cannot apply when the second grant is itself a grant which is not genuine, and, on the analysis which I have made of these matters, the second grant, which in this case is the so-called "licensing" agreement of June 1 1977, was in one crucial respect not a genuine grant. The signing of it, therefore, would not, in my view, result in the surrender by operation of law of the existing grant, namely the statutory G letting from year to year under the Act of 1948.

For these reasons, I have come to the conclusion that, full and careful as was the judgment of the judge below, he arrived at the wrong result. I have come to the conclusion that the statutory letting which was created by the original set of transactions continues, and, that being so, the farmer was entitled to remain on the land after November 30 1977 as the tenant under that letting.

It would, in my view, have been possible for the Short brothers to have given the farmer reasonable notice, which would have been of at least six months and possibly more, before they required him to sign a further written grazing H agreement, that this time the agreement was not like the previous ones, an agreement which was not genuine, but an agreement which was intended to be genuine in all its terms. If they had given such notice to the farmer, and if he, being aware of the true position, had then signed that further agreement, different considerations would apply, and it may well be (it is not necessary to decide the point) that there would have been a surrender by operation of law of the statutory letting from year to year. No notice of that kind was given, however, and accordingly, in my judgment, there was no such surrender. In my opinion, therefore, this appeal should be allowed, the judgment of the deputy circuit J judge should be set aside and there should be substituted for it an order dismissing the claim.

Agreeing, Geoffrey Lane LJ said: There is nothing intrinsically wrong or objectionable in a series of genuine six months' grazing licenses being granted to a farmer, and the fact that there is such a series, or indeed that it is the owner's intention in his own mind, at the outset, to grant such a series will not, on its own, prevent the licence remaining a licence without being converted into a tenancy from year to year by section 2 of the 1948 Act. That much is clear from the decision of this court in *Scene Estate Ltd v Amos* [1957] K 2 QB 205.

In this case, however, the effect of the judge's findings seems to me to be as follows (although it is not the way in which the judge himself expressed those findings). Before the first of the forms of agreement here were signed on December 1 1963 the owner agreed with the farmer that the farmer would have a licence to graze, and would not be disturbed for what was in effect three years certain and thereafter until the owner chose to tell him to go upon reasonable notice or perhaps upon six months' notice. This, despite the fact that the written document said otherwise. Consequently, the document, as the learned judge has found, was not a L genuine record of the true agreement between the parties. It was without doubt the owner's intention that similar forms of agreement should continue to be signed by the farmer ostensibly, but not genuinely, limiting the licence period to six months.

As I said, I do not accept that the learned judge's interpretation of those facts is correct. What he said reads as follows:

> The collateral agreement was that in return for the defendant signing the written agreement the Short Bros would allow him to remain in occupation for several years. The defendant did not then expect or contemplate that he would remain in occupation M forever, and for more than a limited number of years. That collateral agreement was discharged by performance after the Short Bros had allowed the defendant to remain in possession for the appropriate number of years. I shall not attempt to define the number of years meant by the words "several," "few," "some," or "many," but I am satisfied that both J G Short and the defendant considered them to mean fewer than the thirteen and a half years that had elapsed prior to the written agreement of June 1 1977. Therefore by that date, and even before it, the written agreement was unfettred by the oral collateral agreement, and the terms of that written agreement brought it within the proviso that would deprive the defendant of security. What then happened

to the tenancy from year to year effected by the operation of section 2(1) of the Act? I agree with the submission of the plaintiff's solicitor that it had been surrendered by operation of law.

Now it is clear, it seems to me, that the object of this exercise performed by the appellants here, which was indeed explained by J G Short when he gave evidence, was this: It was to enable the owner, if he could, to avoid the automatic conversion of the farmer's interest into a tenancy from year to year by virtue of section 2. This attempt failed because the licence was not for a "specified period of the year." Now the owner must be taken to have known in those circumstances—as indeed Mr Wood conceded—that the farmer's true interest had become a tenancy from year to year, yet he continued, nevertheless, to send these non-genuine forms of agreement for the farmer to sign with the object of avoiding what he no doubt regarded as the mischief of the 1948 Act. That sort of behaviour does not, to say the least, become a plaintiff. I have no doubt that it did not occur to the Short Bros that they were doing anything wrong or indeed were doing any more than they were legitimately entitled to do, however deplorable it might be in fact.

There is a fatal flaw in the owner's contention in this case and it is this: In agreeing in effect to allow the farmer to remain for an indefinite period of at least three years, the owners are saying this, that although you are, by these acts of mine, coupled with section 2 of the Act, getting a tenancy from year to year, yet you must sign a series of meaningless documents, and you must further agree to go, to surrender that tenancy when I give you notice so to do, without any of the statutory protection to which you are entitled. It seems to me that that is a gambit, albeit not so subtly contrived, such as that which was attempted by the landlords in the case of *Johnson v Moreton*, to which we were referred, the report of which is to be found in [1978] 3 WLR 538. It seems to me that this gambit of the landlord in the present case deserves and gets the same sort of treatment as the effort made by the landlords in *Johnson v Moreton*. The farmer cannot bargain away in advance his statutory rights in this way.

That is not to say that there are not circumstances—and indeed *Foster v Robinson* [1951] 1 KB 149 is one of them, where there may be, perfectly properly, an actual surrender of a statutory right, a surrender in fact. But a bargain to give such a surrender in the future or, to put it more accurately in the circumstances of the present case, an agreement at the time the protected tenancy is granted not to seek protection by this statute is not an agreement which should be enforced by the court.

It simply remains for me to mention the question of surrender. I read from p 862 of the first volume of *Woodfall's Law of Landlord and Tenant*, 27th ed: "The term 'surrender by operation of law' or 'implied surrender' (there being no distinction) is an expression used to describe all those cases where the law implies a surrender from unequivocal conduct by the parties which is inconsistent with the continuance of the existing tenancy." It would to my mind have been a strange thing if a surrender by operation of law could take place by the signing of a document which was, at the time it came into existence, designed to conceal the true nature of the parties' rights. Such an act cannot, as I see it, by any stretch of imagination, be described as unequivocal. I can see no basis for the judge's conclusion that the circumstances here have in some way changed so as to turn the latter document, or if you like, the final document of June 1977, from what up to that stage had been an ungenuine document into a genuine one. The signing of this document could not in any event, as I see it, amount to a surrender in the circumstances of this case.

For those reasons, as well as those expressed by my Lord, Brandon LJ, I would allow this appeal with the result that he has indicated.

Also agreeing STEPHENSON LJ said: Out of respect to the admirably clear and thorough judgment of the learned deputy circuit judge, I would wish to add something to what my Lords have said. He has held that the written agreement of December 1 1963 was not a genuine agreement in so far as it provided for the farmer's use of the owners' land for a period of less than a year. The real bargain was that the period was to be several years, and that period was agreed orally and collaterally with the farmer before the written agreement was signed by J G Short on behalf of the owners. It does not seem to me to matter whether there were two agreements, as found by the judge, or one agreement, partly oral and partly reduced (incorrectly) to writing. The reason is that it is common ground that the effect of what was agreed in 1963 was an agreement for the letting of the land from year to year by reason of section 2(1) of the Act of 1948. That statutory tenancy from year to year was outside the exemption of the proviso to the subsection, and would not then be subject to an agreement not to rely on it. But the judge went on to say that that statutory tenancy from year to year was surrendered by the farmer's acceptance of the written agreement of June 1 1977 after discharge of the collateral agreement by performance. The appropriate number of years contemplated by that agreement had expired by then, if not earlier. The farmer's acceptance thereafter of a new written agreement operated in law to estop him from asserting his tenancy from year to year and reduced the period of his lawful occupation to six months.

I agree with Mr Wood that there may be circumstances in which a tenancy created by section 2 can be surrendered by a new agreement inconsistent with it, just as a tenancy under the Rent Acts can be surrendered by such a new agreement—see *Foster v Robinson* [1951] 1 KB 149. But I also agree with Mr Pill that the new agreement must be clearly proved by the owners and must be not only inconsistent with the old agreement but valid. If the new agreement is consistent with the old, or if it is not genuine and misrepresents the true agreement in a material respect, it cannot operate as a surrender; and that is, I think, conceded by Mr Wood. Here the new form of agreement was made collateral not to the same oral agreement but to the tenancy which section 2(1) created from it, and in exactly the same terms as the old. It was sent by the owners and signed by the farmer, as all the similar agreements from first to last had been, without comment or qualification on either side. Those agreements were not genuine in that they contained a false statement about the period of use and occupation, which was not six months but from year to year. Yet implicit in the judge's findings that the collateral agreement was discharged is, I think, a finding that thereafter those forms of agreement became genuine and stated the agreed period truly. Such a finding is essential to the success of the owners' claim to possession—see *Scene Estates v Amos* [1957] 2 QB 205, and I see the logic of the learned judge's reasoning which led him to uphold the claim. But I agree that the collateral agreement which the learned judge found is too uncertain to be capable of being a legally effective contract, and I would give it validity by interpreting it, as do my brothers, as a licence to occupy the land for grazing, to be determined on notice and not before three years from December 1 1963.

The owners then cannot, in my judgment, get rid of the fetters imposed by the oral agreement of 1963 as the judge got rid of them. The agreement continues, and continues to be relied on and acted on until further notice. Or if that is wrong, and it is spent, as the learned judge in effect found and as Mr Wood submits he was entitled to find, I do not think that the owners have got rid of the statutory tenancy by getting rid of the oral agreement. The effect produced by section 2 is not spent. That effect might have been ended if the owners had given notice, when they sent the form of agreement for signature, that the real 1963 bargain was over

A and the farmer had then accepted that termination by signing; but as no notice was given, the agreement had not been discharged or spent by June 1 1977, and there was nothing to indicate to the farmer that in signing the last agreement he was signing away his contractual rights under the licence or his statutory rights under the Act. We should be enforcing an agreement to surrender the tenancy if we were to hold that the farmer was estopped, by signing the agreement of June 1 1977, from disputing its genuineness or from asserting his tenancy from year to year.

I therefore find no surrender by the farmer of the major asset of a protected tenancy and I am of opinion that the B learned judge fell into error in holding that a surrender had been proved and that the farmer was not, in 1977 as in 1963, a tenant from year to year.

I agree that the appeal should be allowed with the results stated by my Lord, Brandon LJ.

The appeal was allowed with costs in the Court of Appeal and below.

COURT OF APPEAL

December 21 1978

C (Before Lord DENNING MR, Lord Justice LAWTON and Lord Justice BRANDON)

JACKSON v HALL AND ANOTHER

Estates Gazette March 10 1979

(1978) 249 EG 955

Agriculture (Miscellaneous Provisions) Act 1976—Date for determination of "eligibility" of applicant for tenancy under family succession provisions—Appeal from decision of Sir Douglas Frank QC that eligibility was to be determined as at date of deceased tenant's death—Held by majority of Court of Appeal that eligibility requirement has to be satisfied at date of application and at date of D hearing, not at date of death—Nothing objectionable in arranging matters after the death, eg by divesting oneself of interest in a commercial unit, so as to achieve eligibility at date of application to agricultural land tribunal—Dissenting judgment by Brandon LJ—Leave to appeal to House of Lords granted

This was an appeal by Graham Christopher Jackson from a decision of Sir Douglas Frank QC, sitting as a deputy judge of the Queen's Bench Division, holding that the appellant was not eligible to apply to the Agricultural Land Tribunal for the Yorkshire and Lancashire Area for a new tenancy of Grange Farm, Keyingham, under the E provisions of the Agriculture (Miscellaneous Provisions) Act 1976 following the death of the tenant, his father. The landlords were William Horner Hall and Geoffrey Alan Marr, the trustees of Geoffrey Edwards Marr deceased. Sir Douglas Frank had decided that the appellant did not satisfy the conditions of eligibility because he was an occupier of a commercial unit, White House Farm, otherwise than as a licensee, at the date of his father's death.

E C Evans-Lombe QC and C Joseph (instructed by Warren, Murton & Co, agents for Stamp, Jackson & Proctor, of Hull) appeared on behalf of the appellant; G O A F Sebestyen (instructed by Collyer-Bristow, agents for Chambers, Thomas & Williamson, of Hull) represented the respondents.

Giving judgment, LORD DENNING MR said: Under the Agricultural Holdings Act 1948 the tenant of a farm was given security of tenure during his life: but it ceased on his death. When he died, the landlord could turn out his widow and family. He could give them notice to quit so long as he did it within three months after the tenant's death—see section 24(2)(g) of the 1948 Act. In 1976 Parliament extended this protection so as to give security to a member of his family after his death. Take for instance a son. G If he had been working on his father's farm for the last five years, and had no other farm of his own, he could apply for a new tenancy to himself. But he had to show that he was an "eligible person." The question in this case is whether he has to be an "eligible person" at the date of his father's death, or whether it is sufficient if he is eligible at the date of his application to the tribunal. The point has given rise to differences of opinion between the various agricultural land tribunals. So we have to settle it.

Such being the question, I will state the facts which give rise to it. The farmer here was James Jackson of Keyingham, north of the Humber. He died on March 22 1977, leaving two sons, Graham and Martin Jackson. H The father had farmed two holdings, which I must describe. One was Grange Farm of 519 acres. It was owned by the Marr trustees. The father was the legal tenant. But he and his two sons were partners in a farming business and they carried on Grange Farm as partners. The other was White House Farm of 400 acres. It was next door to Grange Farm. It was owned by the father himself, but he had let it on a legal tenancy to the partnership of himself and his two sons; and they carried on White House Farm as partners. White House Farm was a "commercial unit" within section 40 of the Agriculture Act 1967 because it was large enough to give full-time employment for an occupier and J one man. No doubt the partners carried on the two farms together as one unit of agriculture. That was obviously the most efficient and economical way of running the business.

The father died, as I have said, on March 22 1977. The two sons were, of course, secure in White House Farm because of their succession to their father, who was the owner. But they were not secure in Grange Farm because that was owned by the Marr trustees. These trustees, on the father's death, gave them notice to quit. As a result, the sons would have had to quit Grange Farm unless they could take advantage of the 1976 Act so as to claim a family succession. K To do this the two sons, or one of them, had to be an "eligible person."

"Eligible person" is defined in section 18(2) of the 1976 Act. It has three requisites (a), (b) and (c). Each of the two sons satisfied (a) the family relationship of being a "child," and (b) the length of service, not less than five years on the farm. But there was this third requisite: "(c) he is not the occupier of a commercial unit of agricultural land within the meaning of Part II of the Agriculture Act 1967 or, if he is, occupies it as a licensee only."

Now, *at the date of the father's death,* neither of the sons could satisfy that requirement because at that date they were the occupiers of White House Farm, L which was a "commercial unit" and they occupied it as tenants and not as licensees only. But, *soon after the father's death,* the two sons took legal advice and in consequence they took steps to get Graham out of White House Farm. On June 22 1977 Graham assigned to his brother, Martin, all his interest in White House Farm. So, thereafter, Martin was the sole owner of White House Farm. If Graham should visit it afterwards, he would do so as a licensee only: see *Harrison-Broadley v Smith* [1964] 1 WLR 456.

Having taken those legal steps, on the very next day, June 23 1977, just within the permitted three months from March M 23 1977, Graham applied to the tribunal for a tenancy of Grange Farm. He said that *at that date*—the date of his application—he was an "eligible person." He satisfied all the requirements (a), (b) and (c). In particular he satisfied (c) because he was no longer the occupier of a commercial unit at White House Farm, or, if he was, he occupied it as a licensee only.

The tribunal seems to have regarded the assignment by Graham to Martin as a "contrivance." The judge regarded it as a "subterfuge." So regarding it, they made short work of it. They said that the material date was the date of the

A death and not the date of the application to the tribunal. I do not so regard the conduct of the two sons. This was a legitimate transaction done so as to keep the two farms together. To keep them in the family and to farm them as one efficient and economic unit. If they had to give up Grange Farm and be confined to White House Farm only, it might be disastrous for their business. The transaction may or may not have achieved the desired result—that I will soon consider—but I would not subscribe to the view that there was anything discreditable in it—any more than there was anything discreditable in the Duke of Westminster paying his gardener his wages by way of a deed of B annuity: see [1936] AC 1. It was a legitimate way of avoiding a stern law.

So I turn to the question of law in the case stated. In order to qualify as an "eligible person," what is the material date? Is it the date of the death of the deceased? Or is it the date of the application? To my mind this question is solved by reference to the actual words of section 20(1) and (2). These read: "(1) Any eligible person may within the relevant period apply to the Tribunal for a direction entitling him to a tenancy of the holding. (2) Where only one application is made under this section the Tribunal, if satisfied that the applicant is an eligible person, shall determine C whether he is in their opinion a suitable person to become the tenant of the holding." Note the present tense. Under subsection (1) the words refer to the date of the application. Under subsection (2) they refer to the date of the hearing by the tribunal. Taking the two subsections together, it seems to me that the applicant must, in the ordinary way, satisfy the requirement of being an "eligible person" both at the date of the application and at the date of the hearing. But not at the date of death. This solution is similar to that reached by the court in cases under the Rent Restrictions Acts: see *Kimpson v Markham* [1921] 2 KB 157; *Benninga (Mitcham)* D *Ltd v Bijstra* [1946] KB 58; *Zbytniewski v Broughton* [1956] 2 QB 673: and under the Landlord and Tenant Act 1954, see *Betty's Cafés v Phillips Furniture Stores Ltd* [1959] 1 AC 20.

On the other side emphasis was placed on the words "survivor of the deceased" in the definition of "eligible person" in section 18(2). But that seems to me to be neutral. At the date of the application the son is a "survivor of the deceased." Likewise, at the date of the hearing he is a "survivor of the deceased." In order to insist on the date of death, you would have to insert the words "at the date of the death of the deceased" after the words "in which case" and before the words "the following conditions are satisfied E —(a) . . ., (b) . . . and (c). . . ." I see no reason to insert those words.

Other matters were raised which were said to point in favour of one or other construction. But these all seem to me to be neutral. One thing, however, which does influence me is that, before the date of the death, the family could make an arrangement so as to ensure the succession of one son. So why should they not do it after his death? The clever far-sighted families will consult their lawyers before the death. The nice good natured ones will wait till after the old man has died. I would not let them lose on that account. I would hold that a family can, by appropriate arrangements, F ensure that a son shall be an "eligible person" so long as the transaction is carried out before the date of the application.

I would allow the appeal and hold that eligibility is to be determined, not at the date of death, but at the date of the application and hearing.

Agreeing that the appeal should be allowed, LAWTON LJ said: By the Agriculture (Miscellaneous Provisions) Act 1976, Parliament gave four classes of persons, who were "eligible" as defined in the Act, the right to apply to an agricultural land tribunal for a tenancy of an agricultural holding of a deceased tenant. These tribunals have a legally qualified chairman and two members who are not lawyers. It was such a tribunal, sitting in the Yorkshire and Lancashire area, which had to construe a definition section (section 18(2)) G which has thrown up for the consideration of this court niceties of grammar and of the meaning of one word. Mr Sebestyen for the freeholder argued for grammatical purity and a restricted meaning; Mr Evans-Lombe for the applicant for a wider meaning.

The grammatical purity for which Mr Sebestyen contended is said by him to come from the correct phrasing of the words

> eligible person means . . . a survivor of the deceased in whose case the following conditions are satisfied—(a) he falls within paragraphs (a) to (d) of subsection (1) above; (b) in the seven years ending with the date of death his only or principal source of H livelihood throughout a continuous period of not less than five years, or two or more discontinuous periods together amounting to not less than five years, derived from his agricultural work on the holding or on an agricultural unit of which the holding forms part; and (c) he is not the occupier of a commercial unit of agricultural land . . . or, if he is, occupies it as a licensee only.

If a survivor of the deceased means, and can only mean, a person who is to be identified as going on living at the moment of the tenant's death, that person to be eligible must then have all the qualifications set out in paragraphs (a), (b) and (c); and for the purpose of specifying those qualifications at that moment the use of the present tense in paragraph (c) is both appropriate and correct. J Mr Sebestyen went on to submit that this construction of a statutory definition slots easily into the sections of the Act in which it has to be applied, the most important being section 20. Eligibility is fixed by reference to the past and must continue up to the dates of application and determination (see section 20(1) and (2)).

There is much force in Mr Sebestyen's suggested construction. It is a possible one. But is it the only one? If there is another construction, which should be preferred? The alternative construction is derived from the concept that a survivor of a deceased lives on. His state as such continues until he himself dies. K In the ordinary usage of the English language this is a common meaning as is shown by such phrases as "he is a survivor of the First World War." If being a survivor is a continuing state, then the qualification set out in paragraph (c) is one which an otherwise eligible person can acquire within any relevant period and it need not be one acquired on the tenant's death. The use of the present tense in paragraph (c) ensures that the person applying under section 20(1) then has the necessary qualification and continues to have it when the tribunal considers his application. This construction finds some support in the use of the passive tense in section 18(1) in the words L "the sole tenant of an agricultural holding dies and *is survived* by any of the following persons. . . ."

In my judgment the alternative construction is the one which would the more easily be given by literate persons, especially as paragraph (c) omits any reference to the time of death which might well have been there in the interests of clarity. This seems to have been the construction put upon paragraph (c) by the draftsman of the statutory form which has to be used when an application is made to a tribunal under section 20—see the Agricultural Land Tribunals (Succession to Agricultural Tenancies) Order 1976 (SI 1976 No 2183), M rule 2 and the appendix to that order. Paragraph 11 of the form is as follows: "(a) I occupy as owner-occupier/tenant/licensee the following agricultural land [*give particulars of the land occupied, including area*]: (b) I do not occupy any other agricultural land." The form does not require the applicant to state what agricultural land he occupied when the tenant died, as it would have done had the draftsman thought that section 18(2) should be construed as Mr Sebestyen submitted it should be. The form cannot, of course, be a guide to the construction of section 18(2). Such value as it has lies only in showing what a draftsman, used

A to looking closely at words, thought the ones used in section 18(2) meant.

I can see nothing in the policy of the Act which should lead me to prefer Mr Sebestyen's construction to that of Mr Evans-Lombe. The broad policy of the Act is clear—anyone taking over a deceased tenant's holding must not be an occupier of another "commercial unit" when he applies for a direction entitling him to do so. Anyone having another "commercial unit" may divest himself of it in expectation of the tenant's death—and probably will do so if he takes legal advice. Such a divesting would not be contrary to the policy of the Act. In my opinion a divesting between death B and an application under section 20 would be no more contrary to the policy of the Act than one before death. If applicants enter into colourable transactions which are mere sham divestings, the tribunals can disregard them.

I would allow the appeal.

Dissenting, BRANDON LJ said: Part II of the Agriculture (Miscellaneous Provisions) Act 1976 ("the Act") provides for the succession to tenancies of agricultural holdings by survivors of deceased tenants after their deaths. Applications for such succession are made to and decided by the agricultural land tribunal for the area concerned.

Under section 20(1) of the Act the right of a survivor of C a deceased tenant to apply to the tribunal for a direction entitling him to become the tenant of the relevant holding depends on his being an "eligible person." Further, under section 20(2) the tribunal to which the application is made must first of all, before deciding whether the applicant is suitable to become the tenant, be satisfied that he is an "eligible person."

Under section 18(2) an "eligible person" means a survivor who possesses three specified qualifications. Qualification (a) is that he should have been either the spouse or a sibling or a child (actual or by treatment) of the deceased. Qualification (b) is that his only or principal income should, during a D specified period or periods before the death, have derived from his agricultural work on the holding or on an agricultural unit of which the holding forms part. Qualification (c) is that "he is not the occupier of a commercial unit of agricultural land within the meaning of Part II of the Agriculture Act 1967 or, if he is, occupies it as a licensee only."

The question for decision in this appeal is at what time or times an applicant has to possess qualification (c) above in order to have the right, as an "eligible person," to apply under section 20(1), and in order to enable the tribunal to be satisfied that he is such a person under section 20(2). The E Agricultural Land Tribunal (Yorkshire and Lancashire Area) and the High Court (Sir Douglas Frank QC sitting as deputy judge) on a case stated by that tribunal, have both decided that the relevant time is the deceased's death. The appellant, who does not, if that view is correct, possess qualification (c) in this case, appeals against the latter decision, contending that there are two relevant times and that these are, firstly, the time of the application to the tribunal, and, secondly, the time of the hearing before it.

In support of the appellant's contention reliance has been placed on the use of the present tense, firstly, in the phrases F "he is not the occupier" and "or, if he is, occupies it as a licensee only" in paragraph (c) of the definition of eligible person in section 18(2), and, secondly, in the phrase "that the applicant is an eligible person" in section 20(2). The appellant's interpretation of the Act was further said to be supported by *Kimpson v Markham* [1921] 2 KB 157, a decision on the Rent Acts. In my judgment the contention for the appellant is wrong, and the decisions of the tribunal and Sir Douglas Frank QC are right, for the following reasons.

First, "eligible person" is defined in section 18(2) as "a survivor of the deceased in whose case the following conditions are satisfied," the conditions concerned being those G set out in paragraphs (a), (b) and (c) following. It seems to me that this subsection, by making the test of eligibility whether certain conditions are satisfied in the case of a survivor, that is to say (to use the related language of section 18(1)) a person by whom the deceased is survived, must by necessary implication be providing that the test should be applied at the time of survivorship, that is to say the time of the deceased's death.

Second, the question whether conditions (a) and (b) are satisfied in the case of a survivor clearly fall to be decided by reference to the time of the deceased's death. It is, therefore, a natural interpretation of the definition of "eligible H person" as a whole that the question whether condition (c) is satisfied should also be decided by reference to the same time.

Third, the use of the present tense in the phrases "he is not the occupier" and "or, if he is, occupies it as a licensee only" in paragraph (c) of the definition of "eligible person" in section 18(2) is no more significant than the use of the same tense in the phrase "he falls within paragraphs (a) to (d)" in paragraph (a), or in the phrase "agricultural unit of which the holding forms part" in paragraph (b). The latter phrase, despite the use of the present tense, must surely relate to the time of death. J

Fourth, section 20(1) provides that an "eligible person" may apply to the tribunal within "the relevant period," which is defined in section 18(2) as being (subject to certain exceptions which are not material) the period of three months beginning with the day after the date of the deceased's death. It seems to me that, if an "eligible person" can apply at any time from one day after the deceased's death to 91 days or thereabouts after it, the test of his eligibility must be a test as applicable on the first day as on any subsequent day up to the last. This will only be so if the question whether condition (c) is satisfied in his case is decided by reference to the time of the deceased's death. K

Fifth, the use of the present tense in the phrase "that the applicant is an eligible person" in section 20(2) is not significant. "Eligible" means the same as "qualified." The question whether a person is presently qualified in one respect or another will, more often than not, depend on whether certain conditions were satisfied in his case in the past. For instance a barrister is presently qualified if he was called to the Bar in the past, and a doctor is presently qualified if he acquired the necessary medical qualifications in the past. That being so, there is no difficulty in interpreting the phrase "that the applicant is an eligible person" in section 20(2) as meaning "is eligible because he possessed the necessary qualifica- L tions, including qualification (c), at the time of the deceased's death."

Sixth, the contention for the appellant that there are two relevant times for the satisfying of condition (c), namely the time of the application and the time of the hearing before the tribunal, itself necessarily involves interpreting the phrase "that the applicant is an eligible person" in section 20(2) as referring, so far as the earlier of those two times is concerned, to the past. If there were any problem, therefore, about the use of the present tense in that phrase (which I do not think, for my fifth reason set out above, that there is) the applicant's M interpretation would not eliminate such problem.

Seventh, it seems to me more consistent with any policy of the Act which can reasonably be attributed to the legislature that the test of eligibility for succession contained in paragraph (c) of the definition of "eligible person" in section 18(2) should be applied by reference to the situation of a survivor as it exists at the time of the deceased's death rather than by reference to some different situation of his which has come into existence between the time of the deceased's death and the time of such survivor's application to the tribunal anything up to three months later.

So far as *Kimpson v Markham*, the authority of the Rent Acts relied on by the appellant, is concerned, I would only say that, since it is a decision on a different statute, I do not think it affords any real assistance in the interpretation of the statute here concerned. In any case it cannot be in any way decisive of such interpretation.

My conclusion on this matter differs from that of the other two members of the court and I express it therefore with diffidence. For the reasons which I have given, however, I would dismiss the appeal.

The appeal was allowed, with costs in the Court of Appeal and below. Leave to appeal to the House of Lords was given.

RICHARDS v ALLINSON

(Before His Honour Judge MEURIG EVANS at Portmadoc County Court on June 6 1978)

Estates Gazette January 6 1979

(1978) 249 EG 59

Arbitration—Special case stated—Paragraph 6 of Schedule 6 to Agricultural Holdings Act 1948—"Within 28 days from the appointment of the arbitrator"—What is effective date of appointment of arbitrator?—Date when all necessary requirements have been fulfilled, sealing by minister (if appropriate), acceptance by arbitrator, receipt by him of notice of appointment and receipt by parties of notice—28 days begins to run on next day—No power for arbitrator to extend time for delivery of statement of case

This was a special case stated by David Thomas Hefin Davies FRICS, the arbitrator appointed by the Minister of Agriculture to determine claims arising on the termination of a tenancy between Lieut-Commander R M Richards, the landlord, and Martin Vernon Allinson, the tenant, of a holding at Gwynedd. The material facts are set out in full in the judgment. The questions of law which were submitted by the arbitrator for the opinion of the court were as follows:

(1) Was the date of appointment of the arbitrator as referred to in paragraph 6 of Schedule 6 to the Agricultural Holdings Act 1948 the date on which the minister sealed the appointment, the date on which the deed of appointment was received by the arbitrator, or the date on which the arbitrator formally accepted the appointment and notice of the appointment was given to the parties?

(2) Was the statement of case and particulars, posted by first-class recorded delivery post by the landlord's agents in accordance with the agreed statement of facts, delivered to the arbitrator within the time limited by Schedule 6, paragraph 6, to the 1948 Act?

(3) Was the landlord entitled to rely on matters alleged in the statement of case and particulars delivered to the arbitrator in accordance with the agreed statement of facts?

(4) If a party to an arbitration is not entitled to rely on matters alleged in a statement of case and particulars delivered to the arbitrator because it has not been delivered within the said period of 28 days as set out in paragraph 6 of Schedule 6, has the arbitrator power to permit him to rely on such matters?

H A C Densham, solicitor, of Burges, Salmon & Co, of Bristol, appeared on behalf of the landlord; Pryce Michael Farmer (instructed by Edward Jones & Son, of Blaenau Ffestiniog) represented the tenant.

Giving judgment, His Honour JUDGE MEURIG EVANS said: The dispute in the present case arises out of a landlord's claim against a tenant for damages for dilapidations. Such claims have to be determined by arbitration by an arbitrator appointed (in default of agreement between the parties) by the Minister of Agriculture, Fisheries and Food, to whom particulars of the claim and of the defence have to be submitted.

The Agricultural Holdings Act 1948 governs the procedure to be complied with in such arbitration, and provides, *inter alia*, a time-limit, from the appointment of the arbitrator, within which the claim has to be submitted. The relevant statutory directions are to be found in Schedule 6, paragraph 6, to the Act (as amended) as follows:

6. The parties to the arbitration shall, within twenty-eight days from the appointment of the arbitrator, deliver to him a statement of their respective cases with all necessary particulars and—

(a) no amendment or addition to the statement or particulars delivered shall be allowed after the expiration of the said twenty-eight days except with the consent of the arbitrator;

(b) a party to the arbitration shall be confined at the hearing to the matters alleged in the statement and particulars delivered by him and any amendment thereof or addition thereto duly made.

The *main* question raised for determination in this present case is: What is the proper date to be taken as the date of "the appointment of the arbitrator"? The deed of appointment under the seal of the Ministry of Agriculture, Fisheries and Food is dated August 23 1976. *Prima facie*, that would be the date of appointment of the arbitrator, and time would begin to run from that date. That at any rate was the view of the arbitrator in the present case. On September 29 1976 he accordingly wrote to the parties the letter of that date pointing out that the landlord's statement of case had not reached him until September 22, whereas it should have been delivered within 28 days from August 23—ie not later than September 21. It was suggested that it was therefore out of time. Correspondence followed between the arbitrator and legal adviser. The arbitration was deferred and the matter was eventually referred for the opinion of this court.

The hearing took place at Portmadoc on June 6 1978, nearly two years after the issue had first been raised by the arbitrator.

It may be convenient to recall the relevant dates, which are as follows:

August 12 1976—Minister informs arbitrator that he proposes to appoint him.

August 18 1976—Arbitrator replies he is prepared to accept.

August 23 1976—Minister posts letter and deed of appointment of that date to the arbitrator.

August 24 1976—Deed of appointment received by arbitrator.

August 24 1976—Minister posts notice of arbitrator's appointment to landlord's agent.

August 26 1976—This notice of appointment of arbitrator received by landlord's agent from the minister (though dated August 23, not posted from Ministry until August 24 and not received until August 26).

August 30 1976—Notice of arbitrator's appointment received by landlord *from arbitrator* (though dated August 25, not posted until August 27 and not received until August 30).

No point was taken by either party that this letter dated August 25 warns the landlord that the statement of case must reach the arbitrator by September 21.

The submissions by Mr Densham for the landlord were as follows:

(1) That the *effective* date of "appointment of the arbitrator" was *not* August 23 (the date on his deed of appointment) but the date by which each of four requirements had been satisfied, namely:

(1) The minister had signed and sealed the deed of appointment (August 23).

(2) The deed had been delivered to the arbitrator (August 24).

(3) The arbitrator had accepted the appointment (prior to August 23).

(4) notice of the appointment had been received by the landlord and tenant (August 26);

A (2) that the 28 days from whatever is the proper date of the appointment must be construed as 28 clear days, and that time for delivery of the pleading would therefore run, not from the day of appointment itself but from the day following the day of appointment;

(3) that the statement of case required by paragraph 6 of Schedule 6, having been posted by recorded delivery on September 20, would normally have been delivered on September 21 and was therefore delivered within the 28 days although in fact it had not been delivered until September 22;

(4) that in so far as the landlord's statement of case had been delivered outside the 28 days time-limit, the arbitrator never-B theless would have been entitled to grant and should have granted an extension of time and/or allowed the landlord to rely on its contents as the basis of his claim.

For the tenant, Mr Farmer's submissions may be summarised as follows:

(1) that the date of appointment of the arbitrator was, and was stated to be in the deed, the date of its execution, viz August 23;

(2) that neither delivery to the arbitrator nor notice to the landlord affected the matter;

(3) that neither was a prerequisite to the validity of the appointment of August 23; C

(4) that the court ought not to attempt an artificial construction of the words of the statute, thus causing delay to the arbitration, one of the objects of the arbitration processes and of Schedule 6 in particular being to bring about an expeditious settlement of differences between landlord and tenant;

(5) that the 28-days limit expired, as the arbitrator had stated, on September 21;

(6) that the landlord could not therefore rely on the contents of the statement of case, having regard to the specific term of paragraph 6(8);

D (7) and that the arbitrator had no jurisdiction to extend the time or to allow the landlord to rely on the contents of the late delivery of statement of case.

In my view the question of the effective date of the appointment has to be determined by reference to the principles laid down by the Court of Appeal in *Tradax Export SA v Volkswagenwerk AG* [1970] 1 QB 537. The question there in issue was whether there could be an effective appointment of an arbitrator in a commercial arbitration when the arbitrator designated by the one party had (a) not been asked whether he was willing to act in the particular matter and (b) had not consented so to act. It was held that there could be no effective appointment E of an arbitrator until both these conditions, and also a third condition, viz that his name and appointment had been communicated to the other side, had been fulfilled. At p 544 Lord Denning MR states the position in unambiguous terms:

> So we have to decide [he states] what is necessary to constitute the appointment of an arbitrator? I think the answer is this: First, it is necessary to tell the other side. That is plain from *Tew v Harris* (1847) 11 QB 7. Second, it is necessary to tell the appointee himself. That is obvious because he often has to start acting at once. Third, it is necessary that he should be willing to act and have intimated his willingness to accept the appointment. In *Russell on Arbitration*, 17th ed (1963), at p 160, it is said: "Acceptance of the office by the arbitrator appears to be necessary to perfect his appointment." There F is a passage in *Ringland v Lowndes* (1863) 15 CBNS 173, 196, which gives some support to that statement.

Mr Densham by great diligence has discovered the case of *Cox v Johnson* (1914) 14 SR(NSW) 240, where it is said at p 250: "In my view all that is required by the section"—a similar section—"is nomination by each party to the other of the person whom he has selected to act on his behalf."

I think that statement may have been right in relation to the facts and evidence in that case. But I do not think it is of general application. I think in general it is essential not only that the other side should be told, but also that the arbitrator himself should be told."

And again Salmon LJ (as he then was) said in the *Tradax* case at p 545: G

> In my judgment, three things are necessary in order to make an effective appointment under this clause: first of all, the arbitrator must be communicated with and asked if he is willing to act. Secondly, he must express his willingness and be clothed with authority to act. And thirdly, as is apparent from *Tew v Harris* (1847) 11 QB 7, in order to perfect the appointment the other side must be notified of the name of the arbitrator. Sometimes these steps are taken by letter or by cable. There is nothing to prevent them being taken over the telephone or by telex. I am not saying that there may not be a special case in which an arbitrator has informed a party in advance that he is always prepared to receive instructions to act as arbitrator in any dispute in which that party may be concerned. If consent has been given in advance, it is enough to communicate the appointment to the arbitrator, and then give the other party the appropriate notice. H

By reference to the time-table which is set out in the earlier portion of my judgment it will be seen that though the arbitrator's deed of appointment was dated August 23 he was not informed of it until the following day, August 24. Consequently the date of appointment could not have been prior to August 24.

There still remains the third request to be satisfied, namely communication of the name of the arbitrator to the parties. Although a letter to the landlord's agent giving this information was written on August 23, it was not posted until August 24 and not received by the agent until August 26. It follows that J the third condition not having been satisfied until that latter date, the effective appointment of the arbitrator was not completed until that date, namely August 26.

Reference to the judgments in *Sclater v Horton* [1954] 2 QB 1, and in particular to that of Evershed MR at p 10, shows clearly that there may well be a valid distinction between what might be termed a "bare" appointment of an arbitrator, even when appointed by a minister, and an effective appointment, so far as the parties to the arbitration are concerned.

As regards Mr Densham's third submission, this was not, of course, necessary to the determination of the present case, having regard to my decision on his first point and was not K accordingly pressed. In my view, nevertheless it is not well founded in law. The presumption of delivery of a letter by recorded post delivery applies only "unless the contrary is proved." In this case the contrary was proved. The letter is agreed to have been delivered on September 22 although actually posted on September 20.

Mr Farmer for the tenant submitted that different principles applied in the case of an arbitrator appointed by a minister under the Agricultural Holdings Act 1948 from those applicable to a commercial arbitration under a "Centrocon" clause as in the *Tradax* case. I cannot accept that. If it were so, absurd L results might follow. For example, if as he contended, it were unnecessary for notice of the appointment to be communicated to both parties before the appointment could become effective, the 28-days' limitation for pleadings could expire or at least a significant portion of that period could have expired before one or both parties knew anything about the appointment, and grave injustice could result. I find that proposition equally inconsistent with authority as with commonsense. In my opinion the principles of the *Tradax* case are as applicable to the appointment of an arbitrator under and by virtue of the provision of the Agricultural Holdings Act 1948 as they are to the appointment of an arbitrator under the Centrocon clause in a private M charter party. Moreover there is direct authority on this point: see *Russell on Arbitration*, 18th ed p 96 and the cases there cited, in particular the judgment of Lord Denman, "Neither party can be said to have chosen an arbitrator until he lets the other party know the object of his choice. Nomination implies notice": See also *Russell* p 213, footnote 5.

Accordingly I consider that the effective date of the appointment of the arbitrator was August 26 1976—ie the date on which the landlord in fact received notice of the appointment as stated in the special case. In any event the date could not be earlier than August 24, the date on which the appointment was

A notified to and received by the arbitrator. If, as I have held, August 26 is the effective date of the appointment it is common ground that the delivery of the landlord's statement of case to the arbitrator was within the 28-days' time-limit. The same result follows if the correct date was August 24.

I have been asked however to deal with the other points raised and argued at the hearing and in particular with the date for which the time of 28 days for delivery of the statement of case began to run. In my view and on the authority of *Stewart v Chapman* [1951] 2 KB 792 and the authorities there cited, the day of the appointment of the arbitrator, whether August 26 or August 24, has to be excluded and the 28 days computed B as from the following days, August 27 in the one case, or alternatively August 25 in the other case. In either event the statement of case was delivered within the 28 days.

Finally, as regards the power of the arbitrator to extend the time for delivery of the pleadings. This point has been the subject of decision in various county courts, and with one exception the consensus of judicial opinion seems to have been in favour of a strict interpretation of the statutory time-limit. I take the same view. Had the legislature intended that the arbitrator should be at liberty to extend the time-limits laid down by the Act, the legislature would have made provision to that effect—or a statutory discretion given by some other statute. C That does not appear to have been done, though a discretion is given to the arbitrator by Schedule 6 to allow amendments to the statement of case.

Moreover the legislature itself extended the time-limit by another Act of Parliament, from the original 14 days in the 1948 Act, to 28 days by the amending Act in 1963 (section 20). This would have been unnecessary if the arbitrator himself could have extended the time-limit. Reference was made by Mr Densham to the judgment of Denning LJ, as he then was, in *Seaford Court Estates Ltd v Asher* [1949] 2 All ER 155 at p 164, but I do not consider that has any relevance to the present D case. The various decisions relating to time clauses in leases suggest that time-limits in that class of case can be regarded as guide-lines only. In my view, however, statutory limits depending for their validity not on agreement between the parties but by legislative decree have to be observed. It may be arguable whether or not such limits can even be waived by the parties (*Kammins Ballrooms Co Ltd v Zenith Investments* (*Torquay*) *Ltd* [1971] A C 850) but waiver was not an issue in this case.

I answer the questions submitted to the court as follows:

(1) The date of appointment is not the date upon which the minister sealed the appointment;

Nor is it necessarily the date on which the appointment is E received by the arbitrator;

Nor is it necessarily the date on which the arbitrator formally accepted his appointment;

Nor is it necessarily the date on which notice has been given to the parties. The date of appointment is the first date by which each and all of the four requirements set out in paragraph (1) of Mr Densham's submissions (*supra*), have been fulfilled, ie in this case August 26 1976.

(2) Yes—August 26 1976 was the date on which the final requirement was fulfilled, which made the appointment effective.

F (3) Yes.

(4) No. Unless possibly there had been waiver or consent.

TUMMON v BARCLAYS BANK TRUST CO LTD G

(Before His Honour JUDGE GOODALL at Bodmin County Court on February 27 1979)

Estates Gazette June 9 1979

(1979) 250 EG 980

Agricultural arbitration—Special case stated by arbitrator—Rental arbitration under section 8 of Agricultural Holdings Act 1948—Important points of law—Part of land comprised in holding used as a developed caravan site—How to value a holding where the caravans constitute a more lucrative enterprise than agriculture—Whether road works, drains, erection of toilet block and shop for the caravan site are tenant's improvements within section 8(2)(a)—Whether, if H **so, "latent value" released by works should be disregarded —Distinction between potential and realisation of potential —Problem of "relief from payment of rates" in section 8(2)(b)—"Appalling difficulties" faced by arbitrators—Arbitrator must ensure that tenant gets the benefit of the relief and as a judicial person must decide the matter according to the evidence—Arbitrator must do his best to specify the quantum of rate relief if requested by a party—"Relief" means full relief—Judge's practical advice to arbitrators in dealing with a "terrifying" problem**

This was a special case stated by Thomas Elwyn Thomas FRICS, the arbitrator appointed by the agreement of the J parties to determine, in accordance with section 8 of the Agricultural Holdings Act 1948, the rent properly payable in respect of the agricultural holding known as East Trevose Farm, St Merryn, in the county of Cornwall. The specific questions raised in the stated case for the opinion of the court are set out in the course of the judge's opinion.

Bruce Mauleverer (instructed by Ralph & Co, of Truro) appeared on behalf of the landlord; H A C Densham, solicitor, of Burges, Salmon & Co, of Bristol, represented the tenant.

JUDGE GOODALL said that the holding comprised in all some 300 acres, of which 202 or thereabouts consisted of viable farmland, and 9½ acres represented the caravan site on which K there was planning permission to station 190 caravans. The holding also contained some acreage of cliff-land and waste and there was a dwelling-house, cottage and yard. At the commencement of the tenancy there were no caravans, but subsequently the site was established, at the expense of the tenant, with roads, toilet facilities and a shop. It was common ground that at the present time the revenue from the caravan site was substantially greater than that from the agricultural activities. The questions set out in the arbitrator's very helpful special case fell into two parts, those relating to problems arising from the caravan site and those concerned with the application of section 8(2)(b) of the 1948 Act, which required the arbitrator not to take into account the relief from the payment of rates. L

Valuation of holding containing caravan site

Logically the first question which had to be answered was the one posed by the arbitrator in the following terms:

Should I value the holding as an agricultural holding, accepting the following extract from *Muir Watt on Agricultural Holdings* (12th ed) p 23?

"The rent to be determined is the rent at which the holding might reasonably be expected to be let in the open market as an agricultural holding, which means, *inter alia*, as land used for agriculture as a trade or business, not as a purely non-commercial M hobby, nor for a profitable but non-agricultural purpose such as a riding school, nor for mere amenity or sport. But the reflection in the rent of an element of profitability from an incidental non-agricultural purpose (not inconsistent with the terms of the tenancy), such as the letting of rooms with board in the farmhouse to summer visitors, or sporting rights not reserved to the landlord, is not excluded, provided that the holding remains in substance an agricultural holding."

All that the above-quoted passage was saying was that a farm should not be valued on the basis that it could be put to a more valuable use than its actual use—in other words the

A well-known rule in rating valuations that properties should be valued *rebus sic stantibus*. The passage did not throw light on the present problem. Mr Densham, on behalf of the tenant, citing *Godfrey v Waite* [1951] EGD 9 and *Howkins v Jardine* [1951] 1 KB 614, submitted that in order to determine whether a property was an agricultural holding one had to look at the situation at the commencement of the tenancy and apply the "predominant user" test. This test had to be applied to value, not to acreage, and if it were to be applied at the date of the hearing the effect would be that the farm would not be an agricultural holding. The solution suggested on behalf of the tenant was that the arbitrator should value as if the
B predominant user test was still satisfied and that no more than the agricultural rent should be attributed to the caravan site. That argument could not be accepted. Suppose it were the landlord who carried out all the work to provide a lucrative caravan site on a farm. Could the tenant agree a larger rent and then after three years turn round and say that the rent should not reflect the value of the caravan enterprise? The task of the arbitrator was to determine the rent properly payable. That could only relate to the whole of the holding, caravan site and all. The answer to the arbitrator's question under this heading was that the holding should be valued in its present state, ie as part farm and part caravan site.

C

Are the works for the establishment of the caravan site improvements within section 8(2)(a)?

The next question to be considered was the following:

Are the works carried out by the applicant during the current tenancy in the establishment of a caravan and camping site, namely, the laying out by the applicant of roads, the provision of drains and the erection of a toilet block in the fulfilment of a condition precedent to the establishment of such site, and the provision subsequently of a shop thereon, tenant's improvements, or are any such works individually improvements which fall to be disregarded by me when awarding the sum of rental to be charged for this agricultural holding within the meaning of section 8(2)(a) of the
D 1948 Act?

It was said on behalf of the landlord that improvements within the meaning of section 8(2)(a) were confined to agricultural improvements, ie improvements which would improve the farm as a farm and so improve its agricultural output. That argument would create great injustice to the tenant. The broad principle underlying the provisions as to the tenant's improvements both under the 1948 Act and the Landlord and Tenant Act 1954 was that a tenant who improved the property should not pay an increased rent because of it. The word used in section 8(2)(a) was "improvements" *simpliciter* and it would be wrong to cut it down. The answer to the arbitrator's
E question was, therefore, that all the work carried out by the applicant during the current tenancy in the establishment of a caravan and camping site, namely, the laying out by the applicant of roads and the provision of drains, the erection of a toilet block and the provision of a shop, and indeed any other improvements he might have carried out, fell to be disregarded by the arbitrator when awarding the rental to be charged.

Should release of latent value be disregarded?

The arbitrator then asked:

F If the works listed above are tenant's improvements which fall to be disregarded, does this mean that I have to disregard merely the rental value attributable to such improvements or should I also disregard the latent value released by the carrying out of the said works and the consequent implementing of the planning consent for change of use?

A supplementary question was the following:

Was the latent rental of the land which is used as the caravan and camping site released by the original granting of the planning permission rather than by the implementation of the permission by the carrying out of the works listed above?

It was submitted on behalf of the tenant that the whole of the increase in the rental value of the holding attributable to the caravan site was due to the tenant's improvements. It G
should therefore be disregarded. That argument was too facile. The increase of rental value was due to a combination of three elements, namely, the attractive situation of the farm on the holiday coast of Cornwall, the existence of the planning permission, and the works carried out by the tenant. The tenant's submission that nothing should be awarded to the landlord in respect of his ownership of this very valuable site must be rejected.

A very similar question had been put to R E Megarry QC (as he then was) in 1959 when he addressed the Third Conference of Agricultural Arbitrators. The question and the answer which he gave were as follows: H

"The rental value of a farm is often enhanced by an improvement beyond the annual equivalent of the cost of the improvement, the enhancement being due to the effect of the improvement in releasing 'potential value' which was in the farm. If the improvement was effected at the tenant's expense, is the arbitrator entitled to include this potential value in the rent he fixes or should he ignore it in the same way as he must ignore the cost of the improvement?"

You have, of course, to consider the consequence of what has been done by the tenant. If he has carried out an improvement you must consider not only what it has cost him but also the consequences of putting in the improvement, which produces certain results and makes for better farming. The Act might have J
said that you should leave out of account the cost of the improvement but take into account the enhancement of the potentialities of the farm, but that would be most unfair to the tenant.

Section 8(2)(a)(i) of the Act of 1948 says—putting it shortly—that the arbitrator is not to take into account "any increase in the rental value of the holding" which is due to the tenant's improvements; and "any increase in the rental value of the holding" means not only the actual cost of the improvement but its consequences in increasing the value of the farm. They are all to be left out of account, so that you must disregard them.

In assessing the value of the farm, however, you will take into account this fact. If it is a farm which, as it is run at the moment and disregarding the particular improvement, has a low value, but K
by spending a relatively small amount on improvements it can be made much more valuable, one of the things which you will take into account is that if the farm were let to a new tenant he could, by doing this improvement himself at a small capital outlay, vastly increase his return from the farm. You will regard it as a farm capable of easy and cheap improvement, and therefore a valuable farm. You have to leave out of account the fact that the improvement has been done by the tenant already, but take into account the possibility that a new tenant might himself carry out that improvement and so get much better value out of the farm.

As one would expect, the problem posed in the present case was correctly answered by Mr Megarry, as he then was. The words "latent value released" in the arbitrator's question L
cover two quite separate matters, namely, latent value or potential and the release of that value or the realisation of the potential. The latter, the release or the realisation, was due solely to the work of the tenant, but the potential was not in any way due to the work of the tenant. Mr Megarry was dealing with an agricultural improvement, but there seems no reason for applying a different rule to non-agricultural improvements. The proper approach in the present case was that the farm should be valued on the basis that the caravan site had not been established, and that the farm was as it was at the beginning of the tenancy so far as its physical state was concerned, but that there was this potential for a caravan M
site and there was planning permission to carry it out. It had been argued that as planning permission had been granted many years ago it would have lapsed if the work had not been carried out and, therefore, the arbitrator should value the holding as if there was no planning permission. This argument was fallacious. The planning permission had been acted upon and had been kept alive by the action of the tenant. It was extant today. The fallacy lay in postulating a planning permission given in 1961 and then assuming that it had lapsed.

The answer to the supplementary question quoted above should have become apparent from what had already been said. The latent value was released by the carrying out of the works, but the latent rental, the potential, was a matter which had to be taken into account. Attention had already been drawn to a possible misconception as to the meaning of "release of latent value."

" Relief from payment of rates ": general effect

It was argued that section 8(2)(b) of the 1948 Act would have no effect. That argument could not be accepted. The Scottish Court of Session in *Guthe v Broach* 1956 SC 132 had decided that the provision did not mean that the arbitrator should ignore rating relief—it meant that he should secure to the tenant the benefit of that relief, however difficult it might be to do so. It had been argued on behalf of the landlord that, in the absence of evidence of open market lettings, arbitrators used as comparables the rents determined by other arbitrators. If those other arbitrators had correctly applied section 8(2)(b), there was a built-in compliance with the provision. It was suggested that an arbitrator was bound to assume that his colleagues had obeyed the law in this respect. This, however, could not be assumed. Indeed it seemed from Mr Walmsley's *Agricultural Arbitrations* that it was highly probable that they had not. If they had not, and if the starting point was a true open market rent, then a deduction fell to be made in applying section 8(2)(b).

Reliance on professional experience rather than evidence?

The arbitrator in his case stated asked the following question:

> Am I entitled to give effect to section 8(2)(b) in the manner in which I propose and to rely upon my own professional expertise rather than determine the matter in accordance with the evidence that would otherwise be proffered?

The answer was that an arbitrator did not act as a valuer. He performed judicial functions. He was bound to listen to the evidence which was offered. He could not refuse to hear witnesses and decide the matter from his own professional experience.

Must the arbitrator quantify rate relief?

The next question was:

> Am I entitled to deal with the matter without defining in cash terms the quantum of rate relief to be deducted from the rent which would otherwise be payable in the absence of the provisions of section 8(2)(b)?

It was clear that the arbitrator did not need to quantify the rate relief in cash terms unless a party asked him to do so. If a party did ask him was he obliged to do so? Paragraph 16 (b) of Schedule 6 to the 1948 Act required the arbitrator, on the application of either party, to specify the amount awarded in respect of any particular improvement or any particular matter. Was the deduction for rate relief an "amount awarded" to the tenant? This was more difficult, but in the end the conclusion was that it could properly be said to be such an amount. If a party applied the arbitrator could not refuse to specify it.

Is it full relief which arbitrator must not take into account?

The final question asked by the arbitrator was:

> Is the amount of relief from payment of rates which I am obliged not to take into account by the provisions of section 8(2)(b) the full relief afforded by all the Rating Acts for which Part V of the Local Government Act 1929 was the authority (now repealed but preserved by the General Rate Act 1967) or only the relief afforded *de novo* by the 1929 Act?

The question asked under this heading was whether the relief which the arbitrator had not to take into account was only the relief given by the Local Government Act 1929 itself —ie the *additional* relief given by that Act and not the full relief given by that Act together with the previous Acts which had partially derated agricultural land. The Acts had now, of course, been repealed and replaced by the General Rate Act 1967. There was no doubt, and both sides agreed, that the rating relief which the arbitrator was directed not to take into account was the full rating relief.

Comments on arbitrators' problem in applying section 8(2)(b)

At the close of his judgment, Judge Goodall said that the application of section 8(2)(b) could involve arbitrators in appallingly complicated calculations. The unfortunate arbitrator in the Scottish case of *Guthe v Broach*, mentioned earlier, embarked on the herculean task of, in effect, preparing a new rate for the whole county of Dumfries. This would mean, in the case of Cornwall, preparing complete valuations for the county and working out on the new total of rateable values what rate in the pound would be required to produce the sum required for the local authority's budget. In addition, if the total rateable value were increased it could well affect the rate support grant from central government. It might be helpful to express a view on how the matter might be approached in a practical way instead of attempting an impossible task. The Layfield Committee recently suggested that agricultural land should be rerated and the National Farmers' Union got out figures showing the increased cost of food which was likely to result. Such figures could be very useful. If it were said that the rating of agricultural land would increase the cost of food by x per cent, then one had a starting figure. This was only a suggestion, which was no part of a judge's function, but which was mentioned in case it might be found of some assistance to arbitrators faced with this problem.

ARBITRATION

COURT OF APPEAL

October 4 1978

(Before Lord DENNING MR, Lord Justice SHAW and Lord Justice BRANDON)

PLINTH PROPERTY INVESTMENTS LTD v MOTT, HAY & ANDERSON

Estates Gazette March 24 1979

(1978) 249 EG 1167

Arbitration—Case stated under Arbitration Act 1950—Rent revision clause in letting of floors in an office block—Provision for arbitration in case of dispute—" Fair rack rent rental value " to be determined—Rent had to be fixed having regard to rental values then current and to the provisions of the lease—The lease contained restrictions on change of user and on assignment and underletting—Question whether arbitrator should have regard only to strict terms of lease or whether account could be taken of the possibility that the lessors might agree to relax the restrictions—Held by Court of Appeal, upholding decision of Slynn J, that the arbitrator must fix the rent on the basis of the strict terms of the lease and accordingly at the lower of the two figures mentioned in the case stated

This was an appeal by the lessors from a decision by Slynn J on May 30 1977 on a case stated by an arbitrator appointed under the lease by the President of the RICS. The question raised by the arbitrator was whether in determining the rent of part of an office block called St Anne House in Wellesley Road and Walpole Road, Croydon, he was required to value in strict accord with certain restrictions in the lease as to user and assignment and underletting or could take account of possible relaxations. The arbitrator was P R S Downham FRICS. The lessors were Plinth Property Investments Ltd, the present appellants, and the lessees Mott, Hay & Anderson, a firm of consulting engineers, the present respondents.

G Godfrey QC and C Priday (instructed by Berwin Leighton) appeared on behalf of the appellants; M Essayan QC and O Albery (instructed by Slaughter & May) represented the respondents.

Giving judgment, Lord Denning MR said: This is a case involving the construction of a rent review clause. A sum of over £300,000 is involved. It is only a short clause in an underlease. It is an underlease of an office block called St Anne House in Croydon. The lessors let the ground, third, fourth and eighth floors of the building to the firm of Mott, Hay & Anderson, who are civil engineers. The underlease was to run from April 1 1966 to June 21 1984. For the first 10 years the rent was £32,914 a year. In these days of inflation the lessors naturally put into an underlease a rent review clause: and, for the last eight years, from April 1 1976, there was to be a rent review. An arbitrator was to be appointed by the President of the Royal Institution of Chartered Surveyors to fix the rent from April 1976. The underlease says—and these are the important words—" . . . the amount to be determined by the arbitrator shall be of such amount as shall in the opinion of the arbitrator represent the fair rack rental market value for the demised premises." Pausing there for a moment, the position is that, if there had been no special clauses in this underlease affecting the matter, at the present time the rent would have been fixed at a sum of over £130,000 a year for these premises, as against the figure in 1966 of £32,914 a year. The underlease goes on to say that the rent is to be fixed " having regard to rental values of property then current and to the provisions of this Underlease." That is the crux of the case—" having regard to . . . the provisions of this Underlease."

One of the important provisions of the underlease is a restriction on the tenants " not to use the demised premises or any part thereof or suffer the same to be used otherwise than as offices . . . in connection with the Lessee's business of Consulting Engineers." It seems to me that that restriction would affect very seriously any rent which the lessee would pay for the premises. He would not pay anywhere near the full market rent if his use of the premises was restricted to the business of civil engineers. For instance, he could not underlet them to a chartered accountant or any other professional person or to a business firm.

I must also read two other clauses about assigning or underletting. Clause 18(a) says: " That there shall not be any assignment underletting or parting with possession of part only of the demised premises nor any underletting or parting with possession or control of the demised premises." That clause was concerned with assigning *part* of the premises. Then clause 18(b) goes on to deal with the assignment of the *whole* of the premises. It says: " That there shall not be any assignment of the whole of the demised premises without the previous written licence of the Lessors and the Superior Lessors such licence not to be unreasonably withheld in the case of a responsible assignee or underlessee."

Although the underlessees could assign the premises with the written licence of the lessors, that did not affect the clause about user. If they assigned to anybody else, the assignee would still be subject to the clause that they were not to be used otherwise than as offices in connection with civil engineering business.

The point before the arbitrator was this: if there was no restriction and the full market rental was awarded, the rent would be £130,455, but if there was a restriction confining it to people who could only use the premises for the business of civil engineers, the rent would be £89,200. There was the conflict. Was the rent for the last eight years of the term to be £130,455 a year or only £89,200? Evidence was given before the arbitrator by distinguished surveyors on each side and the arbitrator stated a case for the opinion of the court. It was decided by Slynn J. Both the arbitrator and the judge ruled in favour of the lower sum of £89,200. The lessors appeal to this court. They say the correct amount should be the full £130,455.

I will not read the detailed questions stated by the arbitrator. No doubt they were framed by counsel on each side. Mr Godfrey submitted that it is wrong to take the restriction on user as having much effect on the rent because it was always open to the lessor to relax it or waive it; and no doubt he would relax it or waive it in his own interests so as to get a higher rent from time to time. Mr Godfrey said that he would relax it or waive it as a matter of business, and he would get a higher rent accordingly. Accordingly the rent should be fixed at the high sum. I will read the one important sentence in the case stated: ". . . whether the arbitrator can take into account the fact that the landlord might consent to a change of user and/or to an underletting of the whole or part of the premises." As Brandon LJ pointed out in the course of the argument, it does not say "with or without consideration." Looking at it as a matter of practical business, one has to remember that, owing to this restrictive clause, if the lessees wanted to use the premises for any other purpose than the business of civil engineers they would have to get the landlord's agreement to do so. The landlord would immediately say, "I'm not going to give my consent unless you make it worth my while. If you pay me a substantial sum, then I'll give my consent for these premises to be used for some other business, such as the business of a chartered accountant; but I am not going to give my consent unless it is made worth my while." That seems to me obvious as a matter of commercial practice. So this clause (restricting the user of the premises to the business of consulting civil engineers) obviously must have a very depressing effect on the rent. No one is going to take premises of this kind with such a depressing restriction on them unless there is a considerable discount on the rent. That is the way in which the arbitrator viewed it. He could not possibly assess the rent on the basis that a landlord will always relax or waive a clause of this kind wholly gratuitously. A landlord would never do that unless he was paid a considerable sum—and how is one to work that sum out? It is all hypothetical. It is so intangible that, as a practical matter, it cannot be done. I think the arbitrator was quite entitled to say that this restrictive clause has such a depressing effect on the rent that the proper rental value of these premises under the review clause for the succeeding eight years should only be the figure of £89,200 and not the figure of £130,455.

We have been referred to some of the cases, and some of the observations in them are quite interesting, but I do not think it is necessary to go into them. It seems that the arbitrator and the judge were quite right; and I would dismiss the appeal.

Agreeing, SHAW LJ said: Mr Godfrey has, if he will allow me to say so, put forward on behalf of the appellant lessors an argument as spirited and attractive as the terms of the lease in question permit. But those terms inexorably defeat his submission that the arbitrator should have taken account of the possibility that the lessors might have relaxed the restriction on user. The lease says what matters are to be taken into account in determining the fair rack rental. They do not include highly speculative factors, which are incapable of estimation on any basis such as the chance that the landlord might waive the restriction on user and, if he does, how much he would require to be paid for so doing.

Slynn J arrived at the correct conclusion, and I would dismiss the appeal.

Also agreeing, BRANDON LJ said: I am of the same opinion. It seems to me that when a lease provides that the arbitrator is to have regard to the provisions of the lease, it means the provisions that give rights and impose obligations on each of the parties to it. What the arbitrator has to consider is what those rights and obligations are on either side and assess the rent in the light of them. He is not to say to himself "Those who have rights may not enforce them and those who have obligations may not be required to perform them." He is to assume that the rights will be enforced and the obligations will be performed. He is to look at the legal position of the parties and nothing else.

I therefore agree with the judgment of the learned judge.

The appeal to the Court of Appeal was dismissed with costs. The order as to costs made by Slynn J stood, ie that the defendants (the present respondents) were to have the costs of the hearing before Slynn J of the case stated and the question of the costs of the arbitration to be remitted to the arbitrator. Leave to appeal to the House of Lords was refused.

COMPULSORY PURCHASE AND COMPENSATION

A **COURT OF APPEAL**

November 23 1978

(Before Lord Justice ROSKILL, Lord Justice GEOFFREY LANE and Sir David CAIRNS)

WAKERLEY AND OTHERS v ST EDMUNDSBURY BOROUGH COUNCIL

Estates Gazette February 17 1979

(1978) 249 EG 639

Appeal from Lands Tribunal on reference by consent—Compensation payable on surrender of yearly agricultural tenancy to acquiring authority—Terms of surrender agreement—Express incorporation of section 48 of Land
B **Compensation Act 1973 and implied incorporation of normal statutory basis of compensation, but additional provision that tenants were to be granted a gratuitous licence for three years and three months to farm part of land surrendered—Main question whether benefit of licence should be set off against compensation or whether tenants should retain benefit—A difference between £14,425 and £24,005—Held by majority of the Court of Appeal, reversing decision of Lands Tribunal, that value of licence should not be deducted from the tenants' compensation**

This was an appeal from a decision of the Lands Tribunal,
C reported at (1977) 241 EG 921, in a reference by consent as to the amount of compensation payable to Joseph Wakerley and the personal representative of his father, the late Arthur Wakerley, in respect of the acquisition by St Edmundsbury Borough Council of the appellants' interest in part of Eldo House Farm, near Bury St Edmunds. The main issue in the appeal was whether the value of a gratuitous licence to farm, granted to the tenants as part of the surrender arrangements, should be set off against the compensation due to them. The licence, for three years and three months, was granted in order to prevent the land from lying idle until it was needed by the council for a town development scheme. The Minister of Agriculture had given his consent to the licence under the
D provisions of section 2(1) of the Agricultural Holdings Act 1948, thus ensuring that it did not take effect as a tenancy from year to year.

L Read QC and J Denbin (instructed by Ellis & Fairbairn) appeared on behalf of the appellants; A Fletcher (instructed by the solicitor to the St Edmundsbury Borough Council) represented the respondent council.

Giving the first judgment at the invitation of Roskill LJ, GEOFFREY LANE LJ said: This is an appeal from a decision of the Lands Tribunal (R C Walmsley FRICS) dated January 28 1977. The tribunal in a reference by consent were charged
E with the task of assessing the amount of compensation properly due to the appellants (hereinafter called the tenants) from the compensating authority, the St Edmundsbury Borough Council (hereinafter called the council) in respect of the acquisition of the tenants' yearly tenancy of part of a farm. The tribunal judged that the proper sum was £14,525. This sum was made up of the figure of £14,425, being the capitalised value of the farming land the tenants had lost by the acquisition less a deduction for profits earned from it since acquisition, together with a nominal sum of £100 in respect of the open-market value of the tenants' interest in F
the land. The farm in question is Eldo House Farm close to Bury St Edmunds. It comprised 600 acres and was at all material times owned by the Orttewell Estate. From 1941 it was farmed by Arthur Wakerley. He was joined in 1953 by his son, Joseph (hereinafter called Mr Wakerley). From April 26 1954 these two held a joint yearly October tenancy in pursuance of a written agreement. The agreement contained the following provision in clause 34:

> That it shall be lawful for the landlords at any time or times upon giving four calendar months' notice in writing, to enter upon . . . and to resume possession of . . . any part or parts of the farm . . . for purposes of building . . . or for any other purpose G
> not being the use of the land for any agriculture.

Arthur retired in 1958 but the joint tenancy continued. He died in May 1976 after proceedings had started. They continued thereafter in the names of Mr Wakerley and of Arthur's personal representatives.

In 1971 the council's predecessors in title (the Borough Council of Bury St Edmunds), decided to buy certain land which was shown as a town development area in the development plan in accordance with the provisions of the Town Development Act 1952. This area included 155 of the 600 acres of Eldo House Farm. There were two phases. We are H
concerned here only with the second. This involved 126.6 acres of the farm. The remainder of the farm, that is to say, some 445 acres, is still held and farmed by Mr Wakerley. It was in January 1973 that the council bought the freehold of the 155 acres (subject to the tenants' tenancy) by private treaty from the estate. Thereafter negotiations were put in hand with a view to the council acquiring the tenants' interest in the land.

There seemed to be no reason why matters should not proceed expeditiously, and the original plan was that the yearly tenancy should be surrendered in January 1973. However, delay was caused by local opposition to the council's J
plans, and it was not until July 14 1973 that surrender finally took place. The agreement was very largely negotiated between two valuers engaged respectively by the council and tenants. They were Mr H O Wilkin for the council and Mr F A Laws for the tenants. Each was widely experienced and highly regarded. Their task was to some extent complicated by two events. It was not envisaged that all the 126.6 acres would be required at once, and it was therefore necessary for special provisions to be included in the agreement to ensure that good agricultural land was not left idle for longer than necessary. In order to meet that difficulty the agreement embodied provisions whereby upon surrender K
of the yearly tenancy the council would grant to the tenants a licence in respect of the 126.6 acres for 3 years and 3 months, that is, until October 1976, which would cover the period up to the 1976 harvest. The council reserved the right to enter upon any part of the land on 10 days' notice. They were obliged to pay compensation in respect of any loss of crops in 1973 and 1974, but the tenants had to crop during 1975 and 1976 at their own risk.

As far as compensation was concerned, it seems to have

been originally agreed as part of the negotiations between valuers that the sum should be computed on what up to that time had been the statutory basis under section 20 of the Compulsory Purchase Act 1965 and sections 9 and 12 of the Agriculture (Miscellaneous Provisions) Act 1968 for land compulsorily acquired. The effect of these provisions was to provide compensation for the value of the unexpired term and also for any loss or injury sustained, the loss in the present case being five times the annual rent to which the landlord would be entitled, in this case agreed at a notional £12 per acre, making in all a sum of £60 per acre. However, before a binding agreement had been reached, there was a change in the law. On February 23 1973 there was introduced into the House of Commons the Land Compensation Bill, which was to become law on May 23 of that year. This had the effect of altering the basis of compensation and increasing in certain respects the rights of dispossessed occupiers, as will hereafter appear.

The two valuers realised that the new Act would affect the situation which they were handling, and would require to be mentioned in the eventual form of agreement between the parties. They were *ad idem* as to the basis of the agreement (or at least they thought they were) and the deed was eventually drawn in its present form by a solicitor instructed jointly by both parties. The deed was executed on March 26 1974, a few days before, under the reorganisation of local government, the present council took over the powers and duties of the Mayor, Aldermen and Burgesses of the Borough of Bury St Edmunds. The deed contained *inter alia*, the following provisions:

> 1. In consideration of the Covenants to be observed and performed the Tenants hereby acknowledge to the Council as follows:
>
> (i) That they will surrender to the Council their tenancy of the land described in the First and Second Schedule hereto with effect from the 14th day of July One Thousand Nine Hundred and Seventy Three.
>
> (ii) That all claims to compensation which might otherwise have been made by either the Council or themselves are replaced by this Deed and Agreements marked "A" and "B" attached and that the Tenants have no further claim against the Council in respect of any statutory or Common Law Provisions arising out of their surrender of the land set out in the First and Second Schedule hereto or otherwise.
>
> 2. In consideration of the Tenants surrendering their tenancy of the said land conveyed to the Council as described in the First and Second Schedule hereto the Council hereby covenant that they will:
>
> (a) Forthwith pay to the Tenants the compensation to which they may be entitled under the provisions of section 48 of the Land Compensation Act 1973 or the sum of Sixty Pounds (£60) per acre whichever shall be the greater in respect of the land hereinafter described in the First Schedule hereto together with interest at the rate of 15 per cent per annum from the fourteenth day of July One Thousand Nine Hundred and Seventy Three until paid or from the first day of January One Thousand Nine Hundred and Seventy Three in the event of the Tenants not being entitled to compensation under the provisions of the said Section 48 of the Land Compensation Act 1973 . . .
>
> (g) As from the fourteenth day of July One Thousand Nine Hundred and Seventy Three grant to Joseph Arthur Wakerley in accordance with the approval of the Minister of Agriculture a Gratuitous Licence to farm the land set out in the first Schedule hereto in the form of the Agreement attached hereto and marked "A."

The licence and tenancy agreements referred to in the main deed were duly executed as therein provided. The minister's consent was obtained under the provisions of section 2(1) of the Agricultural Holdings Act 1948. This had the effect of preventing the licence taking effect as a tenancy. The licence was expressed to run from July 14 1973 to October 11 1976. The tenancy agreement is not material to the matters in issue; it relates solely to the "phase 1" land, with which we are not concerned.

There remained what may well then have seemed to be the formality of the two valuers agreeing the proper figure which the council should pay to the tenants by way of compensation under the terms of the agreement. Some of the items were agreed at the outset, some have since ceased to be in issue but there is still a broad area of disagreement. Before turning to those matters, the situation now that the licence has expired is that, as a practical arrangement, Mr Wakerley has agreed to act as caretaker on behalf of the council of such of the 126.6 acres as remain capable of being farmed. The council pay him compensation for cultivating and reimburse him for any expenditure he may make on seeds and fertiliser. The council have offered him a fresh licence from October 11 1976 to October 10 1979 at £15.50 per acre per annum, which he has declined for reasons which will become apparent.

The following matters are agreed between the parties:

(1) That had the council not acquired his interest in the land Mr Wakerley could have expected to farm the land for the rest of his life, and that that expectancy has been destroyed by the acquisition.

(2) That Mr Wakerley was at the material times a man of 38 years of age with a life expectancy of 33.5 years.

(3) That the net profit per acre per annum over the material period was £38.

(4) That the tenants are not entitled to the equivalent of one year's rent for disturbance.

(5) That the tenants do not have to bring into account as a set-off against any compensation to which they may be entitled the value of the fresh licence which they have declined to accept.

These last three matters were all in dispute before the tribunal but the parties have accepted the tribunal's findings in relation to them. The issues argued before this court were as follows:

(1) What is the correct number of years' purchase to be applied by way of multiplier to the annual profit of £38 per acre so as to capitalise that profit? Put in another way, what risk rate is properly applicable to a man in Mr Wakerley's position? Is it 15 per cent as the tenants, or 20 per cent as the council, maintain? The tribunal decided that 20 per cent was the appropriate figure, giving a multiplier of 4.989 and a capitalised loss of £24,003 against the tenants' rival contention of £30,808.

(2) Was the tribunal right in deducting from the tenants' loss as capitalised the capital value at July 14 1973 of the profit earned by the tenants from 1974 through 1976 as a result of the farming of the land under the licence?

This was the main point of the appeal, the tenants' contention being that on a true construction of the agreement nothing fell to be deducted from the capitalised loss, whatever that might be. The tribunal held that the value of the licence must be set off. That value the tribunal assessed at £9,581. The resulting net figure for compensation was therefore £14,422, which was rounded up to £14,425.

(3) There was a further subsidiary matter argued before us, namely whether the tribunal, assuming that they were correct in making any deduction at all in respect of the value of the licence, were right to deduct the gross profit before tax rather than the net profit after tax.

As to the question whether the proper discount was 15 per cent or 20 per cent, this is eminently a matter for the tribunal,

A and it cannot be suggested that there was anything wrong in principle or in law in selecting the higher of the two figures.

As to the question whether any deduction in respect of profit should be made gross or net of tax there are two answers, both fatal to the tenants' proposition. First, the profit itself was calculated on a " gross " basis before tax and accordingly the deduction must, one would have thought, be on the same basis. Secondly, no evidence was called by the tenants before the tribunal as to the incidence of tax and it would accordingly be impossible for this matter to be determined without sending the case back to the tribunal. I can see no justification for taking such a course in the circumstances.

B That brings one to the main problem in the case, namely what do the words of the deed mean, and in particular the words in clause 2 (a) and 2 (g) of the deed as expanded in the supplemental deed? What is the compensation to which the tenants " may be entitled " under section 48 of the Land Compensation Act 1973?

That section (so far as material) runs as follows:

Compensation in respect of agricultural holdings

48.—(1) This section has effect where in pursuance of any enactment providing for the acquisition or taking of possession of land compulsorily an acquiring authority—

C (*a*) acquire the interest of the landlord in an agricultural holding or any part of it; or

(*b*) acquire the interest of the tenant in, or take possession of, an agricultural holding or any part of it.

(2) In assessing the compensation payable by the acquiring authority to the landlord in connection with any such acquisition of an interest as is mentioned in subsection (1)(*a*) above——

(*a*) there shall be disregarded any right of the landlord to serve a notice to quit, and any notice to quit already served by the landlord, which would not be or would not have been effective if—

(i) in section 24(2)(*b*) of the Agricultural Holdings Act 1948 (land required for non-agricultural use for which D planning permission has been granted etc) the reference to the land being required did not include a reference to its being required by an acquiring authority; and

(ii) in section 25(1)(*e*) of that Act (proposed termination of tenancy for purpose of land's being used for non-agricultural use not falling within section 24(2)(*b*)) the reference to the land's being used did not include a reference to its being used by an acquiring authority; and

(*b*) if the tenant has quitted the holding or any part of it by reason of a notice to quit which is to be so disregarded, it shall be assumed that he has not done so.

E (3) In assessing the compensation payable by the acquiring authority to the tenant in connection with any such acquisition of an interest or taking of possession of land as is mentioned in subsection (1)(*b*) above (hereafter referred to as " the tenant's compensation "), there shall be disregarded any right of the landlord to serve a notice to quit, and any notice to quit already served by the landlord, which would not be or would not have been effective if the said sections 24(2)(*b*) and 25(1)(*e*) were construed in accordance with subsection (2)(*a*)(i) and (ii) above.

(4) Section 42 of the Agriculture (Miscellaneous Provisions) Act 1968 (tenant's compensation to be assessed without regard to his prospects of remaining in possession after contractual F date) and section 15(1) of that Act (effect on tenant's compensation of provision enabling landlord to resume possession for non-agricultural use) shall cease to have effect.

It is apparent from its terms that section 48 does not entitle a tenant to compensation as clause 2 of the deed suggests. It is designed to prevent the tenants' rights to compensation being diminished in certain respects. Its effect is first to deprive the landlords of the advantages given to them by section 42 of the Agriculture (Miscellaneous Provisions) Act 1968, which provided that in cases of compulsory acquisition of land which was an agricultural holding the tenants' compensation was to be assessed " without regard to the tenants' prospects, if any, of remaining in possession G of the holding after the relevant date," and, secondly, to alter the ruling of the House of Lords in *Rugby Joint Water Board v Foottit* [1973] AC 202. That decision enabled a landlord to take advantage of the provisions of section 24(2)(b) of the Agricultural Holdings Act 1948 (that is to say to give a valid notice to quit if the land is required for certain non-agricultural purposes) even though the person so requiring the land was not the landlord himself but an authority with compulsory purchase powers.

The landlords' compensation is therefore to be calculated in disregard of any right in him to serve a notice to quit on the strength of the council's intention to use the land for H building purposes. The tenants' compensation in its turn is to be calculated on the same basis and with the further protection provided by subsection (3). Clause 2(a) only makes sense if it is construed as importing the whole statutory method of assessing an agricultural tenant's compensation for compulsory purchase of his interest, including the provisions of section 48. Mr Read suggested the following paraphrase as fulfilling what the parties really meant by the compressed phraseology of the clause:

forthwith to pay to the tenants the compensation to which they might have been entitled if the council had acquired the yearly tenancy or taken possession of the land in pursuance of any J enactment providing for the acquisition or taking of land compulsorily and there was disregarded any right of the council to serve a notice to quit which would not be effective if sections 24(1)(b) and section 25(1)(e) of the 1948 Act were construed so that the reference to land being required for a non-agricultural use did not include a reference to its being required by an acquiring authority.

That paraphrase seems to me to reflect accurately what the parties must have intended. It makes it clear that the relevant statutory enactments prior to 1973 are necessarily incorporated and that effect is to be given to section 48 of the 1973 Act. Indeed there is little if any dispute between the K parties on this aspect of the case. That interpretation of this clause does not help to solve the problem of whether or not the profit from the licence has to be taken into account. Thus the area of dispute is small. The proper figure for compensation calculated according to the various statutory enactments is, if to be taken gross, £24,003 (rounded up to £24,005): if net after the deduction of the value of the licence (£9,581) the sum is £14,422 (rounded up to £14,425).

Which figure is correct depends on the meaning of clause 2 of the deed. If 2(a) had stood on its own unqualified by the remaining subclauses then without doubt, on the authorities, the tenants would have had to deduct any profit made under L the licence from their compensation. But enough has already been said about clause 2 to show that on any view it does not mean what it says, and it is both permissible and desirable to examine in more detail the factual background against which it came to be drawn. The two valuers had reached agreement in principle on the proper basis for compensation; hence the jointly instructed solicitor. The tribunal accepted as accurate the evidence of Mr Laws, the tenants' valuer, on this aspect. He described how in the back-end of 1972 and the early part of 1973 the council were anxious to expedite their development scheme. There was then a property boom. Everyone thought that by about 1975 there would be little M of the 126.6 acres left to farm. Mr Laws said that no suggestion had been made by anyone until recently about setting off the farming profits against compensation, and that such a suggestion was a major departure from the whole basis upon which his negotiations with Mr Wilkin had been conducted. If he had had any suspicion of such a possibility he would have never allowed his clients to accept the licence at all. It is true (as the council are not slow to point out) that what was passing through Mr Laws' mind is not a relevant consideration, and that the jointly-instructed solicitor may nevertheless by his choice of words have succeeded in producing the unforeseen result found by the tribunal. It is true that if he has that is the tenants' misfortune. The extent of their misfortune is clear. By their hard work and farming skill they have succeeded in reducing the council's liability to them by £9,581, a figure which might have been substantially increased if the council had persisted in their claim that the notional value of the fresh licence, offered by the council, but refused by the tenants, should also be set off. So, if the council's interpretation of clause 2 is correct, they have received surprising benefits from the agreement. They have been relieved of the burden of expensive and time-consuming compulsory purchase proceedings and of the risk that such proceedings might in the end be unsuccessful. They have avoided the expense of farming the land themselves or alternatively the possibility of public criticism for failing to make proper use of good farming land not immediately required for building. Have they had to pay for those advantages? No, they have not. If the tribunal was correct they have got them for nothing. That is a surprising result.

I do not believe that that is the meaning of clause 2. The clause is expressed to contain the various things which the council are promising to do for the benefit of the tenants as a reward for the voluntary surrender of their tenancy and for abandoning any claims they might have against the council as a result of that surrender. That being so it is plain that clause 2(g) is intended to convey a benefit and not a detriment. The only way in which it can convey the intended benefit is if it read as giving to the tenants and not to the council any profit which may accrue from the farming activities under the licence. That I believe to be the intention and effect of the clause. I would allow the appeal and increase the tenants' compensation to £24,005.

Dissenting on the main issue before the court, SIR DAVID CAIRNS said: In this judgment my references to "the claimant" mean Mr J A Wakerley or both him and his father as may be appropriate in the context. The claimant's right to compensation is a contractual right. The amount to which he is entitled depends on how clause 2(a) of the main deed of March 26 1974 is to be construed and whether its express terms are to be supplemented by any implied term. It has not been suggested that clause 1(b) of the other deed of the same date adds to or detracts from what the former clause provides.

It is common ground that the words "the compensation to which they may be entitled under the provisions of section 48 of the Land Compensation Act 1973" cannot be taken literally. The section does not create an entitlement to compensation; it merely lays down some fresh rules to be applied in assessing the compensation to which a landlord or tenant is entitled under other statutes. There is no dispute that the intention was that the claimant should have such compensation as he would be entitled under what has been called "the compensation code." The specific reference to section 48 of the Act of 1973 is easy to understand. It made a great difference to the scale of compensation payable and was in the forefront of the minds of the valuers who were negotiating about the compensation.

The main bone of contention is whether the licence granted on the same day as the deeds were executed is an element to be taken into account in assessing the compensation payable or whether it is to be treated as a separate transaction giving the claimant rights additional to his right to compensation.

The compensation payable to the tenant of an agricultural holding when an authority compulsorily acquires his interest or takes possession of the holding or part of it consists of various elements derived from a succession of statutes.

It is necessary to refer first to the Agricultural Holdings Act 1948. Section 24(1) of that Act imposed restrictions on the operation of a notice to quit given by the landlord of an agricultural holding to his tenant. Section 24(2), however, provided that the restrictions should not apply when the land was required for use other than agriculture for which planning permission had been given, and that fact was stated in the notice. Section 34 provided for compensation for disturbance when the tenant of such a holding left it in consequence of a notice to quit. The compensation was to be the cost of removal etc but not less than one year's rent.

The general rules as to the assessment of compensation for compulsory acquisition are contained in section 5 of the Land Compensation Act 1961 and the relevant rules are as follows:

> (2) The value of land shall . . . be taken to be the amount which the land if sold in the open market by a willing seller might be expected to realise.
>
> . . .
>
> (6) The provisions of rule (2) shall not affect the assessment of compensation for disturbance or any other matter not directly based on the value of land.

Section 20(1) of the Compulsory Purchase Act 1965 provided that if a tenant from year to year of land the subject of compulsory purchase was required to give up possession before his term expired he should be entitled to compensation for the value of the unexpired term and for "any loss or injury he may sustain." Section 20(2) provided that if any part of his land was taken he was to be entitled also to damage for severance or other injurious affection.

Section 9 of the Agriculture (Miscellaneous Provisions) Act 1968 provided for an agricultural tenant who was entitled under the 1948 Act to compensation for disturbance to have in addition a sum to assist him in the reorganisation of his affairs: the sum was to be four times the annual rent or, if only a part of his holding was taken, the appropriate fraction of the rent. Section 12 provided that section 9 was to apply to compulsory acquisitions or taking possession as if the acquiring authority were the landlord and compensation for disturbance were payable to the tenant. Section 15(1) provided that when there was an early resumption clause in the contract of tenancy of an agricultural holding (such as clause 34 of the claimant's tenancy agreement) it should be treated, in the case of acquisition, as if it provided for resumption of possession at the expiration of 12 months from the end of the year of the tenancy current at the date of the service of the notice to quit and, in the case of taking possession, should be disregarded.

Section 48 of the Land Compensation Act 1973, in relation to tenants, provided, by subsection (3), that in assessing the compensation payable to a tenant in connection with compulsory acquisition or taking possession any right of a landlord to give notice to quit was to be disregarded in cases where the notice would not be effective if section 24(2)(b) and section 25(1)(e) of the Agriculture Holdings Act 1948 were construed in accordance with section 48(2)(a) of the 1973 Act. By subsection (4), section 15(1) of the 1968 Act was repealed. By subsection (5) the tenant's compensation was to be reduced by the amount which the acquiring authority was liable to pay under section 12 of the 1968 Act. But by subsection (6) if the effect of subsections (3) to (5) was to reduce the compensation then it was to be increased by the amount of the deficiency.

The effect of all these statutory provisions in relation to the present case was, in my view, accurately summarised by the tribunal as follows:

> Under the general compensation code the compensation in the present case falls, as I understand it, under three heads, *viz*:
>
> 1. An amount equal to five years' rent, under section 12 of the Act of 1968, equating to four years' rent under section 9 of the Act of 1968 and importing one year's rent from section 34 of the Act of 1948.
>
> 2. The value of the "unexpired term or interest in the land," under section 20 of the Compulsory Purchase Act 1965, assessed in accordance with rule (2) of section 5 of the Land Compensation Act 1961 at open market value.

A 3. "Disturbance or any other matter not directly based on the value of land," assessed as referred to in rule 6 of the Act of 1961.

As to the first head above, section 48(5) of the Act of 1973 requires the amount payable to be deducted from the total amount payable under the last two heads, in order to avoid a double payment.

The third head might also have referred to the words "any loss or injury he may sustain" in section 20(1) of the Act of 1965.

It is to be noted that whereas before the passing of the 1973 Act compensation for a yearly tenant of an agricultural holding containing an early resumption clause was assessable on B the basis that the tenant had security of tenure for not less than two years; the effect of the 1973 Act was that the right of the landlord to serve a notice to quit, under that clause or otherwise, was to be disregarded in certain circumstances. The result was to enable the tenant to receive compensation based on his expectation of continuing to work for the farm for a longer period, in this case 33½ years.

A logical approach to the assessment of compensation in this case would have been to calculate the appropriate figures under each of the three heads and then to assess the compensation as the sum of that due under heads 2 and 3. C Assuming that this would exceed the amount due under head 1, that head could be disregarded.

The two valuers in this case did not assess any value for the interest in the land. They both simply worked out a figure intended to measure the total loss or injury which the claimant would have sustained if there had been a compulsory purchase of his interest. The tribunal took the view that, as no evidence had been given of the value of the claimant's interest in the land, that value should be assessed at the nominal figure of £100. The tribunal added "The whole of the claim as presented falls to be looked at under the heading of disturbance." That was an accurate statement D on the basis that "disturbance" is here taken to include "any other matter not based directly on the value of the land." The claimant's valuer had included in his valuation one year's rent for disturbance, using the word "disturbance" in the sense in which it is used in section 34 of the Act of 1948. That was disallowed by the tribunal and has not been pursued on the appeal.

I see no objection to assessing the compensation in the way the valuers and the tribunal adopted. If a separate substantial figure had been calculated for the value of the interest in the land it would merely have split the total compensation into two parts.

E In capitalising the loss of profit the claimant's valuer took a risk rate of 15 per cent whereas the local authority's valuer took 20 per cent. The tribunal accepted the latter figure. Mr Read sought to persuade us that the tribunal erred in law in adopting 20 per cent. He accepted that all the matters in addition to year to year farming risks which the tribunal took into account in reaching the rate of 20 per cent were properly taken into account but he contended that the tribunal attached too much weight to them. In my view no question of law was involved here but only a question of valuation, to be decided in the light of the evidence given. The tribunal F accepted the evidence of the local authority's valuer and in my opinion it is impossible for this court to say that he was wrong to do so.

No further issue remains about the capitalised loss of profit, assessed at £24,003. But from that figure the tribunal deducted £9,581, being the value of the licence for rather more than three years from July 14 1973. The last-mentioned sum was based on an assessment of the capitalised gross profits for the term of the licence. The issues that arise are whether any deduction should be made in respect of the licence and, if it should, whether it should be based on gross profits or on the profits after deduction of tax.

It was contended for the claimant that the meaning of G clause 2(a) is that compensation is to be assessed as if the claimant's interest in the 126.6 acres had been acquired on July 14 1973 and he had then ceased to farm that part of his land. The licence is not referred to in clause 2(a). It *is* referred to in clause 2(g) as being an additional part of the consideration for which the claimant surrendered his tenancy of the 126.6 acres. If this compensation is reduced by reason of the licence the licence was valueless to him—he was farming this land for three years for the benefit of the local authority. Indeed, if his compensation has to be reduced by his gross profits made under the licence there was a detriment and not a benefit to him from the grant of the licence. H

Mr Read made calculations to show how unprofitable the licence was to the claimant. Using the same basis as that applied by the tribunal he calculated that if the claimant had farmed the whole 126.6 acres under that licence for three years the deduction would have been over £14,000. If he had accepted a licence for a further three years and farmed the whole acreage throughout, the total deduction would have been near to £23,000. Thus in the latter case his compensation would have been reduced to a little over £1,000, though he would have been deprived of the opportunity to farm the land for another 27 years, the capitalised profits from which could be estimated at nearly £24,000. J Mr Read further calculated that the claimant's tax liability for the three years 1974-76 in respect of the 126.6 acres, or the part of it which he still retained during the last two years, was equivalent to a capital sum of £5,567. Thus the result of the tribunal's decision was to cause him a loss of that magnitude. Mr Read further argued that if the licence was to be taken into account then the compensation should be treated as compensation for dispossession. Then the proper basis of assessment would be that the claimant should have taken the capitalised value of the profits from 58 acres (taken over in 1975) for 31.5 years plus the capitalised value of the profit K from the remaining 68.6 acres (taken over in 1977) for 29½ years, which would come to a total of nearly £24,000. (There would, however, have to be a discount for payment in 1973.)

For the local authority it was contended that clause 2(a) means that the compensation code was to be applied in all respects as if there had been a compulsory purchase on July 14 1973. If there had been such a compulsory purchase the compensation which the claimant could have claimed for his loss of profit would have been his net loss after taking into account the profits he made while operating under the licence. If the premises in which a person carries on business L are compulsorily purchased and he moves into other premises and carries on business there his compensation for loss of profit will be measured by the diminution, if any, in profits, resulting from the change of premises. There is no reason for any difference if the business continues in the same premises under licence. And if gross figures are used for the profits lost then gross figures should be used for the profits which mitigate the loss. So, says the local authority, it is the clear intention of clause 2(a) that the compensation should be calculated in that way.

Mr Fletcher did not accept the relevance of any calculation M based on taking possession. As the payment was to be in consideration of the surrender of the tenancy and the surrender was deemed to take effect on July 14 1973 the compensation envisaged by clause 2(a) was clearly compensation for the acquisition of the claimant's interest in the land, as if such acquisition had taken place at that date, together with the loss or injury he thereby sustained. If the compensation could have been assessed on July 14 1973, or on March 26 1974, it would have had to take account of the expectation of profits under the licence. Once it is known what actual profits accrued then it is the actual profits that have to be considered. As Lord Denning MR said in *Bailey v Derby*

A *Corporation* [1965] 1 WLR 213: "Next you must ascertain the compensation for disturbance as it is called. That must be ascertained by looking at what has in fact happened since the notice to treat." So also in *London County Council v Tobin* [1959] 1 WLR 354. Wynn-Parry J said: "Had (the tribunal) been hearing the case at the time any figure at which they arrived would have been wholly an estimated figure. In fact they were sitting nearly two years after the move. In those circumstances they were bound to have regard to what had actually happened in the interval." In support of that proposition Wynn-Parry J cited from the speeches of Lord Macnaghten and Lord Robertson in *Bwllfa & Merthyr*
B *Dare Steam Collieries (1891) Ltd v Pontypridd Waterworks Co* [1903] AC 426.

Next Mr Fletcher pointed out that the effect of the deduction is to reduce the compensation by a little over one third. This is not unreasonable since profits made during the first three years after acquisition carry more weight in reaching the capitalised figure than the prospects of profits in later years. It cannot therefore be said that the parties would have agreed, if the matter had been raised when the transaction was entered into in March 1974, that the licence was to be disregarded in assessing the payment. The claimant got the benefit of a rent-free licence and there could be no
C implication that this considerable benefit was to be disregarded in assessing the payment.

As to tax, Mr Fletcher pointed out that tax was not deducted in assessing the capital value of the profits which would have been made if his tenancy had continued. It was necessary to treat the profits made during the licence period in the same way. In both cases the risk rate adopted was appropriate to gross profits and not to profits after tax. No evidence was adduced before the tribunal as to the amount of tax which the claimant in fact had to pay. If the claimant made a bad bargain by accepting the licence, it was not for the
D tribunal nor for this court to relieve him of the effects of his bargain.

I accept Mr Fletcher's contentions. I am of opinion that clause 2(a) means that the payment is to be calculated as if there had been a compulsory acquisition on July 14 1973, taking into account the whole of the provisions of the main deed, which by clause 2(g) provides for it to be granted and by clause 3(a) for it to be accepted. I find it impossible to imply a term that the licence was to be disregarded. It may be that the claimant, even if he had realised that he would have deducted from his compensation the value of any profits he made under the licence, would have still considered it
E worth while, for the convenient working of his farm, to retain the use of as much of his land as possible for as long as possible. It may be that the local authority, if the claimant had not been willing to accept the licence, would have deferred the date for acceptance of a surrender. These are matters of speculation. I can find no firm basis for implying that the claimant was accepting the licence only on the understanding that he was entitled to make under it a profit which would not be taken into account in assessing his compensation.

As to tax, I am quite unable to see how it could be right for the deduction to be calculated net of tax while the loss of profits which would have accrued under the continuing
F tenancy agreement were not to be so calculated. If a person carrying on business had his premises purchased compulsorily and then proceeded to carry on the same business in another place, earning the same annual profit, it would be absurd to say that he had lost his gross profits from his original premises but mitigated his loss only to the extent of his taxed profits at the new premises. So here I consider that the tribunal was quite right to make the deduction on the basis of gross profits earned.

For these reasons, I am of opinion that the tribunal's decision was not in any way erroneous in law and I would dismiss the appeal.

Agreeing with Geoffrey Lane LJ in allowing the appeal, ROSKILL LJ said: In my view the solution to the principal problem we have to decide turns upon a very short, but to my mind difficult, question of construction of the deed entered into on March 26 1974. Does clause 2 of that deed upon its true construction provide that the benefits derived by the tenant under the licence which the council agrees to grant under clause 2(g) and which the tenant agrees to accept under clause 3(a) shall enure solely to the benefit of the tenant, or is the effect of clause 2 that what the tenant receives from those benefits shall operate to reduce the council's liability to pay compensation due under clause 2(a)? To put the same point in other and shorter words—if the H
sum payable under clause 2(a) is £X and the benefits derived from the licence are £Y, does the council upon the true construction of the deed pay to the tenant £X minus £Y or does the tenant receive from the council £X and retain £Y and of course any other sums which clause 2 provides shall be paid by the council to him?

I unreservedly accept that clause 2(a) is compressed in its drafting and that is language, though clear, requires to be expanded as proposed by Mr Read and accepted by Geoffrey Lane LJ in his judgment. I further accept—indeed the contrary was not argued before us—that if clause 2(a) had stood alone, then in the case of compulsory purchase the J
relevant statutory compensation for disturbance must be calculated by reference to what the tenant proves that he has lost when the compensation falls to be assessed by the relevant tribunal, and that if between the date of the compulsory acquisition and that assessment of compensation some new benefit has accrued to the tenant which operates to reduce his loss caused by disturbance, the quantum of compensation payable under the relevant statutory provision must be *pro tanto* reduced. If, therefore, clause 2(a) stood alone, the tenant would clearly have to bring into account any benefit that he had received after the date of the surrender of his tenancy which operated to reduce his loss by K
disturbance.

But what is the effect of adding subclause (g) in clause 2? All the covenants in clause 2 are expressed to be covenants by the council in consideration of the tenant surrendering his tenancy. The consideration moving from the promisee—that is the tenant—is that surrender. The clause provides that in return for that consideration the council—the promisor—will confer what are plainly benefits of various kinds upon the tenant, that is to say, payments of cash, payments of compensation for fertiliser and lime, forbearance to claim dilapidations and the grant of the licence in question. As L
Mr Read pointed out, the council derived not inconsiderable benefits from this arrangement by way of voluntary surrender. The council were saved the necessity of seeking a compulsory purchase order with all the possible delay, expense and uncertainty involved. By getting the tenant to agree to surrender the land and then to farm the land for the next three years under licence, they were saved having to take possession sooner than they required and then either risking the opprobrium of leaving valuable farming land unhusbanded or of making other arrangements at their own expense for the husbanding of this land until such time as the council's development plans came to fruition—a course for which they M
would have had to expended public money.

If the contention of the council and the decision of the Lands Tribunal be right, this agreement upon its true construction saved the council all this and more; the council could delay taking possession; the land would be properly farmed at no cost to themselves; there would be no risk of opprobrium for wasting valuable farmland and the tenant was not only to do all this for the council for nothing, but the council's liability for compensation under clause 2(a) would ultimately be *pro tanto* reduced. Small wonder that Mr Laws, the tenant's valuer, expressed himself in evidence

A which the member accepted, saying, as Geoffrey Lane LJ has pointed out, that if he had realised that this position might materialise, he would never have allowed the licence for the years 1974 to 1976 to be taken up. That last point must, however, be irrelevant to the question of construction. Agreements are sometimes incautiously entered into which produce unforeseen, unexpected and maybe unjust results. This is neither here nor there. All we have to decide is whether this agreement provides that £Y has to be added to £X or whether £Y has to be subtracted from £X.

I cannot construe clause 2(g) as operating in diminution of the tenant's right to compensation under clause 2(a). There B is no express language to this effect. I am unable to imply any such right in the council. I think clause 2(g) is a separate and independent covenant enuring solely for the benefit of the tenant just as are the covenants in clause 2(c) and clause 2(e). On that narrow point of construction I regret I must differ both from Sir David Cairns and from the member. In agreement with Geoffrey Lane LJ on that issue I would allow the appeal. On the other points argued before us I agree with both the judgments delivered and have nothing to add.

The appeal was allowed with costs in the Court of Appeal and below, except the costs of an application to the Lands C Tribunal to settle the form of the case, which were to be paid by the appellants.

COURT OF APPEAL

July 3 1978

(Before Lord Justice STEPHENSON, Lord Justice ROSKILL and Lord Justice GEOFFREY LANE)

HONISETT AND ANOTHER v ROTHER DISTRICT COUNCIL

D Estates Gazette March 3 1979

(1978) 249 EG 847

Compensation for compulsory acquisition of leasehold interest in a small area of land with development value—Town and Country Planning Act 1968, section 30—Appeal by acquiring authority against Lands Tribunal's decision in favour of claimants—Valuations of freehold and leasehold interests interdependent—Agreed value of freehold in possession but disputes as to deductions to be made in arriving at value of leasehold—Criticisms of Lands Tribunal's valuation rejected—Freeholder with a defective E title assumed to be in the market—No error of law by tribunal

This was an appeal by case stated from a decision of the Lands Tribunal (J D Russell-Davis FRICS) (1977) 243 EG 1007 relating to the compensation payable for the compulsory acquisition of land, just under half an acre in size, at Church Road, Catsfield, East Sussex. The appellants were the acquiring authority, Rother District Council, and the respondents R L O Honisett and H Kennedy, leaseholders under a lease expiring in 1998.

J Sullivan (instructed by Sharpe, Pritchard & Co, agents for J G Millward, Town Hall, Bexhill-on-Sea) appeared on F behalf of the appellant council; M Horton (instructed by Sheppard & Son, of Battle) represented the respondents.

Giving judgment STEPHENSON LJ said: The decision of the Lands Tribunal was given on April 5 last year. The appellant authority is dissatisfied with the amount of compensation awarded to the leaseholders of some land, just under half an acre in extent, at Church Road, Catsfield, East Sussex. There is no dispute that some compensation has to be paid for the compulsory acquisition of that land by the appellant authority under section 30 of the Town and Country Planning Act of 1968.

The leaseholders claimed £11,750; the authority put forward evidence that they were only entitled to £2,850; and G the member of the Lands Tribunal awarded them £10,900. It is submitted by Mr Sullivan on behalf of the appellant authority (and he did not appear for them below) that although he cannot seek to justify as low a figure as that which was put forward on behalf of the authority before the Lands Tribunal, or to justify at least one of the items that went to make up that figure, nevertheless, this court should allow an appeal against the tribunal's award, and, not substituting its own valuation, remit the matter to the Lands Tribunal for rehearing.

The grounds on which he puts forward that submission are, first of all, that the Lands Tribunal failed to make a H material finding of fact; alternatively, if they did make that finding of fact by implication, that there was no evidence to support it; and thirdly, that in giving its reasons, the tribunal failed to have regard to one fact and to deal with one of the substantial issues in the case.

The trouble in this case really arose, as it seems to me, from the fact that the freeholder—or possible freeholder (the Earl of Chichester)—was not certain whether he was able to make a good title to the freehold. This parcel of land (I suppose with others) was originally let as long ago as the 40th year of the reign of Queen Elizabeth I for a term of 200 years from Michaelmas Day 1598. An option J was exercised to continue it for a further 200 years, and that long period of 400 years expires 27¼ years after the valuation date, namely on September 29 1998. And the rent which the Earl (or whoever is the correct freeholder) derives from the lease is the sum of five pence per annum.

The land has a value because of its development possibilities. An outline planning permission was obtained by the respondent leaseholders (the claimants) in 1965 for five detached bungalows. That planning permission had not been acted upon, but the land has now been developed by the authority with eight units and a lay-by, which was a condition K of the later detailed planning permission.

There was correspondence in 1965 with Lord Chichester's land agents, and in a letter of April 29 1965 they said that they understood that outline permission had been given for five units, and indicated that they would not be able to recommend a sale of the reversion for less than £1,750, subject to various other matters being cleared up. However, on the next day, Lord Chichester's solicitors wrote that it had been decided not to sell the freehold reversion at the present time. Later Lord Chichester, on a reference initiated by the authority for the determination of the value of the freehold interest, indicated by his solicitors that, as it had L not been possible to establish title, he would not be represented.

It became clear during the course of the hearing of the reference in respect of the valuation of the leasehold interest that the values of the two interests were largely interdependent. In the absence of the freeholder the tribunal determined to decide the case for the leaseholders, and adjourned that for the freeholder until the former decision had been published One can understand the authority's anxiety not to have to pay twice over, and not to find themselves in the position of, having given full compensation to the leaseholders, thereafter being compelled to pay the freeholder, and losing a sum M greater than that which they would have had to pay had the value of both interests been determined at one and the same time. Nevertheless, in challenging in this court the award in fact made in respect of the valuation of the leasehold, they have to show an error of law on the part of the tribunal.

A fact which was purported to have been agreed, and from which both valuations put before the tribunal started, was that the value of the freehold in possession of the subject land at the material date—and that was the date when the authority entered into possession, on June 30 1971—was

£15,000. The rival valuations were put before the Lands Tribunal, both based on that agreed value of £15,000.

Mr G J Walker, a Fellow of the Royal Institution of Chartered Surveyors, deducted from that figure the cost of acquisition of the freehold reversion, which he worked out at £3,000, considering it a reasonable uplifting from the figure of £1,750 which had been quoted in the letter from Lord Chichester's land agents of April 1965. He deducted also legal fees on the £3,000 scale, and an agency commission on it, totalling £140, and he also deducted a single premium for indemnity against a third party claimant to the freehold, quoted at from £50 to £70, but he put that up to £110. And there was a letter from an insurance broker informally quoting a premium of £50-£70 from Sun Alliance, a London insurance group, for an indemnity insuring the leaseholder against a third party claim, claiming the freehold. Those deductions together came to £3,250, and left a net value of the land, for the claimants' development, of £11,750.

On the other side Mr Nightingale, the deputy district valuer and valuation officer for the Eastbourne District, put forward a much lower valuation on behalf of the acquiring authority. He considered that a deduction of £750 from the agreed figure of £15,000 ought to be made for the lay-by, and that the Lands Tribunal accepted: the construction of the lay-by was properly deducted from the value of the unencumbered freehold in possession, and a value put upon it, with the benefit (it should have been said) of the planning permission that had been granted. So, starting from £14,250, instead of £15,000, Mr Nightingale went on to assume the freeholder being in the market as the most likely purchaser, and he deducted, first (a) for a profit on venture, then (b) for costs of merging interests, and third (c) for resale of freehold in possession 20 per cent, the sum of £2,850—a deduction which Mr Sullivan found difficulty, I do not say in understanding, but in supporting. That gave a figure of £11,400 and he then went on to assume that the parties would agree that their interests were of equal value, so he deducted the freeholders' share of one-half, £5,700, leaving the value of the leaseholders' interest as £5,700, but went on: "As the freeholder was not available, he assumed that only a speculator would be in the market," and he deducted another half of £2,850 from that £5,700, leaving a net value realisable for the leasehold interest of £2,850.

Now, though the member of the Lands Tribunal preferred (reluctantly) Mr Nightingale's contention that the £750 for the lay-by should be deducted, he did not follow Mr Nightingale's reasoning in making a deduction of 20 per cent for profit. He did not follow it in either sense; I do not think he understood it, and he certainly was not going to accept it. The reason he gave was that he regarded the starting point at £15,000—or £14,250—the market value which would have been paid by a purchaser who had already allowed for profit and expenses, or who would reckon it to come out of the development itself. In reaching his conclusion, he said this—in a passage in his decision which has been criticised by Mr Sullivan:

> A freeholder without a good title would have been negotiating from a position of weakness, and I agree with Mr Walker that in the real world he would have been glad to take a sure £3,000 or thereabouts for his interest rather than to chance holding out for a half share. Mr Nightingale's argument of equal shares would have had more force had the freeholder a good title. Furthermore, by halving the leaseholders' residue, I think he is making a double deduction for the same thing; he has already disposed of the "known" freeholder, and until the hearing when he agreed that insurance was practicable I do not think he had applied his mind to its possibility. For these reasons I have to reject his approach and turn to that of Mr Walker.

He then referred to Mr Nightingale having conceded that a known freeholder is the most likely purchaser, and said that that followed from the *Mountview Estates* case and the judgment of this court in *Lambe v Secretary of State for War* [1955] 2 QB 612. We have looked at *Mountview Estates Ltd v London Borough of Enfield* (1968) 20 P & CR 729.

The member concluded:

> I think it is plain that the freeholder's particular position of weakness would have benefited the leaseholder, and I am persuaded by Mr Walker that in those circumstances he could reasonably have been expected to accept £3,000 so that the two interests could be married. By his further deduction of £140 Mr Walker has gone some way to satisfying the point (b)

—that is the cost of merging interests—

> in Mr Nightingale's valuation. Nevertheless, the evidence of the cost of indemnity insurance is not conclusive. The brokers' letter seems to have been written following a telephone conversation with Mr Walker, not a letter setting out the full facts, and the quotation seems to be in respect of a price of £15,000 up to beyond the end of the lease. I think that the insurers would, however, have had to envisage more than that, in that the benefit would be assignable to the five ultimate purchasers after development had taken place, when something like a capital sum of £60,000 for the cost of five bungalows would have been at risk. I prefer to raise the deduction in this respect to £200.

So he came to his criticised valuation of the leasehold interest: £14,250 as the starting point, net value of the freehold in possession after allowing for the lay-by and from that he deducted the cost of the acquisition of the freehold reversion, £2,950; for fees and costs, he accepted Mr Walker's figure of £140, and the single premium indemnity he uplifted from £110 to £200, for the reasons I have just read from his decision. So that the total deductions from the £14,250 were £3,290, leaving a value of £10,910, which he grossed down to £10,900 and awarded.

For my part I cannot see anything wrong in the decision, either in law or in fact; and were it not for the fact that Mr Sullivan has argued that the member has not found whether or not the reputed freeholder—that is, Lord Chichester, with his weak title—would have been in the market as a purchaser, I would not have thought it arguable from the passages at p 12 of the decision which I have read, because those passages seem to me quite beyond doubt to indicate—not to imply, but to express the view of the member—that the freeholder without a good title (namely Lord Chichester) was in the market. That seems to me to be the basis of his decision.

I cannot agree, secondly, that there was no evidence to support that finding; nor could I draw any such inference or hold against that finding the fact that no steps were taken, either by the claimants or the freeholder, to implement the 1965 planning permission after May 1965. Although the agreement as to the £15,000 in 1975 was an agreement between the leaseholders and the authority, I see no reason to suppose from inactivity that this reputed freeholder abandoned all interest in the development of this property and what might be gained for it. Indeed, his position seems to have been kept open by the way in which the action was taken on his behalf by his solicitors in the communication they made to the Lands Tribunal, and in the adjournment of the valuation of the freehold until the leasehold had been valued by the Lands Tribunal. Obviously, the uncertainty or weakness in the reputed freeholder's title made his position difficult, but I am quite unable to conclude that there was no evidence that he was in the market—or to have any doubt that the tribunal found that he was in the market.

In those circumstances, such possible points of law as are raised in the notice of appeal by Mr Sullivan seem to me to be without substance. I certainly cannot find that the reasons given by the member for his decision are inadequate, or fail to deal with the substantial issues in the case—or that

A there is anything substantially wrong or inadequate in the reasons that he gives.

So we are not entitled in this court, following the decision of Megaw J, as he then was, in *Re Poyser & Mills Arbitration* [1964] 2 QB 467 to quash his decision and remit the matter to the tribunal for another valuation. For these reasons I would dismiss the appeal.

Agreeing, ROSKILL LJ said: The reasons of the member of the tribunal have been assailed as lacking in clarity and not dealing adequately with the points raised before him. With all respect to Mr Sullivan, it seems to me that both submissions are unfounded. Reading the reasons as a whole,
B it seems to me absolutely clear that the member intended to find that Lord Chichester was in the market, and there was ample evidence, to my mind, upon which that finding was justified. I think the reasons given are both clear and adequate, and I do not think that this case raises any point of law.

I would only add that I feel bound to express my regret that a public authority should have spent public money pursuing this matter to this court, for it seems to me that this was essentially a matter of fact for the tribunal and that the facts found are quite adequate in justifying the
C conclusions reached.

The tribunal rejected the argument put forward on behalf of the district valuer, and I do not find that in the least surprising when one sees the argument in the form in which it appears in the case. That being so, it seems to me very difficult to justify this matter having been brought further. Like my Lord, I would dismiss this appeal.

GEOFFREY LANE LJ also agreed.

The appeal was dismissed with costs.

D

CHANCERY DIVISION

June 9 1978

(Before Mr Justice GOULDING)

LONDON TRANSPORT EXECUTIVE v CONGREGATIONAL UNION OF ENGLAND AND WALES (INCORPORATED)

Estates Gazette March 24 1979

(1978) 249 EG 1173

Compulsory acquisition of back land of a church—Section 92 of Lands Clauses Consolidation Act 1845—Whether church could claim benefit of section and require whole church
E **premises to be acquired—Meaning of "part only of any house or other building"—Authorities reviewed—Held that the church was a "house or other building" and that the land to be acquired (with certain minor exceptions) was a "part"—As church was willing and able to sell the whole, section 92 applied**

In this case the acquiring authority, London Transport Executive, sought a declaration as to whether the Congregational Union of England and Wales was entitled to claim the benefit of section 92 of the Lands Clauses Consolidation Act 1845 in respect of the acquisition of back land forming
F part of Streatham United Reformed Church. The Union claimed that the section applied, to that the Executive could be required to take the whole of the church premises, and this claim was opposed by the Executive.

K Bagnall QC and K M J Lewison (instructed by the chief solicitor, London Transport Executive) appeared on behalf of the Executive, the plaintiff to the originating summons; G Ryan (instructed by Kingsford, Dorman & Co) represented the Union, the defendant to the summons.

Giving judgment, GOULDING J said: This is an originating summons by which the plaintiff, London Transport Executive, seeks certain declaratory relief in connection with the compulsory acquisition of freehold property in Streatham G
belonging to the defendant, Congregational Union of England and Wales (Incorporated). The short question is whether the defendant may in the circumstances, as it claims to do, invoke section 92 of the Lands Clauses Consolidation Act 1845. The defendant owns its property in Streatham upon charitable trusts defined by the United Reformed Church Act 1972. I shall use the present tense for convenience in describing the property, but save where I state otherwise the description is to be understood as relating to the date of the plaintiff's notice to treat.

The property in question is an approximately rectangular piece of land fronting on the Streatham High Road, which H
here runs roughly north and south. According to the evidence the property is managed and used as a single unit for the purposes of the Streatham United Reformed Church. The church building itself stands on the front part of the land; it is quite big; it can hold several hundred people. At the rear—that is the west of the church—is a roughly square plot, which I shall call the back land, about the same size as the site of the church building itself. Part of this back land is a car park. At the date of the notice to treat the car park was let to a company at a rent of £500 a year, with a provision for a certain number of parking places to be available for the church congregation on Sunday J
mornings. The tenancy of the car park has since determined and has not been renewed in view of the pending compulsory acquisition. The rest of the back land is an area of grass with a hall standing on it, called the Carmichael Hall, and two small outbuildings.

I shall read a few paragraphs from the affidavit of the church treasurer, describing the church's activities upon the property. He deposes:

> The Church has a number of rooms of varying sizes attached to the main part of the Church and also has some detached rooms and outbuildings. The main part of the Church is used for the K
> purpose of worship save for the occasional rehearsals of religious drama. There is a regular Sunday congregation of over a hundred but considerably more when there is a special service such as a parade service or Harvest Festival.
>
> The various rooms (both attached to the Church and detached) are used partly in connection with services (for example, the Minister's Vestry for robing and the Parlour for tea after services) and to a greater extent for purposes of an educational kind which are available to the community at large. There is a large hall (attached) a small hall (attached) and the Carmichael Hall (detached) plus outbuildings used for storage.
>
> The large hall is in great demand for all kinds of activities and is in almost constant use both during the day (eg Women's L
> Guild Community Organisation meetings, bric-à-brac sales, Sunday School) and in the evenings (eg Brownies' and Guides' meetings, the Streatham United Reformed Church Badminton Club, the Drama Society). The result of the demand for this hall is that there is often an overspill into the small hall or the Carmichael Hall. This is particularly so in the case of the Drama Society, which produces two plays a year and has to rehearse in the Carmichael Hall or the small hall, although the performances are given in the large hall.
>
> The small hall is also in almost constant use. Every morning from 9 am to 1 pm it is used by a Nursery School for children aged 2 to 5. At other times it is used by the Adult Education Institute and for meetings of the Townswomen's Guild and others. M
> Some of the Sunday School children meet here as of a scheme for the separation of the age groups. The Nursery School children have not only had the use of the hall but have also been able to play outside on the open ground adjoining the Carmichael Hall (the blue land) and this has contributed greatly to the success of the school.
>
> The Carmichael Hall has always been put to full use by the Church. For over 20 years up until June 30 1976 it was used every weekday as a ballet school by the Miss Robinson School of Dancing. The ballet school closed down in 1976 because of the threat of compulsory acquisition of the blue land on which the Carmichael Hall is situated.

A I omit two sentences dealing with the closure of the ballet school. The affidavit then continues:

> The Carmichael Hall is the headquarters of the local Scout Group known as the Fourth Streatham Common Group. They meet in the hall on at least three evenings a week. The group includes Venture Scouts, Scouts and Cubs (ie boys). The hall is also used by the Girl Guides and Brownies. These groups also make extensive use of the open land surrounding the hall for their outdoor activities; for example, they practise erecting tents and lighting fires there and also hold sports evenings and football training sessions there. Their equipment is stored in the outbuildings on the blue land.
>
> B Apart from its regular use by the Scout Group the Carmichael Hall serves as an overspill for activities which cannot be accommodated in the large hall or the small hall because of other bookings. As I have said in paragraph 8 above this often applies to the Drama Society but it can also occur in relation to groups with only occasional meetings.

The defendant also owns and uses a dwelling-house for the residence of the caretaker of the church. It is one of a terrace of houses abutting on the rear boundary, that is the west boundary, of the back land.

The plaintiff wants to enlarge its omnibus garage, which is a large erection standing to the south of the defendant's C property. On or about September 22 1975 the plaintiff served the defendant with a notice to treat. The notice does not seek to acquire the church or the caretaker's house, but it does embrace the whole of the back land including the car park and the Carmichael Hall. The notice to treat relies on the plaintiff's powers conferred by the London Transport Act 1975. That Act, as is usual in such legislation, incorporates large parts of the Lands Clauses Acts and of the Railways Clauses Acts. One of the incorporated provisions is section 92 of the Lands Clauses Consolidation Act 1845. It is quite a short section and I will read it at once. It is as follows:

D

> And be it enacted, that no party shall at any time be required to sell or convey to the promoters of the undertaking a part only of any house or other building or manufactory, if such party be willing and able to sell and convey the whole thereof.

The defendant claims the benefit of that section. It asks the plaintiff to take the whole property, not just the back land. By the originating summons the plaintiff prays the court to declare that the church is not a house or other building or manufactory within the meaning of the legislation and that the defendant is not entitled to require the plaintiff to acquire any property other than that comprised in the notice E to treat.

In order to resist the plaintiff's case the defendant must show within the meaning of section 92 of the Lands Clauses Consolidation Act 1845 that: first, the church is a house or other building; it is clearly not a manufactory. Secondly, the land to be taken by the plaintiff, or some of it, is part of the house or other building. Thirdly, the defendant is willing and able to sell and convey the whole thereof.

The third point is not in contest and I need say no more about it. If the section were free from judicial authority I should not have thought the first question difficult. In the ordinary use of language at the present day the expression F "house or other building" is equivalent to building, and the defendant's church is a building. That, of course, is the defendant's contention in this litigation. However, many reported cases show a disposition to give a very wide extension to the term "house" in this context, and some of them tend conversely to restrict the interpretation of the words "other building." The plaintiff therefore asserts that a place of worship on the scale of the defendant's church is too far removed from an ordinary domestic edifice to be within either branch of the statutory phrase "house or other building."

Let me take the word "house" first. As long ago as 1861 in *St Thomas's Hospital (Governors) v Charing Cross Railway Co* (1861) 1 John & H 400 the hospital as it then G existed in Southwark was treated as a house for the purposes of section 92. The report does not record any argument on that point, the real contest having been whether the ground to be taken by the railway company was part of the hospital within the meaning of the section. The case has nevertheless been authoritatively mentioned with evident approval as showing that "house" includes a hospital. It was so cited in a leading case in the Court of Appeal, *Richards v Swansea Improvement & Tramways Co* (1878) 9 ChD 425, where a compound containing several buildings and yards, used mainly for manufacturing and trading purposes, but in part H for the residence of the owner's employees, was held to be a single house within section 92. James LJ, when referring to the hospital case, remarked (at the foot of p 431) that the word "house" beyond all question would include a very large inn or anything of that kind built for one purpose. Similarly, in *Caledonian Railway Co v Turcan* [1898] AC 256, a decision on an identical provision in section 90 of the Lands Clauses Consolidation (Scotland) Act 1845, Halsbury LC and Lord Watson found it natural to speak of a warehouse as a house.

Richards' case, to which I have just referred, is also of interest on the words "other building," but only by way J of dictum, for the decision was on the word "house." Brett LJ, as he then was, at p 434 of the report, explained section 92 thus:

> That which is treated of is premises, and there are three kinds of premises which are described. The first kind of premises described is a house; and that word has been held to mean not merely that which is a house in the ordinary sense, but a house both in the ordinary and legal sense; that is to say, it must be a house in the ordinary sense, but it may include more that a house in the ordinary sense, namely, that which is also a house in the legal sense, that is, the house and the curtilage and garden, and all that is necessary to the enjoyment of the house. The K second thing which is described in that section seems to me that which is not in the ordinary sense a house, but a building which is in the nature of a house, although in ordinary language it would not be called a house. The third thing which is mentioned in the section is a manufactory, which may be a house, or may be a building, but which may be something more; that is to say, it may be more than one house, or more than one building, or it may consist of neither house nor building, but only of land used for a purpose of manufacturing. No part of that section is to diminish any other part. Each thing described is different from the others, and it is not circumscribed by the description of any of the others.

Cotton LJ said at p 437: L

> Then "building" and "house" I also look upon as distinct in this way, that there may be a building which could not be said either in the legal sense or in the ordinary sense to be a house, and that therefore that word was added as something different from that in order to include something not necessarily included in this word "house."

Brett LJ's dictum that the "other buildings" referred to by section 92 are buildings in the nature of a house, was applied by Warrington J, as he then was, in *Regents Canal & Dock Co v London County Council* [1912] 1 Ch 583, where it was claimed that the whole undertaking of the M canal company must be treated as a single unit for the purposes of section 92. The learned judge put the question thus, at p 589:

> Now it is not contended at the Bar that this undertaking comes under the expression "manufactory." It is not even suggested that it comes under the expression "house," although of course it includes a large number of houses. The argument is that it comes within the expression "other building," and no doubt in one sense it is a building. Every wall, whether a retaining wall, such as that which lines this canal, or a wall standing above the surface of the ground, is in one sense a building, but is it a

A reasonable use of ordinary language to say that the whole of Regent's Canal, with all its appurtenances is one building within the meaning of section 92?

He then read an extract from Brett LJ's judgment in *Richards'* case, and concluded thus:

> Now I read that passage as meaning that, no doubt, a building is something different from that which is ordinarily called a house, but the Lord Justice did say—and in my opinion he was applying, if I may say so with all respect, the true principles of construction to the language—that the words "other building" used in that collocation meant something "in the nature of a house,' although in ordinary language it would not be called a house. Applying B that view to the present case, can anybody say that this canal is a building which, though not a house in the ordinary sense, is in the nature of a house? It is perfectly true that the undertaking as a whole comprises a number of structures which may be "buildings" according to the true construction of the Act of Parliament; but is the undertaking of the canal one "building" within the meaning of the Act? In my opinion it is nothing of the kind, and on that ground I hold that the present motion fails.

From the foregoing and numerous other authorities Mr Bagnall for the plaintiff seeks to extract the propositions, first that to be a house within section 92 a structure must have an existing or potential residential use; secondly, that C to be another building within the section a structure must be within what he calls a domestic scale. Neither proposition is established by explicit authority. For the first Mr Bagnall in reply relied very strongly on Bacon V-C's observations in *Barnes v Southsea Railway Co* (1884) 27 ChD 536, where he said at p 542:

> To my mind the clause in the Act is distinct, and the law is perfectly distinct. The Legislature, having to frame in words an expression which would cover the subject included in the clause, adopted this word "house," but not affecting to give any description of what a "house" means, because none was necessary. The word "house" had already acquired a legal sig- D nification. The passage which has been cited from Lord Coke is not new law in itself, although it is a very plain and distinct definition. The meaning of "house" is domus, residence, possession—what a man has when he talks about "having a house." The meaning of the word "house" in the Act of Parliament, therefore, is the meaning which Lord Coke ascribes to it in Coke upon Littleton, and it includes all that which may be called the domus.

In my judgment, however, that passage is of no assistance because the learned Vice-Chancellor was not at all considering what sort of building may be called a house, but whether a paddock adjacent to a dwelling-house and garden was part E of the house for the purposes of the section.

Of more relevance is a passage from the judgment of Cotton LJ in *Richards'* case at p 438, where he says:

> With regard to the word "house," it is impossible to say it is to be confined to a building used wholly, or almost wholly, for residential purposes. In fact it was conceded by Mr Dickinson, and it could not be well disputed, that "house" would include business premises, and, in my opinion, that is perfectly correct, although possibly it might be said that "house" points primarily to a residence. Yet it is impossible, especially after the decided cases, to say that a "house" is confined to that of which the primary object is residential. I refer to a case which was so often quoted in the argument, that of St Thomas's Hospital. No doubt F persons did reside there, and the persons, while they were being treated there for diseases or for accidents under which they were suffering, did some of them live there. Yet it is impossible to say that it was a residential house, but it was held to be a house within the fair meaning of this Act of Parliament. That the word "house" is not to be confined to something used for residential purposes is very apparent from what was said by Lord Hatherley, when Vice-Chancellor Wood, that you could not possibly take the theatre of the hospital and then say you were not taking part of the house.

It is noteworthy that Warrington J did not hesitate to say that the Regent's Canal undertaking included a large G number of houses, even though none of the buildings enumerated in the evidence at pp 585 and 586 of the report appears to have been used for residential purposes.

Then I turn to Mr Bagnall's second proposition: that a building other than a house to be within section 92 must, though not necessarily residential, at least be on a domestic scale. There is nothing in the *Regent's Canal* case to support this, and how is the scale to be defined? A large hotel, which according to James LJ would be a house beyond all question, might be much bigger than the defendant's church.

I am not surprised in the end to find that the authorities H do not constrain me to so astonishing a conclusion as to say that a substantial place of worship is not within the expression "any house or other building," and I hold that it is. I do not find it necessary to distinguish between the two limbs of the expression, but for the purpose of examining the second main question, namely whether the plaintiff was proposing to take part of the defendant's house or other building, I will assume, without deciding, that the church is not a house but is another building.

It has long been well settled what is part of a house for the purposes of section 92. In the *St Thomas's Hospital* case Page Wood V-C said at p 404: J

> The construction of the word "house" in the Act of Parliament has been clearly settled by authority to comprise all that would pass by a grant of a house; and this, according to the old authorities, would include not only the curtilage, but also a garden attached to the house; *a fortiori*, therefore, any buildings forming part of or appertaining to the messuage would also be included.

Lord Romilly is reported as having said a year earlier, in *King v Wycombe Railway Co* (1860) 28 Beav 104 at p 106:

> The principle is this: That, according to the true construction K of the Act, the word "house," includes everything which will ordinarily pass under the word "house" in a conveyance. Consequently, if the company require any part of that which is included under the word "house," they are bound to take the whole. The courts have, I think very justly, determined that the word "house," as used in the statute, is to be construed liberally, and that it includes all that which properly passes under the word "house."

To similar effect is the passage I have already read from *Barnes v The Southsea Railway Co;* and in *Caledonian Railway Co v Turcan*, that I have also cited, Halsbury LC, at p 263, speaking of a courtyard adjoining the respondent's warehouse, said: L

> The piece of land with which we are now dealing is part of the curtilage of the house, is just as much part of the house as the hall. I think Lord Montcree uses the very language that I have adopted, that it is part of the vestibule of the house and just as much part of the house as if it were a separate room in the house.

Mr Bagnall submits that none of that line of authority is applicable when you come to consider the words "other building." Section 92 only applies in such a case if the acquisition takes away part of the erected structure itself M or the ground inside its walls. That seems to me, at first blush, a highly improbable interpretation. The purpose of the section was thus described by Cotton LJ in *Richards'* case once again at p 438. He said:

> . . . the object and intention of this section was evidently to give a certain protection to landowners, to persons whose property was taken away from them against their will, so that no person should be required to sell a fraction only of that which ought to be regarded as a unit, when he might be very materially prejudiced by having left on his hands certain fractions only of that unit, not capable of being used efficiently when one fraction had been taken away from it.

If, with that object and intention, Parliament gives rights to owners of houses, and in the same breath to owners of other buildings, how can one suppose, without the slightest indication in the language, that the appurtenant curtilage is to be included in the one and excluded in the other case? The distinction is, to my mind, capricious and unwarranted.

But no, says Mr Bagnall, it is necessary to avoid the absurdity of forcing an undertaker to buy a vast building because he needs a litle bit of open ground close to it, perhaps of trivial value to the landowner. That effect is, however, inherent in the other limbs of the enactment, if you think of a ducal mansion or a big hospital under the word "house," or of an immense industrial site under the term "manufactory." It long ago became common practice to exclude possible extreme applications of section 92 by special provisions in private and local statutes, and for recent general legislation on the subject see section 8(1) of the Compulsory Purchase Act 1965. Therefore, looking at the matter apart from reported authority, I find no force in Mr Bagnall's submission on this point.

As to authority, I think that the argument for the plaintiff is probably inconsistent with Turner LJ's exposition of section 92 in *Grosvenor (Lord) v Hampstead Junction Railway Co* (1857) 1 De Gex & J 446 at pp 453 and 454. It might, if correct, have been employed with possibly decisive effect in *Caledonian Railway Company v Turcan*, already cited, and in *Genders v London County Council* [1915] 1 Ch 1, a case like the present of a nonconformist church, but there is no trace of its consideration in either of those authorities. Of course an argument is not answered by the mere omission of eminent persons to raise it when they might have done. However, the first appearance of Mr Bagnall's important distinction after the Act has been in use for more than a century can only strengthen my view that it is plainly unsound.

I ask then, applying the test furnished by the authorities, does the land to be taken by the plaintiff include part of the defendant's church? I will not concern myself with the car park or the caretaker's house. Special considerations may perhaps apply to them; to the car park because it was subject to a tenancy at the date of the notice to treat; and to the caretaker's house because of its separate position and structure as part of a terrace of houses. I will concentrate my attention on the Carmichael Hall and the plot of grass, with two small outbuildings, on which it stands.

On the affidavit evidence, some of which I have read, I am of the opinion that at the date of the notice to treat those subjects were an integral part of the church property, and should, in accordance with the authorities on the section, be treated as part of the church within section 92. The contrary, indeed, was not strongly argued on the facts, reliance being placed on the propositions of law which I have rejected. It follows that in my judgment the plaintiff is not entitled to the declarations claimed and the originating summons falls to be dismissed.

The summons was dismissed with costs.

COURT OF APPEAL

January 30 1979

(Before Lord DENNING MR, Lord Justice ROSKILL and Lord Justice CUMMING-BRUCE)

GEORGE v SECRETARY OF STATE FOR THE ENVIRONMENT AND ANOTHER

Estates Gazette April 28 1979

(1979) 250 EG 339

Compulsory purchase of house which had become a "massive housing problem"—House owned jointly by husband and wife—Wife not served with separate notice of the order—Validity of order challenged under the Acquisition of Land (Authorisation Procedure) Act 1946, Schedule 15, para 1—Wife's affidavit—Whether wife could be cross-examined on affidavit—Deputy judge allowed cross-examination "de bene esse," but later ruled that she should not have been cross-examined—General rule that deponents should not be cross-examined in proceedings for judicial review or in applications such as the present—Judge's discretion to allow it should be exercised only in most exceptional cases—Affidavit evidence in present case should have been rejected without cross-examination—No substantial prejudice or breach of natural justice established—Judge's order in favour of applicant set aside

This was an appeal by the Secretary of State for the Environment and Greenwich London Borough Council from a decision of Sir Douglas Frank QC, sitting as a deputy judge of the Queen's Bench Division, on May 25 1978 in favour of Mrs Victoria Oluremi George, who had applied to quash a compulsory purchase order in respect of a house owned jointly by Mrs George and her husband at 15 Brookhill Road, London SE18. Mrs George claimed that she had never been served with the compulsory purchase order. Sir Douglas Frank's decision was reported at (1978) 248 EG 49.

H K Woolf (instructed by the Treasury Solicitor) appeared on behalf of the Secretary of State; Desmond Keane (instructed by A A Child, solicitor to the council) represented the Greenwich Borough Council; M Howard (instructed by Anthony Steel & Co) represented Mrs George, the respondent to the present appeal.

Giving judgment, LORD DENNING MR said: Some time ago Mr and Mrs George came from Nigeria. He came to be a student of law. At one time he was said to be a barrister of Lincoln's Inn, but that was not right. He was a student at Newcastle. In 1967 Mr and Mrs George bought a three-storey terrace house, 15 Brookhill Road, London SE18. They bought with the aid of a mortgage from the local council (now the Borough of Greenwich). The house was registered in their joint names. But they did not manage it properly. The house got very badly out of repair. They filled it with many, many people from overseas. It became overcrowded. They did not pay the rates. It presented (in the words of an inspector later) "a massive housing problem for the local council." The council summoned Mr and Mrs George from time to time for not doing this, that and the other. The house presented such a problem to the council that in November 1974 they decided it was a case where they ought compulsorily to purchase the house because it was so overcrowded and so out of repair. According to his statements, Mr George was earning good money. So was his wife. Nevertheless they looked after the house so badly that the council thought the only solution was to acquire the house compulsorily.

The council passed a resolution under the Housing Act to acquire it. In accordance with the requirements of the Housing Act, they sent out notices so as to find out who were the owners of the house. By this time the local council were not the mortgagees. There had been a redemption of their mortgage. It had been replaced by a mortgage to Barclays Bank. The local council sent out the statutory notices to find out who were the owners or who was interested. They sent out a notice both to Mr George and to Barclays Bank. They received back clear and unequivocal statements, not only from Mr George but also from the bank saying that Mr George was the owner of the premises. So they served only Mr George with the notice of compulsory acquisition.

The council overlooked one source of information. They could have looked up the property register—to see who was registered as the owner—but they did not. It would have shown that both Mr and Mrs George were joint owners. Not having inspected the property register, the council believed quite honestly and, I think, quite reasonably that Mr George was the sole owner. So, after making their compulsory purchase order, they served it only on Mr George. They did not serve it on his wife. They put all the proper notices in the news-

A papers. They did everything in proper order, but they unwittingly made one mistake. They did not serve the compulsory purchase order on Mrs George.

The order was made in May 1975. The notice was served on Mr George only on May 28. Mr George resisted the order. He made objection to it. He said he would repair the house, and he would look after it properly. So the minister decided there should be an inquiry. As it turned out there were three inquiries. The first was in August 1975, but the inspector fell ill and could not make his report. There was another inquiry in April 1976; and, eventually, an inquiry in 1977. As a result of those inquiries, the inspector was quite satisfied that the B only way to solve this massive housing problem was for the compulsory purchase order to be confirmed. On May 19 1977 the Secretary of State in his decision letter accepted the recommendation and confirmed the compulsory purchase order.

Under the statute there is a period of six weeks during which the order can be challenged. Schedule 1, para 15, of the Acquisition of Land (Authorisation Procedure) Act 1946 provides:

> (1) If any person aggrieved by a compulsory purchase order desires to question the validity thereof . . . he may, within six weeks from the date on which notice of the confirmation or making of the order or of the giving of the certificate is first pub- C lished in accordance with the provisions of this Schedule in that behalf, make an application to the High Court, and on any such application the Court . . . (b) if satisfied that the authorisation granted by the compulsory purchase order is not empowered to be granted as aforesaid, or that the interests of the applicant have been substantially prejudiced . . . , may quash the compulsory purchase order....

On the very last day of the six weeks Mrs George came along. She challenged the validity of the compulsory purchase order. She said: "I was the co-owner. I was never served with this compulsory purchase order. It is therefore bad as against me." She put in an affidavit in support of her applica- D tion, in which she said: "15 Brookhill Road, London SE18, which is the subject of the above mentioned compulsory order, is jointly owned by my husband and myself. We bought the property in November 1967 with the help of a loan. . . . My husband and I lived in the property for about 3 years from the time of the purchase but later moved (to another address)." Later on—and this is the crucial paragraph—she said: "Last month it came to my notice by reading a local paper that a compulsory purchase order had been made and confirmed in respect of our property. This surprised me because at no time was I served with any notice in relation to the same." She followed that up by two or three further E affidavits in which she elaborated her complaint and grievance. On March 3 1978 she swore an affidavit in which she said: "In May 1975 I was in Oxford doing my training at the Fairmile Hospital. I only came down to London when I had a long day off which was usually once a month. When I came down at the end of May my husband told me that the council wrote to say that a compulsory purchase order had been made in respect of 15 Brookhill Road SE18, but assured me that he would be going to the council to sort things out. . . . Had I been served with the notice I would not have relied on my husband's assurance that everything was alright. I would have sought for legal advice."

F Her application came before Sir Douglas Frank, sitting as a deputy judge of the High Court. The case took several days. One day in March, two days in April and eventually a judgment in May. One question which arose was whether Mrs George could be cross-examined on her affidavits. The council wished to cross-examine her. They said she kept quiet about being a joint owner. She said nothing to anyone about it until the order was about to be confirmed. Then she came out into the open at last and said: "I have not been served, and therefore the order should be set aside."

It is a practical point of some interest. In such a case as this ought the court to allow cross-examination on affidavits? Only one reported case has been found in the books where G there has been any cross-examination on affidavits in cases such as this or in the parallel cases on judicial review: it was *R v Stokesley, Yorkshire, Justices, ex parte Bartram* [1956] 1 WLR 254. Lord Goddard said: "This is probably the first case in recent history in which application has been made in Crown proceedings for leave to cross-examine on affidavits." He only allowed it in that case because it was a case of a "very remarkable" character. It is to be noticed that in *Franklin v Minister of Town and Country Planning* [1947] 1 All ER 612 both Lord Oaksey and Morton LJJ indicated that if application had been made to cross-examine the minister in that case, it might have been allowed, but no application H was made.

It seems to me that, in general, cross-examination shall not be allowed in prerogative writ proceedings or in proceedings for judicial review or in applications like this for compulsory purchase orders. There are very good reasons for this rule. First, usually the affidavits speak as to what took place before the determining body. It may be before a planning inspector, a magistrate or someone of that kind. He may have to make an affidavit as to what took place before him. It is undesirable that he should be subjected to cross-examination. We said so in a recent case about a certification officer. If he is not to be subject to cross-examination, neither should the applicant. J Secondly, experience shows that on these procedural questions there is very little conflict on the affidavits. Thirdly, one party or the other might, by means of cross-examination, try to undermine the actual findings of the inspector or other officer holding the inquiry.

In general, therefore, cross-examination should not be allowed, but I would stress that there is nothing in law to prohibit it. It is undoubtedly permissible under the rules. It is covered by Order 53, rule 8(1), which brings in Order 38, rule 2(3), which in terms provides for cross-examination on affidavits. It is a matter for the discretion of the court. It seems to me that in these proceedings cross-examination should K not be allowed except where the court believes it is necessary in order that justice may be done between the parties.

I am inclined to think that this was not a case which needed cross-examination. The affidavits are (if I may put it this way) open to suspicion. They are untrustworthy. I would not be prepared to accept Mrs George's statements at face value.

Furthermore, I see no substantial prejudice to Mrs George by reason of her not being served with this order. She was the joint owner. She knew she was the joint owner. On her own statement in the affidavit of May 1975, her husband had told her that a compulsory purchase order had been served. She was looking to him to take care of his own interest. No L doubt that would include her interest too.

I do not mean to suggest that it is unnecessary to serve a joint owner. Every joint owner should be served, when he or she is known. A wife should be served as well as her husband. In this case the council, if they had known that the wife was a joint owner, ought to have sent out two separate letters, one to her and one to him. But the failure to do it does not mean that she is automatically entitled to have the order set aside. Under the second limb of paragraph 15(1)(b), which I have read, the burden was on her to show that her interests had been "substantially prejudiced" by reason of the omission to serve her. It seems to me that she has not M established that in the least. Her husband had been served. He took every objection that was open to either of them. Everything was heard at the inquiry. He looked after her interest as well as his. She was not prejudiced, substantially or at all.

As it happened, Sir Douglas Frank did allow Mrs George to be cross-examined. He took it *de bene esse* as she was going off to Nigeria. But he afterwards ruled that she should not have been cross-examined. So he put it out of his consideration. But if it is taken into account it is plain that she contradicted herself time after time. She gave different dates and different circumstances in several places. Her evidence

A was, as Sir Douglas Frank said, somewhat unsatisfactory. The one thing she seemed able to concentrate on was "I was not served with the notice" as though that was all that mattered. If her evidence is looked at, it shows that she was not substantially prejudiced, or prejudiced at all, by the failure to serve the order on her in the beginning.

That brings me to the second point in the case which Mr Howard put in by his cross-appeal. He relied on the failure to serve the order on her as being a breach of natural justice. He relied on the first limb of paragraph 15(1)(b) where the ground of invalidity is said to be that the compulsory purchase order was "not empowered to be granted" under the Act. It is said
B that an order is not empowered to be granted if the steps which have been taken have resulted in a breach of the rules of natural justice. I agree. Lord Russell said recently in the case of *Fairmount Investments Ltd v Secretary of State for the Environment* [1976] 1 WLR 1255 at p 1263:

> . . . if the true conclusion is that the course which events followed resulted in that degree of unfairness to Fairmount that is commonly referred to as a departure from the principles of natural justice it may equally be said that the order is not within the powers of the Act and that a requirement of the Act has not been complied with. For it is to be implied, unless the contrary appears, that Parliament does not authorise by the Act the exercise of powers in breach of the principles of natural justice, and that
> C Parliament does by the Act require, in the particular procedures, compliance with those principles.

Accepting that a breach of natural justice (which results in unfairness to the applicant) is a ground for challenging the order, it is necessary to consider the meaning of "a breach of natural justice" in this regard. We have been referred once more to the many cases upon this subject, particularly to the case of *Hibernian Property Co Ltd v Secretary of State for the Environment* (1973) 27 P&CR 197, the judgment of Browne LJ at pp 217 to 218, together with the later case of *Performance Cars Ltd v Secretary of State for the Environment* (1977) 34
D P&CR 92. On reading those cases, it seems to me that there is no such thing as a "technical breach of natural justice." In the unreported case of *Lake District Special Planning Board v The Secretary of State for the Environment** given on February 13 1975, Kerr J said at p 15F of the transcript:

> I therefore accept the submission of Mr Woolf that a breach of natural justice is not made out *ipso facto* by the mere receipt by the Secretary of State of some further representations after the close of the inquiry and the failure to circulate these to everyone concerned so as to give them an opportunity to make further representations thereon, and so on. I also accept the submission that there can be no such thing as a "technical" breach of the rules of natural justice, since the concept of natural justice is not
> E concerned with the observance of technicalities but with matters of substance.

The question is whether, as a result of any failure in procedure or the like, there was a breach of natural justice. On this approach, the position under the first limb is almost indistinguishable from that under the second limb. You should not find a breach of natural justice unless there has been substantial prejudice to the applicant as a result of the mistake or error which has been made. In this case there was no substantial prejudice. I do not think there was any prejudice at all to Mrs George by the fact that she herself was not served with this initial order in May 1975. Her husband knew
F all about it: and neither her husband nor the bank ever told the council about her interest. She was in touch with her husband at the time, even if she was away working as a nurse. In all probability she decided not to take any steps of her own, but left it to her husband to do all the fighting, which is a natural thing for a wife to do. It was only after the order had been made, after it had been confirmed, on the last day of the six weeks that she said, "I was not served with a notice." It seems to me that there was no prejudice to her whatever. It is now far, far too late for her to question the validity of the order.

I think the appeal should be allowed and the confirmation G
of the minister upheld.

Agreeing, Roskill LJ said: It is now 4½ years since the respondent authority resolved to acquire this property. Judging from the inspector's report at the third inquiry which had to be held with regard to the proposal compulsorily to acquire this property, compulsory acquisition was the only way in which the disturbing housing conditions which were prevailing there could be remedied.

That compulsory purchase order was made on May 8 1975—3 years and 8 months ago. There were then, as I have said, three planning inquiries. One became abortive because the inspector was ill; and it was not until June 2 1977, some 21 H
months ago, that the department concerned confirmed that compulsory purchase order. The whole of this delay has been occasioned first by an understandable but unfortunate slip which was made by the acquiring authority to which the respondent's husband must undoubtedly have been a party by giving inaccurate information to the authority. He was required by notice given by the authority to state who the owners were. He failed to state the fact that his wife was a joint owner with him. As a result, the authority did not, as my Lord has said, serve both the joint owners with the notice; and the husband and now the wife have been quick to seize upon every possible argument, good, bad and indifferent—and mainly, if I may J
say so, bad—in an effort to avoid the consequences of that compulsory purchase order. I hope I do no injustice to Mr Howard's admirable argument if I say that I find myself unable to agree with the reasoning of the learned deputy judge.

First, as to the question of cross-examination, I respectfully agree with all that has fallen from my Lord, and in general I accept Mr Woolf's submission on that issue. If one looks at Order 94, rule 3, one sees that the relevant rule of the High Court applicable to an appeal of this kind under the statute reads thus: "Without prejudice to the powers of the Court under Order 38, rule 2(3), evidence at the hearing of a K
motion under rule 1 shall be by affidavit." When one goes back to Order 38, rule 2(3), one sees that that subrule expressly gives the court the power to order cross-examination in appropriate circumstances. It has long been the practice in the Divisional Court not to allow—and this is for good and obvious reasons—cross-examination of deponents upon affidavits. There is, so far as we have been told by counsel, only one recorded case in the last 30 or 40 years of this being allowed to happen. So far as the present type of proceeding is concerned, there seems to be no recorded case of cross-examination of the kind now in question being sought. I think it would be most undesirable save in the most excep- L
tional case that this type of cross-examination should be allowed, for it would encourage applicants against whom a ministerial decision has been given to seek to reopen on a statutory appeal (which is a supervisory or controlling jurisdiction of the High Court) matters of fact which are not within the province of the High Court in such proceedings. Nevertheless there is a limited residual jurisdiction in the exceptional case to order cross-examination.

In the present case, when one looks at this lady's affidavit, with the greatest respect to Sir Douglas, I find that it stretches my credulity too far to take them at their face value. I think if I had been in his position I would have rejected this appeal M
out of hand on the ground that those affidavits are quite incapable of establishing on the balance of probabilities that she has suffered any substantial prejudice. If that were so, that would be an end to this appeal. But if I had had any doubt and had regarded this case as exceptional, then I might have permitted cross-examination, as he did, *de bene esse*. Had I done so, my judicial incredulity would have been converted, I suspect, into total judicial disbelief of her evidence, because one only has to read the transcript to see, as my Lord has said, that her evidence is unworthy of belief in any relevant respect. There is nothing but repetitive adherence to the contention that

*Summary report at (1975) 236 EG 417.

A she need not have done anything at all because she had never been served. To borrow my Lord Cumming-Bruce LJ's phrase, it looks to me as if at all material times this lady "kept her head well down" in order to take advantage of the possible benefit which might accrue as a result of this unfortunate failure on the part of Greenwich Council.

That is enough to dispose of this case, but on the question of the construction of the statute I respectfully agree with everything that has fallen from my Lord.

The second point on which I venture to disagree with Sir Douglas is on what he said on the question of onus of proof. One has only to look at the relevant provisions in paragraph
B 15(1) of the First Schedule to the Acquisition of Land (Authorisation Procedure) Act 1946 to see that the onus is, as I have said throughout the case, on the person seeking to oppose the order. The manner in which the onus is discharged will of course vary from case to case, but I respectfully disagree with Sir Douglas when he said: "I think in an ordinary case where it was admitted that the order had not been served that would amount to *res ipsa loquitur* and the burden would then pass to the acquiring authority"; later he said: "I have reached the conclusion that it is not sufficient evidence to establish Mr Keane's"—Mr Keane appeared for the authority—"proposition that the applicant
C made a conscious decision not to object to the order and I think that the failure to serve the order deprived the applicant of the chance of objecting to it." In both those passages, with respect to Sir Douglas, I think he put the onus of proof the wrong way round. I think it was for the applicant, under whichever limb the case was put, to show on a balance of probabilities that she was entitled to relief.

For the reasons my Lord has given, I do not believe for one single moment that she has suffered any substantial injustice or prejudice. I do not believe that she has been in any way prejudiced or has lost the chance of appearing at one of these many inquiries which her husband's conduct has
D necessitated.

For those reasons, and also agreeing whole-heartedly with what my Lord has said, I would allow this appeal.

Also agreeing, CUMMING-BRUCE LJ said: The first issue is the question whether the learned deputy judge was right in deciding that the cross-examination which he had already heard *de bene esse* was admissible. The test which he applied was as follows: "It seems to me that cross-examination should be permitted only to determine a clear-cut issue of fact on a procedural matter, as was the case in *Stokesley*." Like my Lords, I take the view that in relation to the Acquisition of Land (Authorisation Procedure) Act 1946 procedure, which
E is governed by Order 94, rule 3, it is only in extremely rare cases that it is likely to be appropriate to permit or order cross-examination.

The kind of situation in which I would expect cross-examination to be ordered is where the affidavits of one party are so unsatisfactory that, although they cannot be regarded without cross-examination as worthless evidence, they cannot be confidently accepted as evidence of fact without cross-examination.

On the facts of this case, on the affidavits, I share the surprise of Roskill LJ that the learned deputy judge, on reading the affidavit of the applicant, was able to convince himself
F that that evidence, although not altogether satisfactory and although giving ground for suspicion, was yet to be accepted without cross-examination as evidence of fact on which judgment could be founded. I would have thought, on reading the affidavits, that there was an overwhelming inference that the evidence was both ingenuous and ingenious and so suspicious that without cross-examination it should anyway be rejected. But, if the learned judge had not been prepared, as he was not, to go as far as that, it was in my view his duty to admit the cross-examination in order to determine whether the evidence was reliable.

On the substantive question arising on the cross-appeal as G to whether there is shown to be any breach of natural justice, I do not for a moment accept that on the authorities there is any ground for the view that there is such a concept known to the law as a technical breach of natural justice. A breach of natural justice means that because of what has happened —something that has been done or has failed to be done— somebody has either actually suffered injustice or there is a real risk that somebody has suffered injustice. Therefore, the starting point is to look at the evidence of fact and see whether the evidence points to the inference either that the applicant did suffer injustice or that there was a real risk that the applicant did. On the facts of this case—as I would have H thought on the evidence was manifest—there was no injustice suffered by the applicant and no risk that the applicant suffered any injustice.

For those reasons I agree with the orders proposed.

The appeal was allowed, the judgment below set aside and the minister's confirmation of the CPO upheld.

CHANCERY DIVISION

March 8 1979

(Before Mr Justice BROWNE-WILKINSON) J

TRUSTEES FOR METHODIST CHURCH PURPOSES AND ANOTHER v NORTH TYNESIDE METROPOLITAN BOROUGH COUNCIL

Estates Gazette May 19 1979

(1979) 250 EG 647

Compensation for compulsory acquisition—Claim by church for compensation on equivalent reinstatement basis—Purported withdrawal of notice to treat by acquiring authority —Whether authority entitled to withdraw—Whether claimants had served a proper notice of claim in accordance with section 4 of the Land Compensation Act 1961—Claim not K **quantified in terms of money—If section 4 had to be interpreted literally there was no answer to authority's contention that it could withdraw under section 31—Claimants' arguments based on alleged impossibility of quantifying the claim and on anomalies of literal construction rejected— Held that a notice complying with section 4 had not yet been served and that authority could still withdraw**

This was a summons taken out by the custodian trustee and a representative of the managing trustees of the Methodist Church and Sunday School in Burdon Street, Percy Main, North Shields, in order to obtain the determination of the court on the construction of certain statutory provisions. L These provisions concerned the withdrawal of a notice to treat following the confirmation of a compulsory purchase order which included the plaintiffs' church property and a notice of claim served by the plaintiffs. The defendants, the North Tyneside Metropolitan Borough Council, were the housing authority for North Tyneside.

Michael Mann QC and D G Robins (instructed by Herbert Reeves & Co, agents for Ingledew Mark Pybus, of Newcastle upon Tyne) appeared on behalf of the plaintiffs; A D Dinkin (instructed by Darley, Cumberland & Co) represented the defendant council.

Giving judgment, BROWNE-WILKINSON J said: This case M raises a short point on the Land Compensation Act 1961 (which I will call "the 1961 Act"). The defendant authority is the housing authority for North Tyneside and in that capacity on March 27 1974 made a compulsory purchase order affecting a block of houses and other buildings at Percy Main, North Shields. Included in the Order were the Methodist Church and Sunday School in Burdon Street, which was vested in the first plaintiff as custodian trustee. The second plaintiff is joined to represent the 17 managing trustees of the church. The compulsory purchase order was duly confirmed and on January 5 1976 the local authority served on the plaintiffs a

A notice to treat and a notice of entry. On April 14 1976 the plaintiffs gave a notice of claim in answer to the notice to treat claiming compensation on the basis of equivalent reinstatement. Although, so I am told, the local authority has proceeded to acquire and demolish the houses surrounding the church, at some stage it decided not to go ahead with the acquisition of the church and in fact has never entered under the notice of entry. On June 3 1977 the local authority informed the plaintiffs that it was going to withdraw the notice to treat at the first available opportunity. The plaintiffs claim in this action that the local authority cannot now withdraw the notice to treat but is bound to acquire the premises.

B In order to understand the matter in issue I must refer to some of the relevant statutory provisions. Section 5 of the 1961 Act lays down rules to be applied in assessing compensation payable on a compulsory acquisition. Rule (2) is the general rule, ie that the land shall be valued at its open market value. Rule (5) provides as follows:

> Where land is, and but for the compulsory acquisition would continue to be, devoted to a purpose of such a nature that there is no general demand or market for land for that purpose, the compensation may, if the Lands Tribunal is satisfied that reinstatement in some other place is bona fide intended, be assessed on the basis of the reasonable cost of equivalent reinstatement.

C The classic cases to which the rule (5) equivalent reinstatement basis applies are cases such as the present where a church or school is being acquired. Until 1970 it was thought that rule (5) compensation had to be assessed on the basis of values and costs ruling at the date of the service of the notice to treat. But in *Birmingham Corporation v West Midland Baptist (Trust) Association Inc* [1970] AC 874 the House of Lords declared the law to be otherwise and held that in a rule (5) case compensation was to be assessed on the basis of prices ruling at the date when the actual work of reinstatement might reasonably have been commenced.

Provisions for the service and contents of a notice to treat
D are contained in section 5 of the Compulsory Purchase Act 1965, subsection (2)(b) of which requires that every notice to treat "shall demand particulars of the recipient's estate and interest in the land, and of the claim made by him in respect of the land." Section 6 of the 1965 Act then provides that if a person served with a notice to treat does not within 21 days from the service of the notice state the particulars of his claim the question of disputed compensation shall be referred to the Lands Tribunal.

It is common ground that once a notice to treat has been served in default of agreement it cannot be withdrawn unless there is an express statutory power so to do. The power to
E withdraw the notice relied on by the local authority in this case is contained in section 31 of the 1961 Act. But in order to understand that section I must first read section 4 of the 1961 Act. Subsection (1) is as follows:

> Where either—(a) the acquiring authority have made an unconditional offer in writing of any sum as compensation to any claimant and the sum awarded by the Lands Tribunal to that claimant does not exceed the sum offered; or (b) the Lands Tribunal is satisfied that a claimant has failed to deliver to the acquiring authority, in time to enable them to make a proper offer, a notice in writing of the amount claimed by him, containing the particulars mentioned in subsection (2) of this section; the Lands Tribunal
> F shall, unless for special reasons it thinks proper not to do so, order the claimant to bear his own costs and to pay the costs of the acquiring authority so far as they were incurred after the offer was made or, as the case may be, after the time when in the opinion of the Lands Tribunal the notice should have been delivered.

Subsection (2) is as follows:

> The notice mentioned in subsection (1) of this section must state the exact nature of the interest in respect of which compensation is claimed, and give details of the compensation claimed, distinguishing the amounts under separate heads and showing how the amount claimed under each head is calculated.

The local authority claim that the plaintiffs have not served any proper notice containing the particulars specified in section 4(2). They say that if the plaintiffs now serve such a G notice they will withdraw the notice to treat under section 31(1) or, if no further notice is served, they will withdraw the notice to treat under section 31(2) after the Lands Tribunal has determined the claim. The plaintiffs maintain that the notice they served on April 14 1976 did constitute a proper notice for the purposes of section 4 of the 1961 Act and that accordingly the local authority cannot at any stage hereafter withdraw the notice to treat under section 31.

The notice to treat in this case was in common form. After referring to the compulsory purchase order, paragraph (3) reads as follows:

> The Council hereby demand particulars of your estate and H interest in the said lands together with all charges and interest to which the same are subject and of the claims made by you in respect thereof, all which particulars, charges, interests and claims should be shown in the accompanying form of claim and thereafter delivered to the undersigned at the underwritten offices of the Council.

Paragraph (4) states the willingness of the local authority to treat, and paragraph (5) then reads as follows:

> If for 21 days after the service hereof you shall fail to state the particulars of your estate, claims and interest as aforesaid . . ., the Council will forthwith proceed to require the amount of such compensation to be settled in the manner directed for settling cases of disputed compensation by the Acts incorporated with and J applied by the said order.

The schedule to the notice refers *inter alia* to the Methodist Church and Sunday School and land attached thereto. Then at the foot of the notice to treat there is a note which reads as follows: "A form of claim to be filled in and signed accompanies this notice to treat, and must be returned within 21 days from the date of the service of this notice."

The form enclosed was also a common form of claim in answer to notice to treat published by a well-known legal publisher. It was headed: "Claim in answer to notice to treat. The particulars of the claimant's estate and interest in the property comprised in a notice to treat served on the 5th K day of January 1976, and of the claims made in respect thereof are as follows." After a series of questions directed to obtaining particulars of the claimant's interest in the property question 14 reads as follows: "Particulars of claim: (i) For the value of the claimant's interest in the land being taken, including any easements. (ii) For severance or injurious affection of other land of the claimant. (iii) For disturbance." Then there is a note to question 14, which is as follows: "Details should be furnished showing how the amount claimed under any head is calculated. (See sections 4(2) and 31 of the Land Compensation Act 1961)." Question 15 refers to L some deductions that might conceivably be made. Then question 16 says: "Net claim (ie the total of 14 less 15)."

In my judgment it is clear from these questions, and in particular from the note to question 14, that the local authority in this case was demanding of the plaintiffs that within 21 days they should give the particulars specified in section 4(2) of the 1961 Act, that is to say the amount of the claim and how the amount was calculated.

The claim in answer was put in by the plaintiffs' surveyors, who inserted answers on the form supplied by the local authority. They answered question 14 as follows. Opposite the heading "For the value of the claimant's interest in the M land being taken, including any easements" they wrote: "Compensation on a rule (5) basis to represent the cost of acquiring a simple site and erecting an alternative place of worship." Heading (ii) was not applicable. In answer to heading (iii) "For disturbance" they wrote: "A claim is reserved under the head." And opposite the note they said: "On the basis of current market values." Then in answer to question 16 they wrote: "Net claim under 14(i) and 14(iii) above."

The claim in answer was in fact served on April 14 1976, that is well after the stipulated 21-day time-limit had expired.

A In these circumstances the argument for the local authority is very straightforward. They say that it is clear from the use in section 4, subsections (1) and (2), of the words "amount" and "calculated" that for a notice to comply with that section a claim must be quantified in terms of money. The claim submitted made no attempt to quantify the amount claimed. Accordingly, no notice has been served containing the matters specified in section 4(2) of the 1961 Act and therefore the local authority will be entitled under section 31 to withdraw its notice to treat when either a proper notice is served or the Lands Tribunal determines the compensation.

B If section 4 of the 1961 Act is to take effect according to its literal meaning there is no answer to the local authority's contention. But the plaintiffs claim that this is one of those exceptional cases where the court is entitled to depart from the literal meaning of the words, since to give them their literal effect would either require the plaintiffs to do the impossible or produce absurd anomalies which Parliament cannot have intended. The plaintiffs say, and the defendants accept, that before any realistic claim can be quantified in a case to which rule (5) applies the following steps would have to be taken. First, a site for the new building would have to be identified; their architects would have to prepare plans for C the new building on that site, then bills of quantities would have to be got out. Even then, the only figure which could be claimed would be an estimate of the then current costs whereas, as the *Birmingham Corporation* case *(supra)* shows, the compensation falls to be assessed at prices ruling at the date on which the work of reinstatement could reasonably begin. It is common ground that it would be quite impossible for any proper claim to be formulated within the 21 days referred to in the notice to treat. Therefore, say the plaintiffs, Parliament cannot have intended the impossible to be done. Moreover, even if 21 days is not an absolute time-limit, they say it would be ridiculous to require all this cost D and trouble to be incurred by the plaintiffs only to find that the authority is not going to acquire the property after all.

I will consider first the argument based on impossibility. I accept the proposition that if the literal meaning of a statutory provision would impose on the claimant a positive duty to do something which is physically impossible that cannot have been Parliament's intention and the court is entitled and bound to depart from the literal meaning. But this legislation does not impose any such duty. The claimant is not under any obligation to make a claim within 21 days under section 6 of the 1965 Act, nor at any later stage under section 4 of the 1961 Act. The scheme of the legislation is not to E impose a duty to make a claim but to provide that certain consequences follow if no claim is made, ie the local authority is free to go straight to the Lands Tribunal under section 6 of the 1965 Act and if no detailed quantified claim is made the claimant may in certain circumstances suffer by being deprived of costs if he fails to put in a claim when he might reasonably have done so. Therefore the statute does not impose a duty to formulate any claim within 21 days. Moreover, a claim can be amended. It may well be that a claim in the form used by the plaintiffs in this case is sufficient for the purposes of section 6 of the 1965 Act; it can then later be amended by putting in the necessary figures F and calculations. When this amendment is made, it will for the first time become a sufficient notice for the purposes of section 4 of the 1961 Act.

Next it is said that because compensation is to be assessed at prices current at a future date, ie when the work of reinstatement could reasonably be started, it is impossible to quantify the claim before that date is reached. I do not think this argument is correct. Although until the decision in the *Birmingham Corporation* case *(supra)* any difficulty caused by future inflation could not have arisen in rule (2) and rule (5) cases (since values were assessed as at the date of the notice to treat) the difficulty has always been present in relation to claims for disturbance, compensation for which has always G been assessed on the basis of actual cost. Yet Parliament plainly intended the claim for disturbance to be quantified. It seems to me that where the future values or prices are relevant the quantification specified in section 4(2) would be satisfied by stating the claim at values current at the date of claim and stating that the claimant will adjust his claim to take account of changes in prices and values between the date of claim and the date of assessment. Therefore in my judgment there is no reason to depart from the literal meaning of the words used on the ground that they require the claimant to do the impossible.

I turn then to the argument that the anomalies produced H by a literal construction are so absurd that it cannot represent Parliament's true intention and therefore the court must give the words some other meaning. It is established that in some circumstances it is legitimate for the court to take this course: see *Re Lockwood* [1957] 3 All ER 520. On the other hand in *Stock v Frank Jones (Tipton) Ltd* [1978] 1 WLR 231 the House of Lords has recently re-emphasised the very strict limits to this doctrine. The views of the House are I think reflected in a passage from the speech of Lord Scarman at p 239 E, where he says:

> If a study of the statute as a whole leads inexorably to the conclusion that Parliament has erred in its choice of words, eg J used "and" when "or" was clearly intended, the courts can, and must, eliminate the error by interpretation. But mere "manifest absurdity" is not enough: it must be an error (of commission or omission) which in its context defeats the intention of the Act.

So in my judgment it is not legitimate for me to give sections 4 and 31 of the 1961 Act anything other than their plain meaning unless I am satisfied that the literal meaning will defeat the intention of that Act.

This is not a case where I have to choose between two possible constructions of the words. Unless some words are added to the statutory provisions the words used have only one possible meaning. K I confess that I have grave doubts whether the draftsman of the Act and Parliament really intended that in a rule (5) case the claimant would either have to put forward a claim based entirely on guesswork or run the risk that when all preparatory steps necessary to remove a church or school to another site had been taken the local authority might decide not to go ahead. But anomalous as this is, I do not think the strict construction would defeat the intention of the Act. To my mind the obvious intention of section 31 is to ensure that until the local authority knows, either from a detailed claim under section 4 or from a decision of the Lands Tribunal, how much it is going to have to pay it is to be free to withdraw from the acquisition. To give the local authority this opportunity it is necessary that they should have a quantified claim L on the basis of which they can make an open offer. The effect of giving a local authority this opportunity to withdraw is that in a rule (5) case it may be impossible for the claimants to achieve any certainty that their property will be acquired unless and until they have expended much money in getting out plans and quantities so as to be able to quantify their claim. Hard and unreasonable though this seems to me, I do not feel able to say that it defeats the intention of the Act. Parliament may have thought that it was a necessary price to pay to ensure that local authorities should not be bound to purchase at a price far in excess of what they had budgeted for. M

When the notice to treat is withdrawn the local authority has to compensate the claimant for pecuniary loss flowing from the service of the notice: section 31(3) of the 1961 Act. In the ordinary case this provision for compensation on withdrawal will hold a reasonable balance between the interests of the local authority on the one hand and the owner of the property on the other. In a rule (5) case it may not be adequate, since pecuniary compensation cannot make good the upset to sensibilities and the disruption inherent in closing a church, which course may well have been taken if the notice to treat is withdrawn at a late date. But since Parliament appears

A to have intended the local authority to have the right to withdraw at any time up to the moment when there is a quantified claim, I do not think that the possible unfairness in one limited type of case justifies me in saying that the literal construction defeats the intention of the Act.

Accordingly I must hold that a notice complying with section 4 of the 1961 Act has not yet been served and that the local authority still has the right to withdraw the notice to treat in accordance with section 31 of the 1961 Act.

Finally, I should mention two points. First, there was some discussion during the hearing whether the words in section 31(2) of the 1961 Act "Where a claimant has *failed* to deliver
B a notice as *required* by the said paragraph (b)" (my italics) cover every case where such a notice has not been served, whether or not the claimant was at fault. The plaintiffs accepted that they did cover every case and therefore I need not comment further on the point. Secondly, in these proceedings I am asked merely to construe the statutes. Nothing in this judgment is intended to affect the rights or claims, if any, arising from the fact that the notice to treat served by the local authority in this case did in fact demand full particulars of a quantified claim within 21 days.

The plaintiffs' summons was dismissed with costs.

C

COURT OF APPEAL

March 23 1979

(Before Lord Justice MEGAW, Lord Justice BRIDGE and Lord Justice TEMPLEMAN)

SERVICE WELDING LTD v TYNE AND WEAR COUNTY COUNCIL

Estates Gazette June 30 1979

(1979) 250 EG 1291

Compensation for acquisition of factory by agreement—Compulsory purchase basis of compensation—Claimants in order to mitigate loss purchased an alternative site and erected new
D **factory—Main items of compensation agreed but dispute as to claim under heading of disturbance for bank interest and charges on overdraft—Amount claimed as directly flowing from acquisition—Whether interest and bank charges should be regarded on the other hand as part of purchase price of new factory—Whether claimants received value for money in return for such expenditure—Latter view held to be correct and appeal from the Lands Tribunal allowed**

This was an appeal by case stated from a decision of the Lands Tribunal (Victor Wellings QC) holding that the claimants, Service Welding Ltd, were entitled to compensation from the acquiring authority, Tyne and Wear County Council, under
E the heading of disturbance, for expenditure incurred in meeting bank interest and charges on an overdraft used to finance the construction and equipment of a replacement factory. The Lands Tribunal decision was reported at (1978) 245 EG 143.

Alan P Fletcher (instructed by Sharpe, Pritchard & Co, agents for John E Hancock, solicitor to Tyne and Wear County Council) appeared on behalf of the appellants (the acquiring authority); Michael Mann QC and Daniel Robins (instructed by Ingledew, Mark Pybus, of Newcastle upon Tyne) represented the respondents (claimants).

Giving the first judgment at the invitation of Megaw LJ, BRIDGE LJ said: This is an appeal by case stated from a
F decision of the Lands Tribunal given by Victor Wellings QC on October 6 1977. The matter came before the tribunal on an agreed reference to determine the amount of compensation payable by the appellant acquiring authority, the Tyne and Wear County Council, for the acquisition of the claimants', Service Welding Ltd's, factory premises in Newcastle. There had been an agreement between the claimants and the acquiring authority (as I shall call them respectively) to sell the claimants' premises to the acquiring authority on the terms of a notional compulsory acquisition pursuant to a notice to treat deemed to have been served on July 4 1973. Having concluded that agreement, and faced with the prospect of losing their existing factory premises, the claimants, who wished G
to continue to carry on their business and to avoid a disturbance claim for the loss of the goodwill of the business, set about finding themselves alternative factory premises. They were, in the event, unable to purchase a factory ready for occupation, but they found a suitable site where a new factory could be built on an industrial estate known as the Brunswick Estate at a place called Wideopen. They acquired the new site some time in the latter part of 1973, and employed building contractors to erect a factory for them; and the new factory was ready for their occupation in July 1974 when they moved into it. The premises in Newcastle which were being acquired were handed over to the acquiring authority in September H
1974.

The main items of compensation payable by the acquiring authority to the claimants were agreed as follows. It was agreed that the market value of the Newcastle factory on September 19 1974, the date when the acquiring authority took possession and the date taken as the appropriate valuation date, was £38,250. Compensation for losses incurred on disturbance, on a wide variety of miscellaneous items of a kind very familiar in claims of this kind, was agreed at the sum of £14,149. Because there was an interval of time between the vacation of the Newcastle factory and the date when the acquiring authority took possession of it, from which date interest on the compensation would run, a further sum was J
agreed as payable for disturbance, representing interest, I suppose accruing between July 24 and September 19, namely, £1,065. Those amounts together made up an aggregate of £53,464.

The additional sum claimed by the appellants by way of disturbance compensation arises out of circumstances which are thus summarised in the agreed statement of facts which was the basis of the reference to the Lands Tribunal:

> 5. The claimants' business was financed in part by bank overdraft secured by a charge over the subject property.

That is a reference to the Newcastle factory. K

> The bank was willing to provide additional facilities to finance the cost of purchasing the new site

that is, the site on the Brunswick Estate

> and the erection thereon of the new factory which finance was secured by a charge over the new site which was to be redeemed (in part) by the proceeds of the sale of the subject property. In order to facilitate the recording of the finance costs attributable to the purchase of the new site and the erection of the new factory, the claimants opened a separate "reinstatement account" with the bank to which was debited all the costs incurred in the purchase of the new site and the erection thereon and making ready for occupation and use of the new factory. Central government grants L
> were made towards the cost of erecting the new factory equal to 22 per cent of certain qualifying expenditure. These were applied for and paid at appropriate intervals during the course of the erection of the new factory and the sums received were credited to the reinstatement account thus reducing *pro tanto* the finance costs. The first debit to the reinstatement account

which was presumably the date when the claimants acquired the new site

> was made on September 17 1973.

On the basis of that history, paragraph 8 of the agreed facts sets out the claimants' contentions.

> The claimants contend that they are entitled to be paid as part M
> of their loss on disturbance and in addition to the sums totalling £15,214 under paragraph 7(2) above a sum equal to the cost (bank interest and charges) of financing the capital progressively laid out on the new site and factory (after crediting the central government grants) throughout the period between September 17 1973 (when the first debit to the reinstatement account was made) until July 24 1974 when the new factory was ready for occupation and occupied. Because the new factory is larger than the subject property the ascertainment of these finance costs involves a scaling down of the reinstatement account overdraft and of the central government grants to reflect a notional expenditure on a new factory equivalent to the subject property.

A At the end of the agreed statement of facts for the tribunal there were set out three alternative calculations of the appropriate amounts of interest payable on borrowed money for which the claimants would be entitled to compensation according to three different contentions. I need only read the first, because this is the basis on which the Lands Tribunal decided that the claimants were in fact entitled to compensation, namely, on the footing "As if (as the claimants primarily contend) the total expenditure on the new factory (excluding bank interest) had never exceeded a sum of £58,509 being a sum which it is agreed would have been the cost of building and equipping a new factory of equivalent size with equivalent B equipment to" the old Newcastle factory. On that basis it was agreed that the amount of interest claimable would be £4,285.

The issue was described as a preliminary issue for the Lands Tribunal and was referred to the tribunal on the footing that it was purely a question of law. This assumption was challenged by the member in the course of giving his decision. He said this:

> During the argument I expressed strong doubt as to whether the preliminary issue was a question of law at all and threatened to adjourn this matter and fix a date for the hearing of the reference in order to enable the parties to call evidence including, if they so C desired, expert evidence. I was asked by counsel for the parties, in order to save costs, not to take that course but to resolve the matter as best I could. In order to assist me in that course they came to certain further agreements as follows:
>
> (1) That in any building project there are four elements in the cost of the completed building: (a) the cost of the site; (b) payments to the building contractor; (c) architect's and other fees; (d) interest charges or loss of interest on all the first three elements.
>
> (2) That the tribunal (that is to say I) as arbitrator, should be at liberty to draw any inferences it might think necessary from the agreed facts or any of them, including the four elements stated above, and should be treated as empowered by the agreement dated July 4 1973 to use my own expertise and draw such conclusions as D to what a purchaser might take into account with respect to the value of the new factory as the tribunal might think fit and to use judicial knowledge in the widest possible manner. I heard very little argument as to which of the three bases of assessment would be correct if I came to the conclusion that the claimants were entitled to compensation under this head. I heard no argument in respect to the third basis except that Mr Robins said that it was an impossible basis.

Then there is the crucial paragraph:

> There is no doubt that in purchasing the new site and erecting and equipping the new factory the claimants were mitigating their loss and taking reasonable steps to do so. In my opinion, whether E or not the expenditure in question flowed directly from the taking, it, in so far as it was reasonable in amount, flowed directly from the mitigation: see *Halsbury's Laws of England*, 4th ed, vol 12, para 1193. In the circumstances the claimants are, in my judgment, *prima facie* entitled to compensation for at least some of the expenditure. There is, I believe, no presumption, either of law or fact, that the claimants have received value for money expended. I am unwilling to make any assumption as to what a prospective purchaser or vendor would take into account in the matter of bank interest and charges except that I think it is likely that the value of the new factory would be precisely the same whether or not any bank interest and charges had been paid. It appears to me that the preliminary issue is more properly one of fact and degree and F perhaps expert evidence rather than of law. I have not been satisfied that the claimants received value for money for the interest and charges paid away. Accordingly I shall decide the issue in favour of the claimants.

The decision goes on to opt in favour of the first of the three alternative bases of quantification.

The formal case stated for the opinion of the court sets out two questions on which our decision is invited.

> First:
>
> Whether there was in law a presumption that the claimants had received value for money in respect of bank interest and charges G paid by them in respect of a bank overdraft with the aid of which they caused to be erected and equipped a replacement factory: if yes, whether the said presumption was rebuttable or irrebuttable.
>
> Secondly:
>
> If the presumption was rebuttable whether, having regard to the agreed facts and other agreements mentioned in my said decision, and having regard to my findings on p 5 with respect to mitigation of loss, I was justified in holding:
>
> (i) that the claimants were *prima facie* entitled to compensation for at least some of their expenditure on bank interest and charges;
>
> (ii) that the first of the three agreed bases of compensation should be applied.

H The principles on which compensation for a compulsory acquisition in such circumstances as these is to be quantified are very well known and are not in dispute. This is a case governed by rules (2) and (6) of the rules for assessing compensation set out in section 5 of the Land Compensation Act 1961. Rule (2) provides:

> The value of land shall, subject as hereinafter provided, be taken to be the amount which the land if sold on the open market by a willing seller would be expected to realise.

That element, of course, is the subject of agreement in the sum of £38,250 for the acquisition of the old Newcastle premises. Rule (6) provides:

> J The provisions of rule (2) shall not affect the assessment of compensation for disturbance or any other matter not directly based on the value of land.

It is pertinent to point out that this is not a case to which rule (5) has any application. Rule (5) is the rule which provides that in certain circumstances where land is acquired of a kind for which there is no general demand or market the acquiring authority is to compensate the landowner on the basis of the reasonable cost of equivalent reinstatement.

It is well established that in a case governed by rules (2) and (6), in addition to receiving the value of the property which is taken from him, a dispossessed owner is also entitled K to all the costs which reasonably flow from the fact of his disturbance. The principle is clearly stated in a passage, particularly relied on by Mr Mann for the respondents, from the judgment of Romer LJ in this court in *Harvey v Crawley Development Corporation* [1957] 1 QB 485. The passage in question is at p 494:

> It seems to me that the authorities to which our attention was drawn do establish that any loss sustained by a dispossessed owner (at all events one who occupies his house) which flows from a compulsory acquisition may properly be regarded as the subject of compensation for disturbance, provided, first, that it is not too remote and, secondly, that it is the natural and reasonable consequence of the dispossession of the owner. L

Where the premises compulsorily acquired are occupied by the landowner for business purposes, the potential disturbance compensation is, of course, represented by the possible extinction of the business and a large sum which in that event would be payable to a business occupier for the loss of his goodwill. But it is clear that where, as here, a business occupier is in a position to find alternative accommodation in which to carry on his business and prevent its extinction, he is under a duty to mitigate his disturbance compensation by removing his business to the alternative accommodation. What the authorities (to which I need not refer in detail) very clearly establish, M however, is that when an occupier, whether residential or business, does, in consequence of disturbance, rehouse himself in alternative accommodation, *prima facie* he is not entitled to recover, by way of compensation for disturbance or otherwise, any part of the purchase price which he pays for the alternative accommodation to which he removes, whether that accommodation is better or worse than, or equivalent to, the property from which he is being evicted. The reason for that is that there is a presumption in law, albeit a rebuttable presumption, that the purchase price paid for the new premises is something for which the claimant has received value for money. If he has made a good bargain and acquired premises which have a

A value in excess of what he has paid for them, that is not something for which the acquiring authority is entitled to any credit. If the claimant has made a bad bargain and has paid a great deal more for the new premises to which he is moving than they are really worth, that is not something for which the acquiring authority can properly be charged.

As Mr Fletcher, for the appellant acquiring authority, illustrated to us, this presumption is, of course, rebuttable. There may be circumstances in which, for example, the displaced claimant, in order to render the new premises which he acquires suitable for his own purposes, must expend money on adapting them in a way which will not enhance their value.
B In those circumstances, the cost of adaptation would properly be recoverable as part of his disturbance compensation.

The real question which arises in this appeal is whether, in the circumstances described in the agreed facts to which I have made reference, the interest charges payable by the claimants on the cost of acquiring the site and on the stage payments that they made to their building contractors during the building of their new factory are properly to be regarded as part of the purchase price of the new premises, to which the presumption applies that the claimants received value for their money, as Mr Fletcher contends; or whether, as Mr Mann submits for the respondent claimants, this was an item
C of expenditure for which no value was received by the claimants, an item of expenditure which flowed as a natural and reasonable consequence of their dispossession from the Newcastle factory, analogous to interest payable upon a bridging loan by a dispossessed owner who is forced to acquire new premises before he is in a position to move into them and before the acquiring authority acquires his old premises from him and becomes liable to him to pay interest on compensation moneys.

I hope I shall not be thought disrespectful to the arguments of learned counsel on either side if I say that their succinct submissions have demonstrated to me that the point
D is a very short one which admits really of no elaboration of argument either way. In a sense it is a matter of pure impression. Given the agreed facts, in particular the facts which were agreed as supplementary to the original agreement and which are set out in the passage from the Lands Tribunal decision which I have read, as to what are the necessary elements in the cost of erecting a new building, it seems to me perfectly clear that interest charges incurred in the circumstances in which the claimants' interest charges were incurred here are part of the costs necessarily incurred in producing as an end-product factory premises ready for occupation as such, where the intending occupier has built
E the factory for himself. It seems to me that in those circumstances those charges are properly to be regarded as part of the purchase price paid by the factory owner for the factory which he has built for himself.

One can test that view by considering the position which would arise if, instead of acquiring a site for himself, and himself undertaking the cost of building a factory upon it, the claimant, displaced from his old factory, had found a developer who had erected a factory on a site on a speculative basis. There is no doubt, in my judgment, in those circumstances, that the developer, willing to sell his newly completed factory to the would-be occupier, would base his
F asking price on the various elements of cost which he, the developer, would have incurred in producing the factory; and one of those items of cost would have been the interest charges or loss of interest on the other elements making up the overall cost of producing the factory. Of course, he might not succeed in getting his asking price. He might be forced into selling at a loss. On the other hand, he might be fortunate and sell at a profit. But in any view he would include the interest charges as part of the cost of producing the end-product. Where the intending occupier, as here, instead of going into the market and finding a developer ready to sell him a finished product, has himself incurred the various elements necessary to be incurred in the production of that G finished product, I can see no reason whatever why any one of those necessary elements of cost should be treated as other than part of the purchase price of the end-product, to which the presumption—the well-established presumption—in law that he receives value for money in my judgment applies.

Accordingly, I would allow this appeal; and, turning to the questions in the case stated, I would answer the first part of the first question in the affirmative: that there was in law a presumption that the claimants had received value for money in respect of the bank interest and charges paid by them. It is academic, but one must say that, technically, H that presumption was rebuttable. Turning to the second question, the presumption being rebuttable, having regard to the agreed facts and the other agreements mentioned in the tribunal's decision, and the tribunal's findings with respect to mitigation, I reach the conclusion, with respect to him, that the member of the tribunal giving the decision was not justified in holding that the claimants were entitled to compensation for any part of their expenditure on bank interest and charges. Accordingly, the question as to which of the three agreed bases of compensation would have been appropriate had they been so entitled does not arise.

I would allow the appeal accordingly.

Agreeing, TEMPLEMAN LJ said: The Newcastle factory was J owned and occupied by the claimants and it was compulsorily acquired. Under rule (2) of section 5 of the Land Compensation Act 1961, the claimants became entitled to the amount which the Newcastle factory, if sold in the open market by a willing seller, might be expected to realise. Of course, the claimants cannot have it both ways. If they are compensated by being paid the value of the Newcastle factory, they cannot be compensated in addition by being paid part of the price of the new factory. What they are entitled to, in addition to the value of the land under rule (2), is the assessment of compensation for disturbance under K rule (6), the compensation, as I understand it, being the costs and losses caused by their having to get out of the Newcastle factory and having to get into the new factory. The claimants, as they were entitled but not bound to do, looked round and found a site which they acquired, and, as they were entitled but not bound to do, they made arrangements for the site to be developed into a new factory and to pay the costs of that being done. In the course of so doing, as they were entitled but not bound to do, they paid certain moneys on account and borrowed moneys for that purpose. It seems to me that the moneys which they raised and the interest thereon for which they became liable were not the costs of L being disturbed from the old factory and getting into the new factory: they were the costs of acquiring the new factory. Accordingly, in my judgment, they fall neither under rule (2) nor under rule (6); and I would agree with the order proposed by my Lord.

MEGAW LJ agreed with both judgments.

The appeal was allowed with costs. The acquiring authority was also awarded costs of the reference from February 18 1977, the claimants to have the costs of the reference up to that date. Leave to appeal to the House of Lords was refused. The effect of the court's answers to the specific questions stated for their opinion was as follows: M

(1) There was in law a presumption that the claimants had received value for money in respect of bank interest and charges on an overdraft used to aid the erection and equipment of a replacement factory;

(2) This presumption was rebuttable;

(3) The claimants were not *prima facia* entitled to compensation for at least some of their expenditure on bank interest and charges;

(4) The question as to the basis of compensation for such expenditure did not, therefore, arise.

FEES AND COMMISSIONS

QUEEN'S BENCH DIVISION

October 18 1978

(Before Mr Justice PETER PAIN)

HAMPTON & SONS v TRADE & GENERAL SECURITIES LTD (THREE THIRD PARTIES)

Estates Gazette May 5 1979

(1978) 250 EG 451

Estate agents' claim for commission for finding a tenant for site of office premises—Defendants claim against third parties on the ground that defendants acted as agents only—A tangled story, but held by judge that plaintiffs' efforts brought about introduction which was effective cause of grant of lease—Commission on RICS Scale 19 applicable—Defendants' third party proceedings wholly fail

In this action the plaintiffs, Hampton & Sons (a firm) sued Trade & General Securities Ltd for commission for introducing as tenant Johnson Matthey, in respect of a site known as The Grange, Southgate. The defendants denied that the plaintiffs had been retained but claimed that in any case they (the defendants) had acted only as agents and brought into the proceedings three third parties, namely (1) Amalgamated Developers (Enfield) Ltd, (2) Stonegate Securities Ltd, and (3) Property Investment & Finance Ltd, who, the defendants alleged, were the principals in any transactions which resulted in liability to the plaintiffs.

P Sheridan QC and P Ralls (instructed by Fremont & Co) appeared on behalf of the plaintiffs; I Davies (instructed by Landau & Co) represented the defendants; E Goldrein (instructed by Chales Buckley & Co of Manchester) represented the first and second of the third parties; C Lindsay (instructed by Herbert Smith & Co) represented the third of the third parties.

Giving judgment, PAIN J said: This case concerns a claim made by the plaintiffs, who are a firm of estate agents, against the defendants for commission in connection with the finding of a tenant for a site of office premises. It is a site which I shall refer to as " the Grange site." The defence put forward was that the plaintiffs did not have any retainer from the defendants. But the defendants, perhaps apprehensive as to the success of their defence, sought to safeguard themselves by taking third party proceedings against three third parties. The proceedings against all of them are on the basis that after a meeting in July 1970 anything done by the defendants, whether in regard to the plaintiffs or otherwise, was done by the defendants acting as the agents on behalf of their principals, namely the first, second or third third party. So far as the first third party goes, that was a company which was incorporated on October 17 1972. It is now common ground, since the events we are concerned with occurred in the latter part of 1970 and the early part of 1971, that the defendants cannot succeed against the first third party and I need say no more about them.

This is a case which turns largely on questions of fact. It is a case which is concerned with events that took place by and large over seven years ago. I have heard oral evidence for some six days, and I say straightaway that after seven years memory tends to play tricks, and I do not attach all that much weight to the oral evidence. I attach a great deal more to what the contemporary documents, which are fairly full, appear to show. It is common ground that ITT, which of course is a vast concern, approached the second third party, Stonegate Securities Ltd, in 1970 saying that they wanted a new office built and asking whether Stonegate could assist. Stonegate of course naturally felt this to be a valuable introduction, and they got in touch with the third third party, Property Investment & Finance Ltd, who were largely represented by Mr Mitchell. They also got in touch with the defendants. The third third parties were to be financiers. The defendants were the people who were to provide a site. On July 3 1970 they had a meeting of which we have the benefit of some notes that were made. There were present Mr Mitchell for the financiers, Mr Swift for Stonegate, and Mr Luck and Mr Richards and a gentleman who made the notes for the defendants. The purpose was to discuss if a site discovered by Mr Luck was suitable for development by Stonegate. It was decided that it was suitable. Stonegate's client was revealed to be ITT. I think it was said that they already had an office development permit, or that there was hope of an ODP for 120,000 sq ft, but no site and a need urgently to find one in the Barnet area for occupation by September 1971. The defendants' job was to be to discover a site, introduce the various parties, and attend to the initial formalities. The fee for the defendants and their associates was to be £50,000 payable by instalments as the development progressed. There were a number of other matters that were dealt with at the meeting, but I do not think I need to go through all of them.

So far as the £50,000 was concerned which the defendants were to have, the reference to the fee to be split between the defendants and their associates led to a sharing agreement under which half was to be taken by the defendants and half by a company called Amalgamated Developers (Projects) Ltd, which was largely controlled by Mr Shingler because he had the major interest in Amalgamated Developers Ltd, which was the parent company. Mr Shingler has been referred to as wearing various hats. He appeared in this matter in three capacities, all of them highly profitable. Firstly, he was the controller of Amalgamated Developers, who were to get £25,000, their share in the introduction fee. Secondly he was a partner, although he himself was not at that time in practice, with Shingler, Risdon, the architects who, we know, got £5,000 for preliminary plans and possibly a good deal more for later plans, and thirdly he was an estate agent as A J Shingler & Partners. As we shall see in due course he succeeded in extracting an introduction fee in that capacity.

So far as the project was concerned it had four essentials: firstly a site, secondly a tenant, thirdly it required an office development permit and fourthly it required planning permission. At the time the arrangement was made in July 1970 none of these essentials had been fulfilled, and it was clear that if any one of the four were missing the scheme as a whole would founder. So far as the first essential, the site, is concerned, that matter was cleared up by a resolution passed by the council for the London Borough of Enfield on March 16 1971. They awarded the site on architectural merit to this consortium, although they did not of course at that stage agree all the terms upon which it should be acquired, and there was a great deal of very lengthy negotiation about that thereafter. With regard to a tenant, difficulties were much greater. It was clear that for a long time ITT was the front runner, and as long as they remained keen there was no reason why the consortium should look elsewhere. But at some stage ITT began to cool in their ardour, and it is accepted on all hands that by mid-April the possibility of ITT becoming the tenant had disappeared. It is not quite clear when they began to cool in their ardour. Maybe it is difficult to put a date on these things. Mr Mitchell says he suspected from January onwards this had happened, though Mr Swift seemed to be much more hopeful in January. Certainly letters were being exchanged on the basis that ITT would become the tenant right up to April, so it may be that with hindsight over seven years it is very difficult to fix any reliable date, save to say that it was quite apparent by mid-April that the thing had fallen through.

It is put forward by the defendants that in the meantime Mr Shingler had introduced a firm called Johnson Matthey who ultimately became the tenants of the office site to the consortium. Mr Brooker of Johnson Matthey telephoned Mr Shingler asking him about this site, and Mr Shingler went to see him. It would appear from the entry in Mr Shingler's diary that he went to see him on March 8. I accept on the basis of that diary that was produced that there was such a meeting on March 8. Two days later on March 10 Mr Shingler writes to Mr Brooker, and he encloses a plan of an entirely different site on the Harrow Road. He said he would be pleased to discuss the matter further, and then said: "There are other sites in which I am also involved which might be of interest, although discussions are under way regarding these rather like the Grange, Southgate." That on the face of it is not a letter which is trying to dispose of or to introduce even the Grange, Southgate, into the negotiations. Mr Shingler insisted in his evidence that the Grange, Southgate, was really the main topic of conversation, and he had to tell Mr Brooker that it was not available at the moment. But he told him he already had a prospective tenant, and "as far as I was concerned, first come, first served." Then Mr Shingler says he went to see Mr Brooker on two further occasions about it. He got Shingler, Risdon to send plans of it.

Apart from the one meeting I reject Mr Shingler's evidence that he went three times or that plans were sent. In fact so far as Mr Shingler is concerned, really I find myself unable to accept anything he says unless it is corroborated. I accept that there was the one meeting because that is corroborated by the documents. But I listened to him very carefully. He was in the box for some time. He gave me the impression of being a reckless witness. Not only did he tell a story which it seemed to me, looking at the documents, could not be true, but it seemed to me when pressed that he said really whatever came into his head in answer to the difficulty with which he was faced without much regard to the truth. Mr Shingler told us that after he had been to see Brooker he said he was in touch with Luck and Mitchell about this all the time. "It was such an important thing as far as I was concerned that to speak to Luck and Mitchell would be one of the first things I would do." I accept that he mentioned this to Mr Mitchell once, because Mr Mitchell corroborates him on that. I think Mr Mitchell was a more reliable witness. I do not accept that he told anybody else. I do not accept that the information was passed on to anybody else. Mr Mitchell did suggest rather tentatively that he may have told Mr Swift, but I do not accept this. Mr Swift in his evidence did not suggest that he knew anything about this approach. In my view this cannot be described as an introduction at all. Either Mr Shingler is to be regarded as a part of the consortium in which case as a matter of ordinary language it can hardly be said that he had effected an introduction when he as a member of the consortium received a phone call; alternatively, if you regard him as not being a member of the consortium, and I am not quite sure whether he was, he did not introduce Mr Brooker to anybody. He merely mentioned it to Mr Mitchell, and apparently, according to Mr Mitchell, told him there were certain difficulties. In any case if I were wrong about that, if this were an introduction, it is an introduction which was entirely ineffectual. If it had been an introduction that had any effect, it is quite inconceivable that further contacts would not have been made with Mr Brooker through Mr Shingler. Prior to mid-April every member of the consortium must have realised, and they also had suspected beforehand, that the withdrawal of ITT had placed the whole scheme in jeopardy and placed the consortium in a very grave difficulty because they had got the local authority to accept them on the basis that ITT was going to be the tenant. They must have been desperate to find another tenant. Had it been common knowledge among them that there was this contact, it really is impossible to understand why it would not have been exploited.

In the meantime P L Cope of the plaintiffs had begun to get busy. He had approached both the competitors before the council made their award, and then he approached Johnson Matthey. Now Mr Cope is a somewhat thrusting individual. I do not say that by way of criticism. It may be inherent in his occupation, after all he is an estate agent, to make money out of it. But it is clear that he creates that impression and one sees in the correspondence how obviously his manner had alarmed Johnson Matthey a little, because they put it in writing on March 5. Mr Venus writes saying: "I feel I should put on record the fact my company has not retained the services of your firm for the purpose of finding alternative accommodation or for any other purpose." That of course is the right way to deal with the matter, making it quite plain. At the same time as he approached Johnson Matthey Mr Cope had approached Mr Luck, and Mr Luck had in effect put Mr Cope on ice. He said he wrote back on March 11. "I made a diary note. I will certainly come back to you if I need your assistance in connection with the letting of this development." In the latter part of April when it became clear that ITT was no longer available, Mr Luck did come back to Mr Cope and spoke to him on April 26. They had a telephone conversation. It was plain in the course of that that Mr Luck told Mr Cope, he having taken the initiative in having made the phone call, that their tenant had gone sour on them and was asking Mr Cope what he could do to help. It is at this stage that one comes up against an important conflict of fact. Mr Cope's account of the matter is that "Luck telephoned me, told me the present negotiations were failing, asked would my clients be interested. I asked what about fees. I said I might not be able to get fees from Johnson Matthey. Would he try and handle this. He said yes of course. The name of Johnson Matthey was disclosed. Luck was particularly anxious to know who he would be talking to to avoid duplication. Not a word was said about Shingler." That was Cope's account.

Luck's account was: "I said things had gone sour with ITT. Were his contacts still in the market? He said yes. I was under the impression Cope was talking on behalf of his clients. If Cope had asked me to pay fees, I would not have agreed on my own behalf but would have referred it to Stonegate and PIFL [Property Investment & Finance]." There is the conflict which I have to resolve. It is said on the one hand that it is inconceivable that Mr Cope would have disclosed the name Johnson Matthey unless he had first got a promise that his fees would be paid. It is said on the other hand it is inconceivable that Mr Luck would have made himself liable to pay estate agents' fees which could be very substantial without consulting the other members of the consortium. Between those two inconceivabilities I have to make a choice.

Had the matter stood alone I think I would have felt that this conversation by itself did not sufficiently prove a con-

A tract. But the matter does not stand alone. There is a great deal more that I can test it by. There is the fact in the first place that when he wrote his subsequent letter of May 13 Mr Cope purported to "confirm" what was already said about fees. I do not attach too much importance to that. People do not all use English very accurately, but it is something. More importantly, as I go through the story there are several matters about which I do not think Mr Luck is telling the truth. Thirdly in this period Mr Luck's behaviour does not seem to have been what one would expect from a true businessman. That being so, the argument that he would not have engaged to pay fees without consulting the others loses B a great deal of its force. Taking all these matters into account, I prefer the evidence of Mr Cope. I accept that fees were mentioned in this phone conversation, and that Mr Luck said he would pay. Cope then went ahead and at Luck's request he fixed up the meeting which took place on May 21. It is quite plain to me, having heard all the witnesses now, that at this stage neither Mitchell, Lilley or Swift had any knowledge before the meeting of the dealings between Cope and Luck or of the correspondence. Equally Shingler. It is also clear to me at this stage that Luck, Lilley and Swift knew nothing about Shingler's contact with Brooker. And while Mr Mitchell knew of it, he attached little weight to it. It is really hardly C in dispute, there cannot be a dispute on the documents, that the meeting of May 21 took place as the result of Cope's introduction. The contact to be made by Shingler had nothing to do with it. Mr Luck went off on his holidays. Perhaps it may be charitable to think that some of the mistakes he made were due to the fact that the holiday was overdue. However, he went off on holiday and when he was away Mr Cope sent the letter of May 13 to Mr Richards. That deals with the arrangements for their meeting and a few other matters. Then he said at the end of the letter:

> I trust that you will not mind my confirming that in the event of Johnson Matthey or one of their associated companies taking D a lease on your development at the above address, my firm will be looking to you for the payment of our fees in accordance with the scale of charges as laid down by the RICS.

It would be impossible to imagine language that was plainer. Of course Mr Richards might have replied and said: "I do not know about the last paragraph. After all Mr Luck is away. I do not know what you agreed with him. I am not committing myself to that." But he did not think that was a wise precaution. However, Richards was clearly alive to the danger if Cope was, so to speak, pushing himself in when he had not been engaged, because having made a note of the meeting for Mr Luck he said at the end: "PPS. I have not E replied to Hampton's letter. Please do so." So it is clear that he realised the significance of that last paragraph.

Then the meeting took place on May 21. It has been suggested in evidence that Mr Brooker was angry when he found that this was all about the Grange site and sought to say he knew about this already from Mr Shingler. I do not accept that he was angry. If he had been one would have expected some reference to it to be in the contemporary note. He may have been somewhat gruff in his speech compared with the more polished gentlemen who were dealing with him. He may well have said that Shingler had already mentioned this site to him. But the one thing that is clear beyond all F peradventure is that it was this meeting and nothing else that got things going. Prior to this nothing had been happening. After this meeting things were set in train. However gruff, however awkward Mr Brooker may have been, this was the vital meeting which got the thing going, set the negotiations on foot which led to the lease. One really, I think, gets the whole picture from looking at Mr Richards' note of Mr Brooker talking about the hoodoo site and so on, explaining his difficulties about his ODP and being reassured about it and the door being opened to further negotiations. Mr Cope was of course there pursuant to his letter. He suggests that he was asked to push on with things at a short chat on the pavement afterwards. I do not disbelieve that, but it does not G accord with Mr Mitchell's recollection and I do not really attach any weight to a few casual words mentioned on the pavement seven and a half years ago. I attach much more importance to the subsequent correspondence.

Mr Mitchell made contact with the Barnet Council about the Barnet site, and he then wrote and told Mr Luck about this and used the phrase: "We have asked Hamptons to satisfy themselves that this is so," that is that Barnet Council were not happy with the proposed development. Mr Mitchell says that he thought at the time that Cope was acting for Johnson Matthey. That I am quite prepared to accept. Nothing had been said to Mr Mitchell or Mr Swift which would lead me H to suppose that Mr Luck had purported to engage Mr Cope as estate agent on behalf of the consortium.

Then Mr Luck came back from his holiday. There is no dispute on the fact that he did not reply to the letter of May 13, but that he did go on to engage in correspondence with Mr Cope on a basis which created the impression in Mr Cope that he was engaged. Mr Luck, I think, had not been frank about this. He swore an affidavit in Order 14 proceedings and he said this: "The question of fees was first mentioned in the plaintiffs' letter dated May 13 1971. I did not reply to this letter prior to the meeting held on May 21 as to the best of my recollection I was at this time away from J the office and did not see the letter until after the meeting. Richards went to the meeting in my place." So far, so good. Then:

> He informed me as did Mr Mitchell who was also present and I verily believe that it was at that meeting Mr Cope first informed Mr Brooker that the proposed site for office space was in fact at the Grange. Mr Brooker was not pleased and said that it was a pity that the meeting had been arranged to discuss that site as he knew all about it already. I verily believe that such information had been given to Johnson Matthey by Mr Shingler. It was therefore clear at the meeting that the plaintiffs had not introduced Johnson Matthey Ltd so that they could not be entitled to any fee for so doing. I did not reply to the plaintiffs' letter dated the K May 13, because it seemed unnecessary to do so in the light of the new circumstances.

Of course that was sworn to when Mr Luck had not seen the letter which Mr Cope had sent to Mr Brooker making it plain that the site to be discussed was the Grange site. That view Mr Luck formed at the time; one would expect him to stick to it. When he came to give evidence here he gave somewhat different reasons. He said he discussed the meeting of May 21 with Richards and Mitchell.

> I did not specifically consider the defendants' liability to Hamptons but assumed that if there were a liability for fees it would rest with the defendants, ie the parties interested in the equity. I L did not see the documents until the first week in June by which time I had conversations with Mitchell. I regarded the letting of premises since Stonegate came in as not my concern. This was the responsibility of Stonegate and PIFL. My responsibility was to secure a site.

Then in cross-examination a slightly different slant:

> "I did not reply because of (a) pressure of work and (b) that Mitchell had taken up the position of negotiating with Hamptons in conjunction with Lilley. When I swore my affidavit I was not aware of the letter of May 13."

There one gets all these different reasons put forward. The M only common thing one can derive from them is that Mr Luck did give his mind as to whether there was a liability by his company to Hamptons. If he had said from the word go "this was all pressure of work and I overlooked this on return from my holiday," one might have felt some sympathy for him, because it is easy enough to make a mistake. But that cannot be accepted. It is plain from his answers that he applied his mind and that he is thinking out one excuse and then when that will not do thinking out another. I find it impossible to think that he has been honest about this. He may of course have realised, I know not, he did not say he had, that he was in a difficult position because the meeting had gone ahead after the letter of May 13 and that Mr Cope might well object if he sought to say that it was all a mistake and it was not desired to instruct Hamptons at all. Cope might therefore well have said, "that is all very well to say that once I have introduced the client." But he could, were he acting honestly, have raised the matter then and there but he did not. He allowed the matter to go forward. He continued to permit Mr Cope to conduct correspondence on behalf of the consortium or purportedly on its behalf. True, Mr Cope was pressing it, but there was no attempt to stop him. Gradually Mr Cope was frozen out and more and more negotiations were conducted by Mr Mitchell and Mr Swift. But this was always on the footing that Mr Cope was willing to help, but he was not really needed. Then once he had been frozen out he was given no further information and he, and later W R Grainger, had to inquire how things were going before they could get any information which showed that the development had taken place.

In 1973 Mr Mitchell paid £5,000 or rather he advised Mr Swift, I think it is right to say, to pay £5,000 to Mr Shingler on the basis that Mr Shingler had made the introduction. Then a year later when Mr Grainger put forward the claim on behalf of the present plaintiffs it was disputed. In the witness box Mr Mitchell described it as an outrageous claim. Of course if one accepted that the thing was due to the introduction by Mr Shingler I suppose it would have been an outrageous claim. I doubt whether Mr Mitchell would have so described it if he had known the facts. So much for the facts as I have found them.

I now have to come to deal with the legal position on the basis of those facts. First of all I have to consider the case between the plaintiff and the defendants. I have to ask myself I think three questions. Firstly, were the plaintiffs retained by the defendants? Secondly, if so, on what terms? And, thirdly, was the introduction by the plaintiffs the effective introduction of the granting of the lease? So far as the retainer is concerned it is plain that they were retained. I hold they were retained at the time of the telephone interview of April 26. Were there any doubt about that certainly a retainer would have been effective on the letter of May 13 coupled with the conduct of allowing Cope to go on to act in the matter. Secondly, if they were retained, on what terms were they retained? At this stage I think it is convenient to deal with the argument put forward by Mr Davies that even if they were retained they are not entitled to the full amount of their claim. From the letter it is plain enough that it is to be the RICS scale. Had the matter been raised promptly so that there might have been some argument that there was in fact in May 1971 no consensus as to scale, it might well have been possible to suggest that there was a retainer but that it ought to be on a *quantum meruit*. Somewhat reluctantly I come to the view that the RICS scale must apply, and that scale 19 is the appropriate scale. I say somewhat reluctantly because what Mr Cope did, although undoubtedly valuable, would from most people's point of view hardly seem to be £27,500 worth of work. But I think the matter was really put very neatly by Mr Sheridan when he said, "It is no ingredient in my case that the fee is reasonable." That seems to me to be right. This is the time-honoured way in which estate agents assess their fees. The argument as to whether it is a satisfactory way is equally time honoured. It is not for me as a judge to do more I think than recognise the facts that exist, however curious the results may be. But I was greatly assisted by an argument from Mr Sheridan, because I indicated to him that if I possibly could, consistent with my duty, I would prefer to hold this is a case for a *quantum meruit*. He addressed an argument to me as to the law of contract and as to the circumstances in which the courts would interfere with contracts which convinced me, I must say, that really there is no basis here on which I could interfere, and that it certainly would not be open to me to give the defendants something in the nature of a sympathy vote. Mr Sheridan will forgive me if I do not refer to all the authorities. I simply say that it was a very neatly presented piece of argument which I enjoyed listening to and which was most helpful. Therefore I am bound to say that I think the contract was payable on the scale, and in regard to this particular job scale 19 is applicable.

That leaves the third question as to whether this was the effective introduction. I have been referred to several authorities. I think they are all really on the same lines as to the test one has to look at in a case like this, and I prefer Harman LJ's phrase, which is the phrase he used in *Bartlett v Cole* (1963) 188 EG 397 at 399. As to that, there really can I think be no doubt. It is perfectly true that it was a long time before the contract for a lease was signed. I have not had any detail about it, but it is apparent from the evidence before me that that was due not to any defect in the introduction but to some highly complicated negotiations that had to be conducted about ground rent and planning permission and all that sort of thing. So far as the introduction went, and it is the introduction that is the estate agent's responsibility, the effective introduction was made at the meeting of May 21. That had become completely effective by July in the sense that Mr Mitchell wrote on July 8 saying in effect that if you are serious, we are serious and we will not try to let to anyone else. That was met by Mr Brooker, who wrote back and said, yes we are serious and setting out the required particulars of various things they wanted. That was what Mr Sheridan in his romantic phrase referred to as the engagement. But from then on, whatever difficulties emerge, Johnson Matthey were solid and they went ahead until they had taken the lease. It is not without significance that only three days before Mr Luck had written saying: "I do not think there is any prospect of doing business with GUS. I suggest you reply to Mr Gold breaking off negotiations. Continue please to prod Mr Brooker gently." So there we are, within the same week as the engagement is announced Mr Cope is being asked to do some helpful proddings. It is really impossible in those circumstances to suggest that the introduction he had made was not the effective introduction so far as the grant of the lease was ultimately concerned. So there is no escape from the conclusion that the plaintiffs are entitled to their full claim. It has been treated throughout this case as a claim for £27,500. I am quite prepared to deal with it on that basis. It was originally a claim for an account, but it may be that that is now quite unnecessary. Perhaps it is still a reamended claim for an account, but when I have finished I will hear what everyone has to say about that. It may be simplest to make it for the money judgment.

First of all I ought to deal with the third party proceedings. The third party proceedings are put, as I have already said, purely on the basis of agency. Mr Davies has done all he can to wrestle with a proposition which really is impossible. He has elicited evidence that it is normal practice that the developer pays the estate agent. He has elicited answers showing that some people regarded it as necessary to employ an estate agent once ITT had gone. But unless one can go so far as to say in these circumstances that an agent is a necessity (I do not think he went that far) it does not help his case, unless he can show that there was a specific agency, that he got authority for what he did. There is nothing from which an agency could be implied. The point was really this, that the arrangement or agreement of July 1970 had been torpedoed by the defection of ITT. And the parties were in a position where they would have to think again; if they were to get the thing off the ground they would have to agree to some variation of that arrangement. There cannot have been an agreed variation here, because the second and third third parties did not even know what Mr Luck was doing. It cannot be said that they had agreed to it or authorised it in any way, nor can it be said that after the meeting of May 21 they did

A anything to adopt what Mr Luck had done in engaging Mr Cope.

In those circumstances it seems to me the third party proceedings really fail almost *in limine*. One just has not got the basic requirements for a case against the third parties. Accordingly there will have to be judgment for the plaintiffs against the defendants and for the third parties against the defendants.

Judgment was given for the plaintiffs for £27,250 together with interest from the date of the lease at 10 per cent, namely £8,175, and costs. The defendants were also ordered to pay the costs of the third parties.

B

COURT OF APPEAL

June 27 1978

(Before Lord Justice MEGAW, Lord Justice ORMROD and Lord Justice BRIDGE)

LEWIS & GRAVES v HARPER

Estates Gazette June 30 1979

(1978) 250 EG 1287

Estate agents' claim for commission—Dispute as to amount
C **claimed—No prior agreement as to rate of commission, but understanding that basis of remuneration should be by way of commission—Vendor had experience of the old scales of commission from having sold properties previously—What was reasonable rate of commission in all the circumstances?—Estate agents claimed £286.20 on sale of house at £13,500, but offered to reduce this by £10—Agents claimed that commission on basis of old scale was reasonable—Ban on scales does not prevent estate agent and client from agreeing a rate based on old scales—Here no agreement on rate but judge justified in deciding that the amount charged was reasonable—Judgment for £276.20 upheld**

D This was an appeal by the vendor, F Harper, from a decision of Judge Walker at the Bridlington County Court, giving judgment in favour of Lewis & Graves, a firm of estate agents, for £276.20 by way of commission on the sale of the vendor's house at 45 Harewood Avenue, Bridlington.

John Altman (instructed by Collyer-Bristow & Co, agents for Harland, Turnbull & Roberts, of Bridlington) appeared on behalf of the appellant; Rodney Grant (instructed by West & Son, of Bridlington) represented the respondents. The respondents' counsel was not called on.

Giving the first judgment at the invitation of Megaw LJ, ORMROD LJ said: This is an appeal by the defendant from an
E order which was made on September 12 1977 in the Bridlington County Court by His Honour Judge Walker. Under that order he gave judgment for the plaintiffs for the sum of £276.20, plus costs on scale 3. The plaintiffs' claim was for commission as estate agents.

The particulars of claim are very short. They say: "The plaintiffs claim the sum of £286.20 for commission payable by the defendant under an agreement made orally on July 20 1976, whereby the plaintiff was instructed by the defendant to find a purchaser for his house 45 Harewood Avenue, Bridlington." Then in paragraph 2 the plaintiffs allege (and it is not disputed) that they introduced a purchaser, Mr and
F Mrs Hiley, who subsequently purchased the house at a figure of £13,500. In paragraph 3 they calculate their commission by a sliding scale: 5 per cent on the first £500; 2½ per cent on the next £4,500; 1½ per cent on the balance; plus VAT; and that came to the figure of £286.20.

The defence admitted the verbal agreement whereby the plaintiffs were instructed to find a purchaser for 45 Harewood Avenue. The defendant admitted that the plaintiffs had sold the property, and he admitted that he had not paid any commission up to that time. In paragraph 4 of the defence he said: "There was no agreement as to the amount of commission on sale payable to the plaintiffs and the defendant
G denies that the commission claimed is fair or reasonable. A fair and reasonable sum for commission does not exceed £100." Paragraph 5: "Save as hereinbefore admitted the defendant denies each and every allegation. . . .": and so on.

There was a request for further and better particulars of the particulars of claim as to how the figure of £286.20 had been arrived at. The answer in the further and better particulars was: "There was no formal or written agreement as to the commission to be charged. The commission was charged on the price of the house and on the work done by the agents which is the normal practice in the Bridlington area."

That was the state of the pleadings. The evidence, which
H is very briefly noted by the learned judge, follows more or less on those lines—rather less, perhaps, than more.

The point to be observed is that it is conceded by the defendant that there was an oral contract in the terms set out and that, quite clearly, it was an oral contract that the plaintiffs should be remunerated by commission and not by any other form of payment—commission, that is, payable in the event that they were instrumental in selling the property and not otherwise.

In the course of giving evidence Mr Graves, who was a partner in the plaintiff firm, said (though it was in issue) that the plaintiff firm had acted for the defendant before. Mr
J Graves valued the property at between £13,500 and £13,750; took particulars; took photographs, put the house on the market, advertised it in their shop window, and advertised it in the papers. They sold it very quickly. He agreed that they never in fact discussed the actual commission rates, but he said that it was the practice of the plaintiff firm to continue to charge the old scales laid down in the "Estates Gazette," which are the same as those of the Royal Institution of Chartered Surveyors. In the cross-examination, the learned judge noted an answer in this form: "Costing it out on *quantum meruit* £60 would not be unreasonable." We are told by counsel that there was in fact a good deal more detailed
K cross-examination of Mr Graves as to how the figure of £60, which no doubt was put to him, was made up.

The defendant in his evidence said that he had gone to the senior partner, Mr Lewis, about buying a house at 4 Vernon Road, Bridlington. He told Mr Lewis that he was going to advertise his existing house in the *Bridlington Free Press* himself, and Mr Lewis had said to him "Why not let us sell it?". The defendant went on "I asked 'How much would you charge?' and he said 'Not a lot; we are the cheapest in town.' I said 'Fair enough; you sell it'."

After the property was sold, the plaintiffs sent in their
L account, and the defendant queried it and complained about the bill. In his evidence to the judge he said "I said it was a little bit excessive." Mr Lewis said it was the "going rate." "He offered to reduce it by £10." The defendant then argued that solicitors had charged £88 and that the plaintiffs' costs should be £20 plus £25 profit. He admitted that he had previously sold a house in Vernon Road through other agents and had been told what the commission was: so much, as he said, on the first £500 and so much thereafter; and that he did not dispute that fee. He also sold a property called St Kitts Hotel and paid fees on the ordinary commission basis in respect of that.

M The notes of the judgment which the learned judge made for his own purposes read as follows: "The defendant had experience of scale fees from the sale of Vernon Road and hotel. He knew the sort of bill to expect now the scale has been abolished. No agreement to pay less or to pay on time and cost basis. The defendant's real objection is that it was a quick sale unlike St Kitts Hotel and Vernon Road. I find that the only form of remuneration contemplated by both parties was by commission at a rate similar to the old scale." So the learned judge gave judgment for the figure of £276.20, which was the figure claimed less the £10 conceded by Mr Lewis in the course of his evidence.

Mr Altman complains, basically, that the learned judge approached the case in the wrong way altogether. He does not, and cannot, contend on this evidence that the plaintiffs should only be remunerated on a *quantum meruit* basis, whatever that may be. But it is plain—and Mr Altman has not really seriously suggested the contrary—that there was never any sort of agreement here that the plaintiffs would be paid for the work that they actually did, whether or not they were successful in selling the property. It is quite plain, in my judgment, that the parties in this case made their agreement and made it on the basis that the remuneration was going to be by way of commission.

The only question, therefore, that was left for the judge to decide was, what was a reasonable rate of commission in all the circumstances of the case? The plaintiffs not unnaturally say that the old scale rates are perfectly reasonable. I call them the old scale rates because as a result of the enquiries of the Monopolies Commission and their report in 1969 a statutory instrument was made in 1970 called the Restriction on Agreements (Estate Agents) Order 1970 (SI 1970 No 1696) which by regulation 3 makes it unlawful for two or more estate agents to agree to charge a minimum fee or a minimum rate of commission. It does not, of course, in any way interfere with the arrangements which an estate agent makes with his own customers. There is nothing unlawful as between estate agent and customer in charging a rate of commission based upon the old scale fee if that is what the parties agree. In this case it is common ground that there was no agreement on the rate of commission. What, then, is a reasonable commission in all the circumstances?

The plaintiffs' case, as I have said, was that £276 was a reasonable sum by way of commission. There was no evidence at all on the other side as to what would be a reasonable rate of commission in the circumstances. Some figures were mentioned. I have already referred to the figure of £60 as the figure said to have been arrived at costing it out on a *quantum meruit* basis—whatever that means. But, since this was a contract to remunerate the plaintiffs by way of commission, that is quite irrelevant. The only other little bits of evidence before the learned judge were: first, the alleged statement by the defendant that he thought that the commission was a little too high. There was also evidence that after he had received his bill he suggested that the plaintiffs should accept half the commission; and at the same time he seemed to be suggesting in the course of his evidence that £45 would be a fair remuneration. So the judge was left, as far as I can see, with virtually no material at all upon which to base an assessment of what was reasonable other than the plaintiffs' evidence that they had charged on the footing which was in effect common in the trade—the old scale fee. True it is no longer binding as a minimum scale; but they said it still represents a reasonable, in the sense of a normal, sort of charge for this kind of work. There was really no evidence to the contrary at all.

It is said that the learned judge was wrong in approaching the case on the basis of ascertaining what was in the contemplation of the parties. It seems to me, with respect, that it is perfectly open to a learned judge to approach a case of this kind in that way. If the evidence before him enables him to arrive at a conclusion as to what the parties actually had in mind, then it is not necessary to go and look for implied terms or find "reasonable" figures. If, on the other hand, it is necessary to find what was a reasonable rate of remuneration, in my judgment the way the learned judge approached the matter was perfectly reasonable, given the material that was before him.

Mr Altman has suggested that the learned judge might have considered the scale of commission which is fixed for sales under the order of the court. That scale is to be found on p 495 of the White Book in a note to Order 31 rule 2. It gives a lower rate than the scale used by the plaintiffs. On the other hand, I observe from the note that it does not include any advertising cost, which is reimbursed separately and only if authorised by the court before it has been incurred or has afterwards been ratified by the court. So one does not know at all how relevant that scale is to the circumstances prevailing in Bridlington in 1977. However, the matter cannot be taken any further because the learned judge was not referred to the scale at all and no argument was addressed to him to suggest that that would be a more reasonable rate of remuneration than the figure that the plaintiffs in fact charged.

Mr Altman referred us to the case of *Re Wolfe (deceased): Heller v Wolfe* [1952] 2 All ER 545, which is a judgment of Roxburgh J; but, with respect, I do not think it carries the matter any further.

In my judgment there was only one matter here for the learned judge to decide on the evidence before him and that was, what was a reasonable rate of commission in the circumstances of this case? It was open to the judge, in my view, to arrive at the conclusion which he did on the evidence, because there was virtually no other evidence before him.

The only other point which should perhaps be mentioned is that the learned judge, it is submitted, did not reflect in his assessment the remark of Mr Lewis about "We are the cheapest in town." That seems to me very much a matter for him, to decide whether that observation was enough to justify any substantial reduction in the amount claimed; and anyway Mr Lewis agreed to reduce the fee claimed by £10.

For those reasons, in my judgment this appeal must fail and should be dismissed.

BRIDGE LJ agreed.

Also agreeing, MEGAW LJ said: Mr Altman has said everything that can possibly be said in the way of seeking to suggest that the learned judge arrived at a wrong conclusion. It appears to me that this is not a case which gives rise to any question of general principle at all. I certainly would not suggest for a moment that the learned judge in this case intended to establish or accept the proposition that, in any case in which an agreement is made between an estate agent and the client of an estate agent under which the estate agent agrees to sell the client's house and nothing specific is said as to the remuneration, the estate agent is, necessarily and as of course, entitled to recover remuneration on what used to be the scale, or on any other particular basis which he is in the habit of charging. But on the facts of this case—which are what concern us—the judge, as I understand it, on the evidence accepted that it was known to and understood by the defendant, as a result of experience of scale fees on past sales which had been conducted on his behalf, that the sort of bill which was to be expected was the sort of bill which was here submitted. The learned judge, on the evidence, was entitled to arrive at his judgment on that basis; and I agree that the appeal must be dismissed.

The appeal was dismissed with costs.

HOUSING ACTS

COURT OF APPEAL

November 30 1978

(Before Lord Justice MEGAW, Lord Justice BROWNE and Lord Justice WALLER)

DUDLOW ESTATES LTD v SEFTON METROPOLITAN BOROUGH COUNCIL

Estates Gazette March 31 1979

(1978) 249 EG 1271

Housing Act 1957—Appeal to Court of Appeal by local authority against quashing of demolition order by county court judge—Question whether or not house was capable of being rendered fit at reasonable expense—Comparison to which regard must be had under section 39(1) of 1957 Act—Meaning of "value" in that subsection—Judge's decision challenged on the ground that he considered the carrying out of repairs to be justified but did not support this conclusion with figures—Contention that he should have expressly compared the cost of repairs with the difference between the value of the house unrepaired and its value after completion of repairs—Held by Court of Appeal that judge had taken account of the factors in section 39(1)—No error of law—Dicta approving statement by Geoffrey Lane LJ in Hillbank Properties Ltd v Hackney London Borough Council—Appeal dismissed

This was an appeal by Sefton Metropolitan Borough Council against the decision of Judge Edward Jones at Liverpool County Court quashing a demolition order in respect of a house at 50 Beechwood Road, Litherland, Merseyside, owned by the respondents, Dudlow Estates Ltd.

N F Riddle (instructed by Sharpe, Pritchard & Co, agents for J R M Heppell, Metropolitan Borough of Sefton) appeared on behalf of the appellant council; A T Sander (instructed by Benjamin, Kay & Co, of Prescot, Lancs) represented the respondent company.

Giving the first judgment at the invitation of Megaw LJ, BROWNE LJ said: This is an appeal by the Metropolitan Borough Council of Sefton (to whom I shall refer as "Sefton") from a decision of Judge Edward Jones in Liverpool County Court, which was given on November 23 1977, by which he quashed a demolition order made on October 25 1976 by Sefton in respect of a house at 50 Beechwood Road, Litherland, in Merseyside. That house was owned by Dudlow Estates Ltd, to whom I shall refer as "Dudlow." The house is the end house of a terrace. It was built about 1900. It is a two-storey house, with two living rooms and a kitchen on the ground floor, and two bedrooms and a bathroom with a wc in it on the first floor. That bathroom, as I understand, had been converted some years ago from a bedroom. The tenants of the house at all material times were Mr and Mrs Hughes, who paid a rent of £2.50 a week.

I think that in order to make the case intelligible it is convenient to refer straight away to the relevant statutory provisions, which are all contained in Part II of the Housing Act 1957. Part II is headed "Provisions for securing the repair, maintenance and sanitary condition of houses." Section 4 defines the matters which are to be taken into account in determining whether a house is or is not unfit for human habitation. Part II then goes on to provide, so far as relevant for present purposes, for two alternative ways in which local authorities can deal with unfit houses. First, there is a group of sections beginning with section 9, which provides, by subsection (1):

> Where a local authority, upon consideration of an official representation, or a report from any of their officers, or other information in their possession, are satisfied that any house is unfit for human habitation, they shall, unless they are satisfied that it is not capable at a reasonable expense of being rendered so fit, serve upon the person having control of the house a notice —(a) requiring him . . . to execute the works specified in the notice. . . .

That is followed by provisions enabling the local authority to do the work themselves and charge the owner if it is not done, and various other ancillary provisions. That group of sections is dealing with the position where the local authority is not satisfied that the house is not capable at reasonable expense of being rendered fit for human habitation.

The second group of sections, which is the one with which we are directly concerned in this case, begins with section 16. That is dealing with the situation where the local authority considers that the house is not capable at a reasonable cost of being rendered fit for human habitation.

Section 16(1) provides:

> Where a local authority, on consideration of

certain information

> are satisfied that any house—(a) is unfit for human habitation, and (b) is not capable at a reasonable expense of being rendered so fit, they shall serve upon the person having control of the house

or various other persons

> notice of the time . . . and place at which the condition of the house and any offer with respect to the carrying out of works, or the future user of the house, which he may wish to submit will be considered by them.

Subsection (2):

> Every person upon whom such a notice is served under subsection (1) of this section shall be entitled to be heard when the matter is so taken into consideration.

Subsection (3):

> A person upon whom such a notice is served under subsection (1) of this section shall, if he intends to submit an offer with respect to the carrying out of works . . . serve upon the authority notice in writing of his intention . . . and . . . submit to them a list of the works. . . .

Subsection (4):

> The local authority may, if after consultation with any owner or mortgagee, they think fit so to do, accept an undertaking from him, either that he will within a specified period carry out such works as will in the opinion of the authority render the house fit for human habitation, or that it shall not be used for human habitation until the authority, on being satisfied that it has been rendered fit for that purpose, cancel the undertaking.

A Subsection (5) provides that

> Nothing in the Rent Act 1968, or the Rent (Agriculture) Act 1976, shall prevent possession being obtained of any premises by any owner thereof in a case where an undertaking has been given under this section that those premises shall not be used for human habitation.

Then section 17(1):

> If no such undertaking as is mentioned in the last foregoing section is accepted by the local authority . . . then, subject to the provisions of this section, the local authority shall forthwith make a demolition order for the demolition of the premises to which the notice given under the last foregoing section relates: B Provided that if in the case of any house the local authority consider it inexpedient to make a demolition order having regard to the effect of the demolition of that house upon any other house or building, they may make a closing order as respects that house instead of a demolition order.

Then alternatively they have a power of purchase under subsection (2).

Section 20 gives the right of appeal to the County Court. Section 20(1):

> Any person aggrieved by—(a) a demolition . . . order made under this Part of this Act . . . may, within twenty-one days after C the date of the service of the order . . . appeal to the county court within the jurisdiction of which the premises to which the order . . . relates are situate, and no proceedings shall be taken by the local authority to enforce any order . . . in relation to which an appeal is brought before the appeal has been finally determined.

Subsection (3):

> On an appeal to the county court under this section the judge may make such order either confirming or quashing or varying the order . . . as he thinks fit and may, if he thinks fit, accept from an appellant any such undertaking as might have been accepted by the local authority. . . . Provided that the judge shall not accept from an appellant upon whom such a notice as D is mentioned in subsection (1) of section 16 of this Act was served an undertaking to carry out any works unless the appellant complied with the requirements of subsection (3) of that section

which, it will be remembered, required him to give notice to the authority making an offer of what he was prepared to do.

There are further provisions as to purchase notices which can be given by the local authority, to which I need not refer. I think that the only other section that I need read is section 39(1), which provides that:

> In determining for the purposes of this Part of this Act whether E a house can be rendered fit for human habitation at a reasonable expense, regard shall be had to the estimated cost of the works necessary to render it so fit and the value which it is estimated that the house will have when the works are completed.

I do not think it is necessary to go in great detail into the earlier history of this matter. The history goes back to 1973, when the house in question was in the area of another local authority, the Litherland Urban District Council, and Dudlow applied to them for an improvement grant. As a result of the local government reorganisation, Sefton became the local authority responsible. It is enough to say, I think, that at one stage the local authority was prepared to make an F improvement grant, but Dudlow were not prepared to take up that grant, which was a good deal less than Dudlow had asked for, because they took the view that the whole of the grant would be taken up with one particular piece of work on the gable of this house.

I think I can begin with the document which is what counsel called a " time and place " notice: in other words, a notice under section 16(1) of the Housing Act. It is addressed, not to Dudlow, but to another company which was an associated company which acts as managing agent and was therefore a proper person on whom the notice could be served. That notice stated that Sefton

> are satisfied that 50 Beechwood Road is unfit for human habitation and is not capable of being made fit at reasonable cost, and G that the condition of the house, and any offer with respect to the carrying out of works, or the future use of the house, which you may wish to submit will be considered by the Council . . . on December 18 1975.

We were told that no offer was made by Dudlow.

The demolition order is dated October 25 1976.

> Whereas the Borough Council of Sefton being satisfied that the house known as 50 Beechwood Road, Litherland, is unfit for human habitation and is not capable of being made fit at reasonable cost, have complied with the provisions of section 16 of the Housing Act 1957, in relation to the house; And Whereas no H such undertaking in relation to the house as is mentioned in the said section 16 has been accepted; Now Therefore the Council, in pursuance of section 17 of the Housing Act 1957, hereby order as follows: (1) the house shall be vacated within twenty-eight days from the date on which this order becomes operative; (2) the house shall be demolished within six weeks from the expiration of the last-mentioned period, or, if it is not vacated before the expiration of that period, within six weeks from the date on which it is vacated.

Dudlow appealed to the county court against that order. I do not think I need refer to their notice of appeal in detail. The only issue before the county court was whether or not the house was capable of being rendered fit for human habitation J at reasonable expense.

The matter, we were told, first came before the county court on February 4 1977. The matter was adjourned on that day, and negotiations took place between the parties which resulted in an agreement which appears in our papers as " Draft Heads of Agreement " but which we are told in fact did lead to a final agreement. However, it turned out to be impossible for that work to be carried out because, as I understand it, of the attitude of the tenants, who were thoroughly obstructive.

Between February 4 1977 and November 7 1977, when the matter came before the county court again, Dudlow made K three offers of alternative accommodation to the tenants, who refused them all for one reason or another. The local authority also made an offer of alternative accommodation at some stage, and they refused that too. So the appeal came back before the county court. In that interval between the two hearings something else had happened, which was that Dudlow had done some of the work which had been scheduled to the heads of agreement as being work which was necessary to be done. As I understand it, that was all external work, which could be done without the agreement of the tenants; and the evidence was that that work had cost something like £400, although there was also evidence that some of the L work was not satisfactory.

The matter was heard before Judge Edward Jones on November 7 and 8 1977; and he viewed the house. Eventually, on November 23 he gave a reserved judgment allowing the appeal and quashing the demolition order. Having referred to the fact that the order had been made over a year ago and that negotiations had been going on, he went on as follows:

> In particular, it is agreed that the house is capable of repair so as to make it fit for human habitation. The issue for my decision is whether this can be done " at a reasonable expense " M as is required by section 16 of the Act. In deciding this issue section 39 requires that " regard shall be had to the estimated cost of the works necessary to render it so fit and the value which it is estimated the house will have when the works are completed." The authorities lay down that such value must take into account that there is a sitting tenant.

Pausing there, it was also admitted that the house in its present state is unfit for human habitation. So that the position was that it was common ground before the county court that the house was not then reasonably fit, that it was capable of being made fit, and, as will be seen later on, there

A was substantial agreement as to the cost of the necessary work.

The judge goes on:

> The first point taken was that on the appeal I was confined to considering whether the authority had properly exercised a discretion which was theirs, and in particular whether they had omitted to consider any matter they should have taken into account or whether they had acted in bad faith. I cannot agree. The Act, in section 20(3), is clear in my view. I have to make up my own mind, on the evidence the parties adduce before me, as to whether I consider the order should be confirmed, quashed or varied.

B Mr Riddle very properly conceded that he could not renew in this court that submission which he had made to the county court judge. It is enough, I think, in this connection to refer to the decision of this court in *Victoria Square Property Co Ltd v Southwark London Borough Council* [1978] 1 WLR 463. I need not read the passage, but I refer to the judgment of Bridge LJ at p 472 G to 473 E, where he said quite clearly that it is for the county court judge himself to consider all the circumstances and make his decision in the light of his views about the matter.

The judge then went on to reject a suggestion, made, I
C take it, by Dudlow, that he might accept an undertaking from them. He decided that he had no power to do that, in which, if I may say so, he was plainly right, having regard to the proviso to section 20(3), because Dudlow themselves had not made any offer of any undertaking to the local authority when they could have done so under section 16.

The judge then described the property and said that it had been agreed that:

> in its present state it is unfit for habitation. The tenants are quite unco-operative and will not help in getting the necessary repairs done. My view is that they are holding everyone to ransom with the object of getting a council house in the area
> D of their own choice. They seem to have acted unreasonably in respect of offers of alternative accommodation both by the owner and local authority. The cost of putting the house in a state which would make it fit for habitation is agreed at £1,400-£1,500. I consider the costs of demolishing the house with all the works necessarily incidental thereto at about £500. The council would have me accept the house has only a nominal value. I do not agree. Some of the work has already been done, some perhaps not altogether satisfactorily. For an expenditure of no more than £1,000 this house can be made fit for a small family to live in it. It must have some real value.
>
> I am a little surprised that the local authority was not prepared to be more flexible in order to have one more home
> E available in the face of the known serious housing difficulties. It is suggested that their decision stems first from an original estimate, now agreed to be far too high, of the cost of the necessary work, and secondly from the attitude of the tenant which they felt made demolition the only way to dispose of the matter. It is also suggested that originally they took a much more serious view of the bowing in the gable end. Their agreement now that tie-bars will remedy the position is a material concession. It is said that the proposed expenditure is not justified. The owner disagrees. For an expenditure (beyond the cost of demolition) of £1,000 he will have a house, producing a rent which can be reviewed every three years and with the chance, very real in this case, that the tenant may well move out. She
> F has been offered one council house which she refused, and she has very recently had a third baby.
>
> Doing the best I can with the figures before me I consider the expenditure involved is justified even if only the value of the house with a (not even this) sitting tenant is to be taken into account. There are a number of other considerations which it was suggested I should have regard to, such as the fact that the money was not being provided by the local authority, that a commercial landlord thought it worthwhile, as well as the matters adverted to earlier in my judgment. My judgment does not require recourse to those other considerations, but they do reinforce and confirm my view that as a financial proposition, purely and simply, the carrying out of the agreed repairs is justified.
>
> I have visited the house. There are many houses in the immediate vicinity, and in the same terrace, which look very comfortable G and good little homes from the outside. Now that I know that technically the house with which I am concerned can be put into habitable condition I have no hesitation in finding that it has all the possibilities of adding to the area another attractive little home. Much may depend on the tenants, and they are far from the best as I had reason to observe, but a real potential is undoubtedly there.
>
> I had hoped that I should be able by my judgment perhaps to control in some way the doing of the work and to decide what should happen to the house. Much depends, in the first place, on obtaining vacant possession of the property, as it is agreed, and obvious, that no real work can be done with tenants in possession. H I am quite satisfied this demolition order should be quashed.

The powers and duties of this court on an appeal from the decision of a county court judge in a matter under the Housing Act have recently been stated by this court in the case to which I have already referred, *Victoria Square Property Co Ltd v Southwark London Borough Council*. At page 473E of the report Bridge LJ said:

> In this court the judge's discretionary decision can only be attacked if it is shown to be erroneous in point of law. Having regard to the nature of the discretion exercised, this burden could only be discharged by showing that the judge acted on some wrong principle, failed to take account of relevant matters, took account J of irrelevant matters, or reached a conclusion so manifestly unreasonable that it could not properly have been reached by a reasonable judge correctly directing himself in law.

My Lord, Waller LJ, said, at p 475H:

> Whether the grounds on which he

[that is the judge]

> said he would have exercised the discretion in the way he did are clear or not, this is a matter on which the judge has a wide discretion, and, provided that he has not misdirected himself, or taken into account considerations which he ought not to have taken into account, this court cannot interfere.

K

Finally, my Lord, Megaw LJ, at the bottom of p 476H, said:

> The jurisdiction of this court on appeal from the county court is statutory. The only statutory provision conferring jurisdiction which could be relevant in this case is section 108 of the County Courts Act 1959. It provides for an appeal on a point of law. An exercise of a discretion by the judge of a county court may, it has long since been held, involve a question of law. It will do so if the discretion has been exercised on a wrong principle. It will have been exercised on a wrong principle if it appears that the judge in exercising his discretion has taken into account some matter as being material which he ought not to have taken into account, or has failed to take into account some matter which would have L been material and which he ought to have taken into account, or that, although no such error can be specifically identified, the exercise of the discretion is so unreasonable that it can be explained only on the basis that the judge must have erred in principle. This court's jurisdiction, as given by Parliament, in a case such as this, is thus much less extensive than the jurisdiction given by Parliament to the county court, when it is dealing with the first appeal to the courts, by the very wide words of section 20(3) of the Act of 1957.

Mr Riddle frankly accepts that that is the burden which he has got to take upon himself in this court. He submits that he can discharge it. What he described, I think, as his basic M submission was that the judge had failed to take into account, or had failed sufficiently to take into account, the provisions of section 39(1). He submitted that what the judge had to do in every case of this sort was to make in his own mind, even if he did not include a specific statement in his judgment, a mathematical calculation on these lines. He said that the judge ought first to take the value of the property as it stood in its unrepaired state. Then he should determine what the value would be if it was repaired. He should then deduct the present value from the repaired value and see whether the difference was greater or less than the cost of repair. He submitted that

A in the present case the judge had not done this. There are various other grounds of appeal, but I do not think I shall be doing Mr Riddle any injustice if I do not go through the other grounds in detail. The section 39 point, as I understood it, was the fundamental criticism of the judge's decision, although it brings in various other factors.

Mr Riddle said that if one looks at the judgment there is no such calculation, and that when the judge said that he did not accept the view that the house only had a nominal present value, there was no evidence to support that because the only evidence before him was the evidence called by Sefton that it had only a value of £50. That evidence of Mr McClennan, of B Sefton, was, as I understand it, the only evidence of the value of the house with a sitting tenant in it. Mr Allan Jones, the witness called by the owners, took into account the possibility of a vacant possession value. What Mr McClennan said was this. He first of all said: "Value, with regulated tenancy at rent of £2.50 per week, poor state of repair—£50 nominal. Because repairs above normal." Then he said that if the repairs specified in what he called the "February" schedule—that is, the schedule to the agreement which had been reached but which it was not possible to carry out—had been carried out, it would be "£1,000 or thereabouts. Fair rent £4 per C week"—that is to say the £2.50 existing rent increased to £4 —less certain expenses, taking six years' purchase, "£990 rounded up to £1,000." Mr Riddle says that there you have an increase of £950 as a result of repairs; the cost of the repairs was £1,400 or £1,500, and therefore the judge's answer cannot be right.

It seems to me that, although the judge did not give any figures, what he is doing is carrying out the exercise which Mr Riddle says ought to be carried out. As Mr Sander says, the key paragraph is the one in which he says: "Doing the best I can with the figures before me I consider the expenditure involved is justified even if only the value of the house with a D sitting tenant is to be taken into account." So there he is taking the worst possible case against the owner. If he took something more than a nominal value, as counsel points out, it makes the case less favourable to the owner. The only evidence of the repaired value with a sitting tenant was the evidence of Mr McClennan, to which I have referred. The cost of repair was agreed. It seems to me that, approaching it on the basis on which the judge did look at it, that is to say the worst case from the point of view of the owner, the judge cannot be said to have been wrong in law in the result at which he arrived. In my view it is not necessary, in every case certainly, for the judge in his judgment to set out specific E figures of the mathematical calculations suggested by Mr Riddle. It is quite enough, it seems to me, if it is apparent from the judgment that he has taken into acount the factors which section 39(1) requires to be taken into account. Section 39(1) only says that the court shall have regard to those two factors. Mr Riddle rightly concedes that that does not mean that those are the only factors which can be looked at. Indeed I think there is authority of this court to this effect.

What seems to have happened is that the judge in fact took the council's figures and said that, even taking those figures and on the worst assumption against the owner that it must be F valued with a sitting tenant, he still thought that the expenditure was reasonable. On that basis the cost of the repairs, having regard to the fact that £400 worth had already been done, leaving £1,000, and the repaired value of the house of £1,000 very nearly balance. It seems to me that the judge was entitled also to take into account, as tipping the balance, if it needed tipping, the expense of demolition. It seems to me that when a court is considering whether it is reasonable for the owner to spend the money necessary for repairs it must be a relevant factor that by spending £1,000 on repairs he will avoid having to pay £500 for demolition. I am inclined to agree with what I think counsel said in argument, that the cost of demolition does not come in strictly under section 39. G But it seems to me that it is clearly one of the relevant circumstances to which the court is entitled to have regard under section 20.

The judge, as I have said, gave his decision on the basis that he should only consider the value with a sitting tenant. He said: "The authorities lay down that such value must take into account that there is a sitting tenant." That, I take it, was based on what Somervell LJ had said in *Bacon v Grimsby Corporation* [1950] 1 KB 272 at p 282. Somervell LJ said this: "I have come to the conclusion that the county court judge was right in taking, as he did, the value of the house spoken of in subsection (3) to be the freehold value, on the H basis that these houses contained, as they did, tenants who had the right to claim the benefits of the Rent Restrictions Acts."

We were told by counsel that that sentence has been taken as laying down a principle. But in point of fact, when one examines that case and sees what the point was, it was entirely an *obiter dictum*. The problem did not directly arise in that case. That has been pointed out in this court in the recent case of *Hillbank Properties Ltd v Hackney London Borough Council* [1978] 3 WLR 260, by Lord Denning MR.

The judge having approached the matter as he did, that he would only look at the sitting tenant value, and there being no J cross-notice, it is not necessary for us in this case to consider what the position would be if we were considering it afresh or if there had been a cross-notice. I do not think it is desirable to try to lay down any hard-and-fast rule about it. But that is obviously a matter which is going to be of great importance, and I think it would be a mistake for me to say anything about it which would only be an *obiter dictum*, beyond this: that, as at present advised, if the point had to be considered I should agree with what was said by Geoffrey Lane LJ in the *Hillbank Properties* case, at p 272 F. He said:

> There are in the present cases, as there usually will be, two different house values for comparison. The "sitting tenant" value, K which is less than the cost of the repairs, and the "vacant possession" value, which is a great deal more. The reasonableness or otherwise of the local authority's notice to repair may largely depend upon which of those two values one selects for purposes of comparison. It may be that evidence is available to show that the tenancy is unlikely to come to an end in the foreseeable future, in which case the landlords should adduce it; it may be that there is evidence to the contrary, in which case the local authority should do the same.
>
> [That was a case under section 9, not section 16.]
>
> If there is no evidence one way or the other, as was the case here, L what is the judge to do? It seems to me that in such a case it is wrong to take either the tenanted value or the vacant possession value without qualification as the standard of comparison. In those circumstances any disparity between the cost of repairs and the value of the property will have much less weight, because one of the figures is necessarily imponderable.

As I understand it, what Geoffrey Lane LJ is saying there is really in substance what Megaw LJ said in the course of the argument, that when one is talking about value one is talking about open market value, and that when you are considering open market value you must consider all the factors which would influence the mind of the willing seller M and the willing purchaser; and obviously the possibility of getting vacant possession and the strength of that possibility is one of such factors.

To summarise the matter, it seems to me that in this case the judge did approach the case having regard to the provisions of section 39, and that, although he did not set out any mathematical computation, it is impossible to say that he went wrong in law in the decision at which he arrived. Accordingly, in my judgment this appeal should be dismissed.

WALLER LJ agreed.

A Also agreeing, MEGAW LJ said: I would only add two observations. The first is that in the passage which my Lord has read from the judgment which I delivered in *Victoria Square Property Co Ltd v Southwark London Borough Council* I was dealing with a case where the learned judge in the county court had exercised his discretion to accept an undertaking. That, having regard to the provisions of section 16 of the Housing Act 1957, coupled with the provisions of section 20(3), is I think undoubtedly correctly described as being a matter of discretion. It might be said that the sort of case with which we are dealing here, where it is not a question of the acceptance or non-acceptance of an undertaking, is not
B strictly to be described as the exercise of a discretion. However, on the whole I think that, in relation to the issues in the present case, that is a matter of words rather than of substance. It is to my mind clear that, whether it be called exercise of discretion or not, the judge exercising his functions under section 20(3) of the Housing Act 1957 has got to act judicially. If evidence is adduced before him, he has to consider that evidence, and he has to arrive at a conclusion on the basis of his assessment of any relevant evidence that is adduced before him on the one side or the other. If it were shown to this court that the county court judge had reached a decision to found which there was no evidence at all, or evidence which
C was so plainly insufficient that no reasonable court could have acted on it, this court would be entitled to interfere, the question then being a question that properly would be described as one of law. This court also is entitled to interfere, on the basis of error of law, if it appears that the judge has taken into account some material factor which he should not have taken into account, or has taken into account as being material some factor which he ought not to have taken into account. This court is also entitled to interfere if the decision is so unreasonable that the court can only reach the conclusion that some such error must have occurred. One of the matters which the judge by statute has got to take into account is the
D matter referred to in section 39(1).

For the reasons given by Browne LJ, it appears to me that
E it cannot be said here that the learned judge arrived at a conclusion in respect of which there was no evidence on which he could act or that the conclusion which he reached is plainly inconsistent with evidence which was before him. On the contrary, when the figures which were given in the evidence are looked at, bearing in mind that they cannot purport to be accurate figures but are merely assessments with a greater or lesser degree of accuracy, it appears to me that what the judge decided was well within the principles in accordance with which he had to act.

The other matter on which I desire to say something is that the learned judge in his judgment in the present case adopted
F the basis that he had to look at the question posed in section 39(1) on the assumption that the house had in it a sitting tenant; and, as I understand it, that was the basis on which he did approach his decision. The view which I take on that matter is that which was expressed in the decision of this court in *Inworth Property Co Ltd v Southwark London Borough Council* (1978) 76 LGR 263. As is set out in the headnote in that case, this court there decided that "value" in section 39(1) of the Housing Act 1957 meant "open market value." In the assessment of the open market value, those facts and future possibilities, including facts and future poss-
G ibilities as to a sitting, protected tenant, would be relevant—which would be likely to be relevant in the minds of a willing buyer and of a willing seller in negotiating for the purchase and sale of the property. It is unnecessary in this case to go further into that matter, for the reasons given by Browne LJ. But, like him, I take the view that the manner in which Geoffrey Lane LJ dealt with the question in his judgment in *Hillbank Properties Ltd v Hackney London Borough Council* [1978] 3 WLR 260 at p 272, is, if I may say so with great respect, entirely correct. It is, however, unnecessary, on the facts of the present case, to say anything further as to any difference of opinion that may exist on that question.

H I agree that the appeal should be dismissed.

The appeal was dismissed with costs.

INSURANCE

COURT OF APPEAL

February 28 1979

(Before Lord Justice MEGAW, Lord Justice GEOFFREY LANE and Mr Justice DUNN)

LEPPARD v EXCESS INSURANCE CO LTD

Estates Gazette May 26 1979

(1979) 250 EG 751

Claim on insurance policy following destruction of cottage by fire—Question as to amount which the claimant could recover—Insurance a contract of indemnity—Rule in Castellain v Preston applied—Submission by claimant that parties had agreed that basis of computation was to be replacement value even if actual loss was less—Submission rejected by Court of Appeal—Not a " valued policy "—Claimant entitled only to recover his actual loss to an amount not exceeding cost of reinstatement—Loss equal to difference between value of cottage immediately before fire and the site value (£3,000, being £4,500 less £1,500)

This was an appeal from a judgment of Mars-Jones J holding in favour of the plaintiff that he was entitled under a fire insurance policy to the full reinstatement cost following the total destruction by fire of a cottage called " Janor " at Higher Pennance, Lanner, Redruth, Cornwall. The plaintiff, the respondent to the present appeal, was Anthony Charles Leppard and the defendants, the present appellants, were Excess Insurance Co Ltd.

E A Machin QC and Timothy Preston (instructed by Barlow, Lyde & Gilbert) appeared on behalf of the appellants; P J Millett QC and C Gosland (instructed by Fulwell & Partners, of Bristol) represented the respondent.

Giving judgment, MEGAW LJ said: This is an appeal from the judgment of Mars-Jones J delivered on August 19 1977. The claim of the plaintiff, Anthony Charles Leppard, is on a policy of insurance against fire and other risks, issued to the plaintiff by the defendants, Excess Insurance Co Ltd. The property insured was a cottage known as " Janor", at Higher Pennance, Lanner, Redruth, owned by the plaintiff. The cottage was destroyed by fire on October 25 1975. The defendants denied liability, on the basis of an alleged nondisclosure and an alleged breach of conditions of the policy. The learned judge, on the evidence, accepting the evidence of the plaintiff, rejected those defences and held that the defendants are liable on the policy. From that part of the judgment there is no appeal. The appeal is only as to the amount which the plaintiff is entitled to recover.

The judge describes in his judgment how the plaintiff came to own the cottage. He says: " It was the home of his father- and mother-in-law until 1972, when they left to move into a council house. Father-in-law had had a leg amputated and could no longer get up and down the hill to the cottage. When the cottage was put up for sale a neighbouring farmer by the name of Bernard Maitland offered £1,500 for it. The plaintiff rightly thought it was worth much more than that; and, as I find, decided to buy it himself for that sum in order to ensure that his in-laws should have the capital to make the move and furnish their new home, and thereafter to share with them any profit made on a subsequent sale. The plaintiff had to borrow the money from the bank, and they insisted upon the deeds being in his name. They also insisted quite naturally that the premises should be insured in the meantime."

The first insurance of the cottage was with other insurers, not the defendants. In September 1974, however, the plaintiff changed his insurers and took out a policy of insurance with the defendants. This he did through a firm of brokers whose representative in this transaction was a Mr Excell. The learned judge's account of the plaintiff's evidence as to the discussion with Mr Excell leading up to the issue of the policy is as follows: " The plaintiff gave details of that discussion so far as it affected the cottage in Cornwall. He told me that he informed Mr Excell that he had a cottage in Cornwall which he wanted to sell. He told him that he wanted it insured. It was in fact already insured with the Co-operative Insurance Society. He told Mr Excell it was an empty cottage that he had for the purpose of sale. Mr Excell said he would take it on when the current policy lapsed."

The judge accepted that evidence in its entirety. He said: " I was impressed by the plaintiff's evidence. Despite strenuous but quite proper cross-examination by Mr Preston, he emerged with his credit unscathed. He was a patently honest man who, in my judgment, was trying to tell me the truth to the best of his recollection. What he told me about this initial conversation is what I would expect to happen between an assured and a broker in these circumstances. I believe that the plaintiff has told me accurately the substance of that conversation."

I come to the documents. At a meeting with Mr Excell, the plaintiff filled in a proposal form. Against " Sum insured " the figure filled in was £10,000. The declaration signed by the plaintiff on the proposal form, contained these words: " Declaration. I declare that: (1) the sums to be insured represent not less than the full value (the full value is the amount which it would cost to replace the property in its existing form should it be toally destroyed)." It is unnecessary for me to read declarations (2) and (3).

The evidence which the judge accepted, given by the plaintiff, as to the completion of the proposal form was summarised by the judge as follows: " He "—that is Mr Excell—" already knew the cottage was in Cornwall. I gave him particulars. I told him that it was insured for £10,000, but I was not asking that for it as the selling price. Mr Excell explained that it was not the selling price that the plaintiff should insure for but the cost of reinstatement of the premises in the event of total destruction."

There is nothing in the pleadings or in the submissions before Mars-Jones J or before us to suggest, assuming that Mr Excell was acting at this point as agent for the defendants, that there was any oral collateral contract as a result of this conversation, or that what was then said can properly be used to determine the true construction of the policy of insurance which followed on that proposal by the plaintiff. I therefore say no more on that matter. That is not to say that the contents of the proposal form itself are irrelevant. They are incorporated into the contract—the policy of insurance—by express reference. I come to that policy, on the construction of which the first question in this appeal depends. I set out those parts

A of the words of the policy which are or may be said to be relevant.

The first three paragraphs of the text of the policy contain the following:

> The Insured by the proposal which shall be the basis of and incorporated in this contract has applied to the Company for the insurance contained herein and has paid or agreed to pay the premium.
>
> Subject to the terms and conditions contained herein or attached hereon the Company agrees to provide insurance and indemnity as hereinafter set out in respect of events occurring during any period of insurance.
>
> The liability of the Company shall not exceed the sum insured B on any one item in any one section nor in respect of any one section the total sum insured.

Then comes "Section 1. BUILDINGS." The first paragraph reads:

> The sum insured is declared by the Insured to represent and will at all times be maintained at not less than the full value of the buildings (including architects and surveyors fees and an amount in respect of debris removal).

Then there are two paragraphs in italic print under the heading "The Property Insured." The first of those paragraphs shows that the policy is intended to relate to "the Insured's private dwelling(s)." I need not read that paragraph. C The second paragraph in italics reads:

> The Company will at its option by payment reinstatement or repair indemnify the Insured in respect of . . .

and then the print changes into block capitals and reads:

> (A) LOSS OR DAMAGE CAUSED BY ANY OF THE UNDERMENTIONED PERILS:
>
> (1) Fire. . . .

The other so-called "peril" to which I ought to refer is paragraph 10, under the main head "(A)." It reads:

> 10 (a) Architect's or surveyor's legal and other fees for estimates plans and specifications necessarily incurred in the reinstatement of D the buildings following damage by an insured peril. . . .

I need not read any more of it. For the plaintiff reliance is placed on that paragraph because it is said to show that the parties were contemplating, in the event of a loss under the policy, that there should be reinstatement of the building.

The schedule to the policy as originally issued in 1974 showed the sum insured as £10,000. When arrangements were made for the renewal of the policy in September 1975 the sum insured, apparently as a result of Mr Excell's suggestion to the plaintiff, was increased to £14,000. In fact that renewed policy, which is the relevant policy for present purposes, has, as "Date of issue," the date October 28 1975. That is three E days after the fire. But that was merely an administrative timelag. The renewal of the policy had been agreed before the fire.

The effect of the fire, counsel agreed before us, was to cause a total loss of the building, not a partial loss. The fact that some part of the walls remained does not affect the position. The building was destroyed. It was so pleaded in the statement of claim, and it was so admitted in the defence. The learned judge was, I fear, in error when he said in his judgment "We are not dealing with a total loss here."

The judge in his judgment then described what had happened regarding the cottage between the time of its purchase F by the plaintiff from his parents-in-law and the time of the fire. I shall read the relevant passage, for these are relevant findings of fact.

> This brings me to the question of damages and the basis upon which those should be calculated. By way of introduction I must set out in summary form what happened to the property after it had been put on the market. At first the plaintiff was asking £12,500 for it. That was within months of his purchasing the property from his parents-in-law for £1,500. However as the months went by the asking price was reduced until by July 1975 it was down to £4,250. Indeed the plaintiff frankly admitted that he would have sold the cottage for £4,000 just before the fire. What appears to have happened is that the neighbouring farmer Bernard Maitland, who gave evidence before me, did everything G in his power to dissuade prospective purchasers from going through with their deals. Having failed to get it himself for a bargain price, he put every obstacle in the way of others who were minded to buy it at a higher price. His land surrounded that owned by the plaintiff. Mr Maitland challenged the plaintiff's right to run a water main through his land and his right of way over the lane leading to the property. He went as far as to ask for the sum of £3,000 for granting such rights to the plaintiff or a prospective purchaser. So, although there were literally hundreds of inquiries from prospective purchasers, nothing came of them. It is only right to add that there were other reasons given for breaking off negotiations, but I am satisfied that Mr Maitland's intransigence was undoubtedly the primary caused for the estate agents' failure H to sell this cottage at a reasonable price. Now it would seem that those differences which had arisen between the plaintiff and Mr Maitland have been resolved, but I am concerned for the purposes of this judgment with the value of the property at the date of the loss.

I agree with the learned judge that the relevant date for the ascertainment of the amount of the loss is the date of the loss —October 25 1975.

Mars-Jones J then records the facts of the helpful agreement made between the parties as to certain figures. I had better read it as the judge set it out: "The parties are happily agreed about figures. If the plaintiff is entitled to the cost of reinstatement, the correct sum after taking betterment into J account is £8,694. Mr Gosland" (who was appearing as counsel for the plaintiff) "contended that he was so entitled. Mr Preston" (who was appearing as counsel for the defendants) "on the other hand argued that the plaintiff was only entitled to recover the value of the loss which he put at the agreed figure of £3,000 being the difference between the realistic value of the property in October 1975 put at £4,500 by Mr Weller minus the site value which is agreed at the figure of £1,500."

The learned judge quoted at length from a text-book on the law of insurance, and he cited passages from cases decided as to the measure of damages in the particular circumstances K of those cases, being actions for tort or breach of contract. However, as I understand his judgment, the judge then accepted the submission of counsel for the plaintiff that those cases "dealt with an entirely different situation from that which arises here. They were concerned with the measure of damages in tort and for breach of contract." I agree with the learned judge that those cases, which were cited to us again on this appeal, are not helpful on the issue before us, even though a claim on a policy of insurance is a claim for damages for breach of contract.

The learned judge then went on: L

> Basically this is an action for specific performance, for a declaration that the plaintiff is entitled to the full reinstatement cost of actual reinstatement. It is not an action for breach of contract as such. The contract here was that in return for the payment of a premium the defendants would indemnify the plaintiff against loss sustained from a peril covered by the policy. The plaintiff is to be put in the position that he was in just before the damage occurred, and this result could only be achieved by paying him the cost of reinstatement minus "betterment." I have come to the conclusion that those submissions are well-founded and that the plaintiff in this case is entitled to a declaration in the terms sought.

I respectfully disagree with the learned judge's analysis. M This is a claim for damages for breach of a contract of insurance. It is not an action for specific performance.

The first question which arises is whether, on the true construction of the insurance policy, the plaintiff is entitled to require the defendants to pay him the cost of reinstatement of the cottage, even assuming—and, to answer the first question, one makes this assumption—that the loss actually suffered by the plaintiff was less than the cost of reinstatement. If the answer to that be "No," then the second question falls to be answered: on the facts of this case, was the amount of the loss actually suffered by the plaintiff the cost of reinstatement

A (agreed at £8,694) or was it the figure of £3,000 for which the defendants contend?

The argument put before us by Mr Millett, who appeared for the plaintiff in this court with Mr Gosland, in support of the answer which the plaintiff seeks on the first question is not, I think, the same argument as was submitted to the learned judge on that question. But there can be no objection to the point being taken in this court as counsel wishes to put it; and no such objection was taken on behalf of the defendants.

Ever since the decision of this court in *Castellain v Preston* (1883) 11 QBD 380 the general principle has been beyond
B dispute. Indeed I think it was beyond dispute long before *Castellain v Preston*. The insured may recover his actual loss, subject, of course, to any provision in the policy as to the maximum amount recoverable. The insured may not recover more than his actual loss. As it was put by Brett LJ in *Castellain v Preston* at p 386.

> In order to give my opinion upon this case, I feel obliged to revert to the very foundation of every rule which has been promulgated and acted on by the courts with regard to insurance law. The very foundation, in my opinion, of every rule which has been applied to insurance law is this, namely, that the contract of insurance contained in a marine or fire policy is a contract of indemnity, and of indemnity only, and that this contract means that the
> C assured, because of a loss against which the policy has been made, shall be fully indemnified, but shall never be more than fully indemnified. That is the fundamental principle of insurance, and if ever a proposition is brought forward which is at variance with it, that is to say, which either will prevent the assured from obtaining a full indemnity, or which will give to the assured more than a full indemnity, that proposition must certainly be wrong.

Mr Millett submits that when one looks at the declaration in the proposal form in this case, it shows that the parties define " the full value " as being the reinstatement cost. The first paragraph of section 1 of the policy, which I have read, with its parenthetical reference to " (including architects and
D surveyors fees and an amount in respect of debris removal)," shows, Mr Millett submits, that " the full value " is to be deemed to be the reinstatement cost. When there has been, as here, as the parties agreed before us, a total loss, that necessarily involves a loss of " the full value." Therefore, the policy provides that the plaintiff, on a total loss, shall be entitled to the cost of reinstatement, irrespective of whether or not it is the actual loss. The parties have, it is argued, by the words used in the declaration and the policy, agreed that this contract, though it is not a valued policy—no specific amount of money representing the value of the thing insured has been agreed—nevertheless shall have the same effect as
E a valued policy, though with the substitution for a specific agreed figure of an agreed basis of assessment of the loss, namely, whatever may prove to be the cost of reinstatement as at the date of the loss. It remains, counsel agrees, a contract of indemnity, with the insured entitled to recover no more than the amount of his loss: but on the words of this policy the parties, as they are entitled to do, have agreed, not what the loss is, but on what basis it is to be computed. That agreement, it is submitted, is overriding. The parties are free to agree how the loss shall be computed, even though the computation may produce a figure different from the actual loss. If they so agree, then their agreement prevails, even
F though it gives what in fact is more than a full indemnity.

I disagree. What the insurers have agreed to do is to indemnify the insured in respect of loss or damage caused by fire. The " full loss " is the cost of replacement. That defines the maximum amount recoverable under the policy. The amount recoverable cannot exceed the cost of replacement. But it does not say that that maximum is recoverable if it exceeds the actual loss. There is nothing in the wording of the policy, including the declaration which is incorporated therein, which expressly or by any legitimate inference provides that the loss which is to be indemnified is agreed to be, or is to be deemed to be, the cost of reinstatement, " the full loss," even though the cost of reinstatement is greater than the actual loss. The G
plaintiff is entitled to recover his real loss, his actual loss, not exceeding the cost of replacement.

There remains the second question. Was the plaintiff's actual loss the cost of the reinstatement of the cottage? Or was it, as the defendants contend, the market value of the property as it was at the time of the fire? The defendants do not rely upon any general principle in support of their submission. They say, rightly in my judgment, that this is a question of fact and that one must look at all the relevant facts of the particular case to ascertain the actual value of the loss at the relevant date. Of course, one is entitled to look to the future so as to bring in relevant factors which would have been fore- H
seen at the relevant date as being likely to affect the value of the thing insured in one way or the other, if the loss of it had not occurred on that date. But on the evidence in this case, and the learned judge's statement of the relevant facts in the passages from his judgment which I have read earlier, it is beyond dispute that the plaintiff himself, at the relevant date, wished to sell the house, and was ready and willing to sell it for £4,500—indeed, on his own evidence, for less. Mr Millett submits that he was not bound to sell it. Of course not. He might thereafter, if the loss had not occurred, have changed his mind. The value of the property might have increased or it might have decreased. But there is no getting away from J
the reality of the case: " It was " (I am quoting again from the judgment) " an empty cottage that he had for the purpose of sale." The learned judge says: " I do not think that this man, the plaintiff, would be put in the same position as he was before this fire merely by being paid the sum of £3,000, the difference between the price that he was prepared to accept for the property at the time of its loss and its site value." With very great respect, I am unable to see why not. If the plaintiff himself was ready and willing, as he plainly was, to sell the property for £4,500, or less, on October 25 1978, just before the fire, how can it be said that that was not its actual value at that time: unless, indeed, some reason K
could be shown why the plaintiff himself should have made a mistake about, or underestimated, its real value. No basis is shown for any such suggestion. The amount of the loss here, in my judgment, is shown by the facts to have been the figure agreed, hypothetically, on this basis, as £3,000.

I would allow the appeal and vary the learned judge's judgment so as to provide that judgment be entered for the plaintiff for £3,000.

Agreeing, GEOFFREY LANE LJ said: Mr Millett, on behalf of the plaintiff in this court, has posed two questions. The first question was this: Were the defendants entitled to contend that the full value of the building was less than the cost L
of reinstatement? No doubt they were not, in the light of the definition of " full value " which is contained in the proposal form as follows: ". . . (the full value is the amount which it would cost to replace the property in its existing form should it be totally destroyed)." But the difficulty facing Mr Millett is this: that is not the problem which is at the root of this case. The question really is: What did the plaintiff lose as a result of the fire? Was it the market value of the cottage at that time, or was it the reinstatement cost?

Mr Millett sought to argue, as indeed he had to if he was going to succeed on this point, that this was in the nature of a valued policy, the defendants undertaking, so he suggested, M
in the case of total destruction to pay the reinstatement cost, whatever that might be, up to the limit of £14,000 contained in the schedule. But the wording of the contractual document does not admit of such a construction. It is to be observed that that construction was never alleged in the statement of claim; nor was the case before the learned judge argued upon that basis at all. This was an ordinary indemnity-only contract. The plaintiff was entitled to be paid his actual loss and no more.

The citation which Mr Machin gave to us from the Australian case of *British Traders' Insurance Co Ltd v Monson*

(1964) 111 CLR 86 was directly in point. The material passage is at p 92. It is in the joint judgment of Kitto, Taylor and Owen JJs; and it reads as follows:

> On appeal to the Full Court of the Supreme Court the respondents' contention was upheld by a majority of the judges. One line of reasoning which was sustained commenced by construing the policy as containing an unqualified promise by the company to pay the respondents in the event that happened the full amount of the sum insured; and on this basis it was said, in answer to the appellant's insistence that a policy of fire insurance is only a contract of indemnity against loss, that according to undoubted authorities on the subject a party insured may in some circumstances recover more than the value of his insurable interest and that the most cogent of all categories of such circumstances must be that in which the parties have by express words so contracted.
>
> It is convenient to deal first with this way of putting the case for the respondents. Its fault lies, we think, not in the fact that it begins with the policy, but in the fact that it gives the policy a meaning which neither general understanding nor the law of insurance will support. It concentrates attention upon the words of obligation: "if the property insured . . . be destroyed . . . by fire . . . the Company will pay to the Insured the value of the property. . . ." If those words be read in isolation from their context, no doubt the obligation is to pay the full value of the property, regardless of the quantum of loss sustained by the respondents by reason of the destruction by fire. But the all-important fact is that they are words in a document possessing unmistakably and on its face a character which flatly contradicts the notion that the obligation of the Company is to pay more than the amount of the respondents' loss. It is issued by an insurance company. It is headed "Fire Insurance Policy." All its provisions, even the very words that are relied upon for their literal meaning, are characteristic of fire insurance policies. It is far too late to doubt that by the common understanding of business men and lawyers alike the nature of such a policy controls its obligation, implying conclusively that its statement of the amount which the insurer promises to pay merely fixes the maximum amount which in any event he may have to pay, and having as its sole purpose, and therefore imposing as its only obligation, the indemnification of the insured, up to the amount of the insurance, against loss from the accepted risk.

Then their Lordships cite a portion of the judgment of Brett LJ in *Castellain v Preston*, to which reference has already been made.

It is plain that the first question posed by Mr Millett must be answered against him.

The second point was this: Was the judge correct in holding that the plaintiff would not be completely indemnified unless the building was in fact reinstated? It seems to me, in agreement with Megaw LJ, that as to that the evidence was all one way. The plaintiff had, from the very moment when he bought this cottage from his father-in-law, wanted to sell it, and it is quite plain (he said so himself) that at the time of or immediately before the fire he would have snapped up the very first offer for this cottage of £4,000. One need only refer to one short passage in the judgment of the learned judge to illustrate that fact. It reads as follows:

> This brings me to the question of damages and the basis upon which those should be calculated. By way of introduction I must set out in summary form what happened to the property after it had been put on the market. At first the plaintiff was asking £12,500 for it. That was within months of his purchasing the property from his parents-in-law for £1,500. However as the months went by the asking price was reduced until by July 1975 it was down to £4,250. Indeed the plaintiff frankly admitted that he would have sold the cottage for £4,000 just before the fire.

The learned judge does not make it altogether clear as to the basis of his conclusion on this aspect. He sets out his reasons in the passage which my Lord has already cited and there is no need for me to cite it again. That is the passage dealing with the question of specific performance. But it is clear, with great respect to the learned judge, that by awarding the sum that he did to the plaintiff, the plaintiff is undoubtedly £5,000 or so better off than if he had succeeded in achieving his ambition of selling this cottage for £4,500 or £4,000. That means, in short, that he has received more than an indemnity against his loss. He has had a bonus; and this policy does not provide for him to have such a bonus. This is an indemnity policy: it entitles him to the amount of his loss, and no more. Accordingly, it seems to me that the amount to which he is entitled in respect of this fire is the £3,000 which is the agreed value of the cottage as it was immediately before the fire. That is all he is entitled to recover. I agree with the proposed order.

DUNN J also agreed.

The appeal was allowed with costs and the judgment below varied by substituting as the amount of the plaintiff's loss the figure of £3,000 for £8,694, with consequential adjustment of interest. Special directions were given as to the costs below.

LANDLORD AND TENANT

GENERAL

A **QUEEN'S BENCH DIVISION**

October 30 1978

(Before Mr Justice FORBES)

RAVENSEFT PROPERTIES LTD v DAVSTONE (HOLDINGS) LTD

Estates Gazette January 6 1979

(1978) 249 EG 51

Landlord and tenant—Important decision on repairing liability—Judge's review of authorities on doctrine of "inherent defect"—Lister v Lane considered—Collins v Flynn doubted—Inherent defect not as such a defence to tenant on repairing covenant—True test a matter of degree—
B **Whether what tenant is asked to do is properly described as repair or whether it involves giving landlord a wholly different thing from that which was demised—Tenants liable under repairing covenant in present case for remedial works caused by stone cladding on concrete frame of maisonettes and flats becoming loose and dangerous—Inclusion of expansion joints not building practice at date of construction**

In this action the plaintiffs, Ravenseft Properties Ltd, claimed as lessors of blocks of maisonettes and flats, known as Campden Hill Towers and Campden Hill Flats respectively, to be reimbursed by the defendants Davstone (Holdings) Ltd,
C the lessees, for the cost of remedial works which the plaintiffs claimed were covered by the defendants' repairing covenants in the lease of these properties. The details of the defects and of the relevant covenants are fully set out in the judgment.

Ronald Bernstein QC and S Burnton (instructed by Forsyte, Kerman & Phillips) appeared on behalf of the plaintiffs; J Colyer QC and K M J Lewison (instructed by Thornton, Lynne & Lawson) represented the defendants.

Giving judgment, FORBES J said: In this case, the plaintiffs, Ravenseft Properties Ltd, sue the defendants, Davstone (Holdings) Ltd, substantially on repairing covenants in a lease in which the plaintiffs are the landlords and the defendants,
D tenants. The building is part of a complex at Notting Hill Gate in London. The part of the building with which we are concerned is firstly a 16-storey block of maisonettes, and secondly a four-storey block of flats, those buildings being known as Campden Hill Towers and Campden Hill Flats respectively. Campden Hill Flats plays only a subordinate part in this matter and I shall concentrate very largely on Campden Hill Towers. That building is a modern, rectangular structure made with a reinforced concrete frame. It looks, if I may be excused the analogy, like nothing more than a set of pigeon-holes, or perhaps to use a marine analogy, a flag
E locker, and the square section pigeon-holes each contain one maisonette on two floors. The edges of the pigeon-holes are veneered, if I may pursue the analogy, with a cladding of Portland stone. This stone is about 3in thick and 8in wide and runs upwards along each of the vertical divisions and horizontally along each of the horizontal ones. At the edges of the structure there are slightly wider bands of stone running vertically. The same treatment occurs on both the eastern and the western elevations. The building was built between 1958 and 1960 and it was let in 1960 to a predecessor of the defendants by a predecessor of the plaintiffs.

In 1973 the stones in the cladding on the eastern elevation were observed to be loose and in some danger of falling, F
and emergency measures were taken to secure the more dangerous of the stones. The plaintiffs notified the defendants of this situation and called upon them to carry out the necessary work to repair and secure the stone work. The defendants, while agreeing that the work was necessary, disputed any liability to pay for it under the repairing covenants. The plaintiffs thought, quite rightly, that the matter was urgent, and the plaintiffs and the defendants accordingly agreed that the plaintiffs should give the orders and assume the primary liability for the remedial work necessary and leave the question of the liability of the defendants over for later determination. G
The work was put in hand. Initially I think the defendants paid half of the earlier bills but later stopped paying, and the sum in dispute is now claimed by the plaintiffs in this action.

I do not think it is necessary to go into great detail about the cause of the difficulty which arose with the stone. The main concrete frame and the stone cladding have different coefficients of expansion. The result was that in certain sections some of the stones were being bowed away from the concrete frame. The potential danger can be realised if it is understood that if the stone cladding is pinned or fixed in some way to the main frame at 10ft intervals, 100th of an H
inch of difference in the extent of expansion in 10ft will result in a bowing of one inch in the centre of that part of the stone work. There were three stones to each floor, six that is to each maisonette, and every third stone was what has been called a "booted" stone; that is that it was, in section, "L" shaped, with the projection of the "L" running back into a channel or piece cut out of the main frame, so that the booted stone was held in that position. The stones, including the booted stones, were intended to be tied back into the main frame by metal straps of some kind, but in fact many of the ties were either not there or misapplied in the eastern elevation. Besides the ordinary shrinkage of the frame J
which occurs over the first few years after construction with a concrete frame building of this kind, there are also other movements which take place in a concrete frame structure during the life of the building and it will be seen, therefore, that a difference in expansion coefficient between the cladding and the frame is likely to cause difficulties where the ends of the stones abut on to the horizontal cladding, where in effect they will be prevented from movement in a vertical direction. In this case the bowing of the stones was principally due to lack of expansion joints, but there was a subsidiary cause of failure and that was the omission to tie in the stones properly. K
As I understand the evidence, if the stones had been tied in properly, any individual stones would have been unlikely to fall, despite the movement of the frame relative to them. The movement would, however, possibly or probably cause cracking and chipping of the stones with, of course, an attendant degree of danger of the fall of smaller pieces of stone, but not the whole stones themselves. When a full inspection was made of the building and the very large number of untied stones were found in the eastern elevation, it was

decided that the only safe way of dealing with that was to take down all the stone cladding on this elevation and replace it with proper ties and also with expansion joints. I should mention here that the top three storeys of the building were slightly out of plumb. The frame was leaning backwards slightly and the stones were set plumb, so that the distance between the back of the stones and the front of the frame increased a little in the top three or four storeys. The expansion joints were effected by cutting off half an inch from the top of each stone which lay under a booted stone, in other words every third stone on the elevation, and filling the gap with a plastic expansion absorbing material. On the western elevation and on Campden Hill Flats the symptoms were not so severe. Nevertheless in view of the potential danger, two out of each lift of three stones were pinnned to the frame by inserting metal dowels through the stones into the frame at a downward angle, and once again the top half inch of every third stone was removed and an expansion joint placed in the gap. It was not necessary, however, as I understand the evidence, on the western elevation of Campden Hill Towers and on Campden Hill Flats, to remove the cladding stones in order to carry out this remedial work.

Throughout the work the defendants' surveyors joined in the inspection of the work from time to time with the plaintiffs' consulting engineers and at no time was there any suggestion that the work being done was unnecessary, though, of course, the defendants maintained their argument that they were not liable under the repairing covenants.

The necessity for placing expansion joints in a construction of this kind was not realised by architects or structural engineers at the time the building was constructed and it was not engineering practice at that time to include expansion joints when a construction of this type was being carried out. It was only in 1961, when the publication of the Building Research Station's Annual Report for 1960 pointed out the difficulty and danger, that the problem was recognised, and that report suggested sensible remedies. By 1973-74 when this remedial work was carried out it could be said to have become standard practice in constructing a building in this way to include expansion joints of this type.

I should turn now to the repairing covenants in the lease. It is, in fact, an underlease, but I do not think I need bother at all about the headlease. There are three sub-clauses to which I need refer and only three. They appear in clause 5 of the underlease, which is the main clause concerned with the tenants' covenants. Subclause 6, to repair, reads in this way:

> when where and so often as occasion shall require well and sufficiently to repair renew rebuild uphold support sustain maintain pave purge scour cleanse glaze empty amend and keep the premises and every part thereof (including all fixtures and additions thereto) and all floors walls columns roofs canopies lifts and escalators (including all motors and machinery therefor) shafts stairways fences pavements forecourts drains sewers ducts flues conduits wires cables gutters soil and other pipes tanks cisterns pumps and other water and sanitary apparatus thereon with all needful and necessary amendments whatsoever (damage by any of the insured risks excepted so long as the Lessor's policy or policies of insurance in respect thereof shall not have become vitiated or payment of the policy moneys be refused in whole or in part in consequence of some act or default of the Lessee) and to keep all water pipes and water fittings in the premises protected from frost and to be responsible in all respects for all damage caused to the premises or to the said buildings or any part thereof or to the neighbouring property or to the respective owners or occupiers thereof through the bursting overflowing or stopping up of such pipes and fittings occasioned by or through the neglect of the Lessee or its servants or agents.

I have read that in full to indicate that it is one of those clauses in which the ingenuity of the draftsman has been given full rein in an attempt to cover every conceivable eventuality.

Then we pass to subclause 11:

> to permit the Superior Lessors and the Lessor at all reasonable times during the said term to enter upon the premises and every part thereof to view the state and condition of the same and to take any measurements plans or sections thereof and of all defects decays and wants of reparation there found to give to the Lessee notice in writing in manner hereinafter prescribed.

And subclause 12:

> within three months next after every such notice as aforesaid (or immediately in case of urgency) well and sufficiently to repair and make good all such defects decays and wants of reparation to the premises at the Lessee's own cost absolutely Provided Always that if the Lessee shall fail to comply with the requirements of any such notice as aforesaid it shall be lawful for the Lessor (but without prejudice to the right of re-entry hereinafter contained) to enter upon the premises at any time after the expiration of such three months or immediately in case of urgency and execute such repairs and works and the cost thereof (including any surveyors' or other professional fees incurred) shall be repaid by the Lessee to the Lessor on demand.

In addition to those three covenants, and subclause 6 and subclause 12 are specifically pleaded in the statement of claim as giving rise to causes of action, there was also an agreement between the parties to which I have already very briefly referred, namely that the work should be done and that there should be an argument, if necessary, about the cost of it later. Now that is pleaded in paragraph 10 of the statement of claim in these terms:

> Further on or about January 24 1974 it was agreed *inter alia* between the plaintiffs and the defendants by their agents that the work of repairing and making good the said defects decays and wants of reparation should be carried out as a matter of urgency by the plaintiffs' contractors and that the cost of such works should be paid by the defendants if they were liable for the sum under the terms of the said underlease. The works of repair and making good were commenced pursuant to the said agreement.

Now that paragraph is admitted in the defence, so that it seems to me it is quite unnecessary for me to look further than at the allegation as set out in the statement of claim and admitted in the defence. There is correspondence of around that period which supports that paragraph, but I do not think it is necessary in view of the admission on the pleadings to advert to that further.

Now despite somewhat lengthy cross-examination of the plaintiffs' witnesses, there is here no dispute on fact and the defendants called no evidence. In these circumstances the plaintiff landlords claim to be reimbursed by the defendant tenants the cost of the work carried out and they put their claim under three headings. First, under the repairing covenant, clause 5(6); secondly under the covenant to pay the cost of repairs carried out by the landlord under clause 5(12); and thirdly under the agreement of January 24 1974. In fact, both counsel agree that the agreement adds nothing new as it merely requires the tenant to pay what it is due under either clause 5(6) or clause 5(12).

The tenants' defence is two-fold. Mr Colyer says, first, that there is in that branch of landlord and tenant law concerned with repairing covenants, a doctrine of inherent defect which is applicable to such covenants to repair. This is that where wants of reparation arise which are caused by some inherent defect in the premises demised, the results of the inherent defect can never fall within the ambit of a covenant to repair. Secondly, he says if that proposition is wrong the covenantor is still not bound to pay for any works which, in fact remedy the inherent defect. The landlords answer that broadly in this way. Mr Bernstein says there is no such thing as a doctrine of inherent defect. The question is simply this, "is what the tenant is asked to do fairly represented by the word 'repair'?" and this question is to be judged as a matter of degree in each case.

The leading cases on the matter have been referred to by both counsel and I need only, I think, list them at this stage

A and then consider at any rate some of them in a little more detail later. They are *Proudfoot v Hart* (1890) 25 QBD 42; *Lister v Lane and Nesham* [1893] 2 QB 212; *Wright v Lawson* (1903) 19 TLR 203, 510; *Lurcott v Wakely and Wheeler* [1911] 1 KB 905; *Calthorpe v McOscar* [1924] 1 KB 716; *Pembery v Lambdin* [1940] 2 All ER 434; *Sotheby v Grundy* [1947] 2 All ER 761; *Collins v Flynn* [1963] 2 All ER 1068 and *Brew Bros Ltd v Snax (Ross) Ltd* [1970] 1 QB 612. All these cases are, I think, very well known and I do not intend to recite the facts of the judgments in detail, though there are one or two matters which will have to be considered more carefully. Of these cases it seems that it is unnecessary B to consider further the cases of *Proudfoot v Hart* and *Calthorpe v McOscar*. They are concerned with questions of the standard of repair required under a repairing covenant rather than what is included in the term "repair."

One should start with the case of *Lister v Lane and Nesham* and the frequently quoted passage from the judgment of Lord Esher MR at p 216.

> If a tenant takes a house which is of such a kind that by its own inherent nature it will in course of time fall into a particular condition, the effects of that result are not within the tenant's covenant to repair. However large the words of the covenant may be, a covenant to repair a house is not a covenant to give a C different thing from that which the tenant took when he entered into the covenant. He has to repair that thing which he took; he is not obliged to make a new and different thing, and, moreover, the result of the nature and condition of the house itself, the result of time upon that state of things, is not a breach of the covenant to repair.

From this passage and the case in general Mr Colyer derives this proposition. If it can be shown that any want of reparation has been caused by an inherent defect, then that want of reparation is not within the ambit of a covenant to repair. Inherent defect he defines as an omission of something D in the original design. A defect in the quality of workmanship or materials is not, he says, an inherent defect. The want of reparation in this present case, Mr Colyer says, was the failure to reconstruct the wall; in *Lister's* case, by providing the whole house with underpinned foundations and that want of reparation was directly due to the omission of proper foundations in the original construction.

I started with *Lister's* case and the well-known passage from Lord Esher's judgment, but Mr Colyer says that *Lister's* case itself was founded on *Soward v Leggatt* (1836) 7 C & P 613, and it is necessary to look briefly at that case. There the floor joists were repaired by laying them on bricks rather E than on mud, as the original floor joists had been laid, and Mr Colyer seeks to found an argument from that, that any work to a building which involves a different method of construction must be regarded as giving to the landlord a different thing from that which the tenant took. But such a broad proposition cannot, I think, survive close consideration of the later case of *Lurcott v Wakely and Wheeler*. It should be remembered that in that case the court was dealing with the rebuilding of the eastern external wall of a house in Hatton Garden and that what had happened was that the premises had been certified as being in a dangerous state by the district surveyor and that later, in compliance with a F demolition order, under the London Building Acts, the plaintiff had taken down the wall to the level of the ground floor, and then, in compliance with a further notice by the district surveyor, had pulled down the remaining wall and rebuilt it with concrete foundations and damp courses in accordance with the requirements of the Act. The question was whether the tenant was liable under the repairing covenant for this work and it is clear that the wall originally lacked concrete foundations and damp courses, and that the insertion of these features in the rebuilt wall was the result of statutory notices by the local authority. Both Sir Herbert Cozens-Hardy MR and Buckley LJ regarded the matter as a question of degree rather than as one of method of construction, and clearly the method of construction was wholly different. In any G event, Mr Colyer's first point depends on causation, and *Soward v Leggatt* was not concerned with direct causation but turned on whether the work was an improvement or not, which is Mr Colyer's second rather than his first point.

Turning to *Wright v Lawson* Mr Colyer explains that in the same way as he explained *Lister's* case. There, he says, the want of reparation complained of by the landlord was the failure to provide a bay window supported on pillars but this failure in turn was directly due to the absence of sufficient stability in the house to support a cantilever bay window. Mr Colyer accepts that the only case in which this doctrine H was argued was *Collins v Flynn*. It is necessary therefore to look at that case but before doing so one should, I think, look at *Sotheby v Grundy* because Sir Brett Cloutman deals with *Lister's* case and *Sotheby's* case in a sense together.

Now the facts in *Sotheby* were not unimportant. The house which was the demised premises in that case was found in 1944 to have bulged and fractured walls and the house was condemned as a dangerous structure and demolished by the council. Expenses incurred by the council were recovered from the landlord who sought to recover them from the tenant as damages for breach of a repairing covenant, so the case itself was not directly concerned with work of repair J but with the cost of demolition due to failure to repair. The evidence showed that in fact the house was built, in defiance of the requirements of the Metropolitan Building Act 1855, entirely without footings or, in some places, on defective footings and in consequence, and because of the defective footings, there was every likelihood that the house would, in fact, fall into a dangerous state in which it did fall. "That very wise and experienced judge," as Sachs LJ in the *Brew Bros* case called him, Lynskey J found that the wants of reparation were caused by what he called "the inherent nature of the defect in the premises" but nevertheless felt it incumbent upon him to consider as a matter of degree whether K the finding that the tenant was liable for the works required would be asking the tenant to give the landlord something different in kind from that which had been demised. This is clearly to reject, or overlook, any argument that a want of reparation caused by inherent defect could not in any circumstances be within the ambit of the repairing covenant.

Now looking at the case of *Collins v Flynn*, Sir Brett Cloutman deals with these points at pp 1073 and 1074, starting at 1073 (f).

> The last case that is really in point is *Sotheby v Grundy*. This was the case of the condemned house built in or about 1861, the L main walls having been built either without footings or defective footings. The foundation had settled and this could have been avoided only by underpinning and substituting a new foundation. On the authority of Lord Esher's judgment in *Lister v Lane and Nesham*, it was held that the tenant was not liable for the cost of demolition. The expenses were incurred because of the inherent nature in the defect of the premises, and, therefore, did not come within the terms of the repairing covenant. Plainly the doctrine of liability for the defects in the subsidiary part could have nothing to do with that case. The case, it seems to me, was on all fours with *Lister v Lane and Nesham*. Oddly enough Lynskey J does introduce it in what I think is an obiter passage. He said, "It may be that the inherent nature of the building may result in its M partial collapse. One can visualise the floor of a building collapsing, owing to defective joists having been put in. I do not think *Lister v Lane* would be applicable to such a case. In those circumstances, in my opinion, the damage would fall within the ambit of the covenant to repair, but, as I say, it must be a question of degree in each particular case";

And the learned Official Referee went on to talk about what he described as "obiter joists," referred to also in *Lister's* and *Lurcott's* cases.

Then at page 1074(d):

> I now come to the crucial point. Do the words "repair" and

A "renew" import a liability to rebuild with newly-designed foundations and footings the pier supporting the girder, which in turn carries a great part of the rear wall and a part of the side wall in addition? This is manifestly a most important improvement which, if executed by the tenant, would involve him in rendering up the premises in different condition from that in which they were demised and on the authority of Lord Esher MR in *Lister v Lane and Nesham* I do not think that the tenant is under any such obligation. Furthermore, although a suggestion of liability for removal of an inherent defect in a subsidiary part seems to have been touched on in *Sotheby v Grundy*, I do not think that the obiter remarks of Lynskey J as to defective joists have any bearing on the present case.

B In these passages it seems to me that Sir Brett misdirects himself on the ratio of *Sotheby's* case. The question of whether the inherent nature of the building might result in its partial collapse was not *obiter* at all. It was part of the ratio in this sense that, treating the question as a matter of degree, a partial collapse, in the view of Lynskey J, would have been of a degree which brought it within the tenant's covenant to repair, whereas a total collapse would put it outside. As, therefore, it was not a matter of part only, but of putting in new foundations in the entire building, the learned judge found it was not within the ambit of the covenant. In so far as he appears to be misdirecting himself on the ratio of C *Sotheby*, the persuasive authority of Sir Brett Cloutman's judgment in *Collins* must be considerably eroded.

The only other cases in the list are *Pembery v Lambdin* and *Brew Bros*. Now *Pembery* was clearly a case of inherent defect, but the court did not there decide that the plaintiff failed because there existed a doctrine such as that put forward by Mr Colyer. One can, I think, epitomise Slesser LJ's judgment in that case in this way. The plaintiff failed because her argument, if correct, would have involved ordering the defendant to give her a different thing from that which was demised. This is clearly, in my view, a decision arrived D at by considering the question as one of degree. *Brew Bros v Snax*, though a case where the defect was not inherent (see Harman LJ's *arguendo* at p 622), was nevertheless a case where at any rate a doctrine of inherent defect such as that suggested by Mr Colyer was put forward in argument by the landlord (see p 618G), and countered for the tenants (see p 622B). But both Sachs LJ at p 640 and Phillimore LJ at p 646 appear to indicate that, whether or not the case of any want of reparation is an inherent defect, the question must still be regarded as one of degree in each case.

This necessarily brief review of authorities indicates quite clearly to my mind that apart from *Collins*, which I consider E of doubtful authority, the explanation of the ratio in *Lister's* case as giving the tenant a complete defence, if the cause of the want of reparation is an inherent defect, has never been adopted by any court, but on the contrary, in *Pembery v Lambdin* and *Sotheby v Grundy*, the court, when dealing with wants of reparation caused by inherent defect, chose to treat the matter as one of degree, which in *Brew Brothers* the court effectively said that every case, whatever the causation, must be treated as one of degree.

I find myself, therefore, unable to accept Mr Colyer's contention that a doctrine such as he enunciates has any place F in the law of landlord and tenant. The true test is, as the cases show, that it is always a question of degree whether that which the tenant is being asked to do can properly be described as repair, or whether on the contrary it would involve giving back to the landlord a wholly different thing from that which he demised.

In deciding this question, the proportion which the cost of the disputed work bears to the value or cost of the whole premises may sometimes be helpful as a guide. In this case the figures have not been finally worked out in complete detail. I have, however, the evidence of Mr Clark, the contracts manager for Stone Firms Ltd, the contractors who actually carried out the work. He was not himself responsible G for any of the work, as he joined that company after the work was completed. He is, however, familiar with that company's methods of charging and so on, and he has studied the drawings and the analysed quotations produced by the company. From these he has been able to give me, not a detailed and accurate costing, but a reliable, broad, estimate of the cost of that part of the remedial work relating solely to the insertion of expansion joints. It is, as he said, an indication of the order of magnitude of the cost. Although cross-examined by Mr Colyer, no rebutting evidence was called and I accept Mr Clark's estimate that the cost would have been in the region of £5,000. The total cost of the remedial works was H around £55,000, the balance of £50,000 being for refixing the stones and other ancillary works which was not, as I find, necessary to cure any defect of design, but to remedy what was originally defective workmanship. For comparison, the cost of building a structure of this kind in 1973 would have been in the region of £3m, or rather more. I find myself wholly unable to accept that the cost of inserting these joints could possibly be regarded as a substantial part of the cost of the repairs, much less a substantial part of the value or cost of the building. Mr Colyer urges me not to consider cost and that may, perhaps, in some circumstances, be right. He argues that the result of carrying out this improvement is J to give back to the landlord a safe building instead of a dangerous one and this means the premises now are of a wholly different character. Further, he argues that because they are of a wholly different character, the work on expansion joints, the work necessary to cure the inherent defect, is an improvement of a character which transforms the nature of the premises demised, and, therefore, cannot fall within the ambit of the covenant to repair. I cannot accept this. The expansion joints form but a trivial part of this whole building and looking at it as a question of degree, I do not consider that they amount to such a change in the character of the building as to take them out of the ambit of the cov- K enant to repair.

I pass to Mr Colyer's second point, namely that the tenant is not liable under the repair covenant for that part of any work of repair necessary to remedy an inherent defect. Again it seems to me that this must be a question of degree. In the case of *Lurcott v Wakely and Wheeler*, the wall was defective in the sense that it had no proper footings or damp course. When it was rebuilt, concrete footings and damp course were provided. The court nevertheless found the tenant liable for the whole cost of the work including these improvements. Mr Colyer seeks to distinguish that case be- L cause, he says, in *Lurcott* the improvements were necessary to comply with the requirements of the statute. Here there was no such requirement and the expansion joints were included merely as a matter of moral duty.

It is quite clear to me from the evidence of both Mr Sculley and Dr Michael, the two expert structural engineers who gave evidence, and the only evidence I have on this point, that no competent professional engineer would have permitted the remedial work to be done without the inclusion of these expansion joints. By this time it was proper engineering practice to see that such expansion joints were included, M and it would have been dangerous not to include them. In no realistic sense, therefore, could it be said that there was any other possible way of reinstating this cladding than by providing the expansion joints which were, in fact, provided. It seems to me to matter not whether that state of affairs is caused by the necessary sanction of statutory notices or by the realistic fact that as a matter of professional expertise no responsible engineer would have allowed a rebuilding which did not include such expansion joints to be carried out. I find myself, therefore, bound to follow the guidance given by Sir Herbert Cozens-Hardy MR in *Lurcott's* case at p 914.

A It seems to me we should be narrowing in a most dangerous way the limit and extent of these covenants if we did not hold that the defendants were liable under covenants framed as these are to make good the cost of repairing this wall in the only sense in which it can be repaired, namely, by rebuilding it according to the requirements of the county council.

Because of the view I have formed, I have not thought it necessary to consider certain further submissions by Mr Bernstein. These were based on the special words used in the covenants in the underlease. I have already mentioned the plethora of words used to describe the obligations of the tenant in clause 5, subclause 6. In clause 5, subclauses 11 and B 12, there are very many fewer words, but they include the important word "defects." The view I have formed is, of course, relative to the use of the word "repair," and that by itself seems to me to be sufficient to render the tenant in this case liable for the whole cost of remedial works. It is not, therefore, necessary to pursue the question of whether, if it had not been so, other words used would have been sufficient to fix the tenant with liability.

I accordingly consider that the plaintiffs are entitled to judgment for the proper cost of the entire works in this case.

Judgment was given for the plaintiffs with costs.

C

COURT OF APPEAL

November 28 1978

(Before Lord Justice MEGAW, Lord Justice BROWNE and Lord Justice WALLER)

SWORDHEATH PROPERTIES LTD v TABET AND OTHERS

Estates Gazette February 3 1979

(1978) 249 EG 439

Action for possession and damages for trespass—Tenant of D flat for a fixed term introduced a person with landlords' consent to live in flat as a licensee and this licensee introduced another licensee—Tenant left before his term ended and the licensees became trespassers—County court judge ordered possession against them but refused an order for damages for trespass—Judge rejected claim for trespass on ground that landlords had not shown that they could have relet during period covered by trespass—Held by Court of Appeal that landlords were entitled to damages without producing such evidence, the damages to be based on ordinary letting value of flat—County court judge's order varied

E This was an appeal by the plaintiffs, Swordheath Properties Ltd, from that part of a decision by Judge Solomon at West London County Court which rejected a claim for damages for trespass against two persons who had occupied Flat 3 at 2 Finborough Road, London SW10, for a period after the tenancy of the tenant of the flat had expired. The tenant, who was a protected tenant, had vacated the flat before the expiry of his tenancy. The two persons against whom the order for damages was sought had no interest in the flat, being merely licensees of the tenant.

P de la Piquerie (instructed by Harold Stern & Co) F appeared on behalf of the appellants (plaintiffs in the county court). The defendants did not appear and were not represented in the Court of Appeal.

Giving judgment, MEGAW LJ said: This is an appeal brought by Swordheath Properties Ltd, who are the owners of a residential property known as Flat 3, 2 Finborough Road, London SW10. The action was brought in the West London County Court, where it was heard by Judge Solomon, who gave judgment in the action on January 19 1978.

The relevant facts, as found by the learned judge, were these. The plaintiffs had let Flat 3 to the first defendant, a gentleman called Mr Tabet, for a fixed term, beginning in April 1976 and ending on July 4 1976. The first defendant, G Mr Tabet, with the permission of the plaintiffs, introduced, to live in the flat as a licensee, a gentleman called Hugh M McGuane, who was the second defendant in the action. Mr McGuane, the second defendant, in his turn introduced, to live in the flat as a licensee, the fifth defendant Miss Spriggs. There seems to have been some confusion there, because she was also joined in the action, under another name, as the third defendant, but in fact she is one and the same person. The first defendant, Mr Tabet, let Flat 3 before the expiry of his fixed term—that is, before July 4 1976. He therefore never became a statutory tenant. The second defendant's right as a licensee and the fifth defendant's right as a licensee came to an end, as a matter of law, when the first defendant's right H thus came to an end by his leaving the premises.

The plaintiffs then brought their claim in possession against all the defendants. So far as Mr Tabet is concerned, he had already left the premises before July 4 1976, so the plaintiffs were not concerned to get an order for possession against him. He was ordered to pay £110.68 for arrears of rent. But so far as concerns the second and fifth defendants, Mr McGuane and Miss Spriggs, they were continuing to reside in the flat and therefore the plaintiffs were anxious to obtain possession against them; and they claimed also damages for trespass for the period from July 4 1976. The fourth defendant can be forgotten. He, it appears, was sued in error. J It is accepted that he never was occupying the flat at all.

By their defence the second and fifth defendants raised two issues, both of which were decided against them and in favour of the plaintiffs by Judge Solomon. The two issues were: first, that Mr McGuane had achieved the status of a tenant by reason of payment of rent by him; and, second, it was contended that the plaintiffs were estopped from denying the existence of a tenancy. Both those issues, as I say, were decided in favour of the plaintiffs by the learned judge. There is no appeal against those findings or the decision of the judge on those defences on the part of any of the defendants. The K learned judge, however, held that, though the plaintiffs were entitled to an order for possession against both the second and the fifth defendants, they were not entitled to any damages for trespass against them. It is against that part of the judge's decision that the plaintiffs appeal.

The plaintiffs in this court are represented by Mr de la Piquerie. None of the defendants has appeared. The only two who would be affected by the appeal are the second and the fifth defendants. Mr de la Piquerie has read to us a letter from their solicitors, dated October 25 1978, in which it is indicated that they do not propose to attend or be represented in order to present opposition to the appeal, though L they do not consent to the appeal being allowed. Mr de la Piquerie, therefore, has had to pursue his appeal and to show this court, if he can, that the learned judge was wrong in his decision.

The learned judge based his decision that there was no valid claim for damages on the ground that the plaintiffs had failed to adduce any evidence that they would have been able to relet the demised premises from the date from which damages for trespass were claimed, namely, July 5 1976; and it was, indeed, the fact that the plaintiffs did not adduce any evidence that they could or would have relet the demised premises M had they not been in the occupation of the two defendants on and after that date.

The judge referred to evidence which had been given by a witness, Mr Sadleir, on behalf of the plaintiffs, to the effect that the plaintiffs "would not have offered the demised premises to anyone other than a person known to the plaintiffs whom the plaintiffs trusted." The learned judge then posed the question: "Could the plaintiffs have found such a person to accept a tenancy of the demised premises either on July 5 1976, or thereafter?" That question the judge answered: "I do not know. There was no evidence about it." So he went on to hold that there was " a special situation of the plaintiffs' own making. In the absence of evidence I am bound to hold that the plaintiffs have failed to prove that as a consequence of the conduct of the second and fifth defendants they suffered damage." Accordingly, while he made the order for possession forthwith, he dismissed that part of the claim which asked for damages.

Mr de la Piquerie submits that the learned judge is wrong. There is indeed, curiously, no authority which directly deals with this question in relation to trespass in residential property. But Mr de la Piquerie has referred us to a passage in the article on " Damages " in *Halsbury's Laws of England*, 4th ed, vol 12, at para 1170. In that paragraph the learned authors of that title say this:

> Where the defendant has by trespass made use of the plaintiff's land, the plaintiff is entitled to receive by way of damages such sum as should reasonably be paid for the use. It is immaterial that the plaintiff was not in fact thereby impeded or prevented from himself using his own land either because he did not wish to do so or for any other reason.

Then in a footnote various authorities are cited in support of that passage in the text. The first authority cited is *Whitwham v Westminster Brymbo Coal & Coke Co* [1896] 2 Ch 538. That is a case which was decided in this court by a court consisting of Lindley, Lopes and Rigby LJJ. It appears to me that the judgments, at any rate, of Linley and Rigby LJJ fully support the proposition which is set out in *Halsbury* in the passage which I have read.

If any further authority were needed, Mr de la Piquerie has referred us to a decision of the Master of the Rolls, Lord Denning (sitting as an additional judge of the Queen's Bench Division), in *Penarth Dock Engineering Co Ltd v Pounds* [1963] 1 Lloyd's Rep 359. In his decision in that case, so far as the question of damages arose, the Master of the Rolls had no hesitation in saying that the plaintiffs, even though they would not themselves have made use, bringing in financial return, of the dock in respect of which trespass was committed, were nevertheless entitled to damages for that trespass, calculated by reference to the proper value to the trespassers of the use of the property on which they had trespassed, for the period during which they had trespassed.

It appears to me to be clear, both as a matter of principle and of authority, that in a case of this sort the plaintiff, when he has established that the defendant has remained on as a trespasser in residential property, is entitled, without bringing evidence that he could or would have let the property to someone else in the absence of the trespassing defendant, to have as damages for the trespass the value of the property as it would fairly be calculated; and, in the absence of anything special in the particular case, it would be the ordinary letting value of the property that would determine the amount of damages.

In the present case, therefore, it appears to me that this appeal falls to be allowed and that the plaintiffs ought to have, not merely judgment for possession, but also damages for trespass for whatever would have been the appropriate amount (on which, no doubt, we can have information from counsel) as being the proper letting value of the property from July 5 1976 to the date of the judgment in the West London County Court.

BROWNE and WALLER LJJ agreed.

The appeal was allowed and the order of the county court judge for possession varied so as to include judgment against the second and fifth defendants for damages for trespass in the sum of £1,276.11. Counsel for the plaintiffs, having taken instructions, did not ask for costs of the appeal and no order for such costs was made. County court order for costs on Scale 2 not disturbed.

COURT OF APPEAL

February 12 1979

(Before Lord Justice MEGAW, Lord Justice ROSKILL and Lord Justice LAWTON)

WEST LAYTON LTD v JOSEPH AND ANOTHER

Estates Gazette April 28 1979

(1979) 250 EG 345

Landlord and tenant—Covenants restrictive of assignment and subletting—" Consent not to be unreasonably withheld "—Important discussion of principles by Court of Appeal—Butcher's shop with living accommodation over—Living accommodation reached only through shop—Covenant restricting subletting of living accommodation (1) to a letting to an employee of tenant in shop, or (2) to a letting on a fully-furnished tenancy, for which landlord's consent in writing was required, such consent not to be unreasonably withheld—Lease granted before Rent Act 1974 extended protection to furnished tenancies—Whether consent by landlord to subletting of living accommodation to two respectable and responsible persons was unreasonably withheld—Criticism by Court of Appeal of the " normal " and " abnormal " assignment doctrine and of the dichotomy of Rent Act and Leasehold Reform Act cases—Impossible to formulate rigid rules—All circumstances and statutory background to be considered—In present case subletting would change the character of the property, resulting in a separate tenancy of the living accommodation, attracting Rent Act protection, subtenants passing through shop, generally a loss of control over the premises—County court judge misled by supposed special rules—Decision that consent had been unreasonably withheld reversed

This was an appeal on behalf of the landlord from an order of Judge Wakley at Brentford County Court declaring that consent had been unreasonably withheld to the grant by the tenants, multiple butchers, to two subtenants of a subtenancy of the living accommodation above a butcher's shop at 278 Bath Road, Hounslow. The special circumstances affecting the title of the landlord are explained at the beginning of the judgment of Roskill LJ.

Michael Brooke (instructed by N Ramsey Murray & Co) appeared on behalf of the appellant; Richard Fernyhough (instructed by W H Hopkins & Co) represented the respondents.

Giving the first judgment at the invitation of Megaw LJ, ROSKILL LJ said: This is an appeal by the landlords against an order of His Honour Judge Wakley, sitting at Brentford County Court, on January 26 1978 in proceedings between the tenants, West Layton Ltd, and the landlords, who, at the hearing before the learned judge, were Louis Joseph and Robin Dannhorn. The tenants sought, and indeed obtained, from the learned county court judge, a declaration that the landlords had unreasonably withheld their consent, under a lease dated June 27 1971, to the granting of a subtenancy by the tenants to would-be subtenants, Mr and Mrs Jack Woodward.

Before I go further it is necessary, so that it is on record, to mention one procedural matter, to which, most properly, counsel for the appellant, Mr Michael Brooke, drew our attention before he opened the substance of the appeal. Mr Dannhorn, the second of the two gentlemen named as landlords, in point of fact had no relevant interest in the premises in question, which are a butcher's shop at 278 Bath Road, Hounslow. The relevant party was Mr Louis Joseph, who held the premises, we were told, as trustees for himself and for another. Unhappily, last September, after the learned judge's decision, Mr Joseph died; and to date probate of the will, in which he named a Mrs Ford as his executrix, has not been obtained. After discussion, the appellant's solicitors, most helpfully, gave an undertaking that they would be responsible for any order for costs which this court might make in favour of the respondents; and they also undertook that as soon as probate has been obtained (we have since been told that it is likely to be obtained in the next five or six weeks, if all goes well) the title of this appeal will be amended by substituting as appellant Mrs Ford, the executrix of Mr Joseph deceased. Meanwhile, it was agreed that, whatever the result of this appeal, no formal order would be drawn up until the title has been so amended. So much for the formalities. I shall deal with the matter as if the executrix of Mr Joseph had been substituted for the late Mr Joseph.

Now, as I say, the landlord appeals. This case has been ably argued by counsel on both sides. The gravamen of the argument for the appellant is that the learned county court judge reached a wrong conclusion and that he ought, instead of declaring that the landlord had unreasonably withheld his consent, to have upheld the landlord's refusal to grant the consent in question.

A few dates are relevant. As I said a moment ago, the term was granted on June 27 1971. On March 4 1974 the benefit of that term was assigned to the respondent tenants, West Layton Ltd, who, we are told, have a number of butchers' shops throughout London. On November 5 1976 the reversion of the lease passed to the late Mr Joseph as trustee. But between the date of the assignment of the term to the respondent tenants and the acquisition of the reversion by the late Mr Joseph, the Rent Act 1974 had become law, on August 14 1974.

The lease contained two covenants which are relevant for present purposes. Mr Fernyhough, for the respondents, invited us to look particularly at clause 2 (9) before we turned to clause 2 (18). Clause 2 (9) provides that the lessee covenants:

> not without the previous consent in writing of the lessor (a) to carry on or suffer to be carried on in or upon the shop portion of the demised premises or any part thereof any trade or business whatsoever other than that of a Butchers and at all times of the year during the usual business hours of the locality to keep the shop portion of the demised premises open as a shop for carrying on the said business and (b) to use the residential portion of the demised premises otherwise than for private residential purposes only.

Clause 2 (19) (a) (i) and (ii) read thus (and this is all-important):

> Not to assign transfer charge underlet or part with the possession of any part of the demised premises except a letting on a service tenancy or occupancy of the living accommodation above the shop to any employee of the lessee or on a fully furnished tenancy for which furnished tenancy the landlord's consent in writing shall first be had and obtained and such consent for such a letting shall not be unreasonably withheld and (ii) not to assign transfer underlet or part with the possession of the whole of the demised premises without the previous consent in writing of the lessor such consent not to be unreasonably withheld in the case of a respectable and responsible person.

Let it be said at once that no one has one single word to say against Mr and Mrs Jack Woodward. Their respectability is beyond question. Their suitability to be subtenants of the residential part of these premises in Bath Road, Hounslow, has not been disputed.

There is attached to our papers a sketch plan of the ground floor of this shop. We have not got a plan of the first floor, where the residential accommodation is contained; but if one looks at the ground-floor plan it is reasonably plain that this is a very simple type of small shop with accommodation at the rear in the nature of a kitchen, a store, a lavatory and so on; then upstairs is a small amount of residential accommodation. It is clear that the intention was that this shop would be used for, and only for, a butcher's shop, and that the tenant should have the right to use the upstairs accommodation without let or hindrance from the landlord, the occupant being any employee whom they might let in on a service tenancy or licence for the purpose of the business of butchery that was being carried on underneath, but that if the tenant wanted to use the upstairs accommodation for any other residential purpose dissociated from the business of butchery, two conditions had to be satisfied. One was that the landlord's written consent had to be obtained and such consent was not to be unreasonably withheld, and the other that such tenancy had to be a furnished tenancy.

This lease, as is obvious from the chronology which I have already given, was entered into before the Rent Act 1974 protected furnished tenancies. It is clear, therefore, that the draftsman of this lease thought that, if there were a letting of the type contemplated by clause 2 (18) (a), when the term of the lease ran out by effluxion of time there would be no difficulty in the landlord regaining possession. It is clear that the position has changed since 1974; and it is feared that if Mr and Mrs Woodward go into possession now and obtain statutory protection under the Rent Act 1974 there will be difficulty in regaining possession when this lease expires in five or six years' time. That is really what this dispute is all about.

The learned judge, in a careful judgment, of which we have a note, said that he found this a difficult case to resolve: he expressed the hope that the matter might be taken—as in the event it has been—to the Court of Appeal. I hope I do the learned judge's careful judgment no injustice if I say that his reason for reaching the conclusion that he did, namely that the landlord had unreasonably withheld his consent, was based upon the fact that he thought that he was following what had been called in argument "the Rent Act cases" and not what had been called in argument "the Leasehold Reform Act cases." He also appears to have taken the view that because there was a substantial period of this lease left unexpired—more than 7½ years, he said in the passage in question—it was almost automatically unreasonable for this landlord to withhold his consent in the circumstances in which he did.

It seems to me that one has to approach this problem not so much by reference to the authorities—of which there are a great number—as a matter of the construction of the covenant which I have already read. The facts are not in dispute. Against those background facts, which I hope I have outlined sufficiently, can it fairly be said that the tenants have shown that this landlord has unreasonably refused his consent?

I do not find it necessary to refer to more than a few of the cases. The cases suggest that there has grown up a practice of asking the question: Is the proposed assignment a "normal" assignment or an "abnormal" assignment? If it be the former, then the consent is said to be unreasonably withheld. If, on the other hand, it be the latter, then the consent is said to be not unreasonably withheld. I respectfully question whether, although that phraseology appears to have been used more than once in the cases, it is very helpful phraseology. One first finds reference to it in the decision of this court in *Swanson v Forton* [1949] Ch 143 in the judgment of Lord Greene MR at p 152, where this court followed the immediately antecedent case of *Lee v K Carter Ltd* [1949] 1 KB 85. Lord Greene, after referring to the judgments in *Lee v K Carter Ltd,* said:

> Nothing that I have said must be taken as applying to a case of what in the argument was sometimes referred to as "normal" assignment. If that phraseology be adopted, the whole basis of fact in the present case is that the assignment proposed was an "abnormal" one in that it was an assignment of the tail end of the term by the lessee not for the purpose of conferring the right to occupy for the few remaining days of the term, but in order to enable the assignee to occupy thereafter under the Rent Restriction Acts. In my opinion, therefore, the appellant's withholding of consent was not unreasonable.

One may deduce from the Master of the Rolls' phraseology in that passage that he was not enamoured of the use of the two adjectives in antithesis to each other—"normal" and "abnormal." I find that view reinforced by two recent judgments in this court—true, both in Leasehold Reform Act and not in Rent Act cases. In the first, *Norfolk Capital Group Ltd v Kitway Ltd* [1977] QB 506, my Lord, Megaw LJ, giving the first judgment, dealt with the unsuccessful argument of counsel for the tenant in this way, at p 512:

> The principle which it is claimed is to be derived from an analysis of those decisions and the developments resulting from one decision to another, is this: that where the effect of the desired

A assignment of the lease is likely to be that the proposed assignee will thereafter gain the protection of the Rent Acts, whereas the assignor either could not or would have been unlikely to have had those benefits in the absence of an assignment, the landlord is not entitled to refuse such an assignment if that assignment is to be regarded as a "normal" assignment: he would be acting unreasonably if he so refuses: he will *not* be acting unreasonably, however, if such an assignment is an "abnormal" assignment.

My Lord went on:

> The definition of "abnormal" which Mr Barnes

—he was counsel for the unsuccessful appellant—

> submits is to be derived from the cases to which I have referred is, for this purpose, that such an assignment is *abnormal* where
> B the sole intention (I think on the part of the intended assignor and the intended assignee) is to give to the intended assignee statutory protection under the Rent Acts which would not be available to the assignor. It is a question of the *intention*. If, however, the effect of the proposed assignment would be to give such a protection, it is nevertheless unreasonable, as a matter of principle, for the landlord to refuse consent, provided only that that effect was not within the intention of the parties to the assignment.
>
> I do not propose, with all respect to Mr Barnes' argument, to go into the authorities on which it is said to be based. It does, in my judgment, involve very many difficulties, both of law and, if the principle as stated be right, of practical application. But I
> C am content, for the purposes of this appeal, because I think it is unnecessary to do otherwise, to assume that the principle to be derived from those cases in respect of what I may call the Rent Act position is as it has been put by Mr Barnes. That indeed, as I understand it, was the way in which it was dealt with by Judge Leslie in the court below. But, like Judge Leslie, it is my opinion that the argument put forward on behalf of the tenants fails at the second stage, even making that assumption in respect of the first stage of the argument.
>
> There is no authority to the effect that, and I can see no valid reason why, that principle of "normal" and "abnormal" assignments, said to exist in relation to the Rent Acts, should be carried into the different sphere with which we are concerned in this case,
> D namely, the case where there is a potentially serious disadvantage to the landlord in the proposed assignment by virtue of the provisions of the Leasehold Reform Act 1967. The principle—if principle it be—is one which, for my part, I think ought not to be extended beyond the sphere to which it has been applied, if it has indeed been so applied.

I find in that passage in my Lord's judgment in that case the same dislike, if I may use that word, for the adjectives "normal" and "abnormal" which I think I detect in the judgment of Lord Greene MR some 30 years earlier.

The matter does not rest there, for within a few days of that judgment, which was given on June 30 1976, a further
E judgment was given, by the Master of the Rolls, Orr LJ and Waller LJ on July 8 1976, in *Bickel v Duke of Westminster* [1977] QB 517. The Master of the Rolls repeated in very similar language what I have already said, though he did not in terms refer to the decision in *Norfolk Capital Group v Kitway*. At the bottom of p 523 he said:

> The other proposition is that, where a house is subject to the Rent Acts, the landlord cannot reasonably refuse his consent to a *normal* assignment during the contractual term, even though it means that the assignee will be able to stay on afterwards as a statutory tenant. Such is the result of *Thomas Bookman Ltd v Nathan* [1955] 1 WLR 815.

F I pause there to point out that the first judgment in that case had been given by the Master of the Rolls (then Denning LJ).

> But he can reasonably refuse it in the case of an *abnormal* assignment of the "fag end" of the contractual term, made for the purpose of giving the assignee the benefit of the Acts. Such is said to be the result of *Lee v K Carter Ltd* [1949] 1 KB 85; *Swanson v Forton* [1949] Ch 143; *Dollar v Winston* [1950] Ch 236. Those decisions are said to be binding on this court. If they are good law, they bear a close analogy to the present case. This is an absolutely normal assignment of the lease for the last seven years of the term. It is not the "fag end" of the lease. There is nothing abnormal about it. The Grosvenor Estate cannot, there-
> G fore, reasonably refuse their consent, even though it means that the lady will be able afterwards to enfranchise the premises under the Leasehold Reform Act 1967.
>
> If those cases can properly be regarded as laying down propositions of law, I would agree that we ought to hold the landlords' refusal to be unreasonable. But I do not think they do lay down any propositions of law, and for this reason. The words of the contract are perfectly clear English words: "such licence shall not be unreasonably withheld." When those words come to be applied in any particular case, I do not think the court can, or should, determine by strict rules the grounds on which a landlord may, or may not, reasonably refuse his consent. He is not limited by the contract to any particular grounds. Nor should the courts
> H limit him. Not even under the guise of construing the words. The landlord has to exercise his judgment in all sorts of circumstances. It is impossible for him, or for the courts, to envisage them all. When this lease was granted in 1947 no one could have foreseen that 20 years later Parliament would give a tenant a right to buy up the freehold. Seeing that the circumstances are infinitely various, it is impossible to formulate strict rules as to how a landlord should exercise his power of refusal. The utmost that the courts can do is to give guidance to those who have to consider the problem. As one decision follows another, people will get to know the likely result, in any given set of circumstances. But no one decision will be a binding precedent as a strict rule of law. The reasons given by the judges are to be treated as propositions
> J of good sense—in relation to the particular case—rather than propositions of law applicable to all cases.

I respectfully agree with and adopt that passage in the judgment of the Master of the Rolls, which seems to me to confirm the doubts expressed by my Lord in the *Norfolk Capital Group* case to which I have already referred.

I also agree with an observation made by Lawton LJ during the argument of this appeal: that it is impossible to see why there should be some special rule applicable to Rent Act cases on one side of the line and some different rule applicable to the Leasehold Reform Act cases on the other side of the line. I think that the right approach, as the Master of the Rolls
K suggested in the *Bickel* case, is to look first of all at the covenant and construe that covenant in order to see what its purpose was when the parties entered into it; what each party, one the holder of the reversion, the other the assignee of the benefit of the relevant term, must be taken to have understood when they acquired the relevant interest on either side. It is plain, when one looks at this covenant, that its purpose was that the lessee should have the benefit of the living accommodation for the use of any of his staff linked with the carrying-on of the business of butchery; but that if he wished to go beyond that user and to use that accommodation for some purpose disconnected with butchery, he must grant no
L more than a furnished tenancy for which he must obtain the landlord's written consent, which was not to be unreasonably withheld.

The landlord has not got to consider anybody else's interests except his own. He is the person who has in all the circumstances to decide whether or not he will grant consent. As the Master of the Rolls said, circumstances may vary endlessly. In the present case one of the matters which has caused a change of circumstance is the passing of the Rent Act 1974: just as, in the cases of leases entered into before 1967, the passing of the Leasehold Reform Act 1967 altered the background.

M It seems to me that the effect of the request which this landlord has had made to him by the tenant is to invite him to agree to alter the nature of the property which was being let from commercial property, namely a butcher's shop with residential accommodation above, to property which would be let on a multiple tenancy—by which I mean to more tenants or lessees than one, because there will be not only the tenancy of the shop but also the separate subtenancy upstairs of the residential accommodation, which would be a tenancy attracting Rent Act protection. That is indeed conceded.

There is this further point. It does seem, if one looks at the downstairs plan, that the entrance to the accommodation upstairs is only either through the front of the shop or through the back of the shop. Either the subtenants, to proceed upstairs, can come in through the door marked "A," walk through the shop, go through the door marked "B" and then up the stairs, or they have to come round from behind, go through the door marked "E," go through the kitchen, which is part of the shop, through the door "D," through part of the door marked "C," and then up the stairs. So that the overall position will be radically altered, because the subtenants will then have a right, if they are granted this subtenancy, to go through the shop. I would have thought that a further consideration would be that if that right were possessed by someone who had no connection with the shop, that must adversely affect the letting value of the shop. Be that as it may, with great respect to the learned county court judge, I feel myself compelled to reach a different conclusion from that which he reached. But even if, contrary to my view, it is still right to ask oneself the question whether this proposed subletting is "normal" or "abnormal," I would answer the question by saying that it is "abnormal," in this sense, that it is proposed that there should be a different user of a part of the premises, namely independent residential user, different from and dissociated from the business the letting for which this lease was primarily intended, namely, the carrying on of a butcher's business at these premises in Bath Road, Hounslow.

For those reasons, which I have given at a little length out of respect for Mr Fernyhough's argument, for which I am indebted, and because we are differing from the learned judge, I would allow the appeal and set aside the declaration which he granted.

Agreeing, LAWTON LJ said: I approach this case with only a dim recollection of the many cases in the last three decades which deal with the meaning of common words in leases, namely, "consent for a letting shall not be unreasonably withheld." But for that dim recollection, I should without any hesitation have regarded the question whether a consent to a letting had been unreasonably withheld as one of fact. That, so it seems to me, was the approach of the Master of the Rolls in *Bickel v Duke of Westminster*, in the passage to which Roskill LJ referred in his judgment. Whether anything is "unreasonable" depends upon all the circumstances of the case; and, in relation to lettings, unreasonableness must be considered against the background of the statutory provisions which are applicable. Those statutory provisions may be the Rent Acts; they may be the Leasehold Reform Act 1967; and they may be the Landlord and Tenant Act 1954. When the trial judge is deciding whether anything is "unreasonable," he must take the existence of the statutory provisions into account. The statutory provisions, however, are just one factor. There may be others. When during the period of the contractual tenancy a landlord is asked to consent to one tenant being replaced by another well able to pay the rent and likely to observe the covenants in the lease, in almost all cases it would be unreasonable for him to refuse his consent. It may be that that sort of case is what in the past has been referred to on occasions as a "normal letting." But cases tend to differ on their facts. This case differs from the kind of case to which I have just referred by the fact that when the landlord, Mr Joseph, took an assignment of the reversion, the residential accommodation on the first floor had been unoccupied for a year. The effect of that emptiness above the shop was that he could look forward with a reasonable degree of confidence to being able to maintain complete control of the premises at the end of the lease or such extension of it as might be granted pursuant to the Landlord and Tenant Act 1954. The consequence of being forced by the court to give his consent would have been that he could no longer look forward at the end of the contractual tenancy or such extension of it as might be granted to being able to maintain control of the premises and have them available for reletting, for sale, or redevelopment as business premises. This, it seems to me, was an important factor, which the learned judge was entitled to take into account and should have taken into account.

Mr Fernyhough reminded us, and rightly so, that this court is dealing with a question of fact, and we ought not to interfere unless the learned judge took into account factors which he ought not to have done. In my judgment the learned judge did take a factor into account which he ought not to have done, namely, he thought that there were some special rules relating to cases which came within the Rent Acts and to cases which amounted to "normal" lettings. In my judgment there are no such rules. For those reasons it seems to me that this court can properly interfere with the learned judge's findings; and I would do so. Like Roskill LJ, I would allow this appeal.

MEGAW LJ also agreed.

The appeal was allowed with costs. The appellant's solicitors undertook, immediately upon grant of probate to Mrs Ford as executrix of Mr Louis Joseph deceased, to amend the title by substituting her name as appellant. The order of the court was not to be drawn up until the title was amended. Liberty to apply on terms of the order.

COURT OF APPEAL

January 17 1979

(Before Lord Justice ORR, Lord Justice ORMROD and Lord Justice GEOFFREY LANE)

COMMERCIAL GENERAL ADMINISTRATION LTD v THOMSETT

Estates Gazette May 12 1979
(1979) 250 EG 547

Landlord and tenant—Action for possession against tenant on ground of alleged breach of contract not to do or permit anything which might be a nuisance to landlords or to tenants or occupiers of adjoining premises—Action dismissed by county court judge—Nuisance caused by over-persistent admirer of actress tenant—Persistent telephoning, use of entry telephone system and shouting from outside the premises—Meaning of "permit"—Means giving leave or not taking reasonable steps to prevent act which it is within one's power to prevent—Tenant had installed an "Ansaphone," had disconnected entry telephone and had sought injunction and obtained undertaking to prevent shouting—Judge correct in holding that she did all that could reasonably be expected to abate nuisance—Appeal dismissed

This was an appeal by the landlords, Commercial and General Administration Ltd, from a decision of Judge Stucley at Bloomsbury and Marylebone County Court dismissing an action for possession against a tenant, Sally Thomsett, alleging breach of a covenant not to do or permit a nuisance etc on the demised premises.

J K Reynolds (instructed by Michael Conn & Co) appeared on behalf of the appellants; C Algar (instructed by P S Levy & Co) represented the respondent.

Giving the first judgment at the invitation of Orr LJ, GEOFFREY LANE LJ said: This is an appeal from an order of His Honour Judge Stucley delivered on July 19 1978 whereby he dismissed a claim for possession brought by the plaintiff company, Commercial General Administration Ltd, against the defendant, a young lady called Sally Thomsett. The basis of the claim was somewhat out of the ordinary. It arises in this way. The defendant is occupant of a flat which is owned by the plaintiff company. They let the flat to her originally in 1976 and over the years the term was extended or fresh terms were granted until finally the letting came to an end in June 1978. By the time the proceedings in the Bloomsbury and Marylebone County Court took place it was agreed on all hands that that term had come to an end.

A The defendant was accordingly holding over as a statutory tenant on the terms, in so far as they were relevant, of the agreement which she had earlier signed.

The claim before the learned judge took a number of forms. There is only one of them with which we are concerned on this appeal and that is the claim that the defendant ought to be ordered to give up possession on the basis that she was in breach of one of the clauses of the agreement; that was clause 2(3) which it is necessary for me to read verbatim. It runs as follows:

> The tenant hereby agrees with the landlords as follows: Not to do or permit to be done on the premises anything which B may be or become a nuisance or annoyance to the landlords or to the tenants or occupiers of the adjoining premises or which may be injurious or detrimental to the reputation of the premises and the building in which they are situated and in particular but without prejudice to the foregoing (a) not to play or permit to be played any musical instrument or to allow any singing on the premises at any time or to allow any wireless set or television set or gramophone or tape recorder to be used on the premises between the hours of 11 pm and 9 am or at any time so as to be a nuisance or annoyance to the other inhabitants of the said building, (b) not to keep any animals, birds or other pets on the premises and (c) not to use the premises or permit the same to be used for an illegal or immoral purpose.

C The complaint made by the landlords was this. It seems that the defendant is by way of being an actress or in some way connected with show business and it seems likewise that a number of people were from time to time entertained by her at this flat. Be that as it may, there was one man, by name Slade, who was more persistent in his attentions to this young lady than the others with whom she was acquainted. There is no doubt that—in the colloquial sense—he from time to time over the months of this tenancy made himself a nuisance. He made himself a nuisance—I repeat, in the colloquial sense—in a number of ways. First of all by persistently ringing Miss Thomsett, the defendant, on the D telephone whether she was there or not, with the result that the adjoining tenants would be disturbed by the continual ringing of Miss Thomsett's telephone when she was not there, particularly at night. Secondly, Mr Slade would persistently call on the premises and persistently use the entry telephone system, whether that of Miss Thomsett or of the other tenants in the building, in order to try to get in touch with this young lady. That, too, was a great nuisance to the other tenants. It was, no doubt, also a great nuisance to Miss Thomsett herself. The third type of nuisance was this. Apparently there were parties from time to time and a good deal more noise E was caused than was necessary by those present. On one occasion, at least, one of the guests—or a guest of a guest—became drunk and it was necessary for Miss Thomsett to call the police to have this person ejected. Finally, the other form of annoyance about which complaint was made was that the man Slade, not content with ringing up on the telephone, and not content with trying to use the entry telephone at the door, would stand outside and shout, presumably to Miss Thomsett, thereby annoying the other occupants of the premises.

It is said that the learned judge was wrong in the conclusion to which he came. Wrong, Mr Reynolds submits on F behalf of the plaintiffs, in law and wrong in fact. It is suggested that he was wrong in law in his definition of the word "permit" which appears in the clause to which I have already made reference. Mr Reynolds submits to us that in these circumstances, and in the light of the words of that clause, there is a burden on the tenant to take effective steps to stop any nuisance if it is in the tenant's power to do so.

The authorities on which Mr Reynolds relies for the definition of the word "permit" are twofold. First of all there is the case of *AG v Tod Heatley* [1897] 1 Ch 560. When one comes to examine that case, one finds that it is nothing to do with a covenant in a lease, but was a case which was G dealing with a public nuisance which had arisen on a piece of vacant land by reason of persons putting filth and refuse on that land. All that case did was to examine the duties of an occupier of that type of land *vis-à-vis* his liability under the laws of public nuisance. That is not a case which assists us here. The citation of that case overlooks the fact that we are dealing here with a covenant and not with what amounts, or does not amount, to a public nuisance at common law. We are not dealing with what an occupier is under a duty to do or to prevent so far as public nuisance is concerned.

The second case cited by Mr Reynolds in support of the proposition to be advanced was *Sedleigh-Denfield v O'Callaghan* [1940] AC 880. Without, I hope, being disrespectful to H the arguments he courageously put forward before this court, it seems to me that that case, like the former one, has nothing to do with this case and cannot assist the court in the circumstances of the present case.

The matter seems to me to be concluded, so far as the definition of "permit" in these circumstances is concerned, by a decision of this court in 1922, *Berton v Alliance Economic Investment Co* [1922] 1 KB 742. The court consisted of Bankes LJ, Atkin LJ and Younger LJ. As far as I have been able, in a very brief reading of the case, to discover, all three learned judges were agreed. At p 759 Atkin LJ set out a J definition by which we are bound in a case which was very similar on its facts and, so far as one can read, similar in the wording of the covenant to the instant case. He said:

> To my mind the word "permit" means one of two things, either to give leave for an act which without that leave could not be legally done, or to abstain from taking reasonable steps to prevent the act where it is within a man's power to prevent it.

It is the second part with which we are concerned here.

What the learned judge had to decide in this case was whether the plaintiffs, the burden being on them, had succeeded in showing that the defendant had failed to take reasonable steps to prevent the acts, it being within her power to prevent K them. The learned judge did not have that authority cited to him, but in fact, if one examines his judgment, one finds that that was precisely the test that he applied, and great credit to him for that. Having applied that test, as I think correctly, can it be said that the learned judge was wrong in coming to the conclusion to which he did come, that Miss Thomsett had taken all reasonable steps that she could to stop these various types of activity? In order to upset the judge on a matter such as this, it must be shown, as I see it, that there was no basis on which he could properly have come to the conclusion which he did, namely that Miss Thomsett had taken all reasonable steps to that end. What was it that L she had done? First of all, so far as the telephone ringing was concerned, she had installed an "Ansaphone" as it is called and that, as the judge correctly found, had succeeded in preventing the incessant ringing of the telephone when Miss Thomsett was not there. End of first complaint. So far as the entry telephone was concerned, she had taken steps, by means of disconnecting, or breaking, to do the same to that, and to that extent she had succeeded there. So far as the parties were concerned, on the occasion when the drunken guest of a guest was making a nuisance of himself, she had sent for the police and had him taken away. That was all that she could have done. M

That leaves us with the problem of Mr Slade standing outside and shouting. The evidence on this matter was very thin indeed. Notes of the judge make it clear that this was not very deeply investigated, but it is plain from what appears in those notes, and what appears in the learned judge's judgment, that steps had been taken by Miss Thomsett to gain legal advice and take action against Mr Slade by way of an injunction to prevent this matter happening. The matter never apparently reached a hearing because cross-undertakings were given. We do not know what those cross-undertakings were. We do not know what undertakings Mr Slade gave, although

A we can guess them; we certainly do not know what undertakings the defendant gave, but the fact remains that she got undertakings which in the circumstances were the equivalent of an injunction. Now it is suggested that she should have gone further and that if, as appears to be the case, there was shouting or other objectionable behaviour after the date of these cross-undertakings, she should have taken steps to have had Slade committed for contempt of court, in that he did not abide by his undertaking. That she did not do because, as she told the judge, she was apparently advised not to.

Can it be said that the judge was necessarily wrong in coming to the conclusion that what she did was reasonable
B and that what she did was all that could reasonably have been expected of her in order to abate this nuisance? Speaking for myself, it seems to me that the judge was entirely justified in coming to that conclusion. It is a conclusion to which I myself would have come had I been faced with the same problem. It was, so far as one can judge from the notes of evidence, never even suggested to this lady that she should have taken further proceedings endeavouring to get this man committed for contempt or in some way enforcing the undertakings. The burden lay fairly and squarely on the plaintiffs to prove their case. They did not. I think that the judge was absolutely right in that conclusion.

C There is only one other matter which remains to be decided. It is suggested, and was suggested, that the defendant was in breach of the covenant which I read in that she gave interviews to certain newspapers in such terms that the value of the premises was diminished by reason of things which she had done on the premises. As a matter of fact, it is very doubtful whether any such detriment could have been established, or any such potential detriment, on the evidence, but there is, as Ormrod LJ pointed out in the course of argument, a more fatal objection to that contention. The obligation was "not to permit to be done on the premises anything which may" and there was no evidence that any of these interviews
D given to the newspapers were done on the premises. Indeed, the suggestion is to the contrary. Consequently, on that simple ground, that particular aspect of the argument fails.

That deals with the points raised before us and, so far as I can see, despite the valiant arguments of Mr Reynolds, there is nothing really on which he could successfully base this appeal. I would accordingly dismiss it.

ORMROD and ORR LJJ agreed.

The appeal was dismissed with costs.

E

COURT OF APPEAL

June 29 1978

(Before Lord DIPLOCK and Viscount DILHORNE)

ROSSVALE LTD v GREEN

Estates Gazette June 23 1979
(1978) 250 EG 1183

**Summary judgment for possession of club premises under Order 14—Appeal to Court of Appeal from judge's decision upholding master's order—Agreement stating expressly and emphatically that occupation of premises used as a licensed discotheque or social club was held by occupier as licensee
F so as not to confer exclusive right of occupation or create relationship of landlord and tenant—Terms consistent only with licence—No triable issue—Appeal dismissed**

This was an interlocutory appeal against an order of Donaldson J dismissing an appeal from an order of Master Warren under Order 14. The order was made in proceedings for possession brought by the plaintiffs, Rossvale Ltd, against Harry Green in respect of premises used as a social club.

N Primost (instructed by David Lewis & Co) appeared on behalf of the appellant, Harry Green; and J Colyer QC and J Harper (instructed by Polden, Bishop & Gale) represented the respondent company. Mr Colyer was not called on.

Giving judgment, LORD DIPLOCK said: This is an appeal G
against a judgment of Donaldson J for possession given on May 8 1978 in which he dismissed an appeal against an order of Master Warren given on January 23 under Order 14 giving possession of premises to the plaintiff, Rossvale Ltd. By the writ, dated March 9 1977, the plaintiff sought possession of the premises alleging that the defendant, Mr Green, was occupying them as licensee. The defence, in the very familiar circumstances of this kind, was the contention that he was not a licensee but was a tenant.

The premises are used as a social club of which the plaintiffs or an associated company were the tenant, and they ran upon
the premises the club of which Mr Green was originally the H
manager. Those arrangements were changed in 1972 when the company decided to run the club by way of granting a franchise to Mr Green. They then entered into an agreement under which such occupation as Mr Green thereafter had of the premises was granted. The agreement was dated June 23 1972, between Rossvale Ltd and Mr Green. It is referred to as a licence, and clause 2 says:

> Nothing in this Agreement is intended to create the relationship of Landlord and Tenant. (b) Each party to this Agreement has given careful consideration to this matter and has had full opportunity of receiving separate legal advice thereon and it is agreed and this term is to be deemed a condition of this Agreement going
> to the root of the Agreement that this Agreement will create a J
> mere Licence and not a Lease.

Clause 3 is the clause which grants the licence, and it appears that it is

> for the purpose of carrying on the business of a Licensed Discotheque and/or Club for a period of three months from the commencement date and thereafter until this Licence shall be determined by either party giving to the other not less than four weeks' notice in writing. (b) Nothing herein is to be deemed to confer upon the Licensee an exclusive right of occupation of the premises or any part thereof. (c) The Company shall continue to occupy so much of the premises as it needs . . . . K

There are a number of other clauses in the licence. We have been taken through them by Mr Primost, and all I need to say is that, with one exception, which is clause 7, each and every one is much more consistent with the relationship of licensor and licensee than with the relationship of landlord and tenant.

Mr Primost has drawn our attention to the necessary cases which say that the label which is given to an agreement is not necessarily conclusive if, upon reading the whole of its terms, it appears that that label, whether that of licence or of tenancy, is incorrect. However, this agreement is, in my view, consistent, and consistent only, with what it purports to
be: a licence and not a tenancy. L
It does not seem to me that the contrary is arguable or that there is any triable issue of the matter, nor do I find any affidavits here which raise a triable issue which calls for further examination and consideration. I quite understand that Mr Green must have a grievance that after a number of years he should be deprived of his business upon four weeks' notice, but that is what he agreed to and he at least has the consolation that some 18 months have passed since that notice expired.

This appeal must be dismissed.

VISCOUNT DILHORNE agreed.

The appeal was dismissed with costs. Leave to appeal to M
the House of Lords was refused and a stay of execution was refused.

A **QUEEN'S BENCH DIVISION**

December 19 1978

(Before Mr Justice FORBES)

BRITISH ANZANI (FELIXSTOWE) LTD v INTERNATIONAL MARINE MANAGEMENT (UK) LTD

Estates Gazette June 23 1979

(1978) 250 EG 1183

Landlord and tenant—Preliminary issue on points of law—Discussion of principles governing equitable set-off to claims for non-payment of rent—Underlessees of warehouses sued for arrears of rent and for mesne profits—Cross-claims by
B **underlessees under agreements preceding underleases for unliquidated damages in respect of defects in floors caused by inadequate design, faulty materials or workmanship—Whether sufficiently close connection existed between claims for rent and cross-claims for damages to satisfy conditions for equitable set-off—Review of authorities back to Taylor v Beal in 1591—Held that underlessees were entitled to defend claim for rent and mesne profits by raising as set-off their claims for damages for breaches of agreements**

This was the trial of a preliminary issue in an action in
C which the plaintiffs, British Anzani (Felixstowe) Ltd, were suing the defendants, International Marine Management (UK) Ltd, for possession, unpaid rent, mesne profits and interest in respect of land on which two warehouses had been built. On an application for summary judgment under Order 14 the defendants were given leave to defend and it was ordered that there should be tried as a preliminary issue the question whether in the circumstances the defendants were entitled to set off against the plaintiffs' claims for rent and mesne profits the sums which the defendants counterclaimed by way of damages. The present report relates to the trial of this preliminary issue only. Following the decision by Forbes J
D on this preliminary issue the trial of the action proceeded. As mentioned in the judgment, there are also two actions by the present defendants against the present plaintiffs for damages for alleged breaches of agreement relating to the warehouses. These actions were consolidated and were before the Official Referee, Judge Stabb QC.

Jeremiah Harman QC and Derek Wood QC (instructed by Norton, Rose, Botterell & Roche) appeared on behalf of the plaintiffs; M Myers QC and C S Thomas (instructed by Markbys) represented the defendants.

Giving judgment, FORBES J said: In 1972 the plaintiffs, in another name, became tenants of a plot of land of about 53
E acres at Walton Avenue, Felixstowe, Suffolk. The landlords are the Master, Fellows and Scholars of Trinity College, Cambridge, and the lease was a building lease for 99 years on terms that the plaintiffs redevelop the land by constructing a number of warehouse and/or factory units thereon in accordance with certain approved plans, specifications and drawings.

On June 7 1973 the plaintiffs entered into an agreement in writing with the defendants under which the plaintiffs agreed to construct a warehouse, Warehouse No 1, on part of the land, and thereafter to grant an underlease thereof to the defendants. In fact the defendants had gone into possession of the warehouse on September 1 1972 and practical completion of that
F building was on December 31 1972. The underlease to the defendants was dated April 24 1974 and was for a term of 21 years from December 25 1971 at the initial annual rent of £109,020 payable quarterly in arrear on the usual quarter days.

There was a supplemental agreement between the parties, dated May 13 1975, relating to the construction of a second warehouse, Warehouse No 2, on an adjacent part of the land. Practical completion of this warehouse took place in December 1973. The defendants went into occupation in March 1974, and the supplemental underlease was made on May 22 1975. The rent reserved by the underlease was £92,500 a year, payable
quarterly in arrear. The term granted was the same as in G
the underlease in respect of Warehouse No 1.

The writ in the present action was served some few days after it was issued on September 30 1977. The claim by the plaintiffs is for possession, for unpaid rent, for mesne profits and interest thereon. At the date on which the parties first appeared before me the total arrears including mesne profits and interest amounted to over £570,000. A small proportion, under £30,000, of this amount was disputed as arrears by the defendants, the balance being admitted subject to set-off and counterclaim.

The defendants' counterclaim is for damages for alleged
breach of express or implied obligation said to be contained H
in the two agreements and underleases. The breaches alleged are concerned with the floors of both warehouses. The defendants maintain that in about April 1973 serious defects appeared in the concrete floor of Warehouse No 1, consisting of cracking, differential settlement and deflection of the ground beams. These defects, they say, are or were due to inadequate foundation design, the building having been erected on reclaimed land. As a result, they say, they had to evacuate the warehouse from June 1975 to December 1976 and after that were only able to make use of part of the floor area of the demised building. In August 1974, they say, similar defects
appeared in Warehouse No 2. J

By an action started in 1975, the defendants in the present action claimed against their landlords, the present plaintiffs, damages amounting to £315,000 in respect of the alleged breaches of the agreement and underlease relating to Warehouse No 1. By further action started in 1978, they claim further damages in respect of Warehouse No 1 which they say arose after the issue of the writ in the first action and also claim damages in respect of the defects which they say occurred in Warehouse No 2. The amount claimed in the second action is £707,000. The first action is due to be heard before an Official Referee beginning on January 11 1979. There is a
summons for directions in the second action, heard, as I K
understand, on December 12. In that the defendants were asking for the second action also to be transferred to the Official Referee.

The plaintiffs are denying liability in both these actions, but it is accepted and agreed between the parties that, so far as liability is concerned, the defendants have a genuine and bona fide cause of action. The matters to be investigated include the contentions put forward by the parties through their respective solicitors in correspondence, and the trial of these actions will involve a substantial and detailed investigation of many issues.

The plaintiffs are now insolvent and on December 16 1977, L
pursuant to its powers under a debenture of April 28 1972 and a mortgage of March 31 1977, Bankers Trust International Ltd appointed one, Andrew Homan, as receiver of certain property and income of the plaintiff company including Warehouses Nos 1 and 2. The receiver duly gave notice to the defendants of his appointment on December 16 1977. On January 10 1978 the plaintiff company changed its name to Field Cross Ltd and on March 23 1978 its members passed an extraordinary resolution for the voluntary winding-up of the plaintiff company.

In the present case there were Order 14 proceedings which
resulted in an order that the defendants should have leave to M
defend, that there should be tried as a preliminary issue the question whether the defendants are in the circumstances of the case entitled to set off the sums counterclaimed against the plaintiffs' claims for rent and/or mesne profits. By agreement between the parties the preliminary issues with which I have to deal have been widened slightly and these issues are as follows.

(1) Are the defendants in the events which have happened entitled in law and/or in equity to deduct and/or to set-off against the admitted liability for rent and/or mesne profits, the sums claimed against the plaintiffs for breaches of the said agreements and underleases in the actions referred to?

(2) Further or alternatively are the defendants entitled to a stay of execution of any judgment for rent, mesne profits or other moneys due and/or possession?

(3) If the answer to both of the above questions is "no" upon what terms should the defendants be granted relief from forfeiture?

The arguments ranged over a great variety of precedents from *Taylor v Beal* (1591) Cro Eliz 222 to *Federal Commerce and Navigation Co Ltd v Molena Alpha Incorporated* [1978] QB 927 decided in the House of Lords on the day this hearing began. In any question involving set-off today the starting point, it seems to me, must be the judgment of Morris LJ, as he then was, in *Hanak v Green* [1958] 2 QB 9. It is unnecessary to rehearse at length what Lord Diplock called "a masterly account" of the subject: *Modern Engineering (Bristol) Ltd v Gilbert Ash (Northern) Ltd* [1974] AC 689. It is clear from this judgment that there are in general three occasions on which set-off is permissible: (1) set-off under the rule replacing the statutes of set-off (2 Geo 2 c 22 and 8 Geo 2 c 24); (2) what might be called abatement in certain cases at common law; and (3) equitable set-off. The first requires that, and I quote, "The claims on both sides had to be liquidated debts or money demands which could be ascertained with certainty at the time of the pleading": Morris LJ at p 17. The second are cases within the principle of *Mondel v Steel* (1841) 8 M & W 858. "Such cases are confined to those concerned with goods sold and delivered with a warranty, goods agreed to be supplied according to a contract and actions for work and labour done": per Parke B at p 871. The principle only operates in such cases if the defendant can show that, by reason of the breach of contract by the plaintiff, the goods or the work are diminished in value. To that extent the defendant can obtain "an abatement of the price": Parke B at p 872, as a form of defence. The third occasion on which set-off is permissible arises "where a court of equity would have regarded the cross-claims entitling the defendant to be protected in one way or another against the plaintiff's claim": Morris LJ in *Hanak v Green* at p 23.

It is thus necessary to consider in what circumstances a court of equity before the Judicature Act would have afforded such protection to a defendant. The *locus classicus* for the principle involved is the judgment of Cottenham LC in *Rawson v Samuel* (1841) 1 Cr & Ph 161 at p 178.

> We speak familiarly of equitable set-off as distinguished from set-off at law, but it will be found that this equitable set-off exists in cases where the party seeking the benefit of it can show some equitable ground for being protected against his adversary's demands. The mere existence of cross demands is not sufficient.

And then at p 179:

> Several cases were cited in support of the injunction but in every one of them except *Williams v Davies* (1829) 2 Sim 461 it will be found that the equity of the bill impeached the title to the legal demand.

It is this feature that the equity must go to the very root of the plaintiff's claim that is the essential attribute of a valid equitable set-off. An obvious example, as was emphasised in later cases, was *Pigott v Williams* (1821) 6 Madd 95, where the fees which the solicitor claimed were in relation to a suit which would never have been necessary but for the solicitor's own negligence. The principle is clear though there may have been difficulty in applying it correctly to different sets of circumstances. A felicitously expressed statement of the principle occurs in the judgment of Parker J in *The Teno* [1977] 2 Lloyd's Rep 289 at p 297:

> Where the cross-claim not only arises out of the same contract as the claim, but is so directly connected with it that it would be manifestly unjust to allow the claimant to recover without taking into account the cross-claim, there is a right of set-off in equity of an unliquidated claim.

This statement was expressly approved by Goff LJ in the *Federal Commerce* case in the Court of Appeal [1978] 3 WLR 309 at p 350 D.

Now originally Mr Harman argued that the principle of equitable set-off did not apply where what was sought to be set off was an unliquidated demand, but *Bankes v Jarvis* [1903] 1 KB 549 is a distinct authority for the proposition that unliquidated damages may be set off against a claim for debt. There is this passage in the judgment of Channell J at p 553:

> The Judicature Act and more especially the Rules distinctly put an unliquidated claim on the same footing as a liquidated claim for the purpose of set-off and consequently the defendant's claim against the plaintiff's son, which, if liquidated, could have been pleaded before the Judicature Act as a set-off to the plaintiff's claim can now, although unliquidated be relied on as a defence to the extent of the claim.

This proposition is recognised in the sentence from the judgment of Parker J in *The Teno* to which I have just referred. But Mr Harman, driven from that point, still argues that at any rate the sum must not be so vague as to be unquantifiable at the time of the claim. He cited no authority for this proposition and I can find no warrant for it in any broad legal principle. It is of the nature of unliquidated damages that they may remain unquantified until an award is made. A set-off of unliquidated damages is a defence to so much of the claim as is represented by the eventual amount of the award made. If the defendant in some way limits his damages to a sum less than that claimed, then it is obvious that there is no defence to the balance over and above the sum so limited and, no doubt, summary judgment could be obtained for this balance. But where the damages claimed to be set-off are at large and it is claimed bona fide, as is here admitted, that they over-top the claim, then even though not precisely quantified, it seems to me that the set-off amounts to a complete defence to the whole claim.

Mr Harman's next point, and it is an important one, is that there can be no set-off, equitable or otherwise, against rent. He starts by asserting that there is in landlord and tenant law a settled principle that a tenant may claim an abatement of rent similar to the abatement in certain other cases as explained in *Mondel v Steel*, where he has either been dispossessed by his landlord or he has paid money for which his landlord was certainly liable. This principle he says is not one of equitable set-off at all but is rather a common law principle similar to that in *Mondel v Steel*. He relies on *Taylor v Beal* (1591) Cro Eliz 222 and *Waters v Weigall* (1795) 2 Anst 575. Now, *Taylor v Beal* was a case in the Queen's Bench in which two of the judges held that where a landlord had broken his covenant to repair a house and the tenant had expended money on the repairs, the tenant was entitled to set-off against a claim for rent the money so expended. One of the two, however, held that as the matter had not been properly pleaded the tenant failed. The remaining judge thought there was no set-off and would have left the tenant to a separate action for breach of covenant. On another point, however, they were all three agreed that discharge by the tenant of the landlord's obligation to pay a rent charge issuing out of the land where this had been requested by the landlord was a good defence to a claim for rent. *Waters v Weigall* was a case in the Court of Exchequer. The tenant was seeking to excuse himself from paying rent because the premises had been damaged by tempest and his covenant to repair the premises excluded damage by tempest. The report of the argument on demurrer shows that the tenant was suggesting that, because there was no express covenant by the landlord in those circumstances to repair, he could not bring a cross-action and so would be without a remedy at law. Accordingly, so it was argued, it was a case for equitable relief. The court refused equitable relief and the judgment is short and to the point. Macdonald CB said:

> I do not see how you entitle yourself to the interposition of this court. If the landlord is bound in law or equity to repair in

A consequence of the accident that has happened, and you were right in expending this sum in repairs for him, it is money paid to his use and may be set off against the demand for rent. If you fail in making out these points your ground of relief is destroyed in equity, as well as at law.

The other Barons being of the same opinion the demurrer was allowed. The Chief Baron had no doubt that in circumstances which made expenditure by the tenant money paid to the landlord's use, a set-off was available both at law and in equity. It is interesting that when the case came before the King's Bench on the landlord's claim for rent, the tenant was unsuccessful, Lord Kenyon CJ holding that there was no B obligation on the landlord to repair and even if there had been there could be no set-off because the damages claimed were uncertain, which in the context must mean unliquidated.

> One objection to the plea is that it does not set off any certain debt for uncertain damages. I do not indeed see by what covenant the landlord is bound to repair damages occasioned by fire or tempest. The exception was introduced in the lessee's covenant for his benefit and to exempt him from particular repairs, but if the defendant can maintain any action against the plaintiff, his landlord, the sum to be recovered is uncertain. It must be assessed by a jury and there is no pretence to say that those uncertain damages may be set-off to the present action. If the plaintiff has fairly laid out money on repairing what he C was not bound to repair, perhaps a court of equity will grant him relief: *Weigall v Waters* (1795) 6 TR 488.

On a consideration of these cases it seems to me that *Taylor v Beal* is authority for the proposition that there are at least two sets of circumstances in which at common law there can be a set-off against rent, one where the tenant expends money on repairs to the demised premises which the landlord has covenanted to carry out, but in breach has failed to do so (at any rate where the breach significantly affects the use of the premises) and the other where the tenant has paid money at the request of the landlord in respect of some obligation of the landlord connected with the land demised. To this D proposition there must be added two riders. First, that as the landlord's obligation to repair premises demised does not arise until the tenant has notified him of want of repair, such notification must have been given before the set-off can arise; and secondly that the set-off must be for a sum which is not to be regarded as unliquidated damages, that is, it is a sum certain which has actually been paid and in addition its quantum has either been acknowledged by the landlord or in some other way can no longer be disputed by him, as for instance, if it is the subject of an award on a submission to arbitration. The latest expression of opinion about this matter is in *Lee-Parker v Izzet* [1971] 3 All ER 1099. In that case Goff J, as he then was, was dealing among other things with a E claim to a lien, the basis of which was laid on an argument that the tenants were entitled to treat a payment of the cost of repairs, for which the landlord was liable, as a payment of rent and reliance was placed on *Taylor v Beal*. Goff J discussed the principle of *Taylor v Beal* and at p 1107 G he said this:

> I do not think this is bound up with technical rules of set-off. It is an ancient common-law right. I therefore declare that so far as the repairs are within the express or implied covenants of the lessor the third and fourth defendants are entitled to recoup themselves out of future rents and defend any action for payment F thereof. It does not follow however that the full amount expended by the third and fourth defendants on such repairs can properly be treated as payment of rent. It is a question of fact in every case whether and to what extent the expenditure was proper.

I do not think there is any difference between the principle as seen by Goff J and that which I have set out above save for this. Goff J took the view that it was money properly expended which could form a subject of this right. My view is that the right was slightly more restricted, namely, that it could only be exercised when the sum was certain and its amount could not really be disputed by the landlord. This restriction which I think should be made arises from a consideration of the G judgment of Lord Kenyon in *Waters v Weigall*, which was not quoted to Goff J. In that case the tenant had in fact paid £30 but Lord Kenyon still regarded the cross-claim as one for uncertain damages. It seems the quantum of the sum must have been either unchallenged or unchallengeable before it could be regarded as deductible.

Mr Harman's argument is that these cases are similar to the abatement case exemplified in *Mondel v Steel* and that what the tenant must prove is that the premises are worth less to him than they should have been, such diminution in value being the measure of the amount which can be deducted. I can find no trace of such a principle. Rather the H principle appears to be that the money paid is regarded either as paid at the request of the landlord or to his use, but in either case can be treated as a defence. I agree with Goff J that this is strictly not a set-off as such, since set-off is a creature of statute and the principle appears to ante-date the reign of George II, but rather a case where at law the payment was regarded as payment *pro tanto* of the rent. Nevertheless, it is clearly a defence to claim for rent. In this case such a defence is not open to the defendants as they had not in fact paid anything. Their cross-claim is for damages. In so far as Mr Harman seeks to argue that the existence of this ancient common-law right is in some way J inimical to the co-existence of any equitable principle of set-off, I cannot accept such an argument. Of course, if the facts were such that this defence was available to a tenant equity would not assist him, because he had a perfectly good defence at law which was the ratio of *Waters v Weigall*, but in other circumstances, where as here, the common-law defence is not available because the defendant's cross-claim is not founded on payment, but on unliquidated damages, one still has to consider whether or not equity would intervene.

Then Mr Harman argues that rent is something special, that it is invested with something almost in the nature of an aura. He says rent issues out of the land. It is a debt due K which has attached to it many special rights such as the right to distrain and the right of forfeiture, without previously serving a notice under section 146 of the Law of Property Act 1925. These two rights, he argues, are unique examples of self-help remedies which it is not surprising ought not to be interfered with by any set-off, and he claims that there can be no set-off against distress, relying on *Bullers Nisi Prius* 7th ed p 1816 and *Townrow v Benson* (1818) 3 Madd 203. But the *ratio* ascribed by Buller to the decision of *Absolon v Knight and Barber* (1743) Barnes 450 is that the debt could not be set off against distress, not because rent is something special, but because distress is something L special—not an action, but a remedy without action. *Townrow v Benson* recognises that there can be no set-off against distress and also asserts that equity will follow the law in this. Neither of these cases goes to the extent of saying that there can be no set-off in actions for rent. It is interesting, however, that the paragraph in *Buller* immediately following that relied on by Mr Harman is a reference to *Gower and Wife v Hunt* (1734) Barnes 290 and 291. At p 290 of Barnes is also reported *Brown v Holyoak*. The latter case was an action of debt for rents upon a parol lease. The defendant had by his plea set off a debt by simple contract. On demurrer it was held that a debt of an inferior nature cannot be set M off against a superior demand. The reason given was that debt for rent is equal to an action upon a bond. In other words, although the lease was a parole lease it was still equivalent to a specialty when suing for debt and the debt under a simple contract was not of equal degree and for that reason could not be set off.

Gower v Hunt itself provides another pointer. In that case the landlord brought an action of covenant upon a lease under seal for non-payment of rent. The tenant wished *inter alia* to raise a set-off under the statutes of set-off for sums due under other covenants in the same deed relating to

A "spurring up land at a certain sum per acre." At the Suffolk Assizes Denton J refused to allow the tenant to give evidence about this, but the Court of Common Pleas held that the evidence should have been received as it was to set off a certain debt of equal degree with the plaintiff's demand. There appears to be no trace in these old cases of any special treatment of rent as such, and indeed, as I understand the decision in *Gower v Hunt*, it appears to allow a set-off against rent under the statutes so long as the debt set up by the tenant is a sum certain and arises out of a covenant in the lease which itself is under seal.

The only other case of antiquity which requires considera- B tion is *Beasley v D'Arcy* (1800) 2 Sch & Lef 403. This case in the House of Lords was, as Mr Harman pointed out, a case about relief from forfeiture and the House upheld the Lord Chancellor of Ireland's decision to grant relief. But the important part of the case arose in the earlier stages. The facts and *ratio* are succinctly set out by Lord Redesdale LC in *O'Mahoney v Dickson* loc cit at p 408:

> There was a question between the parties what damage should be allowed in consequence of the landlord's cutting down timber from the land. The tenant came into Chancery to restrain the landlord's proceeding for non-payment of rent, and insisted that C damages demandable in this collateral way ought to be applied to discharge the rent. There had been an award ascertaining the damage; but it had been made after the time limited for making the award and therefore the matter still rested in damages. At a time when the ejectment was brought, this was the demand which the tenant had against the landlord, an unascertained, unliquidated demand arising from the act of the landlord, affecting the land in the occupation of the tenant. The tenant gave a consent to judgment, that is, he admitted the rent demanded to be due; but he insisted that he was entitled to a certain sum for damages for cutting the timber. On the answer there was an injunction until the hearing, on the terms of lodging the rent in court, which terms were not complied with: The *habere* was executed in 1797; the cause was heard in 1798, and an issue directed to try whether D the tenant had sustained any and what damage from cutting the timber. The jury found damage in the amount of £103 6s. A supplemental bill was then filed praying that the tenant might be restored to possession, and the value accounted for in the usual way. The answer insisted on the statute Geo. I stating that the ejectment was brought for non-payment of rent. The cause came on again in 1799, when the plaintiff was declared entitled to restitution, on paying what was due for rent after deducting the damages. That was a case in which the court proceeded, not on a question whether so much was due for rent or not; but there being a collateral demand on the person to whom the rent was to be paid, and the amount of that demand appearing to have been nearly equal to the rent, but the subject of the cross- E action, and which could not by any possibility be set-off by way of defence in the ejectment whether this afforded a distinct ground for equitable relief.

It seems clear from this passage that in a case of ejectment for non-payment of rent equity was prepared to grant relief by way of injunction, where there was a claim for unliquidated damages against the landlord for breach of covenant in the lease. The damages were unliquidated because the arbitrator's award was out of time and therefore void. The Lord Chancellor referred to the ratio of *Beasley v D'Arcy* again, at p 412 in these terms:

> F In a case in the House of Lords (that is *Beasley v D'Arcy*) there was no question whether so much rent was due or not. There was no doubt on that subject, and therefore the case there must have been decided on a distinct ground, namely that although the rent was due, yet the money that was due to the other party for an unliquidated demand ought in point of conscience under the circumstances to have been paid by the landlord in order to enable the tenant to pay his rent, the nature of the demand for damages supposing a loss by the tenant of the produce of the land arising from the act of the landlord. I must presume that that decision was right.

The footnote in the report at this point refers to another case of which there seems to be no other report.

In the case of *Hamilton v Johnson* (Exchequer 1808) O'Grady G CB said that the case of *Beasley v D'Arcy* authorised a bill by a tenant against his landlord for an equitable set-off of unliquidated damages, arising out of the relation of landlord and tenant; and the subject matter of the contract subsisting between them as such—but only such damages.

Although there was an injunction in *Beasley's* case and not a set-off Cohen LJ, as he then was, in *Morgan & Son Ltd v S Martin Johnson & Co Ltd* [1949] 1 KB 107 at p 114 pointed out that set-off was now the appropriate way of dealing with such a situation:

> Before the Judicature Act, such claims were very often enforced by injunction, but it is plain from section 41 that an injunction H would not be the appropriate way of giving effect to a set-off now and that effect should be given to it, under Section 38, as an equitable defence if so pleaded.

It is instructive to compare and contrast *Waters v Weigall* with *Beasley v D'Arcy*. In *Waters* the tenant had paid a sum certain for the repairs or so it was said. As there was on that premise a remedy at law equity refused to interfere. In *Beasley* the cross-claim was for unliquidated damages and therefore not available as a set-off at law; equity granted relief.

A consideration of all these cases leads me to the conclusion that except in cases of distress or replevin equity has never refused to interfere to protect a tenant whose landlord J was bringing proceedings based on non-payment of rent, if the tenant had a bona fide cross-claim for unliquidated damages against the landlord, provided that he was not covered by an existing common law remedy and that the ordinary rules pertaining to equitable set-off were obeyed. I referred to these earlier in this judgment and they are as succinctly put by Parker J in *The Teno* at p 297:

> Where the cross-claim not only arises out of the same contract as the claim, but is so directly connected with it that it would be manifestly unjust to allow the claimants to recover without taking into account the cross-claim, there is a right of set-off in equity of an unliquidated claim. K

It may well be that today, even in replevin cases, the anachronism of the special remedy of distress would or should not inhibit a court from applying equitable principles in the way suggested by Lord Denning MR in the *Federal Commerce* case at p 338B:

> It is now far too late to search through the old books and dig them out. Over 100 years have passed since the Judicature Act 1873. During that time the streams of common law and equity have flown together and combined so as to be indistinguishable the one from the other. We have no longer to ask ourselves what would the courts of common law or courts of equity have done before the Judicature Act? We have to ask ourselves: what L should we do now so as to ensure fair dealing between the parties?

Fortunately, I have not in this case to accept Lord Denning's invitation any further than this: it must, I think, be one more refutation of Mr Harman's contention that the ancient common law remedies attached to rent should govern in the fourth quarter of the 20th century one's approach to a tenant's claim to equitable relief.

The last case to which I should refer is *Hart v Rogers* [1916] 1 KB 646. In that case the landlord claimed for unpaid rent and the tenant counterclaimed for damages for breach by the landlord of the implied covenant to repair the roof of the premises demised. Scrutton J, as he then was, held on the M authority of *Surplice v Farnsworth* (1844) 7 Man & G 576 that this cross-claim was no defence to an action for rent. Although it requires a certain amount of hardihood to differ from so great a judge, I feel I must do so in this case. *Surplice v Farnsworth* was the only case in point to which Scrutton J was referred. It was, of course, a case at law and not anything to do with equitable set-off, and it seems to me that the decision in *Hart v Rogers* must be taken as one reached without the benefit of the very full argument on equitable set-off which has been put before me and therefore *per incuriam*.

While I am satisfied that it is proper in principle to allow that a cross-claim could be effective as an equitable set-off against a claim for rent, it by no means follows that such a defence is available in all circumstances. The important qualification is that the equity must impeach the title to the legal demand, or in other words go to the very foundation of the landlord's claim. This seems to me to involve consideration of the proposition that the tenant's cross-claim must at least arise under the lease itself, or directly from the relationship of landlord and tenant created by the lease. The landlord's covenant to repair contained in the lease, if broken, might found, as has been seen earlier, the ancient common law defence to a claim for rent if the tenant had been forced to pay for repairs to maintain the premises in a state fit for the purpose for which they were let. If instead of paying for the repairs the tenant cross-claims for damages for breach of the covenant, there is no common law defence, but there must, in my view, be an equitable right to set off the unliquidated damages. But in this case, there is no covenant by the landlord to repair contained in the lease, and it is necessary to see how the defendants put their case on this aspect. This can be exemplified by considering the position in relation to Warehouse No 1. The same case arises *mutatis mutandis* in relation to Warehouse No 2.

The underlease of April 24 1974, contains, as I have said, no covenant by the landlord to repair. It does, however, contain the following among the landlord's covenants. Clause 4(3):

> That in the event of the floor of the demised premises suffering damage other than superficial damage, by reason of the weight of articles placed thereon being less than 1,120 lb per square foot, the landlord will pay the cost of repairs and reinstatement to the floor subject to the landlord first approving an estimate for the work.

There is a factual issue between the parties about the loading of the floor, the plaintiffs maintaining that the damage to the floor was caused by loading in excess of 1,120 lb per square foot and the defendants that the loading was less than that figure. The defendants seek to suggest that their claim for damages is related to a breach by the plaintiffs of clause 4(3) of the lease. It is plain to me that there has been no such breach. The plaintiffs' liability is subject to the condition precedent, that they must first approve an estimate and no estimate has even been submitted, and even if it had been submitted and approved, this clause does not oblige the plaintiffs to do anything other than pay for the cost of repairs, a phrase which seems to me to involve discharge of a debt which has in fact already been incurred. In other words, the work of repair having been carried out, the landlord is to pay the cost. No work has in fact been carried out. It follows that the tenant cannot show that there has been any breach of any covenant contained in the lease, which is of course the document on which the landlord founds his claim for rent.

The agreement of June 1973 contained this provision:

> So far as the floor of the new building is concerned, the lessors will make good at their own expense any defects which may occur within two years of the completion date and which may be caused by inadequate design, or faulty materials or workmanship. Notwithstanding completion of the underlease, this clause to remain in full force and effect.

It is alleged by the defendants that the defects in the floors arose within the two-year period and were due to inadequate design or faulty workmanship. It is further alleged that the defects in the floors resulted in the premises or parts of them becoming unusable as a warehouse, the purpose for which they were let. For the purpose of these preliminary issues I must of course assume that these allegations are made good. There is a further provision of the agreement which obliges the landlords to cause the works to be carried out in a good and workmanlike manner with good and sufficient materials, but this adds nothing to the points at issue.

Clause 11 of the agreement is in these terms:

> This agreement shall remain in full force and effect after the date of the grant of the underlease in respect of any act, matter, or thing remaining to be performed or effected thereunder after that date and notwithstanding such grant.

For completeness I should add that the agreement required the landlord to grant and the tenant to accept the underlease, a draft of which was annexed to the agreement.

In these circumstances, Mr Harman argues that as the cross-claim arises not under the underlease itself but under the agreement, there is no such close connection between the claim for rent and the cross-claim as is required to support an equitable set-off. Mr Myers contends to the contrary, that the agreement and the underlease are so closely interwoven that a cross-claim under the former can be regarded as impeaching the plaintiff's title to his demand under the latter. Now, there is no provision in the underlease for the terms of the agreement to be incorporated in it. If there were it would be plain that the landlord's breach could properly be said to arise out of the same instrument as his claim for rent. In such circumstances I should have no doubt that the requirements for an equitable set-off were fulfilled, for the breach of covenant relied on by the tenant is of the type to which equity has in the past allowed relief. (See, for instance, *Beasley v D'Arcy*.) But here not only is there no such incorporation of the terms of the agreement but at the end of the special provision in the agreement relating to the floor, there is this sentence, " Notwithstanding completion of the underlease this clause shall remain in full force and effect." Clause 11 is to the same general effect. Although these provisions are relied on by Mr Myers as showing the close connection between agreement and underlease, they seem to me rather to insist on the separate nature and existence of the agreement. There is to be no merger or incorporation; there is to be no extinction of the rights and duties arising under the agreement by reason of the grant and acceptance of the underlease; the agreement is to continue as a separate but co-existent entity.

This at once raises, in an acute form, the question I posed earlier and in general terms: does the requirement that the equity must impeach the title for the legal demand mean necessarily that the tenant's cross-claim must at least arise under the lease itself or directly from the relationship of landlord and tenant, or is it sufficient that it should arise out of some transaction closely connected with the lease, and if so how closely? The passage I quoted from the judgment of Parker J in *The Teno* assumes that the claim and cross-claim arise out of the same contract, but that is no doubt because on the facts before him they did so and this question did not arise for consideration. It is clear that in *Bankes v Jarvis* [1903] 1 KB 549 the claim and cross-claim arose under different contracts. In *Hanak v Green* at p 24 Morris LJ referring to *Bankes v Jarvis* averred that there was a close connection between the dealings and transactions which gave rise to the respective claims. In *Federal Commerce* Lord Denning MR at p 338D said: " It is only cross-claims that arise out of the same transaction or are closely connected with it." In *The Brede* [1974] QB 233 at p 248F he said much the same thing:

> It is available whenever the cross-claim arises out of the same transaction as the claim or out of a transaction that is closely related to the claim.

In view of these passages and in particular having regard to the facts in *Bankes v Jarvis* it does not seem possible to conclude that it is in all cases necessary that claim and cross-claim must arise out of the same contract. Where, as in this case, they do not, it still therefore remains for consideration whether in any particular case the two matters are so closely connected that the principles affecting equitable set-off can be said to apply. It is very necessary when seeking to reach a conclusion on this question in a case such as this to drive from one's mind the insidious prompting of generations of common lawyers that there is something special about rent. As I have already indi-

A cated I can find no trace of such a principle in equity and there is no reason why there should have been or should be now.

In this difficulty as with others concerned with set-off, the best guide is, I think, to be found in *Hanak v Green*. In explaining and approving *Bankes v Jarvis*, Morris LJ at p 24 had this to say, and I think I may already have quoted it in another context.

> The plaintiff, suing as agent or trustee for her son, claimed £50 from the defendant. The defendant had a perfectly good claim for £51 damages against the plaintiff's son. It was held that the defendant could set up as a defence to the claim against him that the plaintiff's son (the *cestui que* trust of the plaintiff) was B indebted to the defendant in a sum for unliquidated damages exceeding the amount of the claim.
>
> The conclusion seems to me to be clearly correct and obviously fair. It would have been manifestly unjust if the defendant had had to pay £50 to the plaintiff (who was an agent or trustee for her son) at a time when the defendant had an unquestioned claim of £51 against the plaintiff's son who had left the country. There was a close relationship between the dealings and transactions which gave rise to the respective claims. If the case had been brought before the Judicature Act it would appear that the defendant would have had strong equitable grounds for asking a Court of Chancery to restrain the plaintiff from proceeding with her case. But since the Judicature Act the position is that matters C of equity on which such injunctions might formerly have been obtained, may now be relied on by way of defence.

In other words, in considering questions of this kind it is what is obviously fair or manifestly unjust that will determine the solution. This is because today, while it is necessary to look back before the Judicature Act to discover the broad principles upon which equity would grant relief, it may not be helpful to seek to find out from the cases what a court of equity would have done in a similar case. The principle may be derived from the older cases. The application of that principle should be reached by a consideration of what today would be D regarded as fair or just. This is but a reflection of the passage I have already quoted from the judgment of Lord Denning MR in *Federal Commerce* at p 338D.

Applying these principles in the light of what is fair dealing between the parties, I have come to the conclusion that despite the insistence on preserving the agreement as an entity separate from the underlease, there is nevertheless here that close connection between claim and cross-claim which equity requires. The agreement was *inter alia* an agreement to enter into the underlease, the terms of which were set out in the form annexed to the agreement. The special provisions relating to the floor were obviously as much in the minds of the parties when E making the agreement as any of the other terms. It would in my view be manifestly unjust to allow the landlord to recover his rent without taking into account the damages which it is alleged the tenant has suffered through failure by the landlord to perform his part of the agreement. Not only is there in my view an adequate connection between the transactions giving rise to claim and cross-claim, there is also the fact that the breach by the landlord is said to render the premises unfit at least in part for the purpose for which they were let. For both these reasons, it seems to me that the defendants' cross-claim can be said to impeach the title to the plaintiffs' legal demand.

I therefore answer the first of the preliminary issues, and F the only one which so far there has been argument, in this way: the defendants are entitled to defend the plaintiffs' claim for rent and mesne profits by raising as set-off or defence a like sum of the moneys claimed by the defendants against the plaintiffs as damages for breach of the agreements (but not of the underleases) referred to in paragraphs 4, 6 and 8 of the amended defence.

Judgment was given for the defendants on the first preliminary issue with costs. The other preliminary issues did not arise for decision.

LANDLORD AND TENANT
BUSINESS TENANCIES

A

CHANCERY DIVISION

November 29 1978

(Before Mr Michael WHEELER QC sitting as a deputy judge of the division)

VICTOR BLAKE (MENSWEAR) LTD v WESTMINSTER CITY COUNCIL

Estates Gazette February 10 1979

(1978) 249 EG 543

Landlord and Tenant Act 1954, section 24A—Question of commencement date for payment of interim rent under section 24A—Rent held to be payable as from date of commencement of proceedings by lessors under section 24A until date of cessation of continuation of tenancy under B **section 64—Stream Properties Ltd v Davis followed—Dictum by Lord Denning MR in a subsequent case that the court could determine the interim rent to take effect from the expiry of the old lease not followed**

This was an application by the lessors, Westminster City Council, brought by way of counterclaim following discontinuance of proceedings by the lessees, Victor Blake (Menswear) Ltd, for a new lease, to determine the interim rent payable under section 24A of the Landlord and Tenant Act 1954 in respect of premises at 61-69 Charing Cross Road, London WC2. In fact the amount of the interim rent had C been agreed at the rate of £17,000 per annum and the date up to which such rent should be payable had also been agreed. The sole question remaining was the date of commencement of the interim rent.

M Rich (instructed by Paisner & Co) appeared on behalf of the plaintiff company; B Green (instructed by E Woolf, solicitor to Westminster City Council) represented the defendant council.

Giving judgment, MR MICHAEL WHEELER QC said: In the form in which they have arrived before me the present proceedings are concerned solely with a short point of law. D The plaintiff company originally commenced proceedings for a new lease under Part II of the Landlord and Tenant Act 1954, but those proceedings were subsequently discontinued on terms that an application which the defendant, the City of Westminster, had commenced in the meantime under section 24A of the Act for the determination of the interim rent should proceed by way of counterclaim; and it is that so-called counterclaim which now comes before me. The rent payable during the interim period—and both sides will understand what I mean by that—is agreed at £17,000 per annum, and the sole question before me is as to the commencement date from which that interim rent is E payable. It is also common ground that the date to which that interim rent is payable is May 3 1978.

Two choices are offered to me for the commencement date. One is June 20 1977, which is the date of the City of Westminster's application under section 24A of the Act. The other is November 14 1975, which is the date of the expiry of the notice given by the City of Westminster under section 25 of the Act. There is thus a difference between the two dates of some 21 months, so that there is a not insubstantial sum in issue.

Very briefly the relevant dates are as follows. On August F 19 1966, the Greater London Council (then the forerunner as lessor of the City of Westminster) granted a lease to a firm of outfitters; it was determinable by either side on six months' notice at any time after December 25 1968 at a rent of £5,250 per annum exclusive of rates. That lease subsequently became vested, on the one hand, in the present plaintiff, Victor Blake (Menswear) Ltd, as lessee and, on the other hand, in the City of Westminster as landlord. On May 7 1975 the council gave notice under section 25 of the Act, determining the tenancy as from November 14 1975. On May 12 1975 the tenant's solicitors served a counter-notice, and on September 2 1975 the tenant issued G an originating summons for a new tenancy under Part II of the Act, which by agreement was adjourned indefinitely in the hope that the matter might be settled by negotiation. As I have indicated, the council's section 25 notice expired on November 14 1975. But the negotiations seem to have lasted for a considerable time and it was not until May 1977 that they finally broke down.

In consequence, on June 20 1977 the council commenced proceedings under section 24A of the Act for the determination of the interim rent which was to be payable. On February 3 1978 the plaintiff company was granted leave to discontinue its originating summons on terms that the council's summons H —that is for the determination of interim rent under section 24A—should proceed by way of counterclaim. On May 3 1978, that is three months later, the tenant vacated the premises and the continuation tenancy under section 64 of the Act came to an end.

I have been taken through the not uncomplicated sections of Part II of the Landlord and Tenant Act 1954 in a very helpful way and I have formed a fairly clear picture of how they are intended to operate. I do not propose to set them out in detail, for this reason. There is in existence a decision of Pennycuick V-C (*Stream Properties Ltd v Davis* [1972] 1 J WLR 645) which directly covers the point which I have to decide, namely the commencement date for the interim rent. I do not think it would serve any useful purpose if I were to set out the facts and the judgment in that case, although both are relatively short. Suffice it to say that I respectfully find the Vice-Chancellor's reasoning and construction of section 24A wholly convincing: and even if I felt free to do so (which I do not) I should not wish to differ from that decision.

Now it is perfectly true that that judgment has come in for what I think might be described as passing criticism from no less a source than Lord Denning MR in *Secretary of State for Social Services v Rossetti Lodge Investment Co Ltd* K (February 18 1975). Mr Green, who appears for the defendant in the case before me, has very helpfully produced a photostat of the judgment of the Master of the Rolls in that case. And Mr Rich, who appears for the plaintiff, has studied the transcripts of all three judgments in the Court of Appeal. It is, I think, clear that the point which the Master of the Rolls takes as to the construction of section 24A was not one which it was appropriate for either side in that appeal to have argued, at least in any depth. Moreover, this seems

A to me to be underlined by the fact, as both counsel before me accept, that, in one respect at least, the Master of the Rolls was in error in his interpretation of the relevant sections of the Act because, when dealing with the date when the old tenancy came to an end, he had overlooked the effect of section 64 of the Act.

It is true that Orr LJ, having given a reasoned judgment, appears also to have stated that he agreed with everything that the Master of the Rolls had said. Scarman LJ on the other hand certainly does not appear to have commented on the point which is before me or to have concurred in the *obiter dicta* of the Master of the Rolls which was as follows:

B It appears, however, that in *Stream Properties Ltd v Davis* the Vice-Chancellor, Sir John Pennycuick, held that an interim rent could only take effect as from the date when the application for the interim rent was made. I do not, as at present advised, accept that decision. The power to determine interim rent was introduced by the Law of Property Act 1969, which inserted a new section 24A into the 1954 Act. I think that section 24A should be given a liberal interpretation in accordance with the manifest intention of the section. Under it the court can determine the interim rent to take effect as from the expiry of the old lease.

With the very greatest respect to Lord Denning MR I do not find in the above extract any compelling reason for
C not following the clear and cogent reasoning adopted by Sir John Pennycuick in *Stream Properties Ltd.* Moreover, although there are many respects in which the 1954 Landlord and Tenant Act is open to criticism, it is, as I think both counsel before me accept, possible for a landlord who is aware of his rights to avoid the possible hardship which might otherwise arise and to which attention was drawn (prior to the amendment of the 1954 Act) by Wynn-Parry J in an earlier case, *Re No 88 High Road, Kilburn* [1959] 1 WLR 279. In other words, a landlord who knows his rights can put in his notice under section 24A at such a time that,
D even on the construction of section 24A which Vice-Chancellor Pennycuick has adopted, it will enable him to avoid any substantial hardship.

In these circumstances, with the consent of both sides and although faced with an apparent conflict between two distinguished judges, I have not thought it necessary or appropriate to reserve judgment as I might otherwise have done in order to set out my reasons in greater detail.

The judge's order was that the rent would be determined at the rate of £17,000 per annum for the period from June 20 1977 until May 3 1978.

The plaintiff company was awarded the costs of the applica-
E tion to determine the interim rent.

COURT OF APPEAL

November 21 1978

(Before Lord Justice MEGAW, Lord Justice BROWNE and Lord Justice WALLER)

STEVENSON & RUSH (HOLDINGS) LTD v LANGDON

Estates Gazette February 24 1979

(1978) 249 EG 743

F **Landlord and Tenant Act 1954—Application to clause in lease of section 38(1) (restrictions on agreements excluding Part II)—Provision in lease imposing on tenant liability to pay costs, including legal costs, incurred by landlord in planning applications and in proceedings under Part II of the 1954 Act—Whether provision involved a "penalty" under section 38(1)—Landlord under provision could recover costs awarded by a court to tenant—Indirect means of preventing a tenant applying for a new tenancy —Held that tenant was liable to reimburse the cost of a planning application made at the suggestion of his surveyor but the provision in regard to costs incurred in connection with proceedings under Part II of the 1954 Act was void as a penalty** G

This was an appeal by the tenant, Martin Telles Langdon, from a decision of Judge Graham Hall at Croydon County Court in favour of the landlords, Stevenson & Rush (Holdings) Ltd. The judge held that the tenant was liable under the covenant in his lease to pay the landlords' costs both of a planning application and of proceedings under Part II of the Landlord and Tenant Act 1954.

Paul Batterbury (instructed by S Z Kmiecik & Co) appeared on behalf of the appellant; Ronald Bernstein QC and Nicholas Hannah (instructed by Percy Holt & Nowers) represented the respondents. H

Giving judgment, Megaw LJ said: This appeal is from the judgment of Her Honour Judge Graham Hall given in the Croydon County Court on October 21 1977. The action in the county court was one in which the plaintiffs are a limited liability company, Stevenson & Rush (Holdings) Ltd: I shall refer to them as "the landlords." The defendant is Mr Martin Telles Langdon, to whom I shall refer as "the tenant." The landlords are the owners of premises, 29 Park Street, Croydon, which are, for the purposes of the Landlord and Tenant Act 1954, business premises. They were let by the landlords to the tenant from July 6 1967 for a term of seven years at a yearly rent of £375. The covenant J
in the lease which is of relevance in this case is clause 3 subclause (17). That clause, so far as it is relevant for the issues with which we are concerned, reads as follows:

To pay to the landlord all costs charges and expenses including legal costs and charges payable to a surveyor which may be incurred by the landlord: (a) in or in contemplation of any application to any planning authority... (c) of and incidental to the preparation and service of any notice by either party on the other under Part II of the Landlord and Tenant Act 1954 and all negotiations subsequent thereto and of all proceedings thereunder And to keep the landlord fully and effectually indemnified against all costs proceedings expenses claims and demands K
whatsoever in respect of the said applications consents notices negotiations and proceedings.

The first point, which was decided by the learned judge in favour of the landlords and against which the tenant appeals, arise under subparagraph (a) of that subclause—"costs and charges... incurred by the landlord... in or in contemplation of any application to any planning authority." That issue can be dealt with quite briefly. In my judgment the learned judge was plainly right in deciding that issue in favour of the landlords. A sum of £54 was claimed by the landlords against the tenant in respect of certain applications which had been made to the planning L
authority in connection with the grant of a new lease which the tenant claimed from the landlords under the provisions of Part II of the Landlord and Tenant Act 1954. It is unnecessary to go into detail of what those planning applications were. Indeed, the information about them is very scanty.

The learned judge in her judgment said this: "Such an application"—that is, an application to the planning authority—"was made on December 29 1975 whereby planning permission (the requirement for which was raised by the defendant's surveyor) for change of use of part of the premises to offices was sought, and a determination granting M
that permission was made by the London Borough of Croydon on the same day." In those circumstances it seems difficult to see how the recovery of that charge by the landlords from the tenant does not fall within the provisions of clause 3(17)(a).

Mr Batterbury, who has appeared for the tenant in this court, sought to say that the learned judge was wrong in so holding. Some limitation, he submitted, must be imposed on the very wide words "any application to any planning authority." It would be inconceivable that a landlord would

A be able to recover from the tenant expenses which the landlord had incurred in respect of planning applications which had nothing whatever to do with the premises with which the tenant was concerned or with any interest of the tenant therein. Mr Batterbury submitted that the limitation which by implication, has to be made on the words of that clause was that it applies only to an application which has been made by the tenant himself. Mr Batterbury said that it might be that in such circumstances the landlord would incur expenses because, for example, he might wish to oppose the application. To my mind that is not a possible construction of the subparagraph. Plainly, it would be limited B to an application that would be concerned with the premises in question. But it is unnecessary to go further into what precisely the limitation might be, because on the judge's finding in the present case the applications that were made stemmed from a requirement made by the defendant's surveyor. On that issue, I wholly agree with the judgment of Judge Graham Hall.

The more important and more difficult question in this case relates to subparagraph (c) of clause 3 subclause (17). That is the subparagraph which provides that the tenant shall pay the landlord all costs and expenses (and so forth) "including legal costs and charges . . . which may be C incurred by the landlord . . . of and incidental to the preparation and service of any notice by either party on the other under Part II of the Landlord and Tenant Act 1954 and all negotiations subsequent thereto and of all proceedings thereunder. . . ."

In this case the tenant gave notice under Part II of the Landlord and Tenant Act 1954 by way of a request for the grant of a new lease, to follow upon the original 1967 lease when it expired. Steps were duly taken, as required by the various sections of the Landlord and Tenant Act 1954, to bring that matter before the court, in order that the entitle- D ment of the tenant (which I think was not disputed in this case) and the terms on which the new lease should be granted (which were in dispute) might be settled by an order of the court. At some stage of the proceedings the parties came to terms. As a result, an order was made by the court which was, and was expressed to be, by consent. That order gave the tenant a further lease for a term of 10 years from December 25 1975 at a rent which had now gone up to £1,500 a year. There were various other provisions, as to the repairs to be done by the tenant, a rent review after five years, and so forth. The consent order also contained, by paragraph 5 thereof, this clause:

E Save as aforesaid the said lease shall be upon the terms of the lease of the said premises dated July 6 1967, save that in clause 3(17) thereof there be inserted the word "reasonable" between the words "all" and "costs" in the first line.

So there was now in the new lease a clause requiring the tenant to pay to the landlords "all *reasonable* costs charges and expenses."

It was agreed before the learned judge that the word "reasonable" must be deemed by implication to be present in the 1967 lease with which we are primarily concerned. The consent order made no provision whatever as to the F costs of the proceedings which had preceded the making of that consent order.

The landlords in due course presented the tenant with a bill in respect of the costs connected with the proceedings which had led to that consent order under the Landlord and Tenant Act 1954. The bill included substantial sums for solicitors' charges, for counsel's fees and for chartered surveyors' fees, relating no doubt to the preparation of evidence about the appropriate rent, and so forth, for the grant of a new tenancy. The total amount of those items was a little over £1,000. The tenant declined to pay that account.

G The matter came before the court on the basis that the landlords were claiming that they were entitled to payment of those sums by virtue, not of the new agreed terms for the further lease, but of clause 3(17)(c) of the 1967 lease. The tenant denied liability by reference to section 38(1) of the Landlord and Tenant Act 1954. That subsection, so far as is relevant, reads as follows:

Any agreement relating to a tenancy to which this Part of this Act applies (whether contained in the instrument creating the tenancy or not) shall be void . . . in so far as it purports to preclude the tenant from making an application or request under this Part of this Act or provides for the termination or the surrender of the tenancy in the event of his making such H an application or request or for the imposition of any penalty or disability on the tenant in that event.

The tenant, the defendant in the action out of which the appeal arises, contended that clause 3(17)(c) was an agreement which purported to impose a penalty on the tenant in the event of the tenant making an application under Part II of the Landlord and Tenant Act 1954. The landlord said: "No; this is not a penalty; it is a perfectly proper provision which is not a void provision." The tenant not only relied upon the terms of section 38(1) of the Act but also asserted that it would be contrary to public policy that the court should permit effect to be given to such a provision. J

On behalf of the tenant it was said that the inclusion of this clause whereby the tenant undertook to pay to the landlords any costs which the landlords incurred, whether by way of solicitors' fees, surveyors' fees, accountants' charges or otherwise, by reason of the tenant's claim to have a new tenancy, would be a *penalty* imposed, or purported to be imposed, by the agreement in the event of the tenant making such an application; and that therefore it was void. The effect of it, it was said, must be that it would tend to discourage the tenant from making the sort of application which the Act plainly intended should be open to a tenant to make. He would be so discouraged because, whatever the result K and however successful he was in the making of his application, however he might succeed in getting an order from the court that the landlord should pay him the whole or part of his costs of the proceedings, nevertheless the result would be that the landlord would be able, by virtue of this clause, to recover from the tenant, not merely any costs which the landlord had been required to pay to the tenant in respect of the tenant's own costs, but also any costs that the landlord had incurred. That would be subject only to this limitation, which it was agreed at the hearing before the learned judge ought to be implied into the clause, namely, that the costs had to be "reasonable." That presumably would merely L mean, not excessive in amount. Thus, the implied "reasonableness" would not prevent the landlord, if the clause were otherwise valid, from recovering from the tenant costs which the court had awarded to the tenant. That, it is said, would be a deterrent and it would constitute a "penalty" for the purpose of section 38(1).

In the county court the learned judge held that this provision was one which ought to be treated as compensation for the landlords, and not as a "penalty" falling within the words of section 38(1). In this court, Mr Bernstein, for the landlords, has submitted that he might indeed be in serious difficulty M if the tenant had proceeded, both by his pleading, by his submissions in the county court, and by his notice of appeal, on the basis of the word "disability" in the phrase "penalty or disability" in section 38(1). But Mr Bernstein contends that the case has proceeded on the basis of the word "penalty" and that, although there might indeed be a *disability*, there is not here a *penalty*. Mr Bernstein, being the realist that he is, appreciated that that argument might ordinarily create difficulties for him in this court; for example, by this court granting leave to amend to enable the appellant to raise the point. But, he pointed out, steps had been taken

A by letter in advance of the hearing of the appeal to invite the tenant to agree to this appeal being discharged, so that the tenant could proceed afresh in the county court asserting "disability" as well as "penalty"; and, in exchange for that concession, the landlords for their part could put their case differently from the way in which it had been hitherto pleaded and argued: that is, the landlords would found their claim, not on the clause in the 1967 agreement, but on the consent agreement. Whether or not that would have been a permissible and appropriate course in other circumstances, it does not arise if, on the true view, the provisions of clause 3(17)(c) are a penalty. In my judgment the provisions of the B clause are properly to be treated as a penalty.

Mr Bernstein's submission to the contrary is founded on the contrast which he suggests is shown, and deliberately shown, by the collocation of the two words, "penalty or disability." He submits that, whatever wider meaning might justifiably have been given to "penalty," if it had not been collocated with "disability," when it is placed alongside "disability" "penalty" has got to be given a more restricted meaning. The meaning that he would seek to give it is, put in this way negatively, that nothing can be a penalty which does not punish. This provision of the contract, he C said, did not punish, was not intended to punish, and could not be treated as providing punishment; and therefore, though it might well be a *disability*, it was not a *penalty*.

With great respect to Mr Bernstein's argument, I do not think that that subtle linguistic distinction can properly be made in this case. Whatever may be the intended meaning of the word "disability," in my judgment, in the ordinary usage of the English language, the provision that was made here, whereunder the tenant agreed that, whatever costs might be incurred on the part of the landlords, and whatever the county court or this court or the House of Lords might think was the appropriate order for the disposal of those costs as D between landlord and tenant, the tenant would nevertheless by contract be obliged to pay them all, is, in my judgment, fairly to be described as being the imposition of a penalty. It is a threat of a financial sanction which is unrelated in its imposition to any question of the merits of its imposition. For that reason, with great respect to the contrary view taken by the learned judge in the court below, I would allow this appeal.

In the circumstances, I do not think that any useful purpose would be served by purporting to decide the other argument put forward by Mr Batterbury—on which, indeed, we did not encourage Mr Bernstein to address us—to the effect that E even if section 38(1) did not make this clause void it would nevertheless be contrary to public policy. I would merely say that my present provisional view is that Mr Batterbury would have had, at least, a more difficult route to pursue in arriving at success on that argument than the route of "penalty" by which, in my opinion, he makes his way to a successful completion of his journey.

I would therefore allow the appeal.

Agreeing, BROWNE LJ said: As to the appeal in respect of the judge's decision on clause 3(17)(a) in regard to the planning application costs, I agree with my Lord that there is F absolutely no substance in that. The application for planning permission was made on the suggestion of the tenant's surveyor, and permission was obtained for the benefit of the tenant. The tenant, the defendant, was occupying these premises and wanted to use part of them as offices, which involved a change of use requiring planning permission. It is not necessary to decide generally what costs would be covered by this clause. It is enough to say that, in my judgment, it is quite plain that the costs of this particular planning application were covered by the clause, and that the judge came to the right conclusion about that.

As to the other point, the point under clause 3(17)(c), I agree with what my Lord has said. Subsection (1) of section G 38 of the Landlord and Tenant Act 1954 is dealing with three types of provision contained in agreements relating to a tenancy. The first type are provisions purporting to preclude the tenant from making an application or request under this Part of the Act. The second type are provisions for the termination or surrender of the tenancy in the event of the tenant making such an application or request. The third type are provisions for the imposition of any penalty or disability on the tenant in that event, that is, in the event of his making application for a new tenancy under this Part of the Act. It seems to me clear that the object of the subsection is to H prevent contracting out of the provisions of the Act which entitle a tenant to apply for a new tenancy. The first and second limbs of the section are dealing with attempts by what I may call direct means to prevent a tenant from applying for or getting a new tenancy. The intention of the third limb, it seems to me, is to exclude attempts to prevent such an application by indirect means. I should therefore expect "penalty" or "disability" in the last limb to include any provision which would have the effect of deterring a tenant from exercising what would otherwise be his right under this Part of the Act. The effect of clause 3(17)(c) is to impose on a tenant, if he makes an application, an obligation to pay J what may be a very substantial sum in respect of costs. It seems to me clear that such a provision will or may deter a tenant from making such an application. Whatever the word "penalty" may mean in some other contexts (and we have been helpfully shown the *Oxford English Dictionary* definitions) it is my view that, in the context of this provision, clause 3(17)(c), does constitute the imposition of a penalty. I agree with my Lord that in the ordinary use of English it may well constitute a penalty. And, having regard to what seems to me to be the plain object of the section, I have no doubt that it is within the meaning of the word in this context. It is not necessary to consider whether clause 3(17)(c) could K also be described as a "disability." Nor is it necessary to consider the scope of that word. It may well be, as my Lord, Waller LJ, said in the course of the argument, that there is an overlap between the two words in this context.

Like my Lord, I think it is unnecessary, and indeed undesirable, to express any views about the public policy argument put forward by Mr Batterbury, which could only be *obiter*.

I agree that the appeal should be allowed to the extent indicated by my Lord.

WALLER LJ delivered a concurring judgment.

The order made by the Court of Appeal was (1) that the order of the trial judge be set aside and that there be L substituted therefor an order that judgment be entered for the plaintiffs for £54; (2) that the respondents pay to the appellant nine-tenths of the taxed costs (unless agreed) of the appeal and nine-tenths of the costs in the County Court and High Court as particularised in a County Court Order of October 21 1977 and thereunder awarded to the plaintiffs.

CHANCERY DIVISION

December 8 1978

(Before Mr Justice WHITFORD)

M HARVEY TEXTILES LTD v HILLEL

Estates Gazette March 17 1979

(1978) 249 EG 1063

Landlord and Tenant Act 1954—Notice to terminate served on tenants under section 25—Whether notice valid—"I intend to occupy the premises for the purpose of a business carried on by me in them"—Landlord a company—Notice given by individual with controlling interest in company—Section 30(3) considered—Notice ought to state whether landlord's intention is to carry on business personally or

A **through medium of company—Notice in present case held to be valid, however, despite absence of reference to company**

This was a preliminary issue in proceedings by the plaintiffs, Harvey Textiles Ltd, to obtain a new lease of the business premises occupied by them at 82 Middlesex Street, Stepney, London E1, as wholesalers of textiles and fancy goods. The plaintiffs claimed that the section 25 notice served on them as tenants was invalid on the ground that their landlords were a company and not the defendant, Abraham Hillel.

G I Bennett (instructed by J Israel, Arnold & Strange) B appeared on behalf of the plaintiffs; C J Lockhart-Mummery (instructed by Thornton, Lynne & Lawson) represented the defendant.

Giving judgment WHITFORD J said: I have to decide a preliminary point in connection with a summons under which the plaintiffs are seeking the grant of a fresh lease of certain premises in Stepney which they have been occupying in connection with a business as wholesalers of textile apparel and fancy goods.

The defendant, who is the landlord, served a notice under section 25 of the Landlord and Tenant Act 1954 to terminate the tenancy. I may add, although I do not think it is of C particular materiality, that, as is not uncustomary, a counter-notice was served on the plaintiffs' side, and, as I have already said, a summons has been taken out under which the plaintiffs seek to secure the grant of a fresh lease. The only point that I am concerned with is whether the original section 25 notice was a valid notice or whether, as the plaintiffs assert, it was not. It is part of an exhibit to an affidavit which Mr Lever has sworn on behalf of the plaintiffs, being his exhibit 1. The relevant premises are identified in the notice, the defendant's name is given; the plaintiffs are required to give notice whether they will be willing to give up possession of the premises specified. Then, in section 3 of the notice, D which is the relevant part of it for present purposes, the defendant says:

> I would oppose an application to the court under Part II of the Act to grant a new tenancy on the ground that on the termination of the current tenancy I intend to occupy the premises for the purposes of a business carried on by me in them.

That is, of course, a notice expressing opposition to the grant of a new tenancy upon the ground specified in (g) of subsection (1) of section 30 of the Act.

Following the receipt of this notice, apart altogether from any question of a counternotice, there has been a little corres-E pondence which I do not think is of any particular materiality. Certainly on February 2 1978 a letter was sent indicating that the plaintiffs would not be willing to give up possession of the premises and observing that the notice stated, as indeed it did, that it was the intention of the defendant to occupy their premises for the purpose of a business carried on by him, and requesting the fullest possible information relevant to the genuineness of the stated intention.

The only other document which I have, for quite plainly there were in this correspondence certain letters which have not been laid before me, is a letter of March 30 1978, sent on the defendant's side, and there is a reference to at least F one intervening letter after the one I have just read, stating that it is imperative for the defendant to occupy the premises, referring to him as "our client," since they are required as essential storage space, and it continues:

> As our clients have factory and offices on the second floor at 82 Middlesex Street

that is part of the relevant premises

> but that floor does not contain the storage space which our client requires, and for security reasons it is essential that our client should be able to store his stock in the same building as that in which he has his factory.

Then it says he wants part of the ground floor for loading G space for lorries, and so forth. I do not know what reply, if any, was made to that letter, other than one further letter which was very kindly read out to me by counsel which does not deal with the extent to which the plaintiffs may have been aware of what was going on so far as the defendant was concerned at 82 Middlesex Street, but at least I have the affidavit of Mr Lever on the plaintiffs' side which seems to me in the terms of paragraphs 6 and 7, the only paragraphs to which I was referred, to show quite clearly that Mr Lever was well aware that the defendant was not personally carrying on any business. He knew that part of the relevant premises were occupied by a company with which the defendant was H concerned, and, indeed, it would appear from paragraph 7 of the defence that he was perhaps aware of the fact that the defendant had a controlling interest in this company.

The point that he takes, which was said by counsel on the plaintiffs' behalf not in any sense to be a technical point, is that the notice is bad because whereas—and I may say that the defendant accepts this entirely in the only affidavit he has sworn—his real intention was that the premises in question at present occupied by the plaintiffs should be used for the company in which he has a controlling interest, what he said in the notice was, "I intend to occupy the premises for the J purposes of a business to be carried on by me," which he must have known was not the fact.

It is accepted on the plaintiffs' side that a notice under section 25 does not need always to be drafted in particular terms, but it has been pointed out to me that the regulations pursuant to the statute have provided that a notice under the section in question shall follow what is described as a prescribed form. It is, in fact, Form 7 under the Landlord and Tenant (Notices) Regulations 1957, and the printed form which forms part of exhibit 1 to Mr Lever's affidavit is based upon the form which we find in the relevant regulations. There are some notes in the Landlord and Tenant (Notices) K Regulations which draw attention to the fact that the court has no power to make an order for the grant of a new tenancy if the landlord, having stated in his notice that he will oppose an application to the court on one of the grounds specified in this Act, establishes that ground to the satisfaction of the court. It then says "The grounds specified are," and it sets out under Note 4(g) the same grounds as we find in section 30. Subject, however, to this; that in relation to (g) what we find is, as a ground of objection relevant for consideration, "that on the termination of the current tenancy the landlord intends to occupy the premises for the purposes, or partly for the purposes, of a business to be carried on by L him in them, or as his residence," with only this additional note that "the landlord cannot rely on this ground if his interest was purchased or created less than five years before the termination of the current tenancy, and at all times since the purchase or creation of the landlord's interest the premises have been let to a tenant occupying them for the purposes of his business."

Now, what was said was that this notice might very well have been a perfectly good notice if it in fact had been the intention of the defendant personally to carry on a business, but that it is deceptive; and it is said that a notice of this kind M must not deceive; if it does deceive then it will be held bad and the landlord will have to start all over again. Further, it is said that a notice of this kind cannot be amended, it must stand or fall on the terms in which it has been sent and that, in considering whether or not the notice effectively informs the party upon whom it is served of the ground which is being advanced in objection to the grant of a new lease, we must follow the ordinary rules of construction. What was urged upon me was that, looking at this as a matter of language, there is nothing to indicate what the true intention of the defendant was. The plaintiffs would be deceived by the notice. I am bound to say I somewhat doubt myself whether Mr Lever was himself in much doubt as to what the true intention was, but that is a matter that I cannot determine on the evidence as it stands. So it is said that if the plaintiffs in seeing the notice did not know what the true ground was, then it must be bad.

It is interesting, I think, just to consider the way in which the matter which is next most relevant for consideration arose, because it arose out of a case that was decided in 1962 and is, indeed, reported in [1962] 2 QB 593, being the case of *Tunstall v Steigmann*, where the point was that a landlord's notice had been served specifying ground (g), and it was agreed that the business which it was intended should be carried on would not in truth be carried on by the landlord but by a limited company in which the landlord held all the shares, and the Court of Appeal held that in those circumstances the notice was bad, for the landlord and the tenant were different personae. It was following that case, which I think the judgments in the Court of Appeal show in their Lordships' view perhaps brought a little bit of a hardship upon the landlord, that there was an amendment to section 30 of the Act making an express provision that opposition could be entered by not only a landlord intending to carry on business personally, but also by a landlord intending to carry on business through the medium of a limited company in which he might hold a controlling interest. When Parliament chose to make this alteration to the Act, they did it in this way; they added a new subsection, subsection (3) to section 30, in these terms:

> Where the landlord has a controlling interest in a company any business to be carried on by the company shall be treated for the purposes of subsection (1)(g) of this section as a business to be carried on by him.

Then there are certain provisions as to what constitutes a controlling interest, and there is no dispute but that the defendant has got a controlling interest in the company which, it now stands accepted, it was always his intention should carry on the business. It is to be observed that in making this amendment there was not any amendment to subsection (1) of section 30, there was not any amendment touching the prescribed form or the notes which we find in the Landlord and Tenant (Notices) Regulations, and it seems to me to be perfectly plain that the intention of Parliament was that, subsequent to the enactment of what is now subsection (3), a notice should state whether the intention was an intention by the landlord to carry on business personally or through the medium of a company.

Now, as I understood it, counsel for the plaintiffs, who has urged everything that I think could conceivably be urged on the plaintiffs' side, was prepared to accept that if all that had been said in the notice that was served was that the landlord would object to the grant of a new tenancy on ground (g) of subsection (1) of section 30, then it might well be that he would be entitled to rely upon the proposed use of the premises by the company in which he has got a controlling interest, but by using the words he in fact used he has put himself out of court. I do not think that this can be right.

I was referred to the case of *Sevenarts Ltd v Busvine* [1968] 1 WLR 1929. It was a case in which again questions arose as to whether in fact a section 25 notice which had been given was or was not valid in circumstances where the landlord who gave the notice was effectively giving it as trustee for a limited company which would be carrying on the relevant business. The report is, I think, a report of some assistance because there are references to a number of judgments given in earlier cases touching notices under section 25 and the effect of entering grounds of objection pursuant to section 30 (1). In particular, there are quotations from the observations of Denning LJ in *Byles v Caesar* [1957] 1 WLR 156 where he is observing that it is sufficient for a landlord in his notice to specify the particular paragraph upon which he relies, and I think it was in the context of observations such as these that counsel for the plaintiffs was perhaps driven to observe that if only ground (g) had been specified he could not take objection to the notice. It was, in fact, in *Byles v Caesar* that Morris LJ observed in connection with (f), which was the ground that was being considered by the Court of Appeal in that case, "I think that paragraph (f) of section 30 constitutes one ground of opposition, which ground may be satisfied in a variety of ways." I think the same thing can be said of ground (g); it is one ground of opposition which may be satisfied if the landlord intends to occupy for the purpose of carrying on a business of his own, or if he intends that the premises shall be occupied for the carrying on of the business of a company in which he has a controlling interest.

The notice that was here served was, in fact, framed in the terms of the statute, that is in the terms of (g). Those who receive notices of this kind must, unhappily for them, be deemed to understand what the statute is about generally, and in any event most people getting a notice of this kind would seek legal advice. To my mind, anybody seeing the notice must appreciate that although the notice has been framed in the terms of (g), the terms of (g) could be satisfied by the proving of an interest such as is dealt with in subsection (3) of section 30, and the relevant ground having been stated it must, of course, be for the landlord to make good his case under that ground when the matter eventually comes to be determined. I am not impressed with the suggestion that, in fact, there is going to be any hardship upon the plaintiffs, or could be, setting aside all questions of what they may or may not in truth have known as to the true circumstances. Let me presume that they knew nothing and at this date have ascertained as much as they have. When the relevant evidence is put in in this case, the full facts will be before them, they will not be taken by surprise. I think in this case the landlord in issuing the notice that he issued has given just such a notice as is required by the terms of the statute and that the notice is a perfectly valid notice.

CHANCERY DIVISION

December 18 1978

(Before Mr Justice GOULDING)

O'MAY AND OTHERS v CITY OF LONDON REAL PROPERTY CO LTD

Estates Gazette March 17 1979

(1978) 249 EG 1065

Landlord and Tenant Act 1954, Part II—Application for new lease of offices occupied by solicitors in the City of London—Agreement on alternative rents of £10 and £10.50 per sq ft according to whether the court did or did not approved landlords' proposals for a variation of terms under which the lease would be a "clear lease" making tenants liable for comprehensive variable contributions to the costs of repair and decoration of the exterior, the structure and the common parts and management, as well as costs of services—Buildings subject to "clear leases" more marketable—Tenants' contention that landlords of a whole building should accept risks of fluctuations in costs—Observations on tenants' right to challenge surveyors' certificate and certified accounts, subject to restrictions—Held that landlords' proposals satisfied four tests: (1) reason shown for variation, (2) tenants compensated by adjustment of rent, (3) security for carrying on profession not impaired and (4) proposals fair and reasonable between the parties—New tenancy ordered to be granted on landlords' terms

A This was an application by originating summons by a firm of solicitors, Ince & Co, of Knollys House, 11 Byward St, London EC3, for a new lease of offices on the fifth floor of Knollys House. The plaintiff tenants' landlords were the City of London Real Property Co Ltd. The issue between the parties as to the terms of the new lease is set out in detail in the judgment of Goulding J.

D Wood QC and J Gaunt (instructed by Ince & Co) appeared on behalf of the plaintiffs; Ronald Bernstein QC and B Levy (instructed by Nabarro & Nathanson) represented the defendants.

Giving judgment, GOULDING J said: This is an application
B for a new tenancy made under Part II of the Landlord and Tenant Act 1954 by the plaintiff tenants, who are a firm of solicitors practising under the style of Ince & Co. The defendant landlord is the City of London Real Property Co Ltd. It is agreed to be a long-established and perfectly reputable concern. The demised property which is the subject of the application is the fifth floor of a modern office building called Knollys House, together with car-parking space in the basement of the building. The premises are occupied by the tenants as their professional offices. Knollys House was erected in the years 1960-62. It is situated near the Tower of London. The tenants also occupy part of the
C fourth floor of the building, but that occupation is under a different title and has not entered into the argument of this case.

The current tenancy for the purposes of the Act was granted by two leases (one of the fifth floor, the other of the parking space) both dated April 25 1972. It was granted for five years from Lady Day 1972. The area occupied by the tenants was extended, and the rent payable was accordingly increased, by two deeds of variation dated May 27 1976. The total yearly rent payable thereafter was £92,871, which, by my arithmetic, is about £9.39 per sq ft of office space.

I need not take time to state the procedural history of the
D application. It is agreed that the new tenancy to be ordered by the court shall continue until March 24 1982. The rent is not itself the subject of dispute, but its amount will depend on the other provisions to be inserted in the new lease. The parties are agreed on a rent of £10.50 per sq ft if the provisions ordered by the court are those desired by the tenants, that is, if the stipulations in the leases of 1972 are reproduced with certain agreed modifications. The parties are also agreed on a smaller rent, namely £10 per sq ft, if the landlord has its way and the provisions of the new lease relating to service charges are substantially different from those hitherto in
E force. The landlord initially proposed a printed standard form of lease identified as Form 20 C. Late in the hearing a revised draft of the two schedules to Form 20 C was produced, and I gave it the name of Document Z. It makes substantial concessions to the tenants' criticisms of the original form. I shall refer shortly to the provisions contained in Form 20 C, modified as regards the schedules by Document Z, as *the landlord's terms*.

I should add, though it will not (I think) affect the form of the court's order, that the parties have agreed out of court that an interim rent of £90,500, now payable under contractual arrangements between them, shall continue until
F March 25 1979 in lieu of any rent reserved by the new lease, and also (in effect) that no other terms of the new tenancy differing from those of the current tenancy shall be enforced in respect of any period before that date.

The main point of contention arises thus. When an owner of tenanted office premises in London desires to sell them, willing purchasers are most easily found among pension funds and life assurance companies. The nature of their business makes them favour an investment yielding a steady and predictable income in money terms. They do not want to suffer fluctuations of income due to variations from year to year in the costs of maintenance, repairs, insurance, man-
G agement, and the like. It is well known that such items have greatly increased in recent years and are difficult to predict accurately for the future. Hence owners of office property favour leases that throw the responsibility for paying fluctuating outgoings on the tenant. That means, of course, accepting a lower rent than if the landlord bore the outgoings, in return for certainty of annual income. When an entire building is let to one tenant, the owner's requirement is simply met by imposing covenants that oblige the tenant to do and pay for external and internal repairs and all the other things that require a variable expenditure. But when, as at Knollys House, one large building is divided
H between a number of tenants under separate lettings, the landlord cannot well avoid the need to carry out the maintenance, management, and so on, himself. The simple solution here is to require the several tenants to reimburse all the landlord's expenses, bearing the burden in proportionate shares among themselves. The aggregate fixed rent reserved by the several lettings thus remains as a clear profit in the landlord's hands. Leases that attain this result, not always perfectly, have been referred to by witnesses at this hearing as *clear leases* and I adopt that as a convenient term to indicate them. I shall also employ the term *service charge* for a tenant's periodical contribution to some or all of the land-
J lord's outgoings.

The evidence shows that the owner of such a building as Knollys House, on letting vacant accommodation therein, will nowadays insist on a clear lease so that his property may command a readier market if he has to sell it. But what is to happen when the tenancy of a sitting tenant approaches its contractual end, and the tenant applies for a new tenancy under the Act of 1954? Landlords naturally want to introduce clear leases on that occasion also, just as though the premises were vacant. I have been told that many tenants, in such circumstances, have accepted a clear lease in place of one in less modern form, the revised rent being
K of course fixed with due regard to the change. But the tenants in the present case do not want a clear lease. They ask the court to determine the terms of the new tenancy pursuant to section 35 of the Act of 1954. They say that the court, in the exercise of its discretion, should reject the landlord's proposal for a clear lease.

I need not read the relevant provisions of the Act of 1954, for they are now well known. The important thing to remember is that section 35 commands the court, when determining the terms of a new tenancy (other than the duration of the new tenancy and the rent) to "have regard to the terms of the current tenancy and to all relevant circum-
L stances." I must, however, state the landlord's proposal in more detail.

Under the current tenancy the tenants pay, in addition to their fixed rent, a service charge in respect of three matters only, described as follows: "(a) cleaning the public parts of the building and general services of the housekeeper," "(b) lighting the general parts of the building," and "(c) heating the demised premises by hot water radiators." Items (a) and (b) are based upon the landlord's scale of charges, revised annually. The scale specifies an amount payable for each unit of 200 sq ft of floor space. It is estimated on what is
M called a portfolio basis, ie from the landlord's expenditure, not only on Knollys House, but also on other buildings owned by the landlord. Item (c) on the other hand is determined by a formula in which the only variable is the current cost of fuel oil to the landlord. By the terms of the current tenancy the tenants covenant to do internal repairs and decoration of the demised premises; the landlord covenants to provide central heating, to clean and light the common parts of the building, and to keep the exterior and the common parts in repair. The lease contemplates fire insurance by the landlord, who does not, however, expressly covenant to insure.

A Under the landlord's terms the tenants are to pay a comprehensive service charge recoverable as additional rent, and defined as fair and reasonable proportions, to be certified from time to time by the landlord's surveyor, of the actual and/or estimated cost of providing a long list of services. They include water, heating, cleaning, staff; repair and decoration of the structure, the exterior, and the common parts of the building; lighting of the common parts; lifts; rates and other outgoings in respect of common parts; security, and some other matters. Up to 10 per cent of the total is to be added as a management charge, and VAT will also be added. The tenants' proportionate share of premiums for insuring the B building against fire and other risks is separately reserved as additional rent. The tenants covenant to do internal repairs and decoration of the demised premises; the landlord covenants to provide, manage and operate the various services in accordance with the principles of good estate management and to insure.

The strictness of the new form as a clear lease is, however, relaxed, as to certain outgoings, by what I shall refer to for brevity as a *funding provision*. Instead of a share of the actual cost in each year, the service charge is to include an annual amount (to be revised annually by the landlord) towards the estimated cost of replacing lifts, boilers and other C equipment, and towards redecoration of the exterior and common parts. The annual sums are to be calculated on a conventional life expectancy, 30 years for lifts, 15 years for boilers and heating equipment, 5 years for external decoration, 7 years for decoration of interior common parts.

I must also mention what has been conveniently referred to as the *challenge provision* contained in the landlord's terms. The landlord is in each year to serve on the tenants an account showing the amounts of the different items that make up the service charge and the proportions attributable to the demised premises. That account is to be certified by the D landlord's accountants. If the tenants want to challenge either the surveyor's certificate or the certified account, the dispute is to be referred to an independent surveyor or consulting engineer, acting as an expert and not as an arbitrator. The tenants are not entitled to question any item on the sole ground that it could have been provided at a lower cost than that actually incurred or estimated in good faith by the landlord. Subject only to the challenge provision that I have just summarised, the certificates of the landlord's surveyor and accountants are to be final and binding on the parties.

The landlord's reason for desiring to alter the terms of the tenancy clearly appears from the following evidence of E Mr Temple, a valuer who was called on the landlord's behalf. I read from a proof of evidence which he repeated in the witness box. Mr Temple said:

> Major investors in my experience are tending to seek investments where there are clear leases and are less willing to purchase properties where some or all of the leases are not clear leases. Investors are able to evaluate their investment more easily where the clear leases give a net investment return simply calculated as the aggregate of true rents reserved. This attitude in the market has two major effects on the value of properties as investments. First, the marketability of a property is markedly affected by whether or not the lettings are under clear leases. F Clear leases considerably improve the marketability of a property as such investments appeal to major investors. Moreover, value is particularly affected by the prospect of sale on a falling market since it is in periods of financial stringency and recession that a property investment company is likely to require to realise its asset. In those conditions the marketability of the investment is of major importance. Second, the value of a property is directly affected as to quantum by the presence or absence of clear leases. I have inspected Knollys House situate at Byward Street in Seething Lane and 46-48 Mark Lane. If the property were let on clear leases, as compared with being let on leases where the tenants pay service charges but the landlord bears the cost of repairs, insurance and management, there would be a difference in the value in the market of the order of two or more years' G purchase. I am informed that the present annual aggregate rental income of the whole building is approximately £800,000. In my opinion there would be a difference of the order of £1m to £2m in the market value of the whole building if let on clear leases as against leases where the costs of repairs, insurance and management are not fully recoverable.

The tenants' witnesses do not impeach the factual basis of the landlord's case. It is, however, argued for the tenants that, whatever the practice of the open market, it is not easy to justify a substantial departure from the terms of the current tenancy in a new tenancy ordered under the Act of 1954. And here, it is submitted, the landlord's proposal should be H rejected, because the major risks (in the sense of uncertainties) of upkeep and management of a big building are not suitable to be borne by the occupying tenants, each occupying a part only—and it may be a small part—of the property. The risks should fairly be assumed, say the tenants, by the owner of the whole building, who is in the best position to investigate and assess them and to provide for them by charging a proper fixed rent. That is one, and I think the main, objection that I have to bear in mind. I may say at once that the superiority of the landlord's position for estimating future uncertainties is almost obvious, and was frankly accepted by the very experienced estate agent, Mr Perkins, who gave evidence for the J landlord. He went so far in cross-examination as to say that a landlord is better equipped to bear the burden of future uncertainty than the tenant. The point is strengthened by the fact that the landlord in the present case employs a skilled professional staff to look after its buildings.

A second and subsidiary objection to the landlord's terms is that a clear lease may incline a landlord, even unconsciously, to extravagance in maintenance and management, so that the service charge may needlessly increase without the tenants having any opportunity to control it. A less scrupulous landlord, it is argued, might gradually improve the building in the K guise of mere maintenance, thus leading to the unjust result that a tenant, continuing in the same premises, might, under a future tenancy, have to pay a market rental enhanced by works effected at his own expense. That also is an objection that I must bear in mind. The landlord's counsel and witnesses say that neither the landlord nor any probable investor in property in the City of London would behave in such a way, but I attach little weight to that contention. After all, the purpose of legal contracts is to reduce the mutual dependence of parties on one another's morality. The second argument of the tenants is put forward, like their first, as a reason why the court should not order a clear lease against their wishes and L in departure from the terms of the current tenancy. It is used in the alternative as an objection to the challenge provision contained in the landlord's terms. The tenants say they should be at entire liberty to challenge the correctness of the landlord's figures, or the reasonableness of the expenditure, by ordinary legal process in the courts.

The tenants say further that, if a clear lease is to be forced on them at all, they object to the funding provision. They point out that the actual life of a lift or of a boiler may be quite different from its conventional estimated life, and indeed it may never be replaced at all if the building is reconstructed or demolished. Moreover, a short tenancy, such as that now M desired, may well run its whole course before the time for replacement arrives and the landlord incurs any relevant expenditure. The tenants object to paying an additional rent estimated by reference to the landlord's future expenditure for specific purposes, when the landlord is bound by no trust or other obligation to retain and use their contributions for such purposes, nor to account for interest upon them.

The tenants observe generally about the landlord's terms that they are put forward because of the state of the property market or current business usage. But you cannot, they argue, justify a substantial transfer of risk or burden from landlord

A to tenant on those grounds. To do so is to disregard the terms of the current tenancy to an extent not allowed by section 35 of the Act of 1954. In any case, it is submitted, the supposed benefit of the landlord's terms to the landlord is illusory. On the evidence, the value of Knollys House as an investment will not be much enhanced until all, or nearly all, the tenants hold on clear leases, and there is no prospect of getting anywhere near that position in the next three years. So the tenants are asked to suffer a disadvantage that they do not want, for the duration of the new tenancy, without any countervailing benefit to the landlord in that time.

Such, in short, are the contentions of the opposing parties. B Under section 35 of the Act of 1954 the terms of the new tenancy are, in default of agreement, to be determined by the court as there directed. Having regard, in obedience to the section, to the terms of the current tenancy, and having regard likewise to all the relevant circumstances proved or admitted at the hearing, I have come to the conclusion that the terms of the new tenancy shall be the landlord's terms.

In deciding to exercise the court's discretion in that manner, I have sought to give proper weight to all the evidence and arguments that I have listened to. If obliged, as I think I am, to state concise explicit reasons for my decision, I would subject the facts of the present case to four tests, while making C it plain that I do not regard them as a complete scheme of analysis for all similar cases. The first test is this. Has the party demanding a variation of the terms of the current tenancy shown a reason for doing so? Widgery LJ, as he then was, said in *Cardshops v Davies* [1971] 1 WLR 591, at 596: "Section 35, to which I have already referred, shows that the terms of the new tenancy must be fixed having regard *inter alia* to the terms of the current tenancy; and a departure from the terms of the current tenancy is something which in my judgment requires to have a reason attributed to it." *Re No 1 Albemarle Street* [1959] Ch 531 is a good example of a case where the court refused to allow a change in the D terms of the tenancy because the party proposing it simply failed to show any real reason for making the change. Should the first question be answered in the affirmative, my second is whether, if the party demanding a change is successful, the party resisting it will in principle be adequately compensated by the consequential adjustment of open market rent under section 34 of the Act. It is not very difficult to imagine stipulations so inconvenient or humiliating to a tenant, or so restrictive or capricious in relation to a landlord, that a pecuniary retribution ought not to be forced on the unwilling party.

E My third test asks whether the proposed change will materially impair the tenant's security in carrying on his business or profession. If it will, I conceive that the court will ordinarily be hesitant to accept it, since the discretion given by the statute is part of the legislative machinery for protecting business and professional tenants as such. Thus an alteration desired by the landlord was refused in *Gold v Brighton Corporation* [1956] 1 WLR 1291, not only because the landlord had shown insufficient reason for it, but also because it would have narrowed the scope of the tenant's existing trade. Similarly, in *Cardshops v Davies*, already cited, the landlord's proposal failed because it might have prevented F the preservation and eventual sale of the goodwill of the tenant's business. And in *Charles Clements (London) Ltd v Rank City Wall Ltd* (1978) 246 EG 739, the landlord's proposal, which was rejected, would have burdened the tenant's business by a substantial additional rent in return for the relaxation of a restrictive covenant that would have been of no value to the tenant as long as the business was still carried on.

If a proposed change in the terms of a tenancy passes all three of the tests I have suggested, it seems to me that there is a *prima facie* case for allowing it. Reading the provisions of section 34 of the Act, which adopts the open market as G the determinant of rent, together with those of sections 33 and 35, which give the court a wide discretion over other matters and little guidance as to its exercise, I am of opinion that Parliament did not by this legislation intend to petrify the economy of business premises. I think rather that the intention was to leave the market to develop freely, subject only to the overriding protective policy of the Act. Some support for that interpretation is perhaps to be found in *Hyams v Titan Properties* (1972) 24 P & CR 359.

The three tests I have so far formulated, or any others framed with like particularity, may condemn a proposal, but can at the best establish no more than a *prima facie* case for H allowing one. For there must always remain the residual and comprehensive question: taking all the relevant matters into account, is the proposal, in the court's opinion, fair and reasonable as between the parties?

Let me now apply my four questions to the facts in the present case. The landlord has in my judgment proved a definite reason for departing from the terms of the current tenancy, namely a wish to bring them into line with the present practice of itself and other landlords, and thereby to make the reversion a more valuable and readily-marketable asset. I am not persuaded by the tenants' objection that no appreciable improvement of the reversion is on any view J attainable during the short time of the tenancy now to be granted. As Mr Perkins said in cross-examination, you must start somewhere, and the result of the present application may considerably affect the prospects of introducing clear leases on subsequent occasions.

Secondly, the tenants can in principle be adequately compensated for the proposed change of terms by a diminution of the open-market rent to be fixed under section 34 of the Act. The tenants' own expert witness, Mr Baker, found no difficulty in agreeing a figure for that purpose. He was very conscious of the tenants' objection to the uncertainties of the comprehensive service charge and to their lack of K control of the landlord's future expenditure. Nevertheless he was satisfied that the agreed rent reduction of 50p per sq ft, or £4,950 in all, was a fair compensation in all the circumstances. If I have understood correctly the figures produced by Mr Treagus, the landlord's chief surveyor and one of its directors, the acceptance of the landlord's terms would increase the service charge by about £2,700 per annum on the basis of figures relating to a year which ended in March 1978. The remaining £2,250 (approximately) per annum can be regarded as compensation, and is agreed to be a fair compensation, for the transfer of uncertainties from the landlord to the tenants L in respect of the three years to end in March 1982.

I turn to my third test. It is apparent, and it was not disputed by Mr Perkins, that under some circumstances the imposition of a clear lease would materially impair a tenant's security in carrying on his business or profession. Imagine a newly-established, or financially-insecure, or highly-speculative, business carried on in part of an old and already dilapidated building. The replacement of part of a fixed rent by a fluctuating contribution to the actual costs of maintenance and management, even if absolutely fair in purely financial terms, might well in such a case threaten the security of the business by the insupportable risk of a very bad year. But in M the present case the tenants are a well-established firm of solicitors in the City of London, willing to pay a yearly rent of £100,000 for a floor in a modern building less than 20 years old. There is no suggestion in the evidence that it was not well built or that it has not been well maintained. In my judgment the landlord's terms will not in fact adversely affect the tenants' security in carrying on their profession.

So I come to the final and comprehensive question, whether the proposed variation of terms is fair and reasonable as between the parties. One is here in the realm of pure discretion, where it is hardest to condense in words the dialogue of the infra-cranial jury room. On the main question

A of introducing a clear lease I can only say that, weighing the landlord's reasons against the tenants' objections, taking into account the terms of the current tenancy, and remembering the compensating diminution of fixed rent, I think it fair and reasonable to adopt the landlord's terms, and I cannot see that any other elements of the case point to an opposite conclusion.

As to the challenge provision I think it is reasonable to place some limitation on disputes about the amount of a service charge. The landlord's terms have in this respect been considerably improved, from the tenants' point of view, by the concessions embodied in Document Z, and are now more B liberal (so far as my experience goes) than those usually found in modern leases. I do not feel it necessary to modify them further.

Nor do I see any objection to the funding provision. It is indeed a not inconsiderable mitigation, in favour of the tenants, of the very uncertainties that are their main complaint against clear leases generally. There is to my mind nothing unfair in asking a tenant who enjoys the use of equipment during a term of years to pay an annual rent reflecting the estimated cost of replacing the equipment, whether in the event it is replaced in the tenant's time, or afterwards, or for C some reason not at all. Whether the proposed provision for replacement will prove excessive, adequate, or insufficient in the hands of the landlord one cannot tell, because the outcome depends on future rates of interest, on future rates of inflation or deflation of the currency, and on the future physical history of the articles in question, all of which are the subject of uncertainties carried by the landlord. But I can find nothing inherently unfair or unreasonable in the scheme of the funding provision as it stands.

Accordingly, while acknowledging that a different judge might without error have come to a different conclusion, I am of opinion for my part that the landlord's terms should D prevail.

In conclusion I must make it clear that this case has been argued on a tacit agreement that the reservation of a service charge, in the sense in which I have used the expression, is one of the terms of the tenancy other than rent, which fall to be determined under section 35 of the Act. The landlord has not suggested that the rent at which, for the purposes of section 34, the holding might reasonably be expected to be let in the open market by a willing lessor must, on the evidence, consist of a fixed rent plus a variable sum, reserved as additional rent, to reimburse a proportionate share of the landlord's outgoings. Subsection (3) of section 34, no E doubt introduced in 1969 with rent review clauses in mind, might not be immaterial in this connection. There may be good reasons why such a contention is not admissible, but I have not heard the point argued and express no opinion upon it.

Subject to any submissions of counsel about the form of order, I will order the grant of a new tenancy of the holding mentioned in the originating summons, to begin (as it must) on the coming to an end of the current tenancy, and to expire on March 24 1982, at an annual rent of £100,700, the other terms of the new tenancy to be those contained in a draft lease to be prepared and signed by junior counsel for the parties. F The draft is to reproduce the provisions of Form 20 C, as modified by Document Z, with such further modifications if any) as may have been, or may be, agreed between the parties not later than a specified date, possibly December 31 1978 or January 31 1979. I will direct that my order is not to be drawn up before the specified date without the agreement of both parties. The case may, if necessary, be mentioned to me informally pending the drawing up of the order.

Counsel for both parties accepted that the specified date for agreeing the draft should be January 31 1979. The order for costs was that no order should be made as to costs up to and including the date of the adjournment of the originating summons into court, but that the plaintiff tenants should pay half the defendant landlords' costs subsequent to the adjournment of the summons into court. G

COURT OF APPEAL

December 18 1978

(Before Lord Justice STEPHENSON, Lord Justice BRIDGE and Lord Justice SHAW)

CARSHALTON BEECHES BOWLING CLUB LTD v CAMERON AND ANOTHER

Estates Gazette March 31 1979

(1978) 249 EG 1279 H

Landlord and Tenant Act 1954, Part II—Application by tenants for a new tenancy opposed under section 30(1)(g)—Landlord's interest vested in two executors who were also entitled under the will to equal shares in the residue, which included the property in question—One of the two intended to occupy the holding for the purpose of a business to be carried on by him therein—Question whether he could be regarded as "the landlord" for the purpose of section 30(1)(g)—Arrangement that he should purchase his co-beneficiary's share in the property—Effect of section 41(2) of the 1954 Act as interpreted in Frish Ltd v Barclays Bank Ltd—The words "beneficiaries under the trust" in section 41(2) must be limited to persons occupying by virtue of their interest as beneficiaries—In view of arrangement to purchase co-beneficiary's interest this condition was not satisfied—Held that the ground for opposing the grant of a new tenancy under section 30(1)(g) had not been established—County court judge's order allowing application for new tenancy accordingly upheld J

This was an appeal from a decision by Judge Lermon at Epsom County Court in which he made an order for the grant of a new tenancy to the applicants, Carshalton Beeches Bowling Club Ltd, of the first floor of premises at 61a Banstead Road South, Carshalton, and a bowling green occupied therewith. The appellants were Lady Irene Eunice Cameron and Peter Henry, executors of a Miss Henry, deceased, who had been the freehold owner of the whole property. The respondents were Carshalton Beeches Bowling Club Ltd, the successful applicants in the court below. K

B S Green (instructed by Copley, Clark & Co) appeared on behalf of the appellants; R H T Smith (instructed by John Chapman & Co) represented the respondents.

Giving the first judgment at the invitation of Stephenson LJ, BRIDGE LJ said: This is an appeal from a judgment of His Honour Judge Lermon, given in the Epsom County Court on March 17 1978, allowing an application by the respondent tenants for the grant to them of a new lease of certain premises pursuant to the provisions of Part II of the Landlord and Tenant Act 1954. The premises in question are the first floor of a property known as 61a Banstead Road South in Carshalton and a bowling green occupied together therewith, which has been occupied for some time before this matter came to court by the applicants as lessees, the Carshalton Beeches Bowling Club Ltd. L

No 61a Banstead Road South, apart from its first floor and its bowling green, had been occupied for many years (together with another house in the same road) as a school which had been run by a lady named Miss Henry, who was the freehold owner of the property and the landlord of the tenants and who lived to the ripe old age of 94. When she died, she made a will whereby she left the property in question as part of the residue of her estate and she gave that residue in equal shares to her niece and nephew, a Mr Peter Henry, the nephew, and the niece, Lady Irene Eunice Cameron. They were also named as the executors under the will, and they as executors were respondents to this application. They opposed the grant of a new lease to the tenants M

A on the ground afforded by section 30(1)(g) of the Act, which provides, as is well known, that a new lease is not to be granted if the landlord proves that on the termination of the current tenancy he, the landlord, intends to occupy the holding for the purposes or partly for the purposes of a business to be carried on by him therein or as his residence.

Now the evidence was, and this was accepted by the learned judge, that Mr Peter Henry was going to carry on the school and he wished to take over the first floor occupied by the tenant club for the purposes of running a proprietary club for the benefit of parents of children of the school and to take in the bowling green as part of the school amenities,
B so although there was an issue about this, the learned judge undoubtedly accepted that Mr Peter Henry had the necessary intention to defeat the tenants' claim for a new tenancy, provided that he could bring himself within the words "the landlord" in section 30(1)(g). Now, of course, at the time of the hearing Mr Henry was not the landlord, but in order to entitle himself to set up the section 30(1)(g) grounds of opposition, he invoked the provisions of section 41(2) and that provides that: "Where the landlord's interest is held on trust the references in paragraph (g) of subsection (1) of section 30 of this Act to the landord shall be construed as including references to the beneficiaries under the trust, or
C any of them." "So," said Mr Henry, "as one of the beneficiaries entitled to an equal half-share in the residue of the late Miss Henry's estate, of which number 61a Banstead Road South, forms part, I am a beneficiary under a trust and as such entitled to call myself a landlord for the purposes of paragraph (g) of subsection (1) of section 30 of the Act."

The difficulty in the way of that contention and that which was held by the learned judge to be sufficient to defeat it arises from the decided case in this court of *Frish Ltd v Barclays Bank Ltd*, which is reported in [1955] 2 QB 541. That was a case where the property in question was held on
D discretionary trusts for a number of beneficiaries. One of those beneficiaries had agreed with the trustees for the grant to him of a tenancy in reversion to the tenancy occupied by the business tenants, and in virtue of that reversionary tenancy, he claimed to defeat the business tenants' application for a new lease. The decision of the court, Sir Raymond Evershed MR, Jenkins LJ and Parker LJ, was that that opposition to the grant of a new tenancy failed because it was not in virtue of his interest as beneficiary in that trust that the person claiming the rights of a landlord under section 41(2) was proposing to occupy. It was not in
E virtue of his interest as a beneficiary that he would occupy the premises, but in virtue of his interest as a tenant under a tenancy to be granted to him on commercial terms by the trustees.

The matter is put in a number of ways in the judgment. Sir Raymond Evershed said, at p 547:

> What is said (and this prevailed in the court below) is that Mr Parnes happens to be a beneficiary under the trusts on which the property is held by the landlords, and that by reason of section 41(2) (to which I earlier alluded) it therefore follows that occupation by someone who is in fact, or can be shown in fact to be, a beneficiary under the trusts on which the property is
> F held by the landlords amounts to occupation by the landlords for the purposes of paragraph (g), even though the intended occupation is in no substantial sense referable to the fact of the beneficial interest of the proposed occupier.

The Master of the Rolls held that the intended occupation by a landlord claiming under section 41(2) must be referable to the fact of the beneficial interest. At p 548 he says:

> And I think they [the words] must be given, as a matter of fair interpretation of the language, a limitation in this respect that only those beneficiaries are referred to whose interest under the trust (putting it quite broadly and generally) is such as to give them the right, as against the trustees, to occupy the property, or is such that the trustees may properly within the terms of G
> their trust let them as beneficiaries into possession.

Then he added at p 552:

> . . . if the intended occupation is to be that of a beneficiary, it must be shown that it is the intention that he should so occupy by virtue of his quality or right as a beneficiary.

Jenkins LJ gave a judgment substantially to the same effect and said that the beneficiary's interest in virtue of which he claims to occupy in order to qualify under section 41(2) must be such an interest as to justify the trustees in letting the beneficiary into occupation.

In this case, if one asks the question whether the executors H
of the will were entitled to let Mr Henry *qua* beneficiary into occupation of the property, the answer is no, as I think was conceded in the court below and as certainly clearly appears from some voluntary further and better particulars of the landlord's opposition which were given prior to the hearing. By those particulars it was stated: "By consent of both beneficiaries, both 61a and 67 Banstead Road South, which together form the residue, will be conveyed into the sole name of Mr Henry on the winding-up of the estate for a consideration to be agreed between them." In other words, Lady Cameron was for a consideration going to transfer her beneficial interest in this property, as forming part of the J
residue, to Mr Henry.

I can quite see that if the residue of an estate comprises sufficient properties for a single property held under a business tenancy to be appropriated by the personal representatives to one of the beneficiaries, who wish to divide the residue between them, and he can make such appropriation without the consent of the other beneficiaries, the residue being large enough to satisfy the other beneficiaries' shares from other property, that beneficiary would be entitled to claim as landlord under section 41(2), but that is clearly not the case here. From the voluntary particulars given and from the evidence which followed those particulars, it is clear that K
it is necessary for Lady Cameron to agree to transfer her share, or a substantial part of her share, for a consideration, which has not been stated in the evidence or in the particulars, to Mr Henry in order to put him into a position to claim as the sole occupier of number 61a, who is going to carry on the school in succession to Miss Henry and who was going to take over (if his wish prevailed) the first floor from the Carshalton Beeches Club and the bowling green.

Mr Barry Green, who has argued this case as fully and as attractively as well it could be, has urged us to take the view that the extent to which the interest of an intending land- L
lord/occupier who claims under section 41(2) must be that of beneficiary and the extent to which the interest in virtue of which he is to occupy would be his interest *qua* beneficiary is a matter of degree. I accept that proposition to this extent, that if it could be shown that any interest other than that of beneficiary which the intending occupier/landlord had acquired was *de minimis*, that could properly be disregarded, but here on the evidence the only inference we can draw is that the interest of Mr Henry in the property as a beneficiary under the will extends only to half his total interest and the other half of his interest is to be acquired by purchase from Lady Cameron. Looking at the expression used in the *Frish* M
case, it does not seem to me it could possibly be sufficient to satisfy the principle which the court regarded as the governing principle by which section 40(2) of the Act has to be construed.

Furthermore, I think Mr Smith has given us another complete answer to the claim against his clients which is advanced on the appellants' behalf. He points out that before very long the conveyance to Mr Henry which is mentioned in the voluntary particulars will be effected and then it will no longer be a question of Mr Henry holding this property as beneficiary under a trust but he will in fact hold it as to

A half the interest which he will have acquired as a purchaser from Lady Cameron; and of course in that capacity he would be defeated in his claim to oppose the tenants' application for a new tenancy by section 30(2), which expressly provides that the landlord shall not be entitled to oppose an application on the grounds specified in paragraph (g) of subsection (1) if the interest of the landlord was purchased after the beginning of the period of 5 years which ends with the termination of the current tenancy.

For those reasons, I think the learned judge came to the right conclusion, and I would dismiss this appeal.

B SHAW LJ agreed.

Also agreeing STEPHENSON LJ said: I thought at one stage when listening to Mr Green's able argument that it might be possible to regard Mr Henry's occupation which was proposed at the time of the hearing at the county court, and which has now become actual since last May, as in a substantial sense referable to his beneficial interest. In the words of Sir Raymond Evershed MR at p 547 of the report to which my Lord, Bridge LJ has referred, I thought it could be said that his occupation was by virtue of his interest as beneficiary, and not by virtue of a commercial bargain or under the terms of an intended commercial bargain.

C Jenkins LJ does not refer, in his judgment, to a commercial bargain but he agreed with the Master of the Rolls that the interest of the beneficiary, qualifying him as a landlord under section 41(2) must be an interest under the trust on the strength of which he intends to occupy and must be in right of his beneficial interest and referable to it. It seems to me, however, that Mr Smith gave the answer to that; that once Mr Henry does occupy the property which might up to that point be called trust property, the trust in so far as it affects him and that property is at an end. It therefore becomes unnecessary to consider the point whether one can look at D the degree of his previous beneficial interest in the property and the amount of the consideration with a view to seeing how far it turns into a commercial bargain the transaction by which he obtains his occupation of the property.

I too would dismiss this appeal.

The appeal was dismissed with costs.

COURT OF APPEAL

January 29 1979

E (Before Lord Justice ORR, Lord Justice ORMROD and Lord Justice GEOFFREY LANE)

KIRKWOOD v JOHNSON AND OTHERS

Estates Gazette April 21 1979

(1979) 250 EG 239

Part II of Landlord and Tenant Act 1954—Grant of new lease—Option to purchase freehold reversion included in county court judge's order for new lease—Whether jurisdiction to do so by virtue of section 32(3) or section 35 of Act—Original lease had contained an option to purchase but it had lapsed—Held that there was no jurisdiction to include such an option as a term of the new lease F —The purpose of the Act was to protect the tenant in respect of his business, not to place a new saleable asset in his hands—Appeal allowed

This was an appeal by Arthur Johnson, Edward Johnson and William Jess Johnson, lessors of the "Three Ways Filling Station," Burstwick, North Humberside, against part of an order by Deputy Judge Davey at Hull County Court. The part of the order against which the appeal was brought included in the new lease in favour of the respondent, Alwyn Kirkwood, an option to purchase the freehold reversion.

Derek Mallard (instructed by Warren, Murton & Co, agents for Stamp, Jackson & Procter, of Hull) appeared on behalf of the appellants; W D Ainger (instructed by Payne G & Payne, of Hull) represented the respondent.

Giving judgment, ORR LJ said: This is an appeal by the lessors of business premises known as the "Three Ways Filling Station" at Burstwick, North Humberside, against that part of an order made on March 16 of last year by His Honour Deputy Judge Davey in the Hull County Court, whereby, in granting to the respondent under the provisions of Part II of the Landlord and Tenant Act 1954 a new lease of those premises, he ordered that the lease should include an option for the respondent tenant to purchase the reversion expectant on the new lease on terms similar to those, to which I shall later refer, contained in clause 8 of H the original lease, save that the price payable on the exercise of the option was increased from £12,500 (less a percentage of the gross rent paid by the tenant to the landlord during the term) to £20,000 (less a similar percentage of the rent). The sole ground of appeal is that the judge erred in law in holding that he had jurisdiction to order that a fresh option to purchase the reversion be incorporated in the new lease. In the circumstances I do not find it necessary to refer to the facts of the case in any detail. It is enough to say that the original lease of the premises dated September 1 1972 was made between three brothers, Arthur, Edward and William Johnson, as lessors, and the respondent, Mr J Kirkwood, as lessee and was for a term of a little over five years from March 20 1972 until May 31 1977 at a rent increasing in defined stages from £10 per week to £27.50 per week. The judge was satisfied on the evidence that the respondent by entering into the lease had agreed to take on a derelict filling station and to operate it at an increasing rent on the basis that if he could succeed in that venture he would be entitled to acquire the freehold with some of the rent back, and that he had in fact worked very long hours at a small return in order to make a success of the business. K

The option for the respondent to acquire the freehold was, by the terms of clause 8 of the lease, exercisable by notice to be given not less than three calendar months prior to the termination of the term, but by an error, not that of the lessee, that option was not exercised by the due date. There were certain other options contained in the lease with which we are not concerned, since the respondent did not seek their incorporation in the new lease.

A notice to terminate the lease on November 14 1977 was given by the landlords in May 1977 and on September 6 the respondent made application for a new tenancy which he proposed should be a five-year tenancy at a rate of L £1,820 per annum. On September 26 1977 the lessors gave notice that they did not oppose the grant of a new lease but objected to the rent proposed and to the renewal of the option. The tenant's application came before the deputy judge on March 7 last year, when, having heard the evidence of the parties' valuers and the arguments of counsel, he granted a new lease in the terms already mentioned, including the option to acquire the freehold.

The Landlord and Tenant Act contains the following provisions with regard to the terms that may be included in a new lease under that Act.

Section 32(3) provides: M

> Where the current tenancy includes rights enjoyed by the tenant in connection with the holding, those rights shall be included in a tenancy ordered to be granted under section twenty-nine of this Act.

Section 35 provides:

> The terms of a tenancy granted by order of the court under this Part of this Act (other than terms as to the duration thereof and as to the rent payable thereunder) shall be such as may be agreed between the landlord and the tenant or as, in default of

A such agreement, may be determined by the court; and in determining those terms the court shall have regard to the terms of the current tenancy and to all relevant circumstances.

In his judgment the judge, having referred to these provisions and also to the judgment of Upjohn J (as he then was) in *Re No 1 Albemarle Street* [1959] Ch 531, came to the conclusion that the right here in question could be included in a new lease under section 35 of the Act, but not under section 32(3), since it did not in his view touch and concern the lease itself, and before turning to the rival arguments on this appeal it will be convenient to refer to the *Albemarle Street* case, the facts of which were that, by the terms of
B clause 7 of a 21-year lease of business premises expiring on December 25 1957, the lessor authorised the lessees to maintain during the continuance of the lease certain signs in their existing positions on the outside of a building comprised in the lease. By its terms the tenancy expired on December 25 1957 and the lessors gave notice terminating it on that date but it was continued, by virtue of section 64 of the Act, until the expiration of a period of three months beginning with the date on which the tenant's application was finally disposed of. Upjohn J in his judgment expressed the view that section 32(3) deals with terms connected with the holding and requires rights enjoyed by the tenant in connection with
C the holding to be included in any tenancy granted under section 29, but that section 35 deals with other terms of the tenancy as to which it confers on the court what he described as the "widest possible discretion." In the exercise of that discretion he ordered that the new lease should contain a clause in the same terms as clause 7 of the original lease.

That case is therefore authority for the proposition that where a discretion arises under section 35 it is a very wide one, but the question at issue in the present case is whether the court had power to confer, by way of a new tenancy under Part II of the 1954 Act, a fresh option for the tenant
D to acquire the freehold, and it is common ground that the answer to that question must be found in the terms of the Act itself.

Mr Mallard, for the appellants, submits that the object of Part II is to provide security of tenure for business tenants, and that it is implicit in Part II that the court can only include in a new tenancy such terms as will preserve or protect the tenant's enjoyment of the premises and of the business conducted thereon. Alternatively, he advances a narrower argument that the court has no power to grant a fresh option to acquire the reversion collateral to and out-
E side the relation of landlord and tenant; and additionally he submits that it is surprising, if Parliament contemplated a grant of such option under Part II of the Act, that no guidance is given in the Act as to such matters as title to the reversion, the ascertainment of the price payable for the reversion and for the option itself, or the mode of exercise and conditions attached to the exercise of the option.

For the respondent, Mr Ainger submits that it is sufficient compliance with the objects of the Act that the terms should be consistent with security of tenure, and he points to the circumstances of this case in which the tenant, if he was to prosper, would have to invest capital for which the option
F would be an inducement. He also disputes that this option was a fresh one, and he relies on the very wide words of section 35.

In support of his arguments, Mr Mallard relied on the statement made by Denning LJ (as he then was) in the case of *Gold* v *Brighton Corporation* [1956] 1 WLR 1291 that nothing in the Act enables the court in exercising its jurisdiction under the Act to create a new saleable asset for the tenant and on the judgment of this court, delivered by Stamp LJ in the case of *G Orlik (Meat Products) Ltd* v *Hastings & Thanet Building Society* (1974) 29 P & CR 126, where the tenant sought to have incorporated in a new lease under section 32(3) or section 35 of the 1954 Act a right to G
park two vans in an agreed position partly on the landlord's premises, and the county court judge upheld that claim, but this court on appeal concluded that the tenant had not had any right of parking such as was claimed and allowed the landlord's appeal. Stamp LJ in the judgment of the court observed as follows on p 133:

> The object of Part II of the Act is to give security of tenure to business tenants by, inter alia, conferring power on the court to order a new tenancy of the property comprised in "the holding," that is to say (see s 23(3)), of the property comprised in the existing tenancy excluding any part not used for business purposes, and, however widely expressed, section 35 cannot, in H
> our judgment, consistently with the scheme found in Part II, be construed to enable the court to enlarge the holding, for example, by ordering the grant of an easement over the landlords' land or by conferring rights over the landlords' land not hitherto enjoyed.

That reasoning is binding on this court and, considered in conjunction with the terms of the 1954 Act and the observations to which I have earlier referred of Denning LJ (as he then was) in *Gold* v *Brighton Corporation*, has satisfied me that the judge in the present case had no jurisdiction to insert in the new lease an option to acquire the freehold. It is true that in the present case the original lease had contained such an option but, in my judgment, it is impossible J
to say that the fresh option confers on the tenant nothing that he did not have under the original lease. The effect of the order was to extend the right to acquire the freehold for a further five years. In that respect, in my judgment, it confers on him, in the words of Denning LJ, "a new saleable asset" and also, in the words of Stamp LJ, "enlarged the holding."

For these reasons I would allow this appeal and delete the option from the new lease.

Agreeing, ORMROD LJ said: I agree. This case raises a point under the Landlord and Tenant Act 1954 which has K
not hitherto been considered by this court. The question is whether the court has jurisdiction under Part II of that Act to include in an order for a new lease an option for the tenant to purchase the freehold. The fact that this point has not arisen in the 25 years since the Act was passed suggests at least the answer to the question.

Mr Mallard submits that there is no such power in the Act. He says quite simply that such a term was outside the contemplation of Parliament altogether, since the purpose of the Act was to provide security of tenure for business tenants, not to enable them to purchase the reversion of their premises or to obtain additional rights to those which they L
had before the order was made. The force of this submission no doubt explains the absence of any previous decision of this court upon the point.

There is no express provision in the Act relating to options to purchase, so that the power, if any, to create such an option must be found by some process of implication, but, in my judgment, there is no basis for such implication. It is common ground that such power can only be derived either from section 32(3) and section 35. It is unnecessary to decide the precise meaning of the former subsection because, on any view, the current tenancy in this case did not include a right to exercise an option to purchase M
the reversion, such right having lapsed three months before the term expired on May 31 1977. Nor could the tenant be said to be enjoying such a right at the time the matter came before the court. In any event, therefore, the option is not preserved by section 24(1). What the position might have been had there been no time-limit on the option is a question which need not be decided in this case. It is enough to say that the reasoning of Cross J in *Longmuir* v *Kew* [1960] 1 WLR 862, to which Mr Ainger called our attention, by analogy may not be very helpful to a tenant in such a position as the tenant in this case.

A Mr Ainger is therefore left with section 35. To succeed on this appeal he, as he concedes, is forced to argue that the discretion given to the court by this section is wide enough to enable the court to create a new option to purchase where none had previously existed. That is a proposition, in my judgment, which has only to be stated to be rejected. So far-reaching an interference with a landlord's property rights would undoubtedly require express statutory provision. In support of his argument, Mr Ainger relied on the decision of Upjohn J in *Re No 1 Albemarle Street* [1959] Ch 531. In my judgment, however, that decision was based on the fact, or at least it was an important factor in the decision, B that the tenants were still enjoying the right in question, that is, to display certain signs on adjacent property belonging to the landlord at the date of the application to the court. This court in *G Orlik (Meat Products) Ltd v Hastings & Thanet Building Society* [1974] 29 P & CR 126 seems to have so regarded it. Moreover, the judgment in that case, and in *Gold v Brighton Corporation* [1956] 1 WLR 1291, indicates that the purpose of the Landlord and Tenant Act is to protect the tenant in respect of his business and not to put a new saleable asset in his hands.

I therefore agree that this appeal should be allowed.

Also agreeing, GEOFFREY LANE LJ said: It is not in dis- C pute that in the lease of September 1 1972, by clause 8 thereof, the landlords gave to the tenant an option to purchase the holding subject to certain conditions. Likewise, it is not in dispute that that option was never exercised and accordingly lapsed, no notice having been given to the landlords (as the clause required) not less than three calendar months prior to the termination of the terms. So, by March 1977, no option remained alive. Does section 32(3) of the 1954 Act apply in these circumstances? I think not. The current tenancy at the date of the hearing did not include rights enjoyed by the tenant in connection with the holding. The court was being asked to create or re-create a fresh right D in favour of the tenant when none at that time in fact existed. The decision in *G Orlik (Meat Products) Ltd v Hastings & Thanet Building Society* [1974] 29 P & CR 126 is authority binding on us that that is not permissible.

The other possibility advanced by the tenant, namely, section 35, is in my view equally of no avail to him. The decision in the *Orlik* case makes that clear, as the passage already cited by Orr LJ makes apparent. I do not accept Mr Ainger's submission that *Orlik* was decided *per incuriam* on the basis that the court had overlooked the effect of section 62 of the Law of Property Act 1925 on section 32(3). E I agree that the judge's order should be varied accordingly as my Lords have indicated.

The appeal was allowed with costs. The judge's order to be varied by deletion from the new lease of the option to purchase the freehold.

COURT OF APPEAL

June 14 1978

(Before Lord Justice BUCKLEY, Lord Justice ROSKILL and Lord Justice GOFF)

GERMAX SECURITIES LTD v SPIEGAL

F

Estates Gazette May 5 1979

(1978) 250 EG 449

Notice to determine business tenancy—Chapter of errors—Mistake as to date at which notice purported to have been given, not as to termination date—Whether mistake affected validity of notice—When notice read with covering letter it would not have misled any reasonable tenant—Tests generally applicable to slips as to dates in notices to quit—Notice held to be valid

This was an appeal from a decision of Judge Dewar at Bloomsbury County Court holding that a notice given by Germax Securities Ltd, the respondents to the present appeal, G was a valid notice to terminate the tenancy of business premises occupied by Armin Spiegal, the present appellant.

R Yajnik (instructed by Paul Davis & Co) appeared on behalf of the appellant; D M W Barnes (instructed by Paisner & Co) represented the respondents.

Giving judgment, BUCKLEY LJ said: This is an appeal from a decision of His Honour Judge Dewar, given at the Bloomsbury County Court on October 7 1977, when he declared that a notice to quit given by the plaintiffs to the defendant, Armin Spiegal, was a valid notice.

The tenancy under which the notice was given was created by a lease of March 19 1971; it was a lease of business premises H for a term of seven years from June 24 1970, therefore terminating on June 24 1977. The landlords wished to bring the tenancy to an end at the termination of the leasehold term, and on August 24 1976 they served a notice purporting to determine the lease on February 27 1977. The point was taken by the tenant's solicitors that the notice was invalid because it was a notice to determine the lease at the wrong date, and that position was accepted by the landlords' solicitors who, on November 25 1976, served by post another notice determining the tenancy at June 24 1977, the correct date.

The draft of that notice to determine the tenancy was in the appropriate form prescribed by the regulations under the Landlord and Tenant Act 1954 and the draft is dated November 25 J 1977. But, by a most unfortunate and careless slip, whoever copied that notice in the offices of the landlords' solicitors for the purpose of preparing the actual document which was sent to the tenant, dated the notice March 25 1977 instead of November 25 1976. As Goff LJ now points out, the draft also is wrong because the date at the foot of the draft is November 25 1977 instead of 1976. So there was not the desired degree of accuracy in the matter, but it must have been clear to the tenant from the covering letter that the notice was intended to be a notice served as on November 25 1976, or on receipt of the notice by the tenant in the due course of post, K and not on March 25 1977, the date which it bears at its foot.

It is a notice in the prescribed form, giving notice terminating the tenancy on June 24 1977; para 2 of the notice requires the tenant within two months after the giving of the notice to notify the landlords' solicitors in writing whether or not the tenant was willing to give up possession of the premises.

The learned judge took the view that this was an obvious slip and that it would have been clear to the tenant precisely what the landlord was intending to do, that is to say, serve notice in November 1976 and not to serve notice which was to take effect only at a later date; accordingly he held that the slip in the date at the foot of the notice was of no significance L and that the notice was a valid notice.

Mr Yajnik, who I think has said whatever can be said in support of this appeal, has referred us to the case of *Hankey v Clavering*, which is reported in [1942] 2 KB 326. That was a case, decided in this court, where a mistake had been made in specifying the date at which the landlord intended to determine the tenancy. In that case there was a lease for 21 years from December 25 1934, terminable at the end of seven years on six months' notice. The landlord gave the tenants a notice which purported to determine the lease on December 21 1941, but that was a date on which the landlord could not have M determined the lease under its terms, though he might have determined it at December 25 1941. In that case it was held that although the mistake as to the date was obviously due to a slip on the part of the landlord, the notice was invalid because the landlord was attempting to do something which under the lease he had no power to do.

In *Carradine Properties Ltd v Aslam*, reported in [1976] 1 WLR p 442, Goulding J had to consider a case of a lease for 21 years from September 27 1968 which provided for its earlier determination by either party at the end of the first seven or 14 years of the term on 12 months' previous notice in writing. The landlords gave a notice dated September 6 1974, stating

A that they intended to determine the term created by the lease on September 27 1973, a date which was actually before the date of the service of the notice. Goulding J held that there it would have been clear to any reasonable tenant that there had been a mistake and that the intention of the landlords was to determine the tenancy at September 27 1975 and he held that the notice was a good notice notwithstanding the slip. In that case no reasonable tenant could possibly have supposed that the date in the notice, September 27 1973, was the date correctly intended by the landlord; it clearly was a mistake which must have been apparent, to anyone who knew the facts, as a mistake.

B The learned judge formulated the test in this way at p 444 at the letter G: "I would put the test generally applicable as being this: 'Is the notice quite clear to a reasonable tenant reading it? Is it plain that he cannot be misled by it?'. Applying that test, if applicable, to the present case," the learned judge went on to say, "I think the notice would be saved because the tenant receiving that notice and knowing the terms of the lease must have seen there was a mistake, as it would not say '1973' in 1974. Once that is accepted, it is obvious that the notice is for 1975 and not 1973."

Mr Yajnik has submitted that in the present case the notice served by the landlords is open to the construction that it was
C intended by the landlords to take effect only from March 25 1977; but, reading the notice with the covering letter, I do not think it can be said that any reasonable tenant would have been misled here. The mistake in this case was not in the operative part of the notice; it is merely a mistake in relation to the date at which the notice purports to have been given by the landlord. That that was a mistake is clear from association of the notice with the covering letter.

In my judgment there is really no ground in the present case for saying that the tenant receiving the notice would be liable to be misled, and for the reasons given by the learned judge in the court below I am of the opinion that the notice took
D effect as a valid notice, and I would dismiss this appeal.

Roskill LJ agreed.

Agreeing, Goff LJ said: I would add, if I may, just one small point on the facts. As my Lord has observed, the date specified in the notice as that on which the tenancy was to determine was the date upon which, according to the contractual terms, it would expire by effluxion of time. It was necessary, however, to serve a notice in order to secure that it would determine at that date because the tenancy fell within Part II of the Landlord and Tenant Act, and, therefore, must automatically continue beyond the contractual date by virtue of section 24, unless determined in accordance with the pro-
E visions of Part II of that Act, which brings in section 25, and that section gives the landlord power, by a proper notice, to determine on the contractual date.

The appeal was dismissed with costs.

COURT OF APPEAL

February 14 1979

(Before Lord Justice BUCKLEY, Lord Justice SHAW and Lord Justice BRANDON)

LAND RECLAMATION COMPANY LTD v BASILDON DISTRICT COUNCIL

F

Estates Gazette May 12 1979

(1979) 250 EG 549

Landlord and Tenant Act 1954—Whether provisions in Part II of the Act for the grant of a new tenancy applied to the demise for a seven-year term of a right of way to the appellants (a company engaged in waste disposal) over land belonging to the respondent council—Whether an incorporeal hereditament, such as an easement, can be "occupied" within the meaning of section 23(1) of the Act—Review of earlier authorities—Meaning of "premises"—Held, upholding decision of Brightman J, that the right of way, which stood by itself and was not the subject of a more compre- G hensive demise including a corporeal hereditament, was not within the provisions of Part II of the Act for the grant of a new tenancy

This was an appeal from a decision of Brightman J (1978) 246 EG 401 on a preliminary point of law. The issue was whether a demise to the appellants, the Land Reclamation Co Ltd, by the respondents, Basildon District Council, of a right of way over land in the ownership of the council was a tenancy to which Part II of the Landlord and Tenant Act 1954 applied. Brightman J held that it was not.

Anthony Scrivener QC, H A P Picarda and D A Lowe (instructed by Ellison & Co, of Colchester) appeared on H
behalf of the appellants; Ronald Bernstein QC and C B Priday (instructed by J L Knight, Basildon District Council Offices) represented the respondents.

Giving judgment, Buckley LJ said: This is an appeal from a judgment of Brightman J who, on a preliminary issue in an application for a new tenancy under the Landlord and Tenant Act 1954, section 26, declared that the tenancy created by a lease dated August 14 1970 between the respondent council as lessor and the appellant company as lessee is not a tenancy to which Part II of that Act applies. The respondents are the successors of the former Basildon Urban District Council, but I shall describe those two local authorities J
as "the respondents" indifferently.

The appellants carry on a business of disposing of waste, both domestic and industrial, on a large area of land exceeding 1,000 acres at Pitsea in Essex. The respondents are owners of other land adjoining the applicants' land, and known as the Sea Transport Depot; the respondents acquired that land from the Ministry of Defence not long before the date of the lease in question. The only existing vehicular access to the appellants' land is along a private road of some considerable length running over the respondents' land from north to south, from a point marked D on an exhibited plan to a point marked A upon it. When the respondents' land was K
in the ownership of the Ministry of Defence, the appellants were licensed to use this road for the purposes of their business.

On August 14 1970 the respondents made the following grant of a right of way over the road; the document is called a lease and I shall refer to it as "the lease." It is dated August 14 1970; it is made between the respondents of the one part and the appellants of the other part; it recites that the respondents are the owners in fee simple of the land at Pitsea known as the Sea Transport Depot, which includes the road coloured red on the plan attached to the lease, and that the company (that is, the appellants) were L
estate owner in fee simple of land to the south-east of the extreme point of the road, which provided the sole means of vehicular access to and from the highway.

The operative part of the document is in the following terms:

Now this deed witnesseth as follows:—1. In consideration of the rent and covenants hereinafter reserved and contained the Lessor hereby demises unto the Company full right and liberty for them and their assigns their agents servants or licensees to pass and repass with or without vehicles along the private road in common with the Lessor its servants agents tenants and licensees and others having a like right and liberty between the hours of 6 am M
to 8 pm Mondays to Fridays and 6 am to 2.30 pm Saturdays (hereinafter called "the permitted hours") for all purposes in connection with the use of the Company's land for the dumping and disposal of waste materials Subject nevertheless to and reserving unto the Lessor the right at all times hereafter or at any times or time to erect renew and maintain a gate or gates across the said roadway at the points marked "G" on the said plan with all necessary fittings and fixtures but so that the same shall not be locked or so erected or maintained as to impede or obstruct the free use and enjoyment of the right and liberty of the way hereby granted in accordance with the tenor hereof to hold the same unto the Company from the 25th March 1970 for the term of seven years paying therefor unto the Lessor during the said term

the yearly rent of ONE THOUSAND POUNDS (£1,000) by equal quarterly payments in advance on the usual quarter days in every year without any deduction the first of such quarterly payments to be made on the signing hereof for the two quarters ending 28th September 1970.

Then clause 2 contains lessees' covenants, which include a covenant to make up part of the road between the points marked A and B on the plan at the cost of the lessee and to the satisfaction of the lessor's surveyor, and to pay to the lessor the cost to the lessor of putting another part of the road, between the points marked B and C into proper state of repair, and to pay to the lessor three-quarters of the cost to the lessor of maintaining during the first year of the term granted, that part of the road between the points marked B and C, and thereafter to pay in like manner a fair proportion according to user of the lessor's cost of maintaining during the remainder of the term of that part of the road. There is a covenant by the lessee not to use the private road outside the permitted hours without the previous consent of the lessor, and a covenant by the lessee to use its best endeavours to prevent vehicles which call at the company's land from parking on any part of the private road or other part of the land belonging to the lessor at the Sea Transport Depot; and there is a covenant for peaceable enjoyment of the right given by the lessor. I need not read any more of that document.

The use of the road in question by the appellants is, and has always been, entirely for business purposes. It has been a heavy use; we are told that the appellants' lorries go down the road at the rate, on an average, of one every four minutes during the permitted hours. There is no alternative access to the appellants' land readily available; they do own other land adjoining the land on which they carry out their tipping operations, but that land is land which is scheduled as being of special scientific interest. They have made an application for planning permission to construct a road across that other land; that application has been refused by the local planning authority and at the present time an appeal to the minister from that refusal is pending. Industrial waste can only be brought to the appellants' land where they carry out their tipping operations by use of the road; it seems, however, that domestic waste can be brought there by water.

On April 20 1976 the appellants applied, purportedly under section 26 of the Landlord and Tenant Act 1954, for a new tenancy of the right of way; on June 14 1976 the respondents served a counternotice under the Act giving as their grounds for refusing a new tenancy: "(i) That on the termination of the current tenancy the Council or landlord intends to carry out substantial work on the holding or part thereof and that the Council as landlord could not reasonably do so without obtaining possession of the holding; (ii) That on the termination of the tenancy the Council as landlord intend to occupy the holding for the purposes or partly for the purposes of a business to be carried on by them therein." That notice was served without prejudice to a contention by the respondents that the lease was not within the Act.

On the application of the respondents the proceedings were transferred from the county court into the Chancery Division, and the following question was in due course directed to be tried as a preliminary issue, namely, ". . . whether the tenancy created by the lease made the 14th August 1970 is a tenancy to which Part II of the Landlord and Tenant Act 1954 applies. . . ." That matter came before Brightman J in December 1977. The respondents first submitted that the legal relationship between the respondents and the appellants was that of licensor and licensee and not that of landlord and tenant. The judge rejected that argument, saying at H on p 1165 of *Land Reclamation Co Ltd v Basildon District Council* [1978] 2 All ER 1162, where his decision is reported: ". . . it seems to me clear that the local authority granted to the company a legal easement for a term of years and not a mere licence." There is no appeal from that decision, and although I would not wish to be taken to decide that on its true interpretation the lease amounted to a grant of an easement for an interest equivalent to a term of years absolute, I proceed upon the footing, for the purposes of this judgment, that that is its effect.

The judge went on to consider whether upon that basis the lease came within the terms of the operation of the Act. He said: "In ordinary speech, a right of way is not a possible subject matter of occupation as distinct from user, nor is it easy in normal circumstances to say that a road is occupied by a person who has a mere right of way thereover." He went on to discuss certain provisions of the Act and certain authorities, and reached the conclusion, at p 1167 of the report, at the end of the judgment: "In my judgment, the company uses the roadway for the purposes of its business but it does not occupy either the easement or the roadway. I take the view that Part II of the 1954 Act has no application to a mere right of way standing by itself, because such a right is not property or premises capable of being occupied for the purposes of a business, or indeed for any other purpose," and he accordingly made the declaration appealed against.

Part II of the Landlord and Tenant Act 1954 is the part of the Act which deals with security of tenure for business, professional and other tenants; section 23(1) provides that: "Subject to the provisions of this Act, this Part of this Act applies to any tenancy where the property comprised in the tenancy is or includes premises which are occupied by the tenant and are so occupied for the purposes of a business carried on by him or for those and other purposes." Mr Scrivener has contended that the Act should be construed as though "use" and "occupy" were interchangeable terms. The appellants undoubtedly made use of the right of way; therefore, he says, they occupied it for the purposes of the Act. He points out that in the Landlord and Tenant Act 1927, section 17, the holdings to which Part I of that Act (which relates to business premises) applies, are premises "held under a lease . . . and *used* wholly or partly for carrying on thereat any trade or business." He suggests that those words are apt to extend to incorporeal rights used for carrying on a trade or business and that it would be unlikely that the 1954 Act was intended to be more restricted in its operation.

In my judgment that argument misconstrues the 1927 Act. Part I of that Act deals with compensation for improvements and goodwill on the termination of tenancies of business premises. Section 4 deals with compensation for goodwill: the tenant is entitled to compensation "if he proves to the satisfaction of the tribunal that by reason of the carrying on by him or his predecessors in title at the premises of a trade or business for a period of not less than five years, goodwill has become attached to the premises" and so forth. It is clear in my judgment from the use of the words "at the premises" and "attached to the premises" that the Act here relates to physical property—land or buildings—and not to incorporeal rights. Section 5 refers to the tenant removing to or carrying on his trade or business "in other premises" and to the grant of a new lease "of the premises at which the trade or business is carried on." The same observation applies there. And in section 17 itself, to which I have already referred, the use of the word "thereat" has in my opinion a similar significance.

These sections were considered in *Stumbles v Whitley* [1930] 1 KB 393, in this court. In that case an hotel had been let under a lease which also included the fishing rights in a water known as the Lower Ley, and the property demised was described as

"all that messuage or tenement and premises known as the Royal Sands Hotel, Slapton, aforesaid, . . . together with the exclusive right at all times between March 3 and November 1 in each year of the tenancy (such period being hereinafter termed the summer fishing) of fishing the water known as the Lower Ley from Slapton Bridge," within specified limits, "and, secondly, the exclusive right of fishing at all times . . . between November 2 and March 2 in

A each year of the tenancy (such period being termed the winter fishing)" within certain limits "... to hold unto the tenants for the term of seven years from March 25 1922, . . . yielding and paying during the said term in respect of the first hereinbefore described hereditaments the yearly rent of £110 . . . and in respect of the fishing rights secondly hereinbefore described the yearly rent of £55."

The plaintiff, as the tenant under that lease, required the grant of a new lease under section 5 of the 1927 Act. The question for decision was whether the new lease should include the fishing rights and it turned upon the meaning of the word "premises" in section 17. Scrutton LJ stated the B landlord's contention at p 399 in this way: "The argument addressed to us is that the Act applies only to 'premises' held under the lease; that fishing rights are not 'premises,' as the right to extract fish from Slapton Ley is not a building; that 'premises' are buildings or structures. Consequently, it is said, the tenant is only entitled to obtain a new lease of the buildings or structures, and not of the mere right to take something from the soil." He held on construction that contrary to that argument, "premises" meant the subject matter of the lease and that accordingly the Act extended to the whole subject-matter comprised in it, both the hotel and the fishing rights. He also relied on an alternative ground of decision at the end of his judgment, where he said: "It may be put C in another way which will limit it to this case probably. It may be said that the subject matter of this lease is the hotel, but it is the hotel with certain rights attached to it, and that the new grant should be of the same subject matter—namely, the hotel with those rights." That construction also concords with the view I take in considering the purposes of the Act.

Greer LJ decided the case on Scrutton LJ's alternative ground. He said, at p 401: "... but I gravely doubt whether the term 'premises' would cover a separate and distinct lease of an incorporeal hereditament"; and at the end of his judgment he said that he desired to reserve his opinion on D the question which might possibly arise thereafter, namely, whether a bare lease of fishing rights came within the Act.

Slesser LJ decided the case on the ground that the word "premises" should be given what he called its ordinary meaning, by which he evidently meant that it referred to the parcels comprised in the lease, but that even if it should be given the narrower construction, meaning the land and erections upon the land, the incorporeal rights should not be excluded from the new lease. On these grounds the court affirmed the decision of the Divisional Court to the effect that the fishing rights formed part of the premises demised and therefore that the fishing rights should be included in the E new lease.

The case went to the House of Lords and is reported in [1930] AC 544. All the other learned Lords agreed with the speech of Viscount Hailsham. After reviewing certain sections of the Act, Lord Hailsham said at p 547:

> It is conceded—indeed it must be conceded—that the word "premises" does not mean only buildings; it means also at least the land on which the buildings are erected and the land immediately surrounding the buildings, and yet the expression "pull down or remodel the premises" would be wholly inept for such a purpose. It was conceded in argument also that it must include some incorporeal hereditaments such as, for instance, easements. When F one gets that concession, which I think was quite properly and necessarily made, then it is manifest that although the word "premises" is being used in a narrow sense to this extent that the Legislature is at times contemplating rather the buildings in which the trade is carried on than the whole of the subject-matter of the lease, yet it does not intend to exclude other things which are properly described as premises in the strict legal sense when it is appropriate that they should be included. When we look at section 17, the definition section, and we find the reference to "any premises held under a lease," I see no sufficient reason for supposing that the Legislature did not there include not merely the actual buildings in which a trade is carried on, but also the land surrounding them, the easements granted as appurtenant to them, and any other incorporeal hereditaments which may form part of G the premises in the strict legal sense of the term which are the subject-matter of the habendum. Any other construction would, it seems to me, defeat the plain purpose of the Act, which obviously was to provide that in the circumstances defined in the Act the tenant should have a right to continue to carry on his trade or business in the premises in the legal sense in which he was carrying them on under the lease for which he seeks that renewal.

So the case was decided on the ground that "premises" in that context signified what Lord Hailsham called the subject-matter of the habendum, that is to say, the parcels to which the habendum applies as a whole. He went on, however, to say: H

> . . . I desire to add that, while that is sufficient to decide this case, I do not wish to be understood as accepting the view that if the profit à prendre here included had not formed part of the premises within section 17, the referee or the county court judge would have had power to compel the appellant to grant a lease of those rights as a condition of the grant of a new lease, which I understand to be the second ground upon which the judgment of the Court of Appeal is based. It is sufficient for me to say that, in my judgment, the first ground on which Lord Justice Scrutton relies disposes of the case, and that I agree entirely with the conclusion which he has reached on that point.

He was there, I think, reserving his position on the point upon which Lord Justice Greer had also voiced reservations; J that is, whether, if the incorporeal hereditament had been created independently of the corporeal hereditament comprised in the lease, the former would have been within the Act.

Lord Denning MR has by way of *obiter dictum* expressed a similar doubt in relation to the 1954 Act in the case of *Jones v Christy* (1963) 107 SJ 374. The case was decided on April 30 1963, and is also to be found in transcript no 119 of the collection of Court of Appeal judgments for that year. At p 4 of the transcript Lord Denning said this: "If the letting were solely of an incorporeal hereditament, such as the right to fish, I doubt whether the Act would apply to it. When the Act speaks of 'premises' being 'occupied' it may K well refer to corporeal hereditaments and not to incorporeal hereditaments."

I return to consider the terms of the 1954 Act, accepting that the expression "premises" should be construed as meaning all the property demised unless the context otherwise requires. Part II of that Act applies under the terms of section 23, "... where the property comprised in the tenancy is or includes premises which are occupied by the tenant and are so occupied for the purposes of a business carried on by him or for those and other purposes." So it will apply when part only of the whole property comprised in the lease is occupied by the tenant for business purposes. This may occur where the tenant uses L some part of the property comprised in the lease for purposes other than business purposes; or it might occur where some part of the property comprised in the lease was, by its incorporeal nature, incapable of occupation.

Section 32 specifies what property shall be comprised in a new lease where an order is made for the grant of a new tenancy. Subsection (1) of that section, as amended, now reads:

> Subject to the following provisions of this section, an order under section twenty-nine of this Act for the grant of a new tenancy shall be an order for the grant of a new tenancy of the holding; and in the absence of agreement between the landlord and the M tenant as to the property which constitutes the holding the court shall in the order designate that property by reference to the circumstances existing at the date of the order.

To discover what is meant by "the holding," one must go to section 23 (3):

> In the following provisions of this Part of this Act the expression "the holding," in relation to a tenancy to which this Part of this Act applies, means the property comprised in the tenancy, there being excluded any part thereof which is occupied neither by the tenant nor by a person employed by the tenant and so employed for the purposes of a business by reason of which the tenancy is one to which this Part of this Act applies.

A The exclusion in that subsection might operate because some part of the property was in the occupation of somebody other than the tenant or an employee of his employed in the relevant business; or it might operate because some part of the property was by its incorporeal character incapable of occupation. This would not result in the incorporeal rights enjoyed by the tenant in connection with the holding under his current tenancy being omitted from any new tenancy granted under the Act, for section 32 (3) provides that

> Where the current tenancy includes rights enjoyed by the tenant in connection with the holding, those rights shall be included in a tenancy ordered to be granted under section twenty-nine of this
> B Act (except as otherwise agreed between the landlord and the tenant or, in default of such agreement, determined by the court.

That subsection is couched in language equally appropriate to a case in which the rights in question were created by the document creating the tenancy as to a case in which they arose by other means. The word "occupy" is not one, in my opinion, which according to its normal meaning in the English language is apt to describe the enjoyment of some incorporeal right such as an easement of way. It may possibly be more suited to describe the enjoyment of some proprietary right such as a riparian owner enjoys in relation to a
C river which flows past, or through, his property.

In *Brown v Peto* [1900] 2 QB 653, at p 664, Vaughan Williams LJ said: "'Occupy' is a word which in one form or another is not infrequently used of an incorporeal hereditament. Eviction, which presupposes occupation, is a term employed in respect of tithe." He was then considering whether a lease which included a house and land and also sporting rights over other land was an "occupation lease" within the Conveyancing Act 1881, section 18; he was considering an incorporeal right, but not an easement. Nor, of course, are tithes analogous to an easement. In my judgment "occupation" is not an appropriate word to describe the enjoyment of a right of way unless the context demands such a con-
D struction. The provisions which I have so far discussed do not appear to me to do so.

It may be that if the right of way were an exclusive right, the circumstances might result in the exercise of the easement amounting to occupation of the land over which it passes (see *Holywell Union and Halkyn Parish v Halkyn District Mines Drainage Co* [1895] AC 117), but that is not this case. The right of way in the present case is not exclusive; it is not even continuous, being limited to certain hours of certain days.

Other provisions of the Act appear to me to be more consistent with the view that an incorporeal right is not such a
E right as can be "occupied" within the meaning of the Act. Section 63 (2) provides as follows: "Any jurisdiction conferred on the court by any provision of Part II of this Act . . . shall, subject to the provisions of this section, be exercised, (a) where the rateable value of the holding is not over the county court limit by the county court; (b) where it is over the county court limit by the High Court."

An incorporeal hereditament is not rateable *per se*: see *Ryde on Rating*, 13th ed, p 153, where the position is stated thus: "Under the General Rate Act 1967 the person made liable is the 'occupier of lands, houses' etc. Incorporeal rights such as easements, tolls and rights of common, are not, *per se*, rateable. Such rights only become rateable if the exercise
F and enjoyment of them necessarily involves the exclusive occupation of land, or if, as in the case of sporting rights and advertising rights, they are made rateable by the statute." Only if the exercise of the right amounts to the rateable occupation of some physical hereditament can anyone be rateable in respect of it, and he would then be rateable in respect of the physical hereditament and not of the easement or right.

Section 37 of the Act deals with the position when the grant of a new tenancy is inappropriate, and with what compensation should then be paid to the tenant. Subsection (2) provides as follows: "The said amount"—that is, the amount of the compensation—"shall be as follows, that is to say—(a) where the conditions specified in the next following sub-
section are satisfied it shall be twice the rateable value of the holding, (b) in any other case it shall be the rateable value of the holding." Subsection (5) provides: "For the purposes of subsection (2) of this section the rateable value of the holding shall be determined as follows," and then paragraph (a) deals with the position when there is a value entered in the valuation list; paragraph (b) deals with the position when no value is so entered in respect of the holding itself, and provides: "Where no such value is so shown with respect to the holding but such a value or such values is or are so shown with respect to premises comprised in or comprising the holding or part of it, the rateable value of the holding shall be taken to be such H
value as is found by a proper apportionment or aggregation of the value or values so shown." Paragraph (c) is in these terms: "Where the rateable value of the holding cannot be ascertained in accordance with the foregoing paragraphs of this subsection, it shall be taken to be the value which, apart from any exemption from assessment to rates, would on a proper assessment be the value to be entered on the said valuation list as the annual value of the holding." That implies, in my view, that apart from any exemption the hereditament would have an assessable rateable value proper to be entered in the valuation list, but an easement can have no such value.

J There are a number of places in the Act where the expression "use or occupation" is used. Examples will be found in section 57 (1), section 58, section 60; and in section 41 there is a reference to "use, occupation or enjoyment." These might not be incapable of being read as though the two words "use" and "occupation" were interchangeable if the context of the rest of the Act required this; but otherwise in my opinion they point in the contrary direction.

We were pressed by Mr Scrivener with the decision in this court in *Lee-Verhulst (Investments) Ltd v Harwood Trust* [1973] 1 QB 204. That case was concerned with a building which had been adapted for occupation as furnished apartments; it was
owned by a company, and a gentleman who was, I think, the K
sole director of the company, lived on the premises in the basement and supervised all the activities of the business in the house. The apartments were let and services were provided by the plaintiff company to the tenants, which services included the provision of light meals if required, the services of maids to clean the rooms and so forth. The resident director exercised a considerable degree of supervision over the activities of tenants and the way in which the business was conducted in the house, and the question was whether the plaintiff company could be regarded as being in occupation of the house for the purpose of carrying on its business or whether, having regard
to the undoubted fact that the apartments were occupied by L
the tenants for residential purposes, that was inconsistent with the view that the house was in the occupation of the company.

Sachs LJ, at p 213, said this:

> Thus one comes to the critical issue—were the premises "occupied" by the tenant for the purpose of that business?

—by "the tenant" he is referring to the company.

> The tenant company must of course be taken, when considering this question, as being represented on the premises by Mr Lee and the staff he engaged: the position in practice being no different than if Mr Lee had himself been the tenant. By Mr Lee and the
> staff the tenant company were present in the premises for the pur- M
> pose of the business day and night; in the course of their services to the occupants they pervaded every room there; control was exercised by Mr Lee over the manner in which the occupancies were conducted—a control (eg by limitations over the cooking that was permitted and over who could stay in each apartment) of a degree much beyond that usual when a flat is let to a tenant on a normal lease: and in addition the tenant company's furniture was in every room.
>
> Is there anything in the Act of 1954 which precludes the court from giving to the word "occupied" in section 23 its natural and ordinary meaning in the context of the subject matter of that Act —a meaning which would in the set of circumstances above des-

cribed clearly lead to it being held that the tenant did occupy the premises for the purpose of the business? Being unable to find anything in the Act which so precludes the court, I have come to the conclusion that this tenant did so occupy the whole of the premises.

On p 214 the learned Lord Justice referred to the question of whether it was right to adopt a broad approach to the word "occupy;" he said, just below E: "It can, however, be said with confidence that such an approach is consistent with the above-noted changes in the phraseology. It is, moreover, the tendency in recent statutes to use 'occupied' with the express intention of avoiding technical distinctions between leases and licences."

On p 215, just above D, Sachs LJ said:

Having come to that conclusion it seems none the less as well to say that despite the fact that "rent books" were in use, the occupants were in my view licensees entitled to exclusive residential occupation of their apartments—rather than subtenants. In case, however, that is not correct and they were subtenants and (again contrary to my view) it became thus necessary to turn the decision here on whether the tenants and occupants technically shared the occupation of the apartments, I would hold there was here a shared occupation—the latter having exclusive residential occupation, the former having such occupation as was required for purposes of the business.

Stamp LJ (p 217 at D) said: "As Sachs LJ has pointed out and as a glance at *Stroud's Judicial Dictionary* will show, the words 'occupation' and 'occupier' are not words of art having an ascertained legal meaning applicable, or *prima facie* applicable, wherever you find them in a statute, but take their colour from the context of the statute in which they are found."

The facts of that case were very far from the facts of the present case. There is nothing there, I think, to support the contention that such violence should be done to the normal meaning of the word "occupy" as I think is necessary to make it applicable to a right-of-way. The problem to which the court there had to address its mind was whether the company's activities were such as to amount to occupation for the purposes of its business, and whether such occupation was negatived by the concurrent occupation of the apartments by the subtenants for residential purposes, and I do not think, in considering that problem, the learned Lord Justices had at all in mind the kind of problem with which we are confronted in the present case.

Finally Mr Scrivener contended that enjoyment of the right of way could result in occupation of the road over which the right of way exists, within the meaning of "occupied" in section 23 (1) of the Act. An occupier of land may be rateable in respect of that land, his occupation being the result of the exercise of some incorporeal right over or in respect of the land, as was decided in the *Holywell* case to which I referred earlier. But the incorporeal right itself is not rateable. To make the occupier rateable at all, his occupation must, I think, amount to exclusive occupation of some land, as was held to be the case in the *Holywell* decision; and see also *Kittow v Liskeard Union* (1874) LR 10 QB 7; and *Westminster City Council v Southern Railway* [1936] 2 All ER 322. Whether the so-called occupier's occupation is sufficient to make him rateable must be a question of fact and degree. It has not been established in the present case that this intermittent and non-exclusive right of way has been enjoyed in a manner amounting to exclusive occupation of the road. But even if this were so, it would not, I think, avail the appellants. The road is not part of the property comprised in the lease; it is not within the premises which are referred to in section 23 (1). So whether the road is or is not "occupied" by the appellants is irrelevant to the subsection. Mr Scrivener was, I think, eventually constrained to accept that this was so.

For these reasons and in spite of a valiant argument presented to us by Mr Scrivener, I am satisfied that the learned judge reached the right conclusion, and I would dismiss this appeal.

SHAW LJ delivered a concurring judgment and BRANDON LJ agreed.

The appeal was dismissed with costs. Leave to appeal to the House of Lords was refused.

—

COURT OF APPEAL

February 23 1979

(Before Lord Justice STEPHENSON, Lord Justice GOFF and Lord Justice BRANDON)

FAWKE v VISCOUNT CHELSEA

Estates Gazette June 2 1979

(1979) 250 EG 855

Landlord and Tenant Act 1954—Questions as to determination of an interim rent under section 24A in a case where the demised premises were in a serious state of disrepair —Works required to eradicate extensive dry rot—Construction of section 24A—English Exporters (London) Ltd v Eldonwall Ltd preferred to Regis Property Co Ltd v Lewis & Peat Ltd, although no actual decision required on this construction point—Main issue effect of disrepair on interim rent—County court judge in error in holding that disrepair was not to be taken into account—Whether court has power under section 24A to fix a differential rent, eg a rent not to come into operation until works of repair are carried out, or to be increased on completion of works or otherwise takes account of state of repair—Held that court has such power, Goff LJ expressing doubt but not dissenting—No doubts in relation to determination of new rent under section 34—Some observations of importance to valuers in such cases—Appeal allowed

This was an appeal from the determination of an interim rent under section 24A of the Landlord and Tenant Act 1954 by Judge Corcoran at the West London County Court. The appellant, Leslie A Fawke, a solicitor, had a tenancy of offices at 27 Sloane Square, London SW1. The respondent, Viscount Chelsea, was the freeholder. The respondent applied under section 24A for the determination of the interim rent following an application by the appellant for a new tenancy. The judge determined an interim rent of £3,450 per annum and the appellant appealed to the Court of Appeal. The judge had also fixed a rent of £3,950 for the new lease and the appellant sought to challenge this determination likewise in his notice of motion.

Patrick Ground (instructed by Leslie A Fawke) appeared on behalf of the appellant; W Poulton (instructed by Lee & Pembertons) represented the respondent.

Brandon LJ, a member of the court who heard and decided the appeal, was absent when the judgments were delivered.

Giving the first judgment at the invitation of Stephenson LJ, GOFF LJ said: This is an appeal from a judgment given on March 15 1978 by Judge Corcoran in the West London County Court, whereby he determined an interim rent under section 24A of the Landlord and Tenant Act 1954 and the rent and terms for a new lease of offices on the first and second floors of 27 Sloane Square, London SW1.

The applicant for a new tenancy under Part II of the Landlord and Tenant Act 1954 was the tenant, a Mr Fawke, who is a solicitor and carries on his practice at the demised premises. The respondent landlord, Viscount Chelsea, did not oppose the grant of a new tenancy but objected to the terms proposed by the tenant and applied to the court for an order determining the rent which it would be reasonable for him to pay while his tenancy continued by virtue of section 24 of the Act—what is called in the Rules of the Supreme Court, Order 97, rule 9A and in the cases, though not in the Act itself or in the County Court Rules, "an interim rent." The learned judge determined an interim rent of £3,450 and a new rent of £3,950.

A The appellant, Mr Fawke, held the premises under a sub-underlease dated November 2 1970, which I will call "the tenancy," from a company called Soletanche Ltd for a term of two years certain and thereafter from year to year until May 30 1975 at a rent of £2,750 per annum. The respondent acquired the freehold on June 24 1975, on the falling in of the most superior head lease, subject only so far as the demised premises are concerned to and with the benefit of the tenancy, all other relevant under leases having determined.

There was a dispute between the parties whether the respondent was subject to any covenant to repair. The learned judge held that he was, and by a respondent's notice the B respondent sought to reverse that finding. Briefly, the point is that in the grant of the tenancy, Soletanche covenanted to perform the lessee's covenants, which included a full repairing covenant, contained in their underlease, being that immediately superior to the tenancy, and the respondent contends that the liability of Soletanche under that covenant ceased when the underlease to Soletanche expired, as it did on June 7 1975. I need not develop this any further because counsel for the respondent has conceded for the purpose of these proceedings only that we shall proceed on the basis that under the terms of the tenancy the landlord is subject to a full repairing covenant concerning the structure of the whole building including C the demised premises. On the other hand, under the tenancy, the appellant covenanted to keep the interior of the demised premises (but excluding all load-bearing walls, roofs and floor joists) and the sewers, drains and walls thereof in good and tenantable repair and condition.

The tenancy continued, of course, after the contractual expiry date, May 30 1975, by virtue of section 24 of the Act, and on December 3 1975 the respondent served notice to determine it on June 7 1976. On January 23 1976 the appellant served a counternotice stating that he would not be willing to give up possession, and on March 29 1976 he applied for a new tenancy, which he proposed should be for 14 D years from June 7 1976 at the rent of £3,600 per annum, and on other terms as stated in his notice. On April 7 1976 the respondent served notice objecting to the proposed terms and asking the court to determine an interim rent under section 24A.

The expert witnesses for the two parties put forward competing figures, and the learned judge made the assessments I have already mentioned. The major battle, however, turned upon the fact that the building, including that part of it which constituted the demised premises, became infested with dry rot due to many years' neglect of the exterior, so that it was necessary not only for extensive external repairs to be carried E out, but also very substantial internal work to eradicate the rot and replace ruined woodwork. This, it was clear, would cause great inconvenience to the appellant, and indeed while work was being carried out within the demised premises he might not be able to use them at all except for storage. There had been two serious floodings in September 1975, owing to a blocked drainpipe, but having regard to the lapse of time these had probably not caused or resulted in dry rot.

E St John FSVA, who gave evidence for the appellant, and T D Fraser-Mitchell FRICS for the respondent, met at the premises in June 1976 and endeavoured to agree figures. Mr St John proposed £3,000 per annum for the interim rent, and F Mr Fraser Mitchell £4,350. Mr St John made a report at that time which shows that he had not made or had the benefit of any survey, but that because of the damp in the premises he took account of the possibility of dry rot, although as far as can be seen he did not know whether or not it then in fact existed.

I read the following passages from his report. Paragraph 7:

> We were not instructed to make a structural or other survey of the building and we did not do so. In view of the water and damp penetration which has occurred and is occurring (to which we shall presently refer again) and the age of the building, the landlord would be well advised to check on the possibility of rot and to take remedial steps to prevent further damage to the structure. G

Then I pass to part of paragraph 10:

> We observed that rainwater from the roof discharges into the main sewer and if the downpipe is blocked, flooding is caused into the rear room on the second floor and telephone room on the first floor. Also, the cladding tiles to the exterior cause stoppage in downpipe backflow of water into the third floor, and this backflow of water pours through the ceiling into the rear rooms second and first floors.

Then paragraph 11 says: "At the time of our inspection, the skylight at the top of the building was open to the sky and water was coming in therefrom." H

The learned judge himself viewed the demised premises as he said in his judgment:

> On Tuesday the 21st February, 1977 at the request of both counsel I had a view of the premises at 27 Sloane Square. I am entitled to take into account my own impression on a view of the premises as it is a part of the evidence in the case on which I can rely. Of course, when considering my impressions I take full account of all the oral evidence I have heard and particularly the evidence given by the expert witnesses.

The learned judge also found, and here I am reading from p 4 of his judgment: J

> After the inconvenience and disturbance to the two rear rooms caused by the flooding, the applicant carried on his solicitor's practice over this period without complaint and had full use of all the rooms. The interior decorations were damaged to the extent already indicated. I saw the premises and, discounting the exploratory work done in January 1978 to assess the extent of the dry rot discovered shortly before and some damage to the decorations caused by comparatively recent penetration of water, the applicant's premises were in a reasonable state.

In November 1977 dry rot was found in areas around the central light well extending from the fourth floor down to sill level of the windows to the well on the first floor which K was due, as the learned judge found, to seepage of water for a long period from a defective gutter. Notwithstanding the exploratory work done in January 1978, it appears that some further dry rot may still be discovered, because Mr Keen, an expert in the eradication of dry rot and the cure of all forms of dampness in buildings whose evidence the learned judge accepted, said: "the damp areas are areas of potential dry rot as they dry out. After an area affected by damp begins to dry out it takes from six to 12 months before it can be ascertained whether or not dry rot is present."

The learned judge said this:

> L He (the witness) said that he would not be able to commence the job having regard to present commitments until the end of April 1978; it would take 2/3 months to complete the work of eradication; then before redecoration could commence one month would have to elapse for the plastering to dry out. That would mean that redecoration could not commence at the earliest before the end of August 1978, and that is on the basis that no further extensive areas of dry rot were discovered during the remedial work. At the expiration of six months from now further exploratory work could begin to ascertain whether there was dry rot in those damp areas already referred to conducive to dry rot. Those areas are not extensive and if dry rot was not found they would be replastered and could be redecorated four weeks later. M If dry rot was found, the work of eradication would not take very long. It would be early December 1978 before the applicant could expect to have his premises fully redecorated. The exterior work necessary to be done would be completed in the main before Mr Keen commenced work.

Work did not commence as early as anticipated and we understand that much remains to be done. The parties sought leave to adduce further evidence about this, but we did not think it necessary to go into precise details. It was common ground that it would seem likely to be completed before the new lease will commence.

A At the trial Mr St John resiled from his figure of £3,000 for the interim rent. He went so far as to say that the premises were unlettable, and he proposed an interim rent of £1,000 per annum, and further that the rent should remain payable at that rate under the new lease until completion of the necessary works of repair and restoration. At that time it was apprehended that the new lease would commence before then. The learned judge rejected this contention and he determined an interim rent of £3,450 per annum and a rent for the new lease of £3,950 with no period of reduced rent while agreed works of rot eradication and reinstatement were carried out.

B Mr Fawke now appeals, asking for an order substituting an interim rent of £1,000 per annum or such other figure as the court should think fit for the period of the tenancy covered by the said section 24A and used for the carrying out of the said remedial and reinstatement work. Included, or intended to be included, in this notice of motion is a challenge to the new rent also.

I must now read section 24A and the material parts of section 34 of the Act. Section 24A says:

> (1) The landlord of a tenancy to which this Part of this Act applies may, (a) if he has given notice under section 25 of this Act to terminate the tenancy; or (b) If the tenant has made a
> C request for a new tenancy in accordance with section 26 of this Act, apply to the court to determine a rent which it would be reasonable for the tenant to pay while the tenancy continues by virtue of section 24 of this Act, and the court may determine a rent accordingly. (2) A rent determined in proceedings under this section shall be deemed to be the rent payable under the tenancy from the date on which the proceedings were commenced or the date specified in the landlord's notice or the tenant's request, whichever is the later. (3) In determining a rent under this section the court shall have regard to the rent payable under the terms of the tenancy, but otherwise subsections (1) and (2) of section 34 of this Act shall apply to the determination as they would apply to the determination of a rent under that section if a new
> D tenancy from year to year of the whole of the property comprised in the tenancy were granted to the tenant by order of the court.

Then section 34 says:

> (1) The rent payable under a tenancy granted by order of the court under this Part of this Act shall be such as may be agreed between the landlord and the tenant or as, in default of such agreement, may be determined by the court to be that at which, having regard to the terms of the tenancy (other than those relating to rent), the holding might reasonably be expected to be let in the open market by a willing lessor, there being disregarded —(a) any effect on rent of the fact that the tenant has or his
> E predecessors in title have been in occupation of the holding; (b) any goodwill attached to the holding by reason of the carrying on thereat of the business of the tenant (whether by him or by a predecessor of his in that business) . . .

Subsection (3) says:

> Where the rent is determined by the court, the court may, if it thinks fit, further determine that the terms of the tenancy shall include such provision for varying the rent as may be specified in the determination.

Section 24A of the Landlord and Tenant Act 1954 (as inserted by section 3(1) of the Law of Property Act 1969) is, as has been stated in *Regis Property Co Ltd v Lewis & Peat*
F *Ltd* [1970] Ch 695 by Stamp J, as he then was, and by Megarry J, as he then was, in *English Exporters (London) Ltd v Eldonwall Ltd* [1973] Ch 415, extremely difficult to construe and apply because it is not easy to reconcile subsection (1) with subsection (3) or to see the true relationship between the opening words of subsection (3), requiring regard to be had to the rent payable under the tenancy, with the later part of the section, incorporating subsections (1) and (2) of section 34 by reference; because the very general expression "shall have regard to" necessarily imports a large degree of uncertainty; and finally because as Megarry J said in *English Exporters (London) Ltd v Eldonwall Ltd supra* at p 430 E:

> "I would only add that the process of applying section 34 G
> to a hypothetical yearly tenancy is one that, at least under present conditions, may often have an air of unreality about it that would puzzle the most expert of valuers."

In the *Regis Property Company* case Stamp J construed the words "shall have regard to the rent payable under the terms of the tenancy" as merely amending section 34(1), as applied to the determination of an interim rent under section 24A, by in effect striking out the words "other than those relating to rent" and substituting "including those relating to rent." So he held there was only one operation to be performed under subsection (3), that is to say the assessment of the rent as a matter of market value in accordance with section H
34 as so, in effect, amended, and, of course, on the basis of a yearly tenancy. He held that one could have regard to the rent under the existing tenancy if, but only if, it had evidential value for the purposes of ascertaining the market value, and he considered that but for the opening words of subsection (3) it could not have been regarded at all, because of the words of exclusion in that subsection.

Once the valuation exercise has been performed, however, in his view the matter is at an end and the figure arrived at cannot be modified either by the provision in subsection (1) that the landlord may "apply to the court to determine a rent which it would be reasonable for the tenant to pay" or J
by the direction in subsection (3) to have regard to the rent. That is to say he rejected what was described as the cushion argument.

Megarry J, taking a different view, accepted that argument. According to him the direction to have regard to the rent, and the reference to section 34 are two separate things, so that one must first provisionally determine the interim rent in accordance with the terms of section 34 but, of course, as if the court were ordering, not a term of years, but a yearly tenancy, and then review that against the actual rent payable under the current tenancy in the light of the direction in subsection (1) that the rent is to be reasonable. Megarry J con- K
sidered that Stamp J was wrong in thinking that unless his construction were adopted one could not refer to the existing rent even in the examples given by him where it would have evidentiary value. In the result, therefore, both agreed, though for different reasons, that the court and the expert witnesses appearing before it may, and should, have regard to the existing rent if it has evidentiary value, but the difference between them lies in Stamp J's view that, apart from that, it is irrelevant and the matter ends with the valuation, whereas in Megarry J's view it then has in any event to be considered in order to determine whether it is reasonable to adopt the valuation without modification. L

Unfortunately no argument was addressed to us as to which of these views we should accept, or whether, indeed, neither affords the correct solution to the problems posed by section 24A, as both parties were prepared to accept Megarry J's decision. We must, therefore, proceed on that basis, and this case cannot be regarded as a binding decision on this particular question of construction, but I must say that, as at present advised, I prefer the reasoning and decision of Megarry J. With all respect to Stamp J I think he fell into error, because he overlooked the fact that there are two tenancies and two rents involved, not one. The rent to which the court is directed to have regard under subsection (3) is M
the rent payable under the contractual tenancy, which is continuing by virtue of the Act, but when one applies section 34 to the determination of an interim rent, as directed by section 24, subsection (3), one is considering a new hypothetical yearly tenancy, and one can no more consider the rent appropriate to such a tenancy when determining that rent than one can consider the rent under a new tenancy when determining that rent.

I should also make some observations about the question of time, Megarry J's fourth point, p 431. It is possible to argue, as Mr Ibbotson [counsel for the landlords in *English*

Exporters case] did, that the valuation for the purposes of section 24A must be made as at the date of the hearing, or at least of the application to determine an interim rent, because the section applies section 34 as if the court were granting a new tenancy, albeit a yearly one, and, of course, when finally determining the matter by ordering a new lease under section 34 the court must of necessity consider the position at the date of the hearing, or at all events the latest date at which evidence is available. However, the reasons given by Megarry J appear to me to be compelling, and I agree that the values to be applied should be those existing when the interim period begins to run, which having regard to section 34A(2), is not necessarily the same date as that on which the contractual tenancy expires.

In my judgment the valuation should be made not, as was argued by the appellants, upon what knowledge of the state of the premises at the commencement of the interim period a reasonable prospective tenant properly advised and having such survey, if any, as might be reasonable would or should have acquired, for that imports a large measure of uncertainty, nor yet upon the actual knowledge of the actual parties (for that is subjective, while the test ought to be objective) but upon all that is known at the date of the hearing as to the condition that the premises were actually in at the commencement of that interim period. I do not see anything in section 24A which requires the court, in determining what rent it is reasonable for the tenant to pay, albeit by reference to a hypothetical letting, to shut its eyes to known facts.

Before considering the evidence in this case and the learned judge's findings, there are certain other questions of construction to be decided. First, has the court jurisdiction under section 24A to award a rent which is not to come into operation until repairs have been effected or is to be increased when they are, or which commences at once but is later to become subject to a moratorium or reduction for a certain period?

I have had the advantage of reading the judgment which Brandon LJ proposes to deliver, and with which I understand Stephenson LJ agrees, and they take the view that it has. The question is I think a very difficult one, and I confess with all respect to them that I have some hesitation in accepting their views on this point, notwithstanding the referential incorporation of section 34, subsections (1) and (2) in section 24A. It seems to me that the words "a rent which it would be reasonable for the tenant to pay while the tenancy continues by virtue of section 24 of this Act" in subsection (1), and the words in subsection (2) "shall be deemed to be the rent payable under the tenancy from the date" etc may well be too strong to admit of such a construction, especially so when one remembers that section 24A is conferring a power to substitute an interim rent for the contractual rent, which otherwise would continue uninterruptedly until a new lease be ordered or refused, and that in the absence of a provision in that behalf in the contractual tenancy the tenant holding over under sections 24 and 64 would not be entitled to any reduction or suspension of rent during repairs. I am, however, not prepared to dissent on this point and content myself with expressing my doubts.

The next question of construction which was argued, although in the events which have happened it is probably now academic, is whether the court has any such power when determining a rent under section 34 for a new tenancy. Here in my view the answer is free from doubt. I am satisfied that section 34 (3) authorises nothing more than the inclusion of a rent review clause, but what of subsection (1)? The learned judge thought that he could not make any such order because this would be "not a variation of the determined rent for the new tenancy, but the determination of two rents." With all respect I do not agree. If supported by evidence that this would be the manner in which "the holding might reasonably be expected to be let in the open market by a willing lessor" I see no reason why the court should not determine a rent increasing by fixed amounts at specified times. By the same token the court has, in my judgment, power to provide that the rent shall not commence, or shall be at a less rate, until repairs are effected, or shall cease to be payable or be reduced as from the time, albeit later than the commencement of the new lease when they are in fact started and until completed. This does not in any way conflict with subsection (3) or render that subsection otiose, because under a rent review clause the court would not itself be determining the rent delegating that function.

The next question is how the valuers should regard want of repair and breaches of covenant, when assessing an interim rent, and it seems to me they should consider what would be a reasonable rent for the tenant to pay from the date for commencement of that rent as a yearly tenant, having regard to the actual condition and state of the premises at that date, and having regard to the terms of the contractual tenancy so far as applicable to a yearly tenancy. This will mean that the hypothetical tenant will have the benefit of any covenant to repair on the part of the landlord and the burden of any on the part of the tenant, or there may be no covenant to repair on either side.

The learned judge said in his judgment:

"a term of the hypothetical yearly tenancy obliged the landlord to well and substantially repair, maintain and keep the premises in good and substantial repair and condition. In determining the interim rent the court must have regard (inter alia) to that term. Likewise the court must have regard to the tenant's covenant to decorate the premises. When the state or condition of the premises is regulated by such terms, in my judgment, lack of repair or want of decoration albeit a lack of repair, is not an element which the court should take into account in assessing interim rent because the landlord by the term of the tenancy is under an obligation to effect necessary repairs and if he does not the tenant has the right to enforce the covenant to repair against the landlord and claim damages. This is the correct procedure for dealing with lack of repair to the premises and consequential damage suffered by the tenant."

He was, of course, right to take into account the landlord's and tenant's covenants to repair, but, with respect, in my judgment he was not right to treat the lack of repair as an element not to be taken into account in assessing interim rent, on the ground that the tenant will have an action for damages or indeed for any other reason. The valuation being on the basis of a new yearly tenancy it must in my view be assumed that these covenants will be duly performed. There can at this stage be no question of breach.

The fixing of an interim rent will not, however, affect the right of either party to sue the other for any breach of the repairing covenants in the contractual tenancy, which may have already occured or may subsequently occur save that, in assessing damages for diminution in value with respect to any period after the determination of the contractual tenancy, credit must be given for the amount by which the interim rent was reduced on account of the want of repair from what it would have been if the premises had been in the state of repair required by the covenants. This may be considerably less than the actual diminution in value during the period of breach, since the valuation postulates prompt repair. The fixing of an interim rent would also not affect a claim by the tenant for damages for extra disturbance on the ground that the landlord's failure to repair after notice had made more extensive repairs necessary.

In cases where the interim rent so determined is less than the contractual rent the court must then consider whether, in the light of its duty to have regard to the rent payable under the terms of the existing tenancy, it should increase the interim rent up or near to that rent, or should in the exercise of its discretion refuse to determine an interim rent at all, which would leave the contractual rent payable. In determining these matters the court will, of course, bear in mind that section 24A is a section obviously designed by Parliament to

A improve the landlord's position, as is shown, if in no other way, by the fact that he alone can apply under it; and certainly where the low valuation is due to breaches by the landlord of his repairing covenants it would I think have to be a very special case in which the court would determine an interim rent less than the contractual rent. Different considerations will apply if the low valuation is due to a fall in property values.

With these principles in mind I turn to consider the facts of this particular case. The basic rent determined for the new lease is not challenged in the notice of appeal but, as I have said, it was argued below and before us that the low figure B of £1,000 should be continued for the new lease until all necessary repairs have been effected, and as I have already indicated in my judgment, the learned judge was wrong in holding that he had no jurisdiction to order that. This point has, however, been overtaken by events, since this appeal has postponed the date when the new lease will come into force, so that as far as can be seen, I gather, the repairs will all be finished before then. Thus the question now affects the interim rent only. It is, therefore, strictly unnecessary to say anything more about the rent for the new lease. I will just add this, however; Mr Fraser-Mitchell gave evidence in cross-examination as follows: "If repairs are necessary, estate C company's policy is to allow rent-free period for tenant to do the repair. Having regard to the evidence heard it would be a period of 3 or 4 months—3 months fair." There seems to be some confusion here, since in the present case it would be for the landlord to do the repairs, but overlooking that point, this passage affords some evidence which might have supported a postponement of the rent under the new lease, although the company's policy is subjective and does not necessarily reflect objectively what a willing lessor might be expected to do.

However, the learned judge did make an allowance for the want of repair in a different way, for he said "In my judgment, having regard to all the circumstances and background D and to the fact that it will be many months yet before the applicant tenant can settle down to quiet enjoyment of his holding owing to the works of reparation and redecoration which must be carried out, I consider that it will be right for rent review to take place at the expiration of seven years." I would, therefore, not interfere with the rent determined by the learned judge for the new lease even if it is open to the appellants to challenge this upon their notice of motion, and I turn to consider the problem in relation to the interim rent.

As I have said, the evidence of Mr Fraser-Mitchell is some, but by itself quite insufficient, evidence to justify a differential rent and in any case the evidence does not show what the E actual state of affairs was in June 1976. Dr Norman Hickin [a consultant in wood decay and wood preservation] in his report dated December 16 1977 says: "Dry rot is a dangerous fungus and grows swiftly—the quicker work is done the less likelihood there is of any additional work being necessary," so that the position may well have worsened substantially between June 1976 and November 1977, and still more so by January 1978. If a differential provision pending or during repairs is to be ordered it will be necessary to have further evidence on the lines which will be indicated by Brandon LJ to establish clearly that on the actual facts in June 1976 this would have been F the only, or at least a probable, basis for a yearly letting.

Apart from this, the question is whether the learned judge was correct in fixing a rent of £3,450 against Mr St John's figure of £3,000 and Mr Fraser-Mitchell's figure of £4,350 when they met in June 1976. The figure of £1,000 which Mr St John put forward at the trial was based mainly at any rate upon the fact that dry rot was discovered in November 1977 and the learned judge quoted from Mr St John's evidence and commented thereon as follows: "He further said: 'I have considered the interim rent pursuant to section 24A' and 'in the period from June 7 1976 with the knowledge I have now (a) The premises are not lettable, (b) In June 1976 the G premises were not readily lettable, (c) The landlord would be lucky to let them at £1,000 per annum, (d) Had I been advising a prospective tenant I would have advised him not to touch the premises having regard to the possibility of rot.'" The learned judge went on to say: "I think it is important to note in this connection that he also says that while the work of reparation is being carried out the applicant's premises will not be usable by the applicant for office purposes, but only as storage and his valuation of the premises for storage only is £1,000 per annum, ie the same figure he now gives for interim rent over a period when the applicant was using the premises for carrying on his practice without complaint and in premises, H which, on the evidence, I have found were in a reasonable state."

I do not think the learned judge was entitled to take into account the fact that in the events which happened the appellant was in fact able to carry on his practice in reasonable comfort, since that became known only after the relevant date, as at which the valuation has to be made, but he was entitled and bound to take into account the condition in which, assisted by a view, he found the premises to have been at the commencement of the interim period.

I think also, in the paragraph (d) quoted from Mr St John's evidence, he (namely the witness) was purporting to consider J the position as it appeared in June 1976 and not relying on subsequent facts, because he speaks of "the possibility of rot," but the learned judge was quite right in taking a general view that his evidence was coloured by the information obtained in November 1977 that there was not merely a possibility of dry rot but actual dry rot.

With respect, the learned judge also erred, as I have said, in disregarding the state of repair when determining the interim rent. Mr Fraser-Mitchell committed the same error, as he, too, made no allowance for the state of repair of the premises and the tenant's decorations.

In the circumstances, for my part, I cannot regard the K judge's determination of an interim rent as entirely satisfactory. It may be, in view of what counsel told us during the hearing, that the questions of principle having been resolved they will be able to agree to accept or modify the learned judge's determination of £3,450, with or without a differential provision pending or during repairs, in which case we could, by consent, order accordingly, but failing that I would, for my part, refer this part of the case back to him for rehearing with liberty to either party to adduce further evidence.

Brandon LJ, in a judgment read by Stephenson LJ, said: I agree in general with the judgment of Goff LJ. There is, however, one question in respect of which he has expressed L doubts. That question is whether the court has power, not only when determining the rent of a new tenancy under section 34(1) of the Act of 1954, but also when determining an interim rent under section 24A, to determine what I shall for convenience call a differential rent, by which I mean a rent which varies from time to time during the period of the tenancy according to the situation with regard to the state of repair of the premises. Although I have found the question a difficult one, I do not in the end share the doubts which Goff LJ has expressed, and I shall accordingly set out my views on the matter and the reasons on which they are based.

I consider, first, the power of the court when determining M under section 34(1) the rent payable under a new tenancy. The rent to be determined is the rent at which, having regard to the terms of the new tenancy other than those relating to the rent, the holding might reasonably be expected to be let in the open market by a willing lessor, in other words the market rent. It is clear that the market rent of premises may in general be affected not only by the terms of the tenancy concerning the obligations of the lessor and lessee in relation to repair, but also by the actual state of repair of the premises at the commencement of the tenancy.

To take an extreme example, let it be supposed, firstly,

A that under the terms of the tenancy, it is the obligation of the lessor to put the premises into repair; secondly, that the premises are, at the commencement of the tenancy, so seriously out of repair, by reason of previous fire or flood, that they are of only partial use to the lessee; thirdly, that an appreciable period must necessarily elapse before the lessor can begin to perform his obligation of putting the premises into repair; and, fourthly, that, while the lessor is performing that obligation, the disturbance will be such that the premises will be virtually of no use to the lessee at all. In those circumstances it may well be that the market rent of the premises, which would be appropriate after they had been put into repair, B would be appreciably reduced during the period which would necessarily elapse before the work of putting them into repair was begun, and even more reduced during the further period while such work was being carried out.

Similar considerations would, as it seems to me, apply if the premises were seriously out of repair at the commencement of the tenancy, not by reason of previous fire of flood, but by reason of the failure of the lessor or the lessee or both to perform their previous obligations to repair. The cause or causes of the premises being out of repair are not, for this purpose, significant; the fact that they are out of repair, for whatever cause or causes, is so.

C On the footing that the market rent of premises in general may be affected by the situation with regard to their state of repair in the manner which I have described, the questions whether and how the market rent of the premises concerned in any particular case are so affected must depend on the evidence in that case. If the evidence showed that the market rent would be a rent which varied during the period of the tenancy according to whether the premises (a) remained out of repair, (b) were in the course of being put into repair and (c) had been put into repair, then it seems to me that the court, applying the principle of determination prescribed by section 34(1), would not only have power to determine, but would D be obliged to determine a differential rent accordingly. If, on the other hand, the evidence showed that the market rent, though affected in amount by the present and future situation with regard to state of repair, would nevertheless be a normal fixed rent over the whole period of the tenancy, then the court's duty would be to determine a fixed rent in accordance with that evidence.

I consider, second, the power of the court when determining under section 24A an interim rent payable during the continuation of a pre-existing tenancy. The rent which the court has to determine in this case, subject always to its additional duty to have regard to the rent payable under the pre- E existing tenancy, is the rent at which the holding might reasonably be let, having regard to the terms of such tenancy other than those relating to rent, from year to year in the open market by a willing lessor, in other words the market rent on a year to year basis. Here again it seems to me, for the same reasons as those discussed above in relation to section 34 (1), that the market rent may in general be affected by the situation with regard to the state of repair of the premises at the date when the interim rent is to commence, and that the questions whether and how the market rent of the premises concerned in any particular case are so affected must depend on the evidence in that case. So, if the F evidence showed that the market rent would be a differential rent in the sense in which I have used that expression, then the court would have both the power and the duty to determine a differential rent accordingly. On the other hand, if the evidence showed that the market rent, though affected in amount by the present and future situation with regard to state of repair, would nevertheless be a normal fixed rent, the duty of the court would be to determine a fixed interim rent.

In what I have said so far I have been considering whether the court has power to determine a differential rent firstly under section 34(1) and secondly under section 24A, and I have concluded that it has power to do so in either case if the evidence warrants it. I should not, however, wish to G encourage the view that this is a power which the court should exercise at all frequently. The cases in which the evidence may support a differential rent on the basis of the situation with regard to state of repair are likely, in my view, to be limited to cases in which the state of disrepair at the commencement of the new tenancy, or the commencement of the period of interim rent, is of a very serious character. In any other case it seems to me that the evidence will be very unlikely to do so.

A further question could in theory arise as to whether, in a case where the situation with regard to the state of repair was such that the premises would, in effect, be worthless to a H lessee for a certain period, the court would have power to determine a nil rent for that period. That question does not, however, arise for decision in the present case for, although the appellant's valuer gave evidence that the premises were, in their original state of disrepair, incapable of being let at all, the learned judge did not accept that evidence. In these circumstances, I should prefer to reserve this further question until there is a case in which the facts found make a decision on it necessary.

There is one other matter which I would mention. It might be thought that, because a rent determined under section 24A is an interim rent, it can only be applicable for a short time, and ought therefore to be in every case a fixed J rent rather than a differential one. In practice, however, the period for which such a rent may be applicable, having regard to the provisions of section 25 (1) and section 64 (2), may well be of considerable length (by which I mean up to two or even three years), either because of delay in the trial of the proceedings in the county court, or because of prolongation of the proceedings by one or more appeals, or both. It would be wrong, therefore, to approach the interpretation of section 24A on the basis that the period for which a rent determined under it will apply is bound to be short, and to construe the section as not giving the court power to determine a K differential rent in a proper case on that account.

STEPHENSON LJ said: I have had the advantage of considering in draft the two judgments of my brethren. I have had many doubts in the course of the case, but of one thing I feel reasonably certain: namely that Parliament in enacting section 34 in 1954 and in amending section 34 and enacting section 24A in 1969 never considered the problem raised by this appeal: the impact of disrepair in the property comprised in a tenancy on the court's power to determine a rent under either section.

I agree with both Goff LJ and Brandon LJ that on its true construction section 34 as amended authorises, by subsection L (3), what has been called a variable rent—that is a provision for the parties varying the rent by a rent review clause—and by subsection (1), what Brandon LJ has termed a differential rent—that is a provision by the court varying the rent payable at different periods during the term of the new tenancy. But I long shared Goff LJ's doubts whether section 24A, on its true construction, permitted a differential interim rent. However, the reasoning of Brandon LJ has resolved those doubts and driven me to his conclusion that, if the court has the power to give effect to disrepair of the property during the new tenancy and to determine a differential new rent, there is nothing in the language or the purpose of section 24A M which prevents the court from giving effect to the disrepair of the premises while the current tenancy continues and "differentiating" the interim rent in a case where the evidence justifies it.

The cases in which the court could fix a differential interim rent must, I would think, be extremely rare, even where the period when interim rent is payable is prolonged by the operation of section 64. But like Brandon LJ, with whose judgment I am in complete agreement, I would not like to rule out the possibility of admitting expert evidence which could

A
lead the court to take the exceptional course of determining a differential interim rent.

On all other points I am in agreement with the judgment of Goff LJ.

The appeal was allowed and the question of the amount of the interim rent, including the question whether it should be a differential rent, and also the question whether there should be any provision for reducing or suspending the new rent, if the works of repair were not completed by the commencement of the new lease, referred back to the county court judge.

B

CHANCERY DIVISION

November 22 1978

(Before Mr Justice FOX)

ROSS AUTO WASH LTD v HERBERT AND ANOTHER

Estates Gazette June 9 1979

(1978) 250 EG 971

Landlord and Tenant Act 1954, Part II—Application for new lease opposed by landlords—Premises in King's Road, Chelsea, in which a variety of licensees holding concessions
C
sold articles from stands—Tenant company's business concerned with the granting of licences to the concession holders—Question whether the tenant company or another company, both controlled by the same individual, actually carried on the business—Landlords claimed that the business was carried on by the other company, so that section 23(1) of the Act was not satisfied—Held that on the evidence the business was carried on by the tenant company and that it occupied the premises for that purpose

The plaintiffs in this action were Ross Auto Wash Ltd, tenants of ground-floor premises at 85 King's Road, London, whose application for a new lease was opposed by the defen-
D
dants, Diana Herbert and John Robert Turner, the landlords.

D Levy (instructed by Amhurst, Brown, Martin & Nicholson) appeared on behalf of the plaintiff company; R J S Thompson (instructed by Broughton & Co) represented the defendants.

Giving judgment, Fox J said: This is an application for a new lease of ground-floor premises at 85 King's Road, London, under Part II of the Landlord and Tenant Act 1954. The only issue before me is whether the plaintiff is entitled to apply for a new lease at all.

Section 23(1) of the Landlord and Tenant Act 1954, which
E
is in Part II of the Act, is in the following terms:

> Subject to the provisions of this Act, this Part of this Act applies to any tenancy where the property comprised in the tenancy is or includes premises which are occupied by the tenant and are so occupied for the purposes of a business carried on by him or for those and other purposes.

The defendants, the landlords, contend that the premises are not occupied by the plaintiff for the purpose of a business carried on by the plaintiff, and that, therefore, the claim for a new lease fails *in limine*. Whether the premises are occupied by the plaintiff for the purposes of a business which it carries on is a question of fact, upon which I have heard
F
oral evidence. It is not in dispute that on the law and the authorities as they stand the activity which is being carried on upon the premises is a business. The only question is who is carrying it on. The defendants say that it is not the plaintiff, but a company called Doranmoor.

By a lease of June 21 1967, and made between the defendants of the first part and the plaintiff of the second part and a Charles Cooper Ross of the third part, the premises (together with other premises with which I am not concerned) were demised to the plaintiff for seven years from September 29 1970. It appears to have been contemplated at the time of the lease that the premises with which I am concerned would be
G
developed as a shopping precinct, and that, in fact, is what happened.

Notice of the termination of the tenancy on September 29 1977 was given by the defendants on March 3 1977. The plaintiff gave a counternotice on March 11 1977 and the originating summons was issued on June 10 1977.

The plaintiff is a company controlled by a Mr Robin Sutherland. The other company with which I am concerned, Doranmoor, is also controlled by Mr Sutherland. He acquired it early in 1977, and prior to that it plays no part in the case, if indeed it had been formed at all. On January 26 1977 the plaintiff sought consent from the defendants to assign the
H
lease to Doranmoor. The defendants, on February 2 1977 indicated their agreement in principle to the assignment, provided that they were given a suitable guarantee against all the obligations under the lease. No guarantee was offered, and the matter of the assignment went off.

On February 28 1977 a letter was written by the plaintiff to Doranmoor in the following terms:

> I write to confirm that we have appointed you Managers of our premises at 83 Kings Road, London, as from today's date.
>
> You will carry out on our behalf the management and servicing of the business, and in particular you will collect the Concession Licensees' fees and arrange for payment of outgoings.
>
> J
> As agreed your company has been appointed Manager only and full possession and control of the premises remains with the company.

Since then Doranmoor has undoubtedly played some part in relation to the business activities conducted on the premises. The question is precisely what part Doranmoor plays. That brings me to the matter of the activities themselves. The plaintiff's case is that at all times since about September 1970 the plaintiff has carried on upon the premises the business of granting concession licences to other persons to trade from stands thereon not used by the plaintiff. That was certainly
K
the position before Doranmoor became involved early in 1977.

There have, at all material times, been stands on the premises from which persons have conducted various businesses. There are at present about 22 stands from which are carried on a variety of businesses—sales of oriental clothes, jeans, records, bric-à-brac, shoes, handbags, as well as a hairdressing business and a coffee bar. The coffee bar is an important means of attracting business to the site; it is not, it seems, very profitable, and I find that the plaintiff has had to take it over and run it some five or six times during the last five or six years. The plaintiff is, in fact, running it now. The coffee bar is run by the plaintiff and not by Doranmoor.

L
I have heard oral evidence from Mr Sutherland, a director of both the plaintiff and Doranmoor; he was examined and cross-examined before me. Mr Sutherland's evidence is that the negotiation and granting of licences are a matter that he deals with himself, and that when doing so he always regards himself as acting on behalf of the plaintiff. I accept Mr Sutherland's evidence on this. I have no doubt (and I find) that the licences are always granted on behalf of the plaintiff and nobody else. For completeness, I mention here that it is conceded by the defendants that the concession holders, the licensees are indeed licensees and are not subtenants.

Mr Sutherland also said that dealing with the licensees from
M
day to day was one of the most important matters in the whole enterprise. All sorts of problems arise. The following are examples: licensees get behind in rent; problems arise as to the mode of display of goods; there are problems of conflict of interest when licensees deal with the same kind of goods; and there may be disputes as to the user of space—these stalls being mobile. Some matters can be dealt with, as I will mention later, by the manageress on the premises, but Mr Sutherland's evidence is that to deal with licensee problems he goes to the premises nearly every day. He always regards himself as dealing with the licensees and their problems as a director of the plaintiff. I accept that evidence.

A The services which are provided to the licensees are as follows:

(a) the provision of heating and lighting for the premises;
(b) payment of rates, taxes and other outgoings relating to the premises;
(c) toilet facilities;
(d) a sound system and records to be played on the premises;
(e) counters, stalls and display units and cash tills;
(f) advertising;
(g) coin-operated telephone;
B (h) the handling of payment by credit card;
(i) cleaning and decorating;
(j) the running of a licensee's stall for short periods in a day if the licensee has to leave the premises;
(k) insurance and burglary alarm;
(l) the opening up and shutting up of the premises.

As to those matters I find the following additional facts:

(i) The cost of the heating and lighting, and the rates, taxes and other outgoings are paid by the plaintiff.

(ii) As regards the sound system and records, the cost of the C equipment is paid by the plaintiff. The cost of the records is shared between the plaintiff and the stallholder, who sells records.

(iii) The actual stalls are supplied by the plaintiff. The plaintiff leases the stalls themselves and subleases them to the licensees.

(iv) Advertising. This is of two kinds: first, press advertising. That is of fairly small amount, and is mostly about Christmas time. It is arranged by the plaintiff. The cost is shared between the licensees and the plaintiff. Second, a display case at the front of the premises. The licensees can display goods on a rota system. The plaintiff organises that and pays for the window D dressing.

(v) Credit cards. Some licensees are too small for the credit card companies to give facilities to them. Accordingly, an overall arrangement was negotiated with the credit card companies, of which the licensees can avail themselves. The negotiation of that arrangement was handled by Mr Sutherland in the name of Doranmoor.

(vi) Insurance and burglar alarms. Insurance is effected by the plaintiff. The burglar alarm contract was made by Doranmoor on behalf of the plaintiff.

(vii) The premises are open from 10 am to 6 pm for E six days a week. There has, for the one and a half years or so, been a full-time manageress. She occupies an office in the premises and is employed by the plaintiff. If a stallholder has to leave his stall for a period during the day the manageress will normally look after it during his absence. She will also deal with day-to-day problems which arise. Many of those problems, however, are such that she cannot deal with, and they are, therefore, dealt with by Mr Sutherland himself, as I have already mentioned.

I come now to the position of Doranmoor. As I have mentioned, it was acquired by Mr Sutherland early in 1977. Mr Sutherland's evidence is that Doranmoor was acquired F initially on accountant's advice in order to have a separate company for management of part of the premises. In 1977, however, Mr Sutherland was examining the possibility of introducing, in effect, what he called a "partner" into the business to provide further capital, and it was thought that Doranmoor G might be a useful vehicle in connection with that. Application was made to transfer the lease to Doranmoor, but in view of the insistence upon a guarantor the idea was dropped. Mr Sutherland says, and I accept his evidence, that the idea of a transfer of the lease to Doranmoor having been abandoned, Doranmoor continued purely as a management company and nothing else.

I find that Doranmoor receives some of the licensees' fees and pays some of the outgoings, in particular the cost of decorations. Most of the fees, I find, are in fact collected by the plaintiff and not by Doranmoor. No final decision has H been made as to what amount Doranmoor is actually to keep. Mr Sutherland says that will depend upon his accountants' advice. The accounts of Doranmoor have not yet been drawn up, and no accounts of the plaintiff for the last three years have yet been drawn up.

Mr Thompson, on behalf of the defendants, put it firmly to Mr Sutherland in cross-examination that the "management" letter of February 28 1977 is just a sham, and that the consent to transfer of the lease having been refused in the absence of the guarantor, the business was *de facto* taken over by Doranmoor in the guise of a manager. At the highest, it was said by Mr Thompson that the business was being carried on J by both the plaintiff and Doranmoor.

Those suggestions were rejected by Mr Sutherland. He said that the plaintiff conducts the business, and Doranmoor merely assists the plaintiff. I accept Mr Sutherland's evidence as to that. In my judgment the business, from first to last, has been the business of the plaintiff and nobody else. I so find.

The business is a business of granting licences. Those licences, I have found, are granted by the plaintiff. The important matter of dealing with the problems of the licensees is conducted by Mr Sutherland as a director of the plaintiff, in so far as the manageress, who is, in fact, employed by the K plaintiff, is unable to deal with them. The bulk of the services are provided directly by the plaintiff. The management letter itself makes it plain that Doranmoor is a manager only, and that full possession and control of the premises remains in the plaintiff. It is, of course, open to the defendants to demonstrate that that statement is a sham and is simply designed to mislead. (See *Teasdale v Walker* [1958] 3 All ER 307, to which Mr Thompson referred me.) But in my judgment no such state of affairs is demonstrated here at all.

In my judgment the other facts substantially support the reality of the position as stated in the management letter. I L do not accept the suggestion that the management letter entitles Doranmoor to retain the net moneys in its hands. As a matter of construction, consistently with the terms of the letter that Doranmoor is a manager only, it seems to me that Doranmoor must hold the balance for its principal, the plaintiff.

Of course, it may very well be that, for tax reasons, it will suit Mr Sutherland that Doranmoor should receive part of the profits of the business as remuneration for its services. But neither that fact nor the fact that the companies have been slow in preparing accounts leads me to the conclusion that anybody but the plaintiff is carrying on the business.

Accordingly I find that the premises are, and have been, M occupied by the plaintiff for the purposes of a business carried on by the plaintiff.

LANDLORD AND TENANT

LEASEHOLD REFORM

A **COURT OF APPEAL**

November 23 1978

(Before Lord Justice STEPHENSON, Lord Justice BRANDON and Sir David CAIRNS)

BL HOLDINGS LTD v MARCOLT INVESTMENTS LTD

Estates Gazette March 3 1979

(1978) 249 EG 849

Leasehold Property (Repairs) Act 1938—Requirements of valid notice under section 146 of the Law of Property Act 1925 and section 1 of the 1938 Act—Importance of strict observance of the four requirements in section 1(4) of latter
B **Act—Lessor's notice must state lessee's entitlement to serve counternotice claiming benefit of 1938 Act, must specify the time within which, and manner in which, counternotice may be served and lessor's name and address for service—Woodfall's form of notice commended—Practitioners should note that failure to state any one of the requirements may render notice bad—Service of counternotice by lessees held not to constitute waiver of defects in notice—Kammins Ballrooms case considered—Lessees' appeal allowed**

This was an appeal by Marcolt Investments Ltd, the lessees, from the decision of a master of the Queen's Bench Division in favour of the lessors, BL Holdings Ltd. In proceedings
C for leave for the purpose of section 1 of the Leasehold Property (Repairs) Act 1938 the master held that the notice given by the lessors was valid. This ruling was challenged in the present appeal.

M Mark (instructed by Cowan Lipson & Rumney) appeared on behalf of the appellants; E Prince (instructed by Berwin Leighton) represented the respondents.

Giving judgment, STEPHENSON LJ said: This appeal raises a short point, or rather two short points, of considerable interest under the Leasehold Property (Repairs) Act of 1938. The appeal really is as unmeritorious as an appeal can be, in my judgment, and it arises simply from carelessness on the part
D of those who served a notice under section 1 of the Act upon the solicitors for the appellants.

The parties concerned are the appellants, Marcolt Investments Ltd (whom I shall call the tenants) and B L Holdings Ltd, the respondents (whom I shall call the landlords). The tenants are the lessees under an underlease of a block of flats at 11 to 15 Portman Street in this city. Their underlease is dated September 14 1965 and it runs for 19 years from June 13 1965, and they are still in occupation of that building. The landlords are the successors in title of those who granted the underlease and are themselves lessees of head landlords.
E The rent which they get from the underlessees is £4,250 a year. It is not disputed that serious defects in the stone cladding of the block of flats have developed, leading to the serving of a dangerous structure notice on the owner of the premises by the Westminster City Council, and to necessary repairs being done to render this building safe, at the cost of some £20,000—a loss which apparently falls on the landlords because the council had to do the repairs themselves and either have charged or will be charging the landlords the cost of them. We are not concerned whether there is any answer to a claim for damages for breaches of the tenants' F
covenants nos 3 and 6 of clause 2 of the underlease which, for the purpose of the points which we have to decide, it is unnecessary to read.

What happened was that faced with these defects in the cladding, the landlords served on the tenants a notice under section 146 of the Law of Property Act 1925, requiring them to carry out the repairs together with a schedule of dilapidations for repairing in the ordinary way. On March 18 1977 they served upon them what purported to be a notice under section 1 of the 1938 Act, to which I have already referred. That notice, after referring to the underlease and setting out the covenants and the statement that they had been broken G
as contained in the schedule of repairs and dilapidations served with the notice, concluded in this way:

> We Berwin Leighton of Adelaide House, London Bridge, London, EC4, as solicitors for and on behalf of B L Holdings Ltd of 21-22 Grosvenor Street, W1, hereby give you notice to remedy the said breach of covenant to repair within three months from the service of this notice upon you in accordance with clause 2(8)(a)(b) of the said lease and to pay compensation to your landlord in respect of such breach and to pay your landlord's solicitors' costs in connection with the service of this notice. Further take notice that if you wish to rely upon the provisions of the Leasehold Property (Repairs) Act 1938, then such notice thereof must be H
> served upon your landlords care of their solicitors and agents, Berwin Leighton, of Adelaide House, London Bridge, London, EC4, within twenty-eight days from the service of this notice upon you.

The tenants contend that that notice is a bad notice: the notice is invalid because it does not comply with the requirements of section 1(4) of the Act of 1938. The master rejected that contention. That section, section 1, restricts the enforcement of repairing covenants of long leases of small houses and it lays down the procedure of notice and counternotice which must be gone through before breaches of repairing covenants can be enforced and damages can be recovered; J
and put shortly, if a notice is served by the landlords and a counternotice is served by the tenants, then the rights are not enforceable without leave of the court.

Section 1(1) provides:

> Where a lessor serves on a lessee under subsection (1) of section one hundred and forty six of the Law of Property Act 1925 a notice that relates to a breach of a covenant or agreement to keep or put in repair during the currency of the lease all or any of the property comprised in the lease, and at the date of the service of the notice three years or more of the term of the lease remain unexpired, the lessee may within twenty-eight days from that date serve on the lessor a counternotice to the effect that K
> he claims the benefit of this Act.

Then subsection (2):

> A right to damages for a breach of such covenant as aforesaid shall not be enforceable by action commenced at any time at which three years or more of the terms of the lease remain unexpired unless the lessor has served on the lessee not less than one month before the commencement of the action such a notice as is specified in subsection (1) of section one hundred and forty-six of the Law of Property Act 1925, and where a notice is served

under this subsection, the lessee may, within twenty-eight days from the date of the service thereof, serve on the lessor a counternotice to the effect that he claims the benefit of this Act.

So there are two provisions in different circumstances in which a lessee has the right to serve on the lessor a counternotice to the effect that he claims the benefit of the 1938 Act.

Then there is subsection (3):

> Where a counternotice is served by a lessee under this section, then, notwithstanding anything in any enactment of rule of law, no proceedings, by action or otherwise, shall be taken by the lessor for the enforcement of any right of re-entry or forfeiture under any proviso or stipulation in the lease for breach of the covenant or agreement in question, or for damages for breach thereof, otherwise than with the leave of the court.

Subsection (4) is the subsection which deals with what such a lessor's notice is required to contain. It provides:

> A notice served under subsection (1) of section one hundred and forty six of the Law of Property Act 1925, in the circumstances specified in subsection (1) of this section, and a notice served under subsection (2) of this section shall not be valid unless it contains a statement, in characters not less conspicuous than those used in any other part of the notice, to the effect that the lessee is entitled under this Act to serve on the lessor a counternotice claiming the benefit of this Act, and a statement in the like characters, specifying the time within which, and the manner in which, under this Act a counternotice may be served and specifying the name and address for service of the lessor.

Subsection (5) deals with what the lessor must prove before he can obtain the leave of the court, and subsection (6) gives the court power to impose conditions on the grant or refusal of such leave.

Subsection (2) of section 7 of the 1938 Act provides:

> The provisions of section one hundred and ninety six of the said Act [that is the 1925 Act, the Law of Property Act] (which relates to the service of notices) shall extend to notices and counternotices required or authorised by this Act.

Section 196 of the 1925 Act requires that notices shall be in writing; it provides also that notices shall be sufficiently served if left at the last known place of abode or business in the United Kingdom of the lessee, lessor, and so on or other person to be served and further shall also be sufficiently served if sent by post in a registered letter addressed to the lessee, lessor and so on, at his place of business; and that permitted service by registered post has been extended by the Recorded Delivery Service Act of 1962 to recorded delivery.

That Act of 1962 provides:

> Any enactment which requires or authorises a document or other thing to be sent by registered post (whether or not it makes any other provision thereto), shall have effect as if it required or, as the case may be, authorised that thing to be sent by registered post or the recorded delivery service; and any enactment which makes any other provision in relation to the sending of a document or other thing by registered post or to a thing so sent shall have effect as if it made the like provision in relation to the sending of that thing by the recorded delivery service or, as the case may be, to a thing sent by that service.

By subsection (3) of section 2 of the Act reference is made to the sending of documents and the serving of documents or any other similar thing.

Bearing those statutory provisions in mind, I come back to subsection (4) of section 1 of the 1938 Act and note that a lessor's notice under the section shall not be valid unless it contains four things. The first is a statement to the effect that the lessee is entitled "under this Act to serve on the lessor a counternotice claiming the benefit of the Act," and the second is a statement in the like characters specifying the time within which a counternotice may be served; the third is a statement of the manner in which, under this Act, a counternotice may be served and the fourth is a specification of the name and address for service of the lessor.

Mr Mark, on behalf of the appellant, submits that the notice of March 18 1977 served by the landlords' solicitors on the tenants' solicitors does not comply with those requirements in two respects. He relies first on the submission that it did not contain a statement to the effect that the lessee is entitled under the Act to serve on the lessor a counternotice claiming the benefit of the Act; and, at the invitation of the court, he took a second point, that it also was invalidated because it did not contain a statement of the manner in which, under this Act, a counternotice may be served.

An ideal specimen or form of notice complying beyond any doubt with the requirements of the Act, is to be found at p 1209 of the current (28th) edition of *Woodfall on Landlord and Tenant*, and it is Form 12 in Appendix A3. Para 5 says: "The lessee is entitled under the Leasehold Property (Repairs) Act 1938 to serve on the lessor a counternotice claiming the benefit of the said Act." Para 6 says that a counternotice may be served within 28 days and so on. Para 7 says:

> Such counternotice must be in writing and may be served upon the lessor by handing the same to him personally. Such counternotice shall also be sufficiently served if it is left at the last known place of abode or business of the lessor in the United Kingdom. Such counternotice shall also be sufficiently served if it is sent by post in a registered letter addressed to the lessor by name at the aforesaid place of abode or business and if that letter is not returned through the post undelivered; and that service shall be deemed to be made at the time at which the registered letter would in the ordinary course be delivered.

Now solicitors or others faced with the provisions and requirements of the 1938 Act may be forgiven for wondering whether it is necessary to set out the rights and duties of the lessee or tenant to whom the notice is addressed in such detail as that, and they might derive or think they derive some support from a decision of this court in the case of *Middlegate Properties Ltd v Spyridon Messimeris*, which is reported in [1973] 1 WLR 168. There a shorter form of notice had been given in which the tenant had been informed that by virtue of the Act "you are entitled if you so desire to serve upon us a counternotice claiming the benefit of that Act" and that "Such notice must be served upon us within 28 days," "should be addressed to us . . . and may be left at or sent by registered post to our office." It was held that that notice was valid, and on appeal it was argued by the lessors that the plaintiffs had sufficiently complied with the requirements of section 1(4) with regard to the statement specifying the manner in which the counternotice might be served, notwithstanding that the statement contained in the notice did not specify that service of such counternotice might be effected by recorded delivery or alternatively by registered post.

This court held that the notice was valid, and dismissed the appeal. Perhaps the main point was that the requirements in section 1(4) as to the characters to be used in the notice were not complied with. But there was a second point, and on that point the learned Master of the Rolls said this:

> The tenant says that paragraph 5 was bad because it only specified registered post and not also recorded delivery service. He points out that under the Recorded Delivery Service Act 1962 recorded delivery is equal to registered post. I think this point is bad also. The statement is sufficient if it specifies one good manner of service. It need not specify all.

It may be that that statement of the learned Master of the Rolls—and I think both Megaw LJ and Willmer LJ agreed with it—went too far if it is taken to mean that it is unnecessary for a lessor to specify more than one good manner of service in all circumstances. If, however, the words of the Master of the Rolls are read in relation to the question which was before the court, namely, whether it was necessary to add to a statement that the notice could be left at or sent by registered post "to our office," a statement that registered post included recorded delivery by virtue of the provisions of the 1962 Act, which I have just read, obviously that decision does not go as far as the language of the Master of the Rolls might appear to go.

One other thing which one must, I think, bear in mind in considering this notice and the statute which requires it to be given is that in the same case the learned Master of the Rolls pointed out that what one had to consider was whether the notice conveyed to the recipient the information which the statute requires it to convey. If the statute is read with literal strictness, surprising results might be achieved. Therefore the statute has got to be read with its object of remedying the abuse of landlords buying property and facing tenants with exorbitant claims for repairs for breach of covenant in mind. And secondly the learned Master of the Rolls said that not only must the Act be looked at sensibly and reasonably, but so must the notice. This is a fact brought out in his judgment, though not I think in the head-note of the report. He referred to the fact that the court had looked at precedent books and went on to say: " Some give forms with more detail than others. All I would say is that these notices should be construed reasonably. If they tell the tenant with sufficient clearness what the statute requires, they are good."

Now approaching this notice in that way, does it tell the tenants what the statute requires them to know about their right under the Act to serve on the landlords the counternotice claiming benefit of the Act? Does it tell them enough about the manner in which, under this Act, a counternotice may be served?

Mr Prince concedes two things. First of all he concedes that this notice is very indifferently drawn and its language is not only not ideal but is not as clear as it should be; and secondly he concedes that for a notice to be good and valid as satisfying the statute requirements it must satisfy the statutory requirements on the supposition that it is being sent to any reasonable hypothetical tenant, and not necessarily to a tenant or a tenant's agent versed (as the solicitors who received this particular notice were assumed to be) in the niceties of the law of landlord and tenant, and probably acquainted with the detailed provisions of this particular statute.

I doubt very much whether the *first* of the four requirements which I have specified was complied with by this notice. It is said by Mr Prince for the landlords that to tell a tenant " if you wish to rely upon the provisions of an Act, a notice thereof," which must mean " of your wish to rely upon it," " must be served on your landlord " is a good enough statement of the effect of section 1(4) and that is enough under the subsection without using its exact words. You are conveying in effect to the lessee by your statement the information that he is entitled to serve a counternotice under the Act. I doubt very much whether that is so; " Such notice thereof " is an obscure phrase. It is admitted that the word " such " has to be disregarded because it obviously does not refer to the only notice referred to in the earlier part of the sentence but must refer to some other notice, reference to which has perhaps been by a clerical error omitted from the particular notice.

But however that may be, I am satisfied that there is no statement to comply with the *third* requirement, of the manner in which, under this Act, a counternotice may be served. Mr Prince has sought to say that from the statement that it must be served on the landlords' solicitors as agents at this particular address it should be inferred or implied that the manner of service is actual delivery by post or otherwise. I cannot imply any such thing or support that Parliament expected a reasonable tenant to imply any such thing. I have no doubt that the tenants' solicitors knew well at least some of the various manners in which they could serve the counternotice, although a " counternotice " has itself to be implied from this notice because the word " counternotice " is not used; " such notice thereof " is the only reference there could be to the counternotice to which the statute refers. But I am certain it is not an inference or an implication which Parliament would have expected a reasonable tenant necessarily to have drawn from it.

It is clear from the wording of the subsection that by specifying the name and address for service of the lessor you do not tell him *how* the person at that name and address is to be served. There is no statement expressed or implied of the manner in which service can be effected or anything to warn a tenant as in fact and in law is the case, that if he adopts the method of serving his notice of intention or wish to rely on the provisions of the Act or to claim the benefit of the Act by ordinary non-registered post, unless that letter actually reaches its destination his service is unavailing and does not preserve the benefit which he is seeking to claim by a counternotice served specifically in that way.

I have therefore come to the conclusion that this notice did not comply with the Act and it was therefore invalidated. There remains the question whether it can be waived or was in fact waived by what happened afterwards. What happened afterwards was this: less than a week later, on March 23, the tenants' solicitors wrote to the landlords' solicitors this letter—and we have had no evidence about it, but we have no reason to suppose that they sent the letter by registered post. We do know that it was in fact received and that the receipt of it was acknowledged, as asked for at the end of the letter. The letter reads as follows:

> Dear Sirs, Re Marcolt Investments Limited and B L Holdings Limited, 11-15 Portman Street, London, W1. We thank you for your letter dated 18th March and acknowledge receipt of the copy-letter and notice which you sent us. Will you please treat this letter as formal notice on behalf of our clients, Marcolt Investments Limited, that they wish to rely upon the provisions of the Leasehold Property (Repairs) Act 1938. Will you please acknowledge receipt of this letter and the notice contained in it.

Now the learned master whose order is under appeal apparently dealt with the matter very shortly and expressed the view that " the notice was OK." He had before him not only the notice but what is relied on, the letter which I have just read, as a counternotice, and I am not surprised that on reading a notice given by solicitors to solicitors in this form and acknowledged by solicitors in this way, as a request that their letter should be treated as a formal notice that they wished to rely on the provisions of the Act, with which they were very familiar, he dismissed the objections of the tenants and gave the landlords leave to proceed with their action under the section—but without being referred to any authority or going into any real consideration of the question whether these defects in the notice, if established, were capable of waiver or could in fact be waived.

Are such defect or defects as were present in this case capable of waiver? First of all, the answer to that seems to me to be yes. Certainly it has not been submitted to us by Mr Mark that they are incapable of waiver, and that concession of his would appear to be well justified by the decision of four out of five members of the House of Lords in the case of *Kammins Ballrooms Co Ltd v Zenith Investments (Torquay) Ltd*, which is reported in [1971] AC 850. Were the two defects in this notice to which I have referred waived, and waived by this letter which I have just read of March 23? The only further facts or developments relevant to the question of waiver were these: that after that letter in March, nothing further was done until the landlords had taken out a summons on October 7 applying for an order giving them a right of re-entry and then for leave to take proceedings for enforcement of their right of re-entry and for damages for breach of covenant and for the directions of section 2 of the Leasehold Property (Repairs) Act as to costs and expenses incurred in relation to breaches of covenant which they wanted to prove.

On this question of waiver, Mr Prince relies on what is admittedly an isolated observation of Lord Denning MR in the case of *Sidnell v Wilson*, which is reported in [1966] 2 QB 67. There the main issue was whether a valid notice had been given, there having first of all been given an admittedly invalid notice, and subsequently a notice which it was alleged, read together with the earlier notice, complied with the requirements of the Act. And there was also the question whether the learned judge, whose decision was reversed, had approached the matter in the right way. But it is plain from p 73 of the report that counsel for the lessor, appealing from the judge's decision, refusing leave to bring proceedings, had submitted first that a valid notice had been given if the two documents were read together, and alternatively that: " the formalities were waived by the conduct of the tenants in serving the counternotice claiming the benefit of the Act of '938." There was no dispute that a valid counternotice had been served in that case.

At p 77 of the report, the Master of the Rolls came to the conclusion that the second letter would be regarded as a fresh notice under section 146 which contained the statement that the lessee was entitled to give a counternotice; and if one looks at the terms of that letter, it is not as full as the form in *Woodfall*, but it appears to set out all the necessary matters required by the subsection. Then he added this sentence: " Alternatively any defect was waived by the tenant giving a counternotice."

Harman LJ did not deal with the validity of the notice but concurred in allowing the appeal. Diplock LJ did. He said: " I agree, though with some hesitation, that the appeal should be allowed. I think the letter of February 24 1965 may properly be read in conjunction with the earlier notice . . . and those two documents together constitute a notice specifying the requirements of subsection (4) of section 1 of the Act," and then he went on to the other point. He did not specifically agree with the statement of the Master of the Rolls accepting the alternative ground that if, contrary to the opinion in which Diplock LJ concurred that there was a valid notice, there was a defective notice, none the less the defects in the notice were waived by the counternotice.

It is I think clear that the head-note, which makes the alternative ground of waiver of the defect part of the decision, is erroneous, and it is conceded by Mr Prince, quite rightly, that what Lord Denning said was an expression of his own opinion and no more; but it deserves the respect which all expressions of opinion from the learned Master of the Rolls call for, and I would very gladly accept it as applying to this case, if I could. It seems to me quite absurd that the tenants' solicitors, having written that letter clearly showing that they knew what their rights were, and knew what the counternotice involved, should be able to write the letter which I have read, clearly claiming the benefit of the Act and not raising any point on the validity of the notice whose terms were quite adequate for their purposes, and then, for purposes of delay when the matter came to court, to take the technical point that the notice was invalid. But however absurd it may be, the taking of technical points is something which the parties and their legal advisers are entitled to do and we have to enforce the law as it is.

The question of waiver was considered at length by their Lordships in the *Kammins* case to which I have already referred, and a majority of their Lordships, including the learned Lord who dissented, came to the conclusion on the question whether a particular matter was capable of waiver that there had been no waiver, no estoppel, no acquiescence, which prevented the landlords in that case from taking the point at a very late stage that application by tenants for a new tenancy of a dwelling under the Landlord and Tenant Act 1954 was not made in time. There the tenants had applied for a new tenancy and the landlords put in an answer of the kind which the Act permits. The proceedings had gone on and the tenants had no doubt incurred a considerable amount of expense; and it might have been thought that by putting in an answer to the application, and taking further steps which I need not detail because they are all set out in the report, the landlords had either waived the point which they sought to take at a time when the tenants were out of time for making any application and had lost their right to a new tenancy altogether, or were estopped from taking the point; but nevertheless the majority of their Lordships held that the landlords were entitled to take the technical point which they had not taken earlier, and there had been no waiver or estoppel.

Mr Prince has submitted that there is a distinction between that case and this in that what was involved there was a tenancy which continued under the Act until properly determined so that the tenants' estate was at risk Here, he says, the letter of March 23 did amount to a representation—I think he would say a clear and unequivocal representation—that the tenants did not intend to dispute the validity of the notice for the purpose of the 1938 Act, and that they did wish to invoke their rights under the 1938 Act to compel the landlords to get leave from the court before they enforced their rights under the underlease; and if one looks at that letter that is how one should read it, and if one reads it, there is here a clear waiver or an estoppel which made it impossible for the tenants to question the validity of this notice at the hearing before the master in October.

I have, however, felt compelled to conclude that, though no two cases are the same and though these difficult questions of waiver and estoppel are primarily questions of fact and here the question is one of the construction of an apparently clear document, quite different from the answer of the landlords and the rest of their conduct which was considered in the *Kammins* case, nevertheless this document of March 23 cannot be considered as more than a statement of intention to go ahead with the service of a counternotice and to go ahead with the claim for benefit under the Act, and it does not involve any promise or representation never to withdraw that intention or never to take any point on the validity of the notice. It was a wise move to preserve the position created by the serving of the notice, however defective, by putting in a counternotice, and this was done within a week of the notice being sent.

It would, I think, have been better if the tenants' solicitors had given earlier some indication that they might wish to take the point that the notice was defective. I would have thought that intelligent solicitors would have been able to see at once whether the notice was possibly defective, bearing in mind the obscurity of its syntax on any view; but they did not give that indication and the fact that they did not do that does not in my judgment prevent them from taking the point before the master and taking it now.

For those reasons, I have come to the conclusion first that the notice did not comply with section 1(4) of the 1938 Act and was therefore invalid. and secondly that though that invalidity was capable of waiver, it was not in fact waived by the letter of March 23. I therefore would allow this appeal.

Agreeing, Brandon LJ said: I accept the principle that the requirements as to the contents of the notice prescribed by section 1(4) of the Act of 1938 must be interpreted in a reasonable way. It is the substance of the notice that matters, not whether it follows precisely the wording of the subsection. There is no room for technicality in the matter. On the other hand the application of this principle must not be carried to the point where, because the notice is not expressed in terms which would be clear to an ordinary tenant, the protection which the Act was intended to afford such a tenant is significantly watered down. In my judgment, to hold that the notice with which we are here concerned

A satisfied the requirements of section 1(4) in either of the two respects discussed by my Lord, Stephenson LJ, would involve just such a watering down and would therefore be wrong.

With regard to the question of waiver, it seems to me that the letter of March 23 1977 contains on its true construction, by necessary implication, a representation that the solicitors for the tenants were not at that time contending that the notice was invalid. They were, on the contrary, proceeding on the basis that it was valid and that a counternotice was therefore appropriate. I cannot interpret the letter, however, as containing any promise not to rely on invalidity of the notice at some later stage if the solicitors should be advised
B that they had good grounds in law for doing so. The letter was written, I think, six days after the notice was served. It was a matter upon which solicitors would no doubt in due course consult counsel and, if they should be advised to put in a defence in the proceedings, which have not been mentioned in that letter it seems to me that they were perfectly entitled to act on such advice. For my part, I do not feel that there is any reason why the tenant should be C regarded as precluded by that letter from relying on the statutory grounds of invalidity of the notice, and in reaching that conclusion I do not, I think, share the regret which my Lord, Stephenson LJ, feels in the matter. It seems to me that the Act is in simple terms and there is no difficulty in those acting for landlords, who have professional qualifications and skill, from complying with the substance of it, and, if they fail to do so, I think it is not unreasonable that the responsibility should in the end rest with them.

Also agreeing, SIR DAVID CAIRNS said: If I can detect a shade of difference between my Lord, Stephenson LJ, and my D Lord, Brandon LJ, as to the invalidity of the notice, on the first ground I would respectfully agree with Brandon LJ about that.

The appeal was allowed, the order for costs being that the appellants were to have half the costs of the appeal. No order was made as to the costs below.

LANDLORD AND TENANT
RENT ACTS

A **COURT OF APPEAL**

December 12 1978

(Before Lord Justice MEGAW, Lord Justice ROSKILL and Lord Justice BROWNE)

DEMUREN AND ANOTHER v SEAL ESTATES LTD

Estates Gazette February 3 1979

(1978) 249 EG 440

Rent Act—Whether separate "licence" agreements resulted in non-exclusive licences unprotected by the Rent Act or whether in all the circumstances there was a protected joint tenancy—Somma v Hazlehurst and Aldrington Garages v B **Fielder discussed and distinguished—Cases depend on particular facts—"Curious" provisions in present "licence" agreements—Finding of fact by county court judge of oral agreement that the two applicants should together enjoy exclusive possession of whole flat—"Licence" agreements did not reflect real agreement—Whether there could be a joint tenancy with unusual feature that each joint tenant was personally liable only for his share of total rent—Held to be so in circumstances of present case—County court judge's decision that there was a protected joint tenancy upheld**

C This was an appeal by Seal Estates Ltd from a decision by Judge Figgis in Wandsworth County Court in favour of Ayodeji Alunkunle Demuren and Olufemi Adefope who had applied under section 105 of the Rent Act 1968 (now section 141 of the Rent Act 1977) for a declaration that they held a protected tenancy of Flat 6, Spencer Park Court, Spencer Park, London SW18. The facts and submissions are set out fully in the judgment of Megaw LJ.

Stuart Stevens and Miss P McCarthy (instructed by Gershon, Young & Co) appeared on behalf of the appellants; Stephen Sedley and Miss Tessa Moorhouse (instructed by Garratt Lane Law Centre) represented the respondents.

D Giving judgment, MEGAW LJ said: This is an appeal from the judgment of Judge Figgis given in the Wandsworth County Court on October 29 1977. The proceedings in the county court arose out of an application by two applicants, Mr Ayodeji Alunkunle Demuren and his cousin, Mr Olufemi Adefope, against, as respondents, Seal Estates Ltd. The application was made under the provisions of the then relevant Rent Act, section 105 of the Rent Act 1968, in which the applicants were, in effect, asking for a decision that they, the applicants, had the protection of the Rent Act in respect of their occupation of Flat 6, Spencer Park Court, Spencer Park, London SW18, of which property Seal Estates Ltd were the owners.

E The two applicants are both post-graduate students who had come from Nigeria to this country to pursue their studies. They were anxious to obtain accommodation together in London. They saw an advertisement which appeared in the *Evening Standard* on September 13 1976, which said: "Flats and Maisonettes to Let. Sharers: Available now or Oct 1st super s/c"—which means "self-contained"—"Flats and Houses. North, South, East and West London for 2, 4 or 6 single sharers. £7.50 per week to £12.50 per week each. Crawfords": and then a telephone number.

The applicants telephoned the telephone number given as being "Crawfords." Crawfords, it would appear, are estate F agents, who act for, among others, Seal Estates Ltd, or at any rate have some connection with them. There was really no evidence as to their status. Their precise position in this transaction is obscure; but it seems to me that they have got to be treated for present purposes as agents for Seal Estates Ltd who appear to be the owners of Spencer Park Court. The vagueness about all these matters will perhaps be less surprising when I mention that at the hearing no evidence was offered on behalf of Seal Estates Ltd. The applicants, having telephoned, arranged to see the flats at Spencer Park Court, of which they had no doubt been told by whoever, representing Messrs Crawfords, had answered the telephone. They went and saw G the flats. They were taken round them by a lady, Mrs Ainsworth, who lived in no 7. The flats which they saw, including no 6, had three rooms as well as a kitchen and bathroom. They liked flat no 6. It included the largest room, which was furnished as a sitting-room; the second largest room, which was furnished as a bedroom with two beds, two wardrobes, and so forth; and the third, the smallest of the living rooms, was furnished as a dining-room, with a table and chairs. They were told by Mrs Ainsworth, when they said they would like to take the flat, to get in touch with Mr Charles who, it appears, was a caretaker. Mr Charles in his turn told them to go to an address which he gave them, in Addison Gardens H in West London, with their cheque books, and to see somebody whose name, it would appear, was Mr Keating.

The two applicants went on that same day, September 13 1976, to the address in Addison Gardens and they saw there a young man, unidentified, who apparently said that he was Mr Keating's assistant. The applicants, according to the findings made by the judge on the evidence, told Mr Keating's assistant that they were interested in a flat. They were told that no 6 alone was available; and they said that they would take no 6. They were then given certain documents to read over, which they read; and after Mr Keating's assistant had dealt with other persons he turned his attention to the appli- J cants and, it would seem, invited them to sign, each of them separately, forms called "Applications," and also, having signed them, to sign and initial what purported to be licence agreements, each of them being invited to initial and sign a separate such agreement. Having done that, they were asked to give cheques for their proportion of the rent. As the applicant Mr Adefope had no bank account, the man with whom they were dealing suggested that Mr Demuren, who had a bank account, should sign cheques making them payable to Mr Adefope, and Mr Adefope endorsed those cheques over to somebody or other, apparently somebody representing Seal K Estates Ltd. Cheques were demanded and made over for the monthly instalments of the rent, or licence fee, whichever it may be payable by the applicants in connection with their occupation of Flat 6. So they gave post-dated cheques for the rent month by month, to be used monthly in advance for the payment of rent. (Mr Demuren ran out of cheques and had to return the next day to complete the required total of cheques). In addition, the applicants gave cheques for a "deposit," against, presumably, the possibility of non-pay-

A ment of the other cheques, or damage, or suchlike. They also, curiously, were asked to give, and did give, cheques each in the sum of £25 to whoever was the person dealing with them, apparently as being commission for the aid and assistance that that person, or Messrs Crawfords, gave in the transaction. If Messrs Crawfords were the owners' agents, it gives rise to certain queries that they should be taking commission from the applicants. If they were the applicants' agents other queries would arise. We need not seek to answer them.

The applicants asked when they could move in. They were told they could move into the flat at any time. In fact they did not move in until the following Saturday, September 18. B They asked if it was possible for them to stay beyond a year—again dealing with this on the collective basis of the two seeking occupation together. They were told by Mr Keating's assistant that if they wanted to stay beyond a year they would have to sign a fresh agreement and give fresh cheques.

Mr Adefope said in evidence that he thought he was signing to rent a flat. It is clear that they did read the documents—whether in whole or in part is not clear; and it is said that they ought to have understood that what they were signing were separate licences and not tenancy agreements. Neither of the gentlemen, so far as appears, was a lawyer or had any experience in law; and they were not asked any C questions as to what they knew about the technical difference between a tenancy and a licence or its possible effects. When they moved into the flat they changed some of the furniture around, so that two rooms would be furnished as bedrooms, one bed in each of two rooms, and one wardrobe. There were no numbers on any of the rooms in the flat, even though the so-called licence agreements refer to a room "numbered One."

The question that arose in the proceedings in the county court was whether in the circumstances the agreements which existed in law between the applicants, either jointly or severally, and Seal Estates Ltd were licences or were tenan- D cies. The learned judge held that they were tenancies and that accordingly these two applicants were entitled to the protection of the Rent Act. It is against that decision that the respondents in the county court, the appellants in this court, Messrs Seal Estates Ltd, appeal. Their appeal has been presented with great diligence and great perseverance by Mr Stevens of counsel on their behalf.

I would be content to say that I agree in its entirety with the judgment of Judge Figgis in the court below; that his findings of fact are findings which are justified, on the notes of evidence which we have seen; and that his conclusions of law drawn therefrom are unexceptionable. However, in this E court we had cited to us two authorities, cases decided in this court since the decision of Judge Figgis in the present case, and of which, therefore, obviously, he could not be aware because they were not in existence when he delivered judgment. Those two cases are: first, *Somma v Hazlehurst* [1978] 2 All ER 1011, a decision given by this court on March 3 of this year; and the other is a judgment given on July 7 of this year, in *Aldrington Garages v Fielder**, a case which so far as I know is not yet reported. We have been provided with a transcript. Mr Stevens' real point is that the present case is indistinguishable on its facts from those two cases, in which it was held that agreements that in some respects F were similar to the agreements with which we are concerned in the present case created licences and not tenancies.

In the second of the two cases, *Aldrington Garages v Fielder*, my lord, Roskill LJ made it quite clear that in his view that case was a case which depended upon its own particular facts and that neither it nor *Somma v Hazlehurst* was to be treated as laying down principles which could necessarily be applied in any case where there were similar but not identical agreements, or where there were relevant facts which were different from the facts in those two cases. In my judgment the facts in the present case distinguish it G from each of those two authorities, so that neither of them affects the decision of Judge Figgis.

I do not propose to go in detail into the terms of the purported licence agreements here. But there are certain points which I should mention. First, there were two separate agreements, signed separately by Mr Demuren and Mr Adefope. Second, in each of them there was a clause 9 which said "Nothing herein contained shall create the relationship of Landlord and Tenant between the parties hereto and the benefit of this Licence shall be personal to the Licensee." Beyond that there are certain curious provisions in the earlier clauses of the "licence" agreements. H I am using the word "licence" in inverted commas. The agreement itself is described at the beginning as being "This Licence made the 13th September 1976."

By clause 1 it is provided: "The Owner"—that is Seal Estates Ltd—"gives the Licensee permission to reside in common with such other person as shall from time to time be nominated by the Owner in the room numbered One (hereinafter called the room) forming part of the Flat known as Number Six (hereinafter called the Flat) in the building known as Spencer Park Court, Spencer Park, SW18."

Mr Adefope signed one such agreement referring to "the J room numbered One." Mr Demuren signed another agreement, in identical terms except for the name of the so-called "licensee"; and in his agreement also there was a reference to "the room numbered One." Although these were separate agreements, they both applied to the same flat; and both Mr Adefope and Mr Demuren were to "reside" (whatever that means) in "the room numbered One." There was, as the evidence showed, at that time no room numbered "1." A number "1" was stuck on thereafter, after controversy arose between the parties which ultimately resulted in the application to the court.

Although the agreements purported to say that these two K gentlemen would, under their separate agreements, "reside" in that supposedly identified room, clause 2 provided that "The Licensee shall be entitled to use the lounge dining room bathroom toilet kitchen (hereinafter called the common parts) in the Flat in common with all other occupants of any part of the Flat for the time being," It is "all other occupants," in the plural, as distinguished from "such other person as shall from time to time be nominated," which is in the singular, as provided in clause 1. The explanation which Mr Stevens offers (his clients gave no evidence at the hearing in the county court themselves or by their agents, Mr Crawford or Mr Keating's assistant, or anyone else) is that this was L what he described as being a common form agreement which might be used by the owners, Seal Estates Ltd, for flats which had more than one bedroom (not being the case in relation to the flat with which we are concerned). In those circumstances it is said that the owners might wish to provide that there should be two or more people residing in common in one bedroom and two or more people residing in common in another bedroom and that they should each be limited, so far as their exclusive residence is concerned—when I say "exclusive" I mean exclusive to the two or more people in a bedroom—to their particular, defined, bedroom; but that they all should be entitled to use the lounge, dining room, M bathroom, toilet and kitchen as being in common for all the occupants of all the bedrooms. It may be that that is the purpose of using the phrase "in common with all other occupants of any part of the flat." If so, it is an agreement which was not appropriate to the present case.

The so-called "licence" agreement, then sets out an even more curious provision. Clause 3 says: "The permission hereby granted shall continue until the 30th day of September 1977 unless previously determined by the Owner giving to the Licensee not less than one week's previous notice in writing." So far, so good—until one realises that under this

*See (1978) 247 EG 557 where the case was reported.

A agreement there is no corresponding provision for the "licensee," so called, to give one week's notice, or any other notice, in writing. So far as he is concerned the obligation which this agreement purports to put upon him, so far as it is capable of being reconciled with other clauses in the agreement, is that he is under an obligation, not, of course, to remain physically living in the nominated "room numbered One" up to September 30 1977, but, whether he continues to reside there or not and to make use of the common part, to continue to pay for his right to "reside" in room number one during the whole of the time up to September 30 1977. One provision, for the owner, that he can give one week's notice: B another, and very different, provision, for the "licensees," under which they have no corresponding provision for bringing the so-called "licence" to an end.

But it becomes even more curious when one comes to clause 4. I have said that when the agreement was made in the offices of Messrs Crawfords on September 13 1977—which stage was reached, as the judge has found, before these documents were signed—the applicants were both required to give cheques over the period of a year, post-dated month by month, in the sum of £54.17 each, as well as a cheque for "deposit." Clause 4 provides: "The consideration for such permission"—that is, the "permission to reside in common C with such other person" as should be nominated "in the room numbered One"—"shall be the sum of £54.17 per calendar month payable by the Licensee to the Owner in advance on the 1st day of each month the first payment being the appropriate proportion from the 17th September 1976 to the 30th September 1976 next being made on the signing hereof."

So, here are these applicants, who have given post-dated cheques covering the future, month by month, for a year, under an agreement in which their occupation as "licensees," so it is said, can be terminated by the owner at a week's D notice at any time: and they themselves have no corresponding right. So what would happen if the owner exercised his right under this agreement, if it be a right, to give a week's notice after one of the post-dated cheques had been paid? Would the position be that the owner would be in possession of payment for the ensuing month and the "licensee" (so called) would be turned out, possibly within a week of that payment having been made? It is said by Mr Stevens, on the basis of some propositions which I am afraid I am wholly unable to follow, that in those circumstances the "licensee" would be able to get back the proportionate balance of his month's payment by some action on the basis of a consider- E ation which had partly failed.

To my mind the provisions of those two clauses are irreconcilable. There is something which is so badly wrong with this agreement that one is bound to look at it with the gravest suspicion.

The real argument put forward on behalf of Seal Estates Ltd in this court to suggest that the learned judge was wrong was the fact that under these two "licence" agreements each of the "licensees" separately is obliged to pay month by month £54.17 for his shared accommodation. It is said that that is inconsistent with its being a joint tenancy agreement because, if it were a joint tenancy agreement, each of the two F joint tenants would be under a liability to pay the whole of the rent, subject, of course, to the right to reimburse himself, if he did pay the whole of the rent, from the other joint tenant. Judge Figgis was impressed with that point. In his judgment, having come to the conclusion, as I think for wholly unexceptionable reasons, that in all other respects the indications were that the true nature of these arrangements was a tenancy and not a licence, he went on: "That does not necessarily mean that a tenancy and not a licence was created. There is still the question raised by the fact that I am satisfied that it was a term that each applicant should be liable for one half of the money payment. It seems to me to be a matter of some difficulty which side of the line that may make the arrangement taken as a whole. G My conclusion is that it is not necessarily inconsistent with a tenancy for a landlord to agree not to look to each of two joint tenants for more than one half of the rent."

Mr Stevens pointed to what was said in the judgment of the court delivered by Cumming-Bruce LJ in *Somma v Hazlehurst* [1978] 2 All ER 1011 at p 1021, in relation to the two separate agreements which had been signed in that case (as two separate agreements had been signed here). The county court judge in that case, without going into any of the background evidence as to the making of the contract, and without inquiring whether or not there had been an oral H contract to which the agreements purported to give effect, had held that the two agreements were to be treated as joint agreements. Cumming-Bruce LJ said this: ". . . . there was no basis for saying that they were joint agreements at all"—that is, in the absence of any extraneous evidence. Then he went on:

> However, if that is too legalistic a way of looking at the situation, as counsel for the respondents submits, it is impossible to reconstruct the separate obligations into one joint obligation without doing violence to the obvious intentions of the parties, particularly of the two grantees, H and S. For example, if one takes clause 1, which deals with the consideration which the grantees are bound to pay, J it would be necessary, as we see it, either to have a joint interest coupled with a several liability for half the weekly amounts, which is something of a logical inconsistency, or else to re-draw the terms of the agreement to make each grantee jointly and severally liable for the full four-weekly instalment of £77.60, as against the contractual liability resting on each under the agreement of £38.80. This is not construing the agreement between the parties: it is re-writing it in the absence of any claim for rectification or any allegation of fraud. Counsel for the respondents was unable to provide any answer to this problem which came near to satisfying us on this point.

Geoffrey Lane LJ, in delivering the first judgment in *Aldrington Garages v Fielder*, K regarded that part of the judgment of the court in *Somma v Hazlehurst* as being substantially the *ratio decidendi* of the case. There, again, there had been two separate agreements. There, again, the county court judge had not made any findings of fact as to any oral agreement leading up to the committing of it into writing. Geoffrey Lane LJ, at page 12/F of the transcript of the judgment, said this:

> If the two agreements are separate then there is no exclusive possession and no tenancy. If they are joint, how is the obligation upon the occupiers to pay their consideration, £54.17 per month, to be construed? In order to effect the intention of the parties, if it is a joint tenancy, L it would be necessary to rewrite the agreements, as I see it, to make both Mr Fielder and Miss Maxwell jointly and severally liable for the total consideration, namely, £108.34 per month, as opposed to the several liability, under the agreement, of £54.17 each. As Cumming-Bruce LJ said in *Somma v Hazlehurst*: "This is not construing the agreement between the parties; it is rewriting it in the absence of any claim for rectification or any allegation of fraud."

"That, indeed," said Geoffrey Lane LJ, "was one of the *rationes decidendi* in *Somma's* case, and we are accordingly bound by it. In my view it concludes effectively this case."

In the present case, in contrast, the learned judge did make findings of fact as to what had been the agreement between M these two applicants and Mr Keating's assistant at Crawfords' office on September 13 1976. The judge said this:

> There was no evidence as to what the landlords really intended, apart from inferences which may be drawn. Counsel for the respondents elected, as he was entitled to do, to call no evidence. I find that when the parties went to see the person who told them to fill-in these documents, they said they wanted to take one of the flats. They were told only Flat 6 was available and said they would like to take it. In so far as the documents provide that they should reside in common in room no 1, they do not represent what anyone contemplated, including the landlord. If anyone had gone

A into the witness-box on their behalf to say that they expected two applicants to occupy one bedroom and the rest of the flat as a lounge and dining room, I would not have believed him. In my judgment there was a mutual, concurrent intention of all the parties that the applicants should together have and enjoy exclusive occupation of the whole of the flat. Indeed that seems to be borne out by the fact that the advertisement refers to a self-contained flat.

In my opinion there was ample evidence to justify the learned judge in reaching the conclusion that there was such an oral agreement which had been made between the parties before the so-called " licence " agreement came to be signed B on that day by the two applicants and, on some unknown date thereafter, before September 21, by somebody representing Seal Estates Ltd.

There would, therefore, have been, on that basis, strong ground for thinking that there could have been a claim for rectification of those written agreements as not giving effect to what was the express intention of the parties in arriving at their oral agreement. But, however that may be, there would still remain the question about the joint tenancy, having regard to the provision for separate payment of rent by the two licensees or joint tenants, as the case may be. I find it unnecessary to go further into that matter, having regard to C the particular facts of this case as they were, which in my judgment distinguish them from both of the cases which I have cited, *Somma v Hazlehurst* and *Aldrington Garages v Fielder*.

A joint tenancy with each of the two tenants paying only one half of the total amount of the rent due for the joint tenancy could indeed, in certain circumstances, be regarded as creating a logical inconsistency. It would particularly be so, perhaps, if the position were that the landlord, if one of the two joint licensees or tenants walked out, would be left without having any recourse in respect of the rent that was due under the agreement from that person other than the personal action D against him. In this case one would have perhaps less sympathy for the owners than might otherwise have been the case because under clause 1 of their own agreement, assuming that it is valid and effective, they would have the right to nominate some other person to occupy the room numbered " 1," charging such payment as might be appropriate to that person thereafter. Moreover, they would have their remedy against the outgoing licensee, if this licence agreement is capable of being given any meaning at all, because, as I have said, there is no provision in that agreement for the licensee to cease to be liable for payment up to September 30 1977 merely because he chooses to leave his residence in room no E 1. There might indeed be a problem after September 30 1977 in relation to a continuing rent-restricted tenancy; but, again, the provisions of clause 1 would seem to be of some avail there.

However, in the circumstances of the present case there are facts which in my view make it not " logically inconsistent " at all, as it was thought to be logically inconsistent in the two earlier cases. There is here the provision of the post-dated cheques, the effect of which is that the landlord has got his payment in advance up to September 30 1977 from each of the two applicants. He has secured for himself the whole F amount due from the two joint tenants (if that is what they are) for the whole period. It seems to me that here there is no logical inconsistency in the conclusion at which the judge arrived, namely, that the agreement was one which involved that, though it was a joint tenancy, nevertheless, unusually, each of the two should be personally liable for, and only for, his share of the total rent of the flat. That is something which may be unusual but is not, as I see it, such as to prevent it from being a joint tenancy, at any rate in the circumstances of this case. It represents, as I see it, the oral agreement between the parties in this case.

In so far as this curious, and in some respects unintelligible, " licence " agreement purports to provide otherwise, it is not G consistent with the reality of the agreement which was made for the letting of a shared flat, particularly because of the advance payments for a year (inconsistent with the purported contractual right to determine on a week's notice) and because of the absence of any provision which could properly or sensibly be said to preclude exclusive possession of the flat by the two sharing tenants. The reference to " all other occupants," when there could be none such in the intention of the parties, is a strong indication that the " licence " agreement does not reflect the real agreement.

I would, accordingly, dismiss this appeal.

Agreeing, ROSKILL LJ said: Mr Sedley in his submission H this morning for the respondents invited us to decide this appeal in a manner which might give some assistance to county court judges who are, day in day out, having to grapple with this problem where owners of property are seeking, perhaps understandably in the circumstances, to get the maximum financial advantage from their properties and to avoid what they no doubt regard as the irksome fetters of the Rent Act. On the other hand, as my Lord said in *Buchmann v May* [1978] 2 All ER 993 at p 1001, ". . . . the court would, and should, be astute to detect a sham, particularly in a case such as this, where the protection of the Rent Acts is involved." J

I do not think it is possible to deal with these cases save in relation to the facts of each particular case, anxious as one always is to help those courts which have to grapple with these problems even more often than we now do. I draw attention to what this court said in a judgment given by Cumming-Bruce LJ in *Somma v Hazlehurst*: " At the end of the day it was clear to us that the instant case is very much a case peculiar to its own facts. The difficulties, which were real, were largely difficulties in determining the content of the findings of the judge."

In the most recent of these cases, *Aldrington Garages v Fielder*, decided on July 7 1978, but I think (as does my Lord) K unreported, we were at some pains to emphasise that these cases do turn on their facts. I hope I may be forgiven for referring to something I said at p 13 of the transcript: " If one applies the reasoning in *Somma's* case to the words and facts of the present situation, it seems to me that the result is inevitably that the agreement, in the circumstances of this case, and against the factual background of this case, creates a licence and not a tenancy. That is not to say—and I repeat what I said earlier on—that these precise words would necessarily, in every case, amount to the creation of a licence: not in every factual circumstances would they produce the same result as they have here." L

As my Lord has said, the factual background of the instant case is very different indeed from that attending both *Somma v Hazlehurst* and *Aldrington Garages v Fielder*. In those two cases there was no finding comparable with the findings which the learned county court judge has made in this case. Those findings are: first, " In my judgment there was a mutual, concurrent intention of all the parties that the applicants should together have and enjoy exclusive occupation of the whole of the flat. Indeed that seems to be borne out by the fact that the advertisement refers to a self-contained flat." Secondly: " I find that the agreement which was actually come M to, to be inferred from the circumstances indicated, is one of tenancy of the whole flat to the applicants jointly."

Mr Stevens argued that there was no evidence to justify those findings. I am afraid I emphatically disagree. Looking back to the learned judge's very full and careful note of the evidence one finds Mr Adefope giving this evidence in chief: " We said we were interested in one of the flats. He "—that is, the landlord's agent—" said only Flat 6 was available, others had already been taken. We said we would take Flat 6. He gave us some forms to read over and complete while he was attending to two other girls who he said were taking Flat 8."

A Then there were post-dated cheques for the whole of the period in question; and this agreement, whatever its character, was an agreement for a fixed term. It was not even expressly extendable at the end of that term. The learned judge rightly said that the appellants took post-dated cheques to cover every instalment of the alleged licence period.

It seems to me, therefore, that the evidence abundantly justified the conclusion of fact at which the learned county court judge arrived. In my judgment, before those two young men from Nigeria left the agent's office that evening there had been a concluded agreement on the terms that they would take Flat no 6 for the monthly payments which had already been B made by means of these post-dated cheques; and the true nature of that agreement was that of a tenancy. I am quite unable to see that the curious and rather ill-drawn document that these young men had to sign afterwards in any way necessitates reaching a different conclusion of fact from that which the learned county court judge reached. It seems to me clear that the appellants' counterpart of that document was not signed until some days later. That document simply does not begin to reflect the true transaction between the parties.

I would add (I think my Lord has already referred to this) that I find clause 3 the most astonishing clause. Not only C does one have the curious point, mentioned by my Lord, of the interaction between clauses 1 and 2, but one finds in clause 3: "The permission hereby granted shall continue until the 30th day of September 1977 unless previously determined by the Owner giving to the Licensee not less than one week's previous notice in writing." Yesterday Mr Stevens argued that that clause could be effective to determine this agreement within a week notwithstanding that on its face it was an agreement for a fixed term of approximately a year. Today, under pressure from the court, he argued that those last words in clause 3 were irrelevant surplusage. It seems to me plain that those last words cannot be made to fit with an D agreement of this kind for a fixed term where the antecedent agreement involved that the totality of the payments made were made by post-dated cheques. I am driven to the view that these documents, though they may have a different result on some other occasions against a different factual background, on this occasion simply do not begin to represent the true bargain between the parties.

For those reasons, in addition to what my Lord has said, I would dismiss the appeal.

I would end by expressing my appreciation of the work that has been done by the Garratt Lane Law Centre in Wandsworth in seeking that these two young men from Nigeria were E able to get their rights, both in the court below and in this court.

BROWNE LJ said: I agree so entirely with the reasons already given by my Lords that I find I cannot usefully add anything. I should like only to say that I agree with the last remark made by Roskill LJ.

The appeal was dismissed with costs. Legal aid taxation of respondents' costs ordered. Leave to appeal to the House of Lords refused.

F

COURT OF APPEAL

November 15 1978

(Before Lord DENNING MR, Lord Justice BRIDGE and Lord Justice TEMPLEMAN)

QUENNELL v MALTBY AND ANOTHER

Estates Gazette March 24 1979

(1978) 249 EG 1169

Rent Acts—Tenancy created by mortgagor landlord contrary to restriction in mortgage instrument—Rule in Dudley & District Benefit Building Society v Emerson that tenant is not protected against eviction by mortgagee in right of title paramount is subject to control by equity—Legal charge to bank containing prohibition of grant of tenancies without chargee's consent—Tenancy granted in contravention—Bank's loan paid off by wife of mortgagor—Charge assigned to wife—Wife claims possession from tenants on ground that tenancy is not binding upon her as mortgagee—Held that court could look at substance of transaction—Mortgagee's right to recover possession must be exercised bona fide and reasonably to enforce security—Here there was an ulterior motive, namely, to get possession for benefit of husband in order to sell house at a profit—Appeal from county court allowed and action for possession dismissed—Equity evidently not yet past age of childbearing G H

This was an appeal from a decision of Judge Granville Wingate at Lewes County Court in favour of Mrs Joan Marilyn Gillespie Quennell, who sought possession of a house at 6 Wallands Crescent, Lewes; against Peter Jeffery Maltby and Roderick Alexander Lupton, who had been granted a tenancy by Mrs Quennell's husband, the owner of the house. At the time when the tenancy was granted the house was subject to a legal charge in favour of Barclays Bank for a loan to Mr Quennell. The legal charge included a prohibition of the creation of tenancies without the bank's J consent, which was not obtained. The facts are fully stated in the judgment of Lord Denning.

D Lamming (instructed by Donne, Mileham & Haddock, of Brighton) appeared on behalf of the appellants, the defendants below; P de la Piquerie (instructed by Anscombe, Hollinworth) represented the respondent, the plaintiff below.

Giving judgment, LORD DENNING MR said: If the judgment of the judge below is right, it will open a gap in the protection which is offered to tenants by the Rent Act. I will first give the facts.

Mr Quennell is a gentleman who lives in Cheyne Row in London. But he is the owner of a large house in Lewes. K It is 6 Wallands Crescent, Lewes, with about nine bedrooms. He has an agent in Lewes who looks after it for him. The house is very suitable for students. In 1973 the agent let it to some students of the University of Sussex. Two of them became the tenants. They were Mr Maltby and Mr Jack. They were let into possession for a term of one year at a rent of £90 a month, expiring on December 31 1974. They had other students there with them, about nine students in the house. While Mr Maltby and Mr Jack were tenants Mr Quennell borrowed money from Barclays Bank and mortgaged this house to secure the loan. It was only for the L sum of £2,500. He executed a legal charge on August 13 1974 in favour of Barclays Bank to cover any moneys which might from time to time be owing to the bank. In that legal charge there was this clause 4, which is in common form:

> During the continuance of this security no statutory or other power of granting or agreeing to grant or of accepting or agreeing to accept surrenders of leases or tenancies of the Mortgaged Property or any part thereof shall be capable of being exercised by the Mortgagor without the previous consent in writing of the Bank.

That meant thereafter from August 13 1974 so long as M this legal charge subsisted to the bank, Mr Quennell could not let the premises or accept surrenders without the consent in writing of the bank.

The tenancy of Mr Maltby and Mr Jack came to an end at the end of December 1974. The house was then relet to two other students, a Mr Quilter and a Mr Lyeth, again for a year. It was not relet to Mr Maltby because it was thought that he was going to the United States. As it happened Mr Maltby did not go to the United States. In fact he stayed on living in the house. So did several other students. At all events, the important thing to note is that the bank

A did not give its consent to this letting to Mr Quilter and Mr Lyeth. No one asked the bank for its consent. No one realised it was necessary. No one interfered and nothing happened. The year 1975 passed. Then at the end of that year there was a fresh letting. This was between Mr Quennell as landlord and Mr Maltby and a Mr Lupton as tenants. That tenancy lasted until December 1976. Again no one asked the bank for consent. No one realised it was necessary. And from January 1977 onwards the tenants remained as statutory tenants, paying the rent to the agents.

The position then arose that Mr Quennell wanted to get possession of the house. If he could get vacant possession,
B he could sell it at a high price. It might be worth £30,000 to £40,000 with vacant possession. Mr Quennell started proceedings for nuisance and annoyance, but he dropped them. Then he went to lawyers for advice. After consulting them, in October 1977 Mr Quennell went to the bank and told them about the tenants in the house. The bank had not heard before about the various changes in the tenancies. Even when they were told the bank made it clear that they had no intention of taking any proceedings to enter the property or to turn the tenants out or anything of that kind. The bank were not concerned to get possession.

Then Mr Quennell's lawyers in London advised him that
C there was a good way in which possession could be achieved. This is what it was: Mr Quennell's wife, Mrs Quennell, paid off the bank. She paid the £2,500 which was owing to the bank and took a transfer of the charge. The bank transferred it to her by a transfer dated January 17 1978. Then Mrs Quennell brought proceedings against the tenants, Mr Maltby and Mr Lupton, seeking possession. She said that she stood in the shoes of the bank, and, seeing that the tenancy was granted without the consent of the bank, it was void. So she could recover possession.

The judge accepted this submission. He held that the wife, Mrs Quennell, was entitled to possession of the premises
D and could turn Mr Maltby and all the other students out of the house. Now it has been held that, when the bank holds a charge and there is a clause in it whereby there are to be no tenancies granted or surrendered except with the consent of the bank in writing, then in those circumstances, if the mortgagor does thereafter grant tenancies without the consent of the bank, those tenancies are not binding on the bank, and the tenants are not entitled to the protection of the Rent Acts. That was decided in *Dudley & District Benefit Building Society* v *Emerson* [1949] Ch 707. Mrs Quennell relies on that case. She says that, as transferee
E of the legal charge, she stands in the shoes of the bank and can obtain possession.

The judge accepted that submission. His decision, if right, opens the way to widespread evasion of the Rent Acts. If the owner of a house wishes to obtain vacant possession, all he has to do is charge it to the bank for a small sum. Then grant a new tenancy without telling the bank. Then get his wife to pay off the bank and take a transfer. Then get the wife to sue for possession.

That indeed was what happened here. In October 1977, when Mr Quennell went to the bank, he told them about the tenancies. They said that they did not intend to take pro-
F ceedings. So he got Mrs Quennell to do it. In evidence, she said: "I paid £2,500. This was for my husband. I took the charge to make the debt to his bank less onerous. I was aware he wanted to obtain possession of the house to sell it. I merely paid off the charge. These proceedings have been brought to get possession to sell." So the objective is plain. It was not to enforce the security or to obtain repayment or anything of that kind. It was in order to get possession of the house and to overcome the protection of the Rent Acts.

Is that permissible? It seems to me that this is one of those cases where equity steps in to mitigate the rigour of the law. Long years ago it did the same when it invented the equity
G of redemption. As is said in *Snell's Equity*, 27th ed (1973) on p 376:

> The Courts left the legal effect of the transaction unaltered, but declared it to be unreasonable and against conscience that the mortgagee should retain as owner for his own benefit what was intended as a mere security.

So here in modern times equity can step in so as to prevent a mortgagee, or a transferee from him, from getting possession of a house contrary to the justice of the case. A mortgagee will be restrained from getting possession except when it is sought bona fide and reasonably for the purpose of enforcing the security and then only subject to such conditions as the
H court thinks fit to impose. When the bank itself or a building society lends the money, then it may well be right to allow the mortgagee to obtain possession when the borrower is in default. But so long as the interest is paid and there is nothing outstanding, equity has ample power to restrain any unjust use of the right to possession.

It is plain that in this transaction Mr and Mrs Quennell had an ulterior motive. It was not done to enforce the security or due payment of the principal or interest. It was done for the purpose of getting possession of the house in order to resell it at a profit. It was done so as to avoid the protection which the Rent Acts afford to tenants in their occupation. If
J Mr Quennell himself had sought to evict the tenants, he would not be allowed to do so. He could not say the tenancies were void. He would be estopped from saying so. They certainly would be protected against him. Are they protected against his wife now that she is the transferee of the charge? In my opinion they are protected. For this simple reason, she is not seeking possession for the purpose of enforcing the loan or the interest or anything of that kind. She is doing it simply for an ulterior purpose of getting possession of the house, contrary to the intention of Parliament as expressed in the Rent Acts.

K On that simple ground it seems to me that this action fails and it should be dismissed. The legal right to possession is not to be enforced when it is sought for an ulterior motive. I would on this account allow the appeal and dismiss the action for possession.

Agreeing BRIDGE LJ said: The situation arising in this case is one, it seems to me, in which the court is not only entitled but bound to look behind the formal legal relationship between the parties to see what is the true substance of the matter. Once one does that, on the facts of this case it is as plain as a pikestaff that the purpose of the bringing of these proceedings via Mrs Quennell is not for her own benefit to
L protect or enforce the security which she holds as the transferee of the legal charge but for the benefit of her husband as mortgagor to enable him to sell the property with the benefit of vacant possession. In substance she is suing as his agent. That being so, it seems to me inevitably to follow that she can be in no better position in these proceedings than her husband would be if they had been brought in his name. If they had been brought in his name, it is clear that the defendants would have had an unanswerable defence under the Rent Acts.

I agree that the appeal should be allowed.

Also agreeing, TEMPLEMAN LJ said: I agree that the appeal
M should be allowed. The landlord Mr Quennell, finding that he was encumbered by a statutory tenant and not able to reap the benefit of a sale with vacant possession, devised under advice a scheme whereby he might obtain vacant possession. It so happened that the landlord had mortgaged the property to his bank to secure his overdraft and other borrowings, and the mortgage contained a common form prohibition on any lettings without the consent of the mortgagee bank.

The lease to the statutory tenant was made by the landlord after the date of the mortgage without the consent of the bank and was therefore in breach of the landlord's covenant contained in the mortgage. That lease was binding on the land-

A lord but void against the bank. On expiry of the lease the tenant became a statutory tenant as against the landlord but not as against the bank.

The landlord being unable to get possession from his own statutory tenant approached the bank and asked the bank to bring an action against the tenant for possession. This would then enable the landlord to sell the property with vacant possession. The bank very properly declined to take any such action which was not required to protect their position as mortgagee. The amount of the debt owed by the landlord to the bank was £2,500, the rent payable by the tenants exceeded £1,000 a year, and the property was worth in the region of
B £30,000 to £40,000. The bank in these circumstances rightly refused to do for the landlord that which the landlord could not do for himself.

The landlord, again under advice and undaunted, conceived an alternative method of obtaining vacant possession. His wife paid off the debt of £2,500 owed to the bank by her husband-landlord; and the bank, as it was bound to do accepted that payment and transferred the mortgage to the wife. The landlord's wife (then the mortgagee) was owed £2,500 by her husband; and she, at the request of her husband, brings an action against the tenant for possession claiming that the lease made by her husband is not binding on her as
C mortgagee and that she can therefore obtain possession and then sell to the benefit of herself and her husband.

As I say, the authorities establish that as a matter of law a lease made in breach of covenant by a mortgagor is void against the mortgagee and, I assume for present purposes, against the transferee unless the lease is adopted by the mortgagee. Neither the bank nor the wife adopted the tenancy. The estate, rights and powers of a mortgagee, however, are only vested in a mortgagee to protect his position as a mortgagee and to enable him to obtain repayment. Subject to this, the property belongs in equity to the mortgagor. In the
D present case it is clear from the facts and the evidence that the mortgagee Mrs Quennell is not bona fide exercising her rights and powers for her own purposes as mortgagee but for the purpose of enabling the landlord mortgagor (her own husband) to repudiate his contractual obligations and defeat the statutory tenancy of the tenant which is binding on the landlord. Mrs Quennell does not even pretend to be acting in her own interests as mortgagee. She brings this action to oblige her husband. In my judgment the court must therefore treat this action, although in form brought by a mortgagee, as an action brought for and on behalf of the landlord mortgagor. The court should deal with it as though the mortgagor
E landlord were the plaintiff, and on that basis possession will not be ordered.

The appeal was allowed with costs in the Court of Appeal and below. Legal aid taxation ordered for the successful appellants (defendants below).

HOUSE OF LORDS

March 8 1979

(Before Lord WILBERFORCE, Lord DIPLOCK, Lord SALMON, Lord FRASER OF TULLYBELTON and Lord SCARMAN)

F **TILLING v WHITEMAN**

Estates Gazette April 7 1979
(1979) 250 EG 51

Rent Act—House "required as a residence for the owner-occupier"—Case 10 in Schedule 3 to Rent Act 1968 (now Case 11 in Schedule 15 to Rent Act 1977)—House jointly owned but required as a residence by one of the joint owners only—Strict interpretation of Court of Appeal reversed by majority of House of Lords—House of Lords adopts a liberal construction which "might, to a conveyancer, appear loose"—One joint owner may satisfy the conditions for obtaining possession—McIntyre v Hardcastle distinguished—"Unfortunate course" of county court proceedings—"Treacherous short-cut" involved in preliminary point of law—Disturbing position as to respondent's liability for costs—Dissent by Lord Fraser G

This was an appeal by Mrs Tilling from the decision of the Court of Appeal, who had upheld a decision of the county court judge dismissing a claim by Mrs Tilling and her co-owner, Miss G L M Dossett, for possession of a cottage in Staple, near Canterbury, let to the respondent, Miss Whiteman. The claim for possession, based as Case 10 in Schedule 3 to the Rent Act 1968, which was then in force, was on the ground that the cottage was required as a residence by Miss H
Tilling. Her co-owner, Miss Dossett, did not wish to reside in the cottage, having left, before the grant of the tenancy to Miss Whiteman, to live in Oxfordshire. Judgments in the Court of Appeal were reported at (1978) 246 EG 1107.

Ronald Bernstein QC and C Summer (instructed by Williamson & Barnes, of Deal) appeared on behalf of the appellant; the respondent, Miss Whiteman, in person.

Delivering the first speech, LORD WILBERFORCE said: This appeal arises under Case 10 of Part II of Schedule 3 to the Rent Act 1968, which reads as follows:

> Where a person who occupied the dwelling-house as his residence (in this Case referred to as "the owner-occupier") let it on a J
> regulated tenancy and—(a) not later than the relevant date the landlord gave notice in writing to the tenant that possession might be recovered under this Case, and (b) the dwelling-house has not, since 8th December 1965, been let by the owner-occupier on a protected tenancy with respect to which the condition mentioned in paragraph (a) above was not satisfied, and (c) the court is satisfied that the dwelling-house is required as a residence for the owner-occupier or any member of his family who resided with the owner-occupier when he last occupied the dwelling-house as a residence.

If these conditions are satisfied, section 10 of the Act requires an order for possession to be made. The appellant, Mrs K
Tilling, owns a small house in Canterbury jointly with Miss G L M Dossett. She claims to have been in occupation of it immediately before February 19 1975. On that date the joint owners let it to the respondent, Miss Whiteman, for two years. The tenancy agreement contained a clause whereby Miss Whiteman agreed to yield up the premises at the end of the tenancy. There was a statement, signed by the joint owners, addressed to Miss Whiteman that under the Rent Acts 1968 and 1974 the landlord (*sic*) may recover possession of the premises under the provisions of Case 10 (*supra*).

Miss Whiteman did not yield up possession as she had L
agreed, so the owners brought proceedings in the Canterbury County Court for possession and other relief. Pleadings were exchanged, and the case came on for trial in May 1977 with both sides legally represented. The learned judge took what has turned out to be an unfortunate course. Instead of finding the facts, which should have presented no difficulty and taken little time, he allowed a preliminary point of law to be taken, whether Case 10 applies to a case where there are joint owners one only of which requires the house as a residence. So the case has reached this House on hypothetical facts, the correctness of which remain to be tried. I, with others of your Lordships, have often protested against the M
practice of allowing preliminary points to be taken, since this course frequently adds to the difficulties of courts of appeal and tends to increase the cost and time of legal proceedings. If this practice cannot be confined to cases where the facts are complicated and the legal issue short and easily decided, cases outside this guiding principle should at least be exceptional.

My Lords, the legal issue in the present case is not an easy one. Case 10, and section 10 of the 1968 Act upon which it is based, say nothing about joint owners, or joint occupiers. To read, or not to read, the singular expressions ("person," "landlord") as including the plural, gives rise to difficulties,

A as the judgments below well demonstrate. In my opinion our task must be to attribute that reasonably admissible meaning to the language which will best carry out what appears to be the legislative intention.

The two alternative views are clearly and forcefully set out in the judgments of the Court of Appeal.

The first is that, for an order for possession to be made, the house must be required for the residence of both co-owners. This commended itself to Stephenson LJ and Shaw LJ. There is no doubt that a powerful case can be made for it on the language used, and some further support may be derived from the Court of Appeal decision in *McIntyre v*
B *Hardcastle* [1948] 2 KB 82, decided on what became Case 8 in the same Schedule.

The second alternative is that there is no such requirement, and that each of the three conditions stated in Case 10 are on the agreed or assumed facts satisfied in the present case. First, Mrs Tilling occupied the dwelling-house as her residence. Second, she let it on a regulated tenancy. Third, she requires it as a residence for herself. On these facts, the court must make an order for possession. This was the opinion of Eveleigh LJ.

My Lords, I propose to do little more than to say that,
C having to the best of my ability compared the weight of these rival arguments, and having carefully considered Miss Whiteman's printed case, I have come to the conclusion that on balance, the judgment of Eveleigh LJ is to be preferred. The arguments in its favour are so clearly stated in the Lord Justice's judgment that nothing would be gained, and something might be lost, by my restating them in my own language. I will only add two observations.

First, the purpose of this piece of legislation, added to the bulky corpus of rent legislation in 1965, was to induce occupiers of dwelling-houses, who for some temporary reason desired, or had, to reside elsewhere for a time, to make their
D premises available for letting to others, on the basis that on their return they would be able, without dispute, to regain possession. (In cases within Case 10 an order for possession is mandatory.) The emphasis is upon occupation: the person concerned must have occupied the dwelling-house as his residence, and must require it as his residence, or that of a member of his family who resided with him when he last resided there. As compared with this emphasis on occupation and residence, ownership plays a subsidiary part. It enters into the matter only because of the inherent fact that the dwelling-house is let, and letting is effected by a landlord. But Case 10, and the policy behind it, is not, if I may person-
E alise, interested in the landlord: he is not, as such, the key figure: that is the "owner-occupier." This consideration, to my mind, provides justification for avoiding a strict interpretation of the words "let it on a regulated tenancy"—the words which mainly, if not wholly, support the argument that one of two or more joint owners cannot satisfy the Case unless they both also require to reside in the house. I find it therefore possible to say that Mrs Tilling, being at the time the occupier, when she decided to go to live elsewhere for a time, let her house on a regulated tenancy, even though, for the letting to be effective, Miss Dossett had to join in. This
F interpretation might, to a conveyancer, appear loose, but is one which might easily appear in common parlance. For the reasons I have given I do not think that the strict conveyancing meaning is intended to be imposed.

Secondly, as regards *McIntyre v Hardcastle* the wording in Case 8 (which has a separate and much longer history, being derived through the Act of 1933 from that of 1920) is (relevantly) "where the dwelling-house is reasonably required by the landlord for occupation as a residence for (a) himself . . ." Here there is an identity between the person who has let the house and the person who requires it. There must then be great strength in the argument that if for one purpose the plural is deemed to be included in the singular, so it must G
for the other—plural landlords must require the house for themselves, not for one of them. This was the argument accepted by the Court of Appeal in *McIntyre's* case. But assuming the correctness of this (and I agree that the question remains open in this House), the argument under Case 10 is different. There is no imposed identity between occupier and landlord: there may be a plurality of landlords, but only one occupier, and it may be possible to say that one of these landlords has let. I find therefore no necessity, or indeed attraction, in following the earlier case.

I would allow the appeal and remit the case to Canterbury County Court to decide the case on the basis that Case 10 H
applies to the agreed or assumed facts.

As to the costs, while I have sympathy with Miss Whiteman as regards the burden which our expensive legal system places on litigants with small resources, I have to bear in mind that, for whatever reason, she did not pursue an application for legal aid in the Court of Appeal, or make an application in this House. Apart from the Legal Aid Fund there is no other fund of public money out of which the costs she is liable to pay can be provided. I can see no alternative to an order that she must pay the appellant's costs in this House and in the Court of Appeal.

LORD DIPLOCK agreed that the appeal should be allowed. J

In a speech agreeing that the appeal should be allowed, LORD SALMON said: Two elderly ladies, Mrs Tilling and Miss Dossett, who are not related to each other, jointly own a cottage in the village of Staple, near Canterbury. Miss Dossett had once resided in the cottage with her friend Mrs Tilling, but some time prior to the tenancy to which I shall presently refer, she went to live in Oxfordshire where she has ever since continued to reside. Mrs Tilling was still living in the cottage immediately prior to February 19 1975. The learned county court judge found it to be an agreed fact that both these ladies had occupied the cottage as their residence, K
"Mrs Tilling immediately before the tenancy, Miss Dossett some time earlier."

By an agreement in writing dated February 19 1975, Mrs Tilling and Miss Dossett, as joint owners of the cottage, let it furnished on a regulated tenancy to Miss Whiteman for a period of two years from February 21 1975 at a weekly rent of £12.50. At the foot of this agreement the following notice appears:

> To the tenant . . . J Whiteman. Take notice under the Rent Acts 1968 and 1974 that the landlord may recover possession of the premises under the provisions of Case 10 of Part II of Schedule 3 to the Rent Act 1968. L
> Dated 19th February 1975.

This notice was signed by both the landlords and beneath it there was an acknowledgment of its receipt signed by Miss Whiteman. Miss Whiteman however failed to deliver up possession of the cottage at the expiration of the two-year term and still continues to reside there. On March 4 1977 Mrs Tilling and Miss Dossett began proceedings in the Canterbury County Court against Miss Whiteman for possession of the cottage and mesne profits. Paragraph 4 of particulars of claim stated that the cottage, which was jointly owned by the plaintiffs, had formerly been occupied by the M
plaintiffs as their residence and is now required as a residence for the first plaintiff.

The case for the defendant was that the cottage, being required as a residence for only one but not for both of the plaintiffs, their claim for possession must fail. This defence succeeded in the county court. The first plaintiff, who is legally aided, appealed from that decision but her appeal was dismissed by a majority in the Court of Appeal. The first plaintiff now appeals to your Lordships' House against the decision of the Court of Appeal.

The result of this appeal turns entirely upon the true

construction to be placed upon Case 10 in Part II of the 3rd Schedule to the Rent Act of 1968.

Part II is headed:

> Cases in which Court must Order Possession where Dwelling-House subject to Regulated Tenancy.

This heading derives from section 10(2) of the Act of 1968 which, so far as relevant, reads as follows:

> If . . . the landlord would be entitled to recover possession of a dwelling-house which is for the time being let on or subject to a regulated tenancy, the court shall make an order for possession if the circumstances of the case are as specified in any of the Cases in Part II of Schedule 3 to this Act.

Case 10 reads as follows:

> Where a person who occupied the dwelling-house as his residence (in this Case referred to as "the owner-occupier") let it on a regulated tenancy and:
>
> (*a*) not later than the relevant date the landlord gave notice in writing to the tenant that possession might be recovered under this Case, and
>
> (*b*) the dwelling-house has not, since 8th December 1965, been let by the owner-occupier on a protected tenancy with respect to which the condition mentioned in paragraph (*a*) above was not satisfied, and
>
> (*c*) the court is satisfied that the dwelling-house is required as a residence for the owner-occupier or any member of his family who resided with the owner-occupier when he last occupied the dwelling-house as a residence.

In my opinion it is plain from the opening words of Case 10 that the words "the owner-occupier" are only a kind of shorthand for the words "a person who occupied the dwelling-house as his residence [and] let it on a regulated tenancy." Accordingly, the latter words can properly be substituted for the former words where they appear in paragraph (*c*) of Case 10. This would, no doubt, make the language of paragraph (*c*) rather clumsy and may well explain why the shorthand version was introduced.

Mrs Tilling is, in my view, indubitably a person who occupied the dwelling-house as her residence and let it on a regulated tenancy. I entirely agree with Eveleigh LJ when he says: "The fact that two people do a thing together does not, in my understanding of the English language, prevent either one claiming that he himself did it. The argument to the contrary entails reading into the Act words something like 'on his own'." The majority of the Court of Appeal placed some reliance on the use of the words "the landlord" in paragraph (*a*) of Case 10. I will assume that these words may be read as "the landlords" when, as in the present case, there is more than one landlord. This, however, does not, in my view, assist Miss Whiteman. Paragraph (*a*) was complied with, since the notice to the tenant was, in fact, signed by both the landlords. It follows, therefore, that

(1) Mrs Tilling is a person who occupied the dwelling-house as her residence and let it on a regulated tenancy.

(2) The necessary notice referred to in paragraph (*a*) was duly served.

(3) The regulated tenancy expired on February 21 1977. and

(4) Mrs Tilling may well satisfy the county court that she requires the dwelling-house as a residence for herself.

Accordingly, on what I regard as the true interpretation of Case 10, all its conditions may have been complied with.

The majority of the Court of Appeal in rejecting the above interpretation of Case 10 placed great reliance on *McIntyre v Hardcastle* [1948] 2 KB 82 (CA). In that case two sisters sought possession under Schedule I to the Rent Act of 1933 of a house they owned jointly on the ground that the house was required as a residence for one of them only. The court held that that Schedule did not entitle them to recover possession because the house was not required as a residence for both of them.

Case 8 in Part 1 of Schedule 3 to the Act of 1968 strongly resembles Schedule I to the Act of 1933. I shall not recite it as it is fully set out in the judgment of Stephenson LJ. I am afraid that I cannot agree that there is any similarity between the language of Case 8 and that of Case 10. Indeed not only the language but also the object of these two Cases are, in my view, strikingly dissimilar. Accordingly it seems to me that the decision in *McIntyre v Hardcastle* as to the true interpretation of Schedule I to the Act of 1933 (substantially reproduced in Case 8) is of no assistance in interpreting the language of Case 10. It is not necessary for me to say any more about that decision; and I shall refrain from doing so. Although Case 10 cannot be described as a model of clarity, I am satisfied that the true interpretation of its language is that which I have stated.

While I agree with Stephenson LJ that Case 10 is capable of more than one interpretation I cannot, however, agree that the interpretation which I favour is likely to lead to any serious difficulty on the ground that the joint owners are likely to disagree on whether the dwelling-house should be occupied by the tenant or one of the owners. In the present case, for example, if Miss Dossett had considered that the dwelling-house should be occupied by Miss Whiteman rather than by Mrs Tilling after the tenancy expired, it is unlikely that she would have signed the notice under paragraph (*a*) which she and Mrs Tilling served upon Miss Whiteman. I recognise that it is possible that Miss Dossett might later have changed her mind and become hostile to Mrs Tilling. But I do not understand how this possibility can affect what I regard as the true interpretation of Case 10 or even suggest that this construction can be contrary to the policy of the Act.

On the contrary, if the construction of Case 10 is to be affected by the policy which caused its introduction, that policy, in my view, strongly supports the construction of Case 10 which I favour. At the time when the Act of 1968 was passed, there existed, and had for many years existed, a serious shortage of residential accommodation. There were many cases of persons in occupation of homes which they owned jointly who, for one reason or another, had to leave them temporarily, sometimes for considerable periods; they would have liked to let them during their absence, but refrained from doing so for fear of losing them for ever to their tenants. Accordingly, many homes remained unoccupied which would otherwise have been let to persons urgently in need of them. Case 10 was in my view designed to safeguard persons who occupied their homes against the danger of losing them should they let them during their absence; and accordingly enabled more living accommodation to become available to the public than would otherwise have been the case. I think, therefore, the construction of Case 10 which I favour is in accordance with the policy of the Act and with the public good. I am afraid that I cannot accept the other construction which could lead to absurd and unjust results. Suppose, for example, a husband and wife who are co-owners of their home have to live abroad for, say, two years on account of the husband's business activities. They then let their home for those two years and comply with all the provisions of Case 10. Unfortunately, the husband falls in love with another woman and his wife divorces him. They, however, remain comparatively friendly, and the husband is willing for his wife to live in their former matrimonial home. The tenancy has expired but the tenant refuses to leave. An action is brought by both co-owners for possession. The defence is that the action cannot succeed because the house is required as a residence for one only of the co-owners. If the majority decision of the Court of Appeal is correct, the action would fail.

Take another example. Three sisters are living in a house. Two of them are co-owners of the house. The third has no legal or equitable title to it. The two co-owners and their sister leave the house to live abroad temporarily: the house

A is let, say, for two years and all the provisions of Case 10 are complied with. Both the co-owners then get married and decide to live abroad. The third sister returns to England. She is a member of the family of both the co-owners and resided with them when last they occupied the house as a residence. The tenant is sued for possession by the two co-owners on the ground that the house is required as a residence for their sister. There could be no defence to the action and the co-owners' sister would be enabled to reside in the house. Suppose, however, that the first co-owner sister and the third sister had married and decided to live abroad while the second sister who was a co-owner returned to England
B and an action for possession was brought on the ground that the house was required as a residence for her. The action would fail because it could not succeed unless the house was required as a residence for both the co-owners. Yet an action to enable the house to be occupied by the sister who had no legal or equitable title to it would have succeeded. This does not seem to me to make any sense at all. But it is the inevitable result of what might happen unless the construction of Case 10 which I favour is correct.

For the reasons I have stated I would allow the appeal and remit the case to the county court for the learned judge to decide the few issues of simple fact which, most unfortunately,
C he did not dispose of before making his findings on the points of law. I would order with reluctance that the costs in this House and in the Court of Appeal be paid by Miss Whiteman. There is, I think, no other course which your Lordships can take. I am extremely disturbed that such a heavy financial burden should fall upon Miss Whiteman's shoulders, particularly as it would have been avoided had she obtained legal aid: it may be that the Law Society will wish to consider whether there are any circumstances which might justify the order as to costs not being enforced.

In a dissenting speech, LORD FRASER OF TULLYBELTON
D said: I regret that I am unable to agree with the view of my noble and learned friends with whom I heard the argument in this appeal. I can, therefore, have little confidence in my own view, but I shall express it as briefly as possible.

The appeal raises a question of construing words in Case 10, which is in Part II of Schedule 3 to the Rent Act 1968. Part II is headed "Cases in which Court must Order Possession where Dwelling-House Subject to Regulated Tenancy." Case 10 begins by describing the class of person for the benefit of whom, or whose family, it will operate if the other conditions are satisfied. The person has to be qualified in two ways. (1) He must have "occupied the dwelling-house as
E his residence" and (2) he must have "let it on a regulated tenancy." The same person is referred to throughout, and he must have both occupied and let the house. His dual character is appropriately recognised in the statutory description of him as "the owner-occupier." The present case proceeds upon an agreed statement of facts, which shows that (and I paraphrase):

(1) The house was jointly owned by two ladies, Mrs Tilling, the appellant, and Miss Dossett, who are not related.

(2) It was occupied by both of them, though only the appellant was occupying it immediately before it was let to
F the respondent.

(3) It was let to the respondent under a lease to which the joint owners were one party, described in the lease as "the landlord."

(4) It is now required as a residence for only one of the joint owners, the appellant.

In these circumstances it seems to me that the "person" who let the house was the composite person consisting of both joint owners. Neither owner separately let the house. It may be that one of them could have let the house without the concurrence of the other, so far as the rest of the world is concerned, but that is not what happened. It follows G
that the only person who falls within the statutory description of "the owner-occupier" is the composite person and, as the house is not required as a residence for that person, the claim for the appellant alone must fail. Like Stephenson and Shaw LJJ I can see no acceptable answer to that simple argument. With all respect to those who think otherwise, I do not regard it as right or possible to read the word "let" in the first clause of Case 10 as if it meant "concurred in letting." Even if one joint owner could truthfully say in casual conversation that she had let the house, a matter on which I feel serious doubt, it does not follow that the statute, which is expressed in precise language, can be read in the H
same way. The singular "person" can, of course, be read as including the plural: see Interpretation Act 1889. If it is read as meaning the plural here, then so also must "the owner-occupier" be read in the plural, with the result that paragraph (*c*) of Case 10 will only apply where the house is required by both owner-occupiers or a member of their joint family as a residence.

Some support for my view is found in *McIntyre v Hardcastle* [1948] 2 KB 82, though that is not directly in point as it was dealing with what later became Case 8 of Schedule 3 to the 1968 Act, where the wording differs materially from the wording of Case 10. Nevertheless, I would adopt the J
words of Tucker LJ at p 90, where he said:

> I do not think that the legislature really contemplated this situation when this paragraph was framed. Therefore, I feel driven to interpret it merely in the light of the actual language used.

That is what I have tried to do.

I do not find any assistance in the policy of section 10 of the Act, which is the section authorising Schedule 3. No doubt section 10 is intended to benefit owners, and to increase the supply of houses for letting, but it is an exception to the main policy of Part II of the Act, which evidently is to give K
security of tenure to tenants. The question is what are the limits of the exception, and the answer must, I think, be found simply in the words used by Parliament. I would only add that, on the facts as known to me, it appears that the merits of the present dispute are on the side of the appellant rather than of the respondent, but that cannot justify straining the words of the Act in her favour.

I would refuse the appeal.

In a speech agreeing that the appeal should be allowed, LORD SCARMAN said: I have had the advantage of reading in draft the speech delivered by my noble and learned friend, L
Lord Wilberforce. I agree with it. I would allow this appeal for the reasons given by Eveleigh LJ in his dissenting judgment. I also agree that the appellant should have her costs before this House and in the Court of Appeal.

The case presents two disturbing features. First, the decision in the county court was upon a preliminary point of law. Had an extra half-hour or so been used to hear the evidence, one of two consequences would have ensued. Either Mrs Tilling would have been believed when she said she required the house as a residence, or she would not. If the latter, that would have been the end of the case. If the former, your Lordships' decision allowing the appeal would now be final. M
As it is, the case has to go back to the county court to be tried. Preliminary points of law are too often treacherous short cuts. Their price can be, as here, delay, anxiety, and expense. Secondly, it is a tragedy that Miss Whiteman, who has appeared in person in the Court of Appeal and before this House because of the expense of legal representation, was not legally aided. She had an eminently arguable case and, if legally aided, would have been protected against the burden of costs which under our law falls upon the unsuccessful litigant. Perhaps she was above the very modest limits set to eligibility for legal aid. But there are indications that

A she chose not to apply. I wonder whether she fully understood the protection offered a litigant by legal aid. Surely those who helped her with advice (and she did have advisers, legal and lay, at various times) might have persuaded her. Perhaps they tried and failed. But the result, whatever its cause, is a financial disaster for her.

The appeal was allowed with costs in the House of Lords and the Court of Appeal. The case was remitted to Canterbury County Court to decide on the facts in the light of the law as interpreted by the House.

B

COURT OF APPEAL

March 7 1979

(Before Lord Justice MEGAW, Lord Justice WALLER and Lord Justice EVELEIGH)

SKINNER AND ANOTHER v COOPER AND ANOTHER

Estates Gazette June 9 1979

(1979) 250 EG 967

Rent (Agriculture) Act 1976—Scope of transitional provisions —Difference of opinion in the Court of Appeal—Effect of events before January 1 1977—Claim by owner of cottage C for possession against widow of former farmworker—Farmworker and wife permitted by previous and present owners to remain in cottage after termination of worker's employment with previous owner through illness—Move to adjoining cottage by arrangement—Widow remained in cottage for some years after husband's death—No rent paid to previous owner after termination of employment or to present owner —Notice to quit given to widow after January 1 1977 when relevant provisions of 1976 Act came into force—How far back the transitional provisions of Schedule 9 were to be taken to apply in deeming the Act to have been in force— Meaning of "at all material times"—Broader view of D majority and narrower view of Eveleigh LJ, dissenting— Order for possession against widow set aside, but she may face an alternative claim in county court

This was an appeal by Mrs Lillian Cooper, widow of Albert Cooper, from a decision of Judge Deane at Sittingbourne County Court granting an order for possession in favour of the plaintiff in the action, Wilfred James Skinner, in respect of a dwelling-house at 6 Paradise Cottages, Lower Road, Hartlip, Sittingbourne.

Amanda Barrington-Smyth (instructed by Allan Jay & Co, agents for C Stuart Fisher, of Gillingham, Kent) appeared on behalf of the appellant; Sebastian Neville-Clarke (instructed E by Amery-Parkes & Co, agents for Winch, Greensted & Winch, of Sittingbourne, Kent) represented the respondent.

Giving judgment MEGAW LJ said: This appeal is concerned with the meaning of the transitional provisions in the Rent (Agriculture) Act 1976, to which I shall refer as "the Act." It is an appeal from the judgment of Judge Deane, in the Sittingbourne County Court, delivered on September 26 1978 whereby he granted an order, in favour of the plaintiff, Mr Wilfred James Skinner, against the defendant, Mrs Lillian Cooper, for possession of a dwelling-house, 6 Paradise Cottages, Lower Road, Hartlip, Sittingbourne.

The defendant is the widow of Mr Albert Cooper. On F January 1 1977, the date of coming into force of the relevant provisions of the Act, she was the occupant of the cottage, 6 Paradise Cottages. She had been born in 5 Paradise Cottages and had lived all her life either in it or no 6. She is the daughter of Mr Attwood, who had been a farmworker, employed as such by the elder Mr Goodhew, a farmer, who owned the cottages. Mrs Cooper's husband, Mr Albert Cooper, was also a farmworker who was employed as such, first by the elder Mr Goodhew, and on his death by his son, who thereafter owned and operated the farm and owned the cottages. During the whole of the time that Mr Albert Cooper lived in the cottage he did not make, nor was he expected to make, any payment, nor was there any deduction from his wages, in respect of his occupation. G

In 1968 Mr Cooper fell ill. In April 1969, when it became clear that because of his illness he would not be able to work again, Mr Goodhew terminated his employment. But he allowed Mr and Mrs Cooper to continue to live in the cottage, free of payment.

In October 1972 Mr Goodhew sold 5 and 6 Paradise Cottages to Mr Skinner, the plaintiff in the action. Mr Goodhew asked Mr Skinner to allow the Coopers to stay on, particularly because of Mr Cooper's illness. Mr Skinner agreed, by way of what he described as "a gentleman's agreement." Presumably he told Mr Cooper of his consent. The permission which he gave was related to the period of Mr Cooper's illness. Mr H Skinner was a builder, not a farmer, and, of course, Mr Cooper because of his illness no longer did farm work. There was no relation of employer and employee between them.

In October 1973 Mr Albert Cooper died. Mr Skinner renovated and sold 5 Paradise Cottages. Mr and Mrs Cooper and their family had by arrangement removed to no 6. If the Act otherwise applies, this move from one cottage to the other does not matter. Because of a provision of the Act, facts which would have been revelant as affecting the question of protection in respect of no 5 would remain equally relevant in respect of the occupation of no 6. This is common ground. J

After Mr Cooper's death, Mrs Cooper offered payment, but Mr Skinner did not accept her offer. From then, October 1973, until May 1977, when Mr Skinner gave Mrs Cooper notice to quit, Mrs Cooper continued to live in the cottage. Mrs Cooper did not give up possession as a result of the notice. Mr Skinner then brought an action for possession in Sittingbourne County Court. The defence, so far as is relevant for this appeal, was based on the Act. For the plaintiff it was contended that the Act did not give the defendant security of tenure. It was further contended that, if it did, the plaintiff was entitled to rely on Case IX in Schedule 4 to the Act, asserting that the cottage was required as a residence for Mr Skinner's mother. K This latter issue was left undecided in the county court, for reasons which I need not go into. The parties agree that, if this appeal should be decided in favour of the defendant, the plaintiff would be entitled to have that further issue heard and decided in the county court. I say no more about it.

Judge Deane accepted the submissions on behalf of the plaintiff and held that the transitional provisions of the Act did not avail the defendant. From that judgment this appeal is brought.

There is no doubt that the Act, by reason of the transitional provisions, referred to in section 40(3) and contained in Schedule 9, provides that in certain circumstances the right of the L occupant of a dwelling-house to have the protection of the Act against an order for possession at the instance of the owner of the dwelling-house is to be decided, if a dispute arises, on the hypothesis that the Act, which came into force on January 1 1977, had been in force earlier than that date. To put it another way, the Act, in certain cases, is, by its own terms, to be deemed to have been in force before it was enacted.

The question in this appeal is how far back the court is to go in deeming the Act to have been in force. If the Act had not been enacted, or had not been given retrospective effect, Mrs Cooper would have had no security of tenure. M She could have been evicted on the termination of her licence to occupy, which would have required no more than reasonable notice. The Rent Acts would not have applied. The notice actually given in this case in May 1977 would have been effective. Apart from the Act, the legal position would have been the same for Mr Cooper at any time during his lifetime as it was for Mrs Cooper after his death. It would have been the same for the occupant of the cottage while it was owned by Mr Goodhew as it was after Mr Skinner bought it.

I shall now trace the position as it would have been if by reason of the terms of the Act it is to be deemed to have been

A in force at all times at which facts occurred or existed which, if they had all happened after the coming into force of the Act, would have been relevant to the question whether or not Mrs Cooper would have the protection of the Act. For the plaintiff it is conceded that, if that be the right hypothesis on the true construction of Schedule 9, then the claim for possession, so far as this appeal is concerned, fails. But the plaintiff contends that the Act does not require or entitle the court to go back so far in deeming the Act to have been in force. On the true construction, it is contended, the retrospective effect must be treated as being more limited.

It is not necessary to go further back than the year 1968. B Suppose that the Act had then been in operation. In 1968, before his illness, Mr Albert Cooper occupied the cottage under a licence from his employer, Mr Goodhew. Mr Cooper was a "qualifying worker" under Schedule 3 paragraph 1. He fulfilled all the conditions for being a "protected occupier in his own right" under section 2(1) of the Act, and thus would have had the security of tenure provided by the Act. After his employment was determined by Mr Goodhew in April 1969, because of his continuing illness, Mr Cooper had a fresh licence to occupy, which was a "relevant licence." Mr Cooper continued as a protected occupier because of section 2(2) or (3). It does not matter which.

C In October 1972, when Mr Goodhew sold the cottages to Mr Skinner, Mr Cooper obtained a new licence, which was a "relevant licence" from Mr Skinner. The old licence automatically terminated by reason of the change of ownership. But that change of ownership and change of licence in no way affected Mr Cooper's status. It was, for example, not affected by any express or implied condition of the licence, such as that the licence was only to continue for the duration of Mr Cooper's illness. The Act being deemed to be in force at that time (the hypothesis on which I am for the moment proceeding), there could be no question of Mr Cooper contracting out of the protection of the Act, any more than a protected tenant D under the Rent Acts can contract out: *Brown v Draper* [1944] KB 309. Nor are there any facts in this case which could be relied on as bringing into operation Case VII of Schedule 4 to the Act, as counsel for the plaintiff at one stage sought to argue before us.

The final stage, on the hypothesis of the Act having been in force, is as follows. When Mr Cooper died in October 1973 Mrs Cooper, the defendant, became a "statutory tenant" by virtue of section 4(2) and (3) of the Act. She so remained at the "operative date," January 1 1977, and thereafter at the date of the notice to quit, of the hearing of the action and of the judgment.

E We had much argument upon the wording of paragraph 2 of Schedule 9. In my opinion, it is not relevant in this case. That is because it is, by its terms, applicable to the question whether a person is, on or after the operative date, a protected occupier. There is no question here of Mrs Cooper being a protected occupier. If the Act gives her any status, it is that of a statutory tenant. The relevant provision of Schedule 9 is, therefore, paragraph 3.

Paragraph 3 reads:

> *Statutory tenancy arising on operative date.* A person who is occupying a dwelling-house as his residence on the operative date shall become the statutory tenant of the dwelling-house if, on F the assumption that this Act and the provisions of the Rent Act 1968 which are applied by Schedule 2 to this Act, including (where relevant) any amendments to those provisions, were in force at all material times before that date, he would be a statutory tenant of the dwelling-house on that date, and this Act shall thereafter apply to him, and by reference to him, on that assumption.

The question which we have to decide is: what is meant by the phrase "at all material times" in that paragraph? To my mind, the phrase is, in that context, capable of one meaning only. It is not ambiguous. "At all material times" means at all times when any fact existed or event occurred which, if the Act had been in force at the time, and thereafter, would G have been relevant for the decision of the question with which the paragraph is concerned. That question, made specific to the present case, is whether Mrs Cooper, who was occupying the cottage as her residence on the operative date, is the statutory tenant. I am unable to see how any point or period of time can be an immaterial time for the purpose of that question, if the answer to the question would be different according to whether you include or exclude reference to that point or period of time.

Accordingly, in my judgment, the foundation is not laid for the argument here adduced on behalf of the plaintiff. That argument is that where a statute has retrospective effect, H as this Act admittedly has, it should be given as limited an effect as is possible without doing violence to the words used in the statute: *Lauri v Renad* [1892] 3 Ch 402, per Lindley LJ at p 421. I should have thought that it was "doing violence to" the words of the statute if one were to adopt the meaning of "at all material times" put forward on behalf of the plaintiff: that is, that the "material times" are limited to the period of time necessary to take one step back; that you cannot go further back than the beginning of the particular licence under which the occupant was occupying at the moment of the attempted termination of that licence. The submissions to this effect were actually made by reference to paragraph 2(3) J of Schedule 9. As I have said, in my view that paragraph is not applicable here. I am not clear how it was sought to apply the submission which was developed in relation to that paragraph to the paragraph of the schedule with which we are concerned, paragraph 3. I cannot see any rational basis for the suggested limitation of meaning of the phrase "at all material times" in that paragraph.

I am also of the opinion that there is a sound basis for Miss Barrington-Smyth's submissions on behalf of the defendant that the retrospective effect which is created by this Act is not the type of retrospective effect which involves or requires the application of some specially strict criterion of construction K so as to impose the maximum limitation on the scope of the words. In the present case Mr Skinner could have terminated Mrs Cooper's licence by obtaining and executing an order for possession at any time before January 1 1977. If it had been suggested that the Act provided for a reversal or variation of such a lawfully completed transaction, it would have been right to apply a very strict test before accepting that that was the effect of the Act. But it is not suggested that the Act has such an effect. It merely has the effect, which many statutes have, of altering the pre-existing law in such a way as to destroy or vary, upon and after the passing of the Act, some legal right which previously existed.

L However that may be, I would decide this appeal on the basis that "at all material times" in paragraph 3 of Schedule 9 means at all times in the past at which any fact existed or event occurred which, if the Act had been in force at that time and thereafter, would have been relevant for the decision of the question with which the paragraph is concerned: namely whether on and after January 1 1977 Mrs Cooper has been a statutory tenant under the Act.

I would therefore allow the appeal and set aside the order for possession. But, because of the matter which I mentioned earlier, as to a possibly outstanding issue in the county court, the action will have to be restored in that court if the plain- M tiff desires to pursue the alternative basis of his claim for possession.

WALLER LJ delivered a judgment agreeing that the appeal should be allowed.

Dissenting, EVELEIGH LJ said: Mr Cooper worked wholetime in agriculture. He was provided with a cottage by his employer Mr Goodhew. In 1968 he fell ill and did no more work. In 1969 his employment was terminated, but he and his family continued to live in the cottage rent-free. In 1972 Mr Goodhew sold the cottage to the plaintiff, a builder, and asked him to allow the Coopers to stay on, particularly in view of Mr Cooper's illness. In 1973 Mr Cooper died. Mrs Cooper continued to live in the cottage. She offered to pay rent to Mr Skinner but he would not accept it. In 1977 the plaintiff served notice to quit.

The defendant relies upon the protection of the Rent (Agriculture) Act 1976. The first ground upon which protection was claimed was on the basis of section 3(2) of the Act. That subsection comes into play when subsection (1) is fulfilled, that is to say "after the death of a person ('the original occupier') who, immediately before his death, was a protected occupier of the dwelling-house in his own right." After such a death a widow who was residing with him at his death becomes a protected occupier if she has a relevant licence in relation to the dwelling-house (subsection (2)). The relevant licence relied upon in the present case is said to be one from Mr Skinner to the defendant. For the purpose of this case we are not concerned with considerations which make a licence to occupy a relevant licence. Applying Schedule 2, paragraph 1, to the Act, it is clear that if her licence is treated as a tenancy and the Rent Act 1968 were modified as mentioned in Schedule 2, Mrs Cooper would have a protected tenancy for the purposes of that Rent Act. However, the fact that she might have a relevant licence from the plaintiff does not necessarily mean that she comes within the wording of subsection (2) when it says ". . . if the widow has, in relation to the dwelling-house, a relevant licence. . . ." In my opinion those words require her to be a licensee at the time of the death. A licence subsequently granted does not, in my opinion, fall within the meaning of the words I have quoted. Section 3(2) in my opinion is intended to cover the case where husband and wife are joint licensees. To cover this objection it is submitted on behalf of Mrs Cooper that the licence granted by Mr Skinner to her husband was one granted to herself also. If that is correct, then Mrs Cooper did have a relevant licence and could benefit from subsection (2) if her husband was a person contemplated by subsection (1), namely "a protected occupier of the dwelling-house in his own right," and she would have to add "at some time." The defendant's contention that her husband was a protected occupier in his own right depends upon the submission that regard must be had to the licence granted by Mr Goodhew at a time when Mr Cooper was employed by him as a farmworker. It is said that paragraph 2 of Schedule 9 entitles the defendant to treat the Act as being in force at the time of the grant of the licence by Mr Goodhew. Thus the basic question is whether or not we can treat Mr Cooper as being a protected occupier some time in 1968 or before that when the licence was granted and he was first employed as a farmworker.

Reliance is placed on paragraph 2(3):

> In applying this paragraph to section 2(3) or to section 3 of this Act (so that the question whether a person is a protected occupier depends on whether he or another person was, at a time before the material date, a protected occupier or statutory tenant) it shall be assumed that this Act and the provisions of the Rent Act 1968 which are applied by Schedule 2 to this Act, including (where relevant) any amendments to those provisions, were in force at all material times before the operative date.

It is said that the words "at all material times before the operative date" are wide enough to relate back to that time. I do not think that this paragraph has the effect contended for. The whole paragraph comes into play when there is a "question whether at any date which is on or after the operative date a person . . . is a protected occupier": paragraph 2(1). That question does arise and consequently we may continue to the following subparagraph.

Subparagraph (2) reads: "So far as the question depends on prior circumstances, they shall be taken into account even if occurring before the operative date." In my opinion this subparagraph does not entitle the court to go back to any time it pleases for all purposes. It can only do so in so far as the status depends on prior circumstances. That means when status depends on prior circumstances made relevant by the Act. This so far as the position of the defendant *on or after the operative date* "depends on prior circumstances"—but only "so far as" and not further. The circumstances must be those *required* to be proved in order to decide the particular question.

Subparagraph (3) does not specifically by itself state that prior circumstances may be regarded whatever their nature. It must mean prior circumstances which under the Act give rise to protection. We are dealing with retrospective legislation. It must therefore be construed strictly and where possible so as not to deprive a person of his existing rights. Mr Skinner's existing right before the Act came into operation was a right to determine the licence and claim possession.

In my opinion subparagraphs (2) and (3) must be read together. The effect is that when the question whether a person is a protected occupier depends upon prior circumstances, regard may be had to those circumstances, and where section 2(3) or section 3 of the Act is invoked the Act shall be treated as though it were in force at the time of the prior circumstances upon which the question of status depends. In order to see whether the question of status depends upon prior circumstances we have to look at the wording of the relevant parts of the Act which declare what those circumstances are.

Section 2(1) immediately provides an example. That provides:

> *Protected occupiers in their own right* (1) Where a person has, in relation to a dwelling-house, a relevant licence or tenancy and the dwelling-house is in qualifying ownership, or has been in qualifying ownership at any time during the subsistence of the licence or tenancy (whether it was at the time a relevant licence or tenancy or not), he shall be a protected occupier of the dwelling-house if—(a) he is a qualifying worker, or (b) he has been a qualifying worker at any time during the subsistence of the licence or tenancy (whether it was at the time a relevant licence or tenancy or not).

In so far as a person has a licence and the dwelling-house is in qualifying ownership and the occupier is a qualifying worker, he is protected and no reference to prior circumstances is required. The subsection, however, refers to events in the perfect tense also: if the house "has been" in qualifying ownership. In such a case, if he is a qualifying worker he may be protected by virtue of the earlier ownership even though that qualifying ownership no longer prevails. That situation may arise in the future and the qualifying ownership, while antedating the time when the question of protection falls to be considered, will none the less be material. It does not necessarily follow that the only case of earlier qualifying ownership will antedate the coming into force of the Act. If section 2(1) had stood on its own without the provisions of paragraph 2 of Schedule 9, it could be argued that the Act was only dealing with a case of precedent qualifying ownership where that ownership could be found on a date after the commencement of the Act. Paragraph 2 of Schedule 9, however, will allow qualifying ownership before the Act to be treated as an ingredient for protection whether one points to its existence before or after the commencement of the Act. Then, again, section 2(1)(b) equates the position of one who has been a qualifying worker with that of one who is a qualifying worker (section 2(1)(a)). Again, in section 2(2) we encounter the expression "has been in qualifying ownership" etc.

In the examples I have read so far it is possible to recognise the prior circumstances without recourse to paragraph 2 (3) of Schedule 9. They are actual matters of fact. The present status of a person as a protected occupier depends not upon the legal status of himself or anyone else at an earlier date but only upon the nature of his employment and the existence of an employer as a licensor. These matters can be ascertained without the necessity of assuming that the Rent (Agriculture) Act 1976 was in force at an earlier date.

However, when we come to section 2(3) a different problem arises. That section reads:

A A person who has, in relation to a dwelling-house, a relevant licence or tenancy shall be a protected occupier of the dwelling-house if—(a) immediately before the licence or tenancy was granted, he was a protected occupier or statutory tenant of the dwelling-house in his own right, or (b) the licence or tenancy was granted in consideration of his giving up possession of another dwelling-house of which he was such an occupier or such a tenant.

Thus it is necessary for the occupier to establish a protected status (depending upon the facts and law) before the grant of the licence or tenancy now relied upon. In considering that protected status, paragraph 2(3) of Schedule 9 comes into play. It comes into play because the question
B whether he is a protected occupier depends upon whether he or another person was a protected occupier at a time before the material date, that is, the date immediately before the relevant licence under examination was granted to him.

Under section 2(3)(a) it is necessary to discover the status of the person claiming the protection of the Act immediately before his current licence was granted. It may be that the current licence was granted before the Act came into force. And therefore on the wording of section 2(3) alone he would not be entitled to protection because he could not say that he was a protected occupier at that time. Paragraph 2(3) of Schedule 9 comes to his rescue by providing that at the material time, that is to say the granting of the
C licence, it must be assumed that the Act was in force.

In my opinion paragraph 2 does not enable the court to go back and look at the interests of other people unless those other people are specifically named as persons from whose status the person now claiming protection can derive title. The words "or another person" in paragraph 2 (3) are there to cover claims under sections 3 and 4. They have no application to section 2.

Section 3(1) reads:

Protected occupiers by succession (1) Subsection (2) or, as the case may be, subsection (3) below shall have effect for determining what person (if any) is a protected occupier of a dwelling-house
D after the death of a person ("the original occupier") who, immediately before his death, was a protected occupier of the dwelling-house in his own right.

Thus, in order to claim protection by succession it is necessary to establish that the predecessor was a protected occupier immediately before his death. If the death occurred before the commencement of the Act, paragraph 2(3) of Schedule 9 comes to the assistance of the person claiming succession. It does not, however, in my opinion come into play to give the predecessor a status which can only be established by tracing back through past events upon the assumption that the Act has always been in force. The starting point of
E section 3 is the status of the original occupier at the time of his death. There are no artificial provisions in the Act upon which he can go backwards to establish his status.

The defendant has contended in effect that the phrase "at all material times" is equivalent to "at all times" or "at any time in the past." It seems to me that in considering a particular case before the court, only one particular time will be of relevance for the application of paragraph 2(3) of Schedule 9. The plural is used in the paragraph, that is, "at all material times," because paragraph 2(3) can be invoked for the interpretation of different sections or subsections in which different events are made material.

F I now turn to consider the application of paragraph 3 of Schedule 9 to section 4 of the Act. Paragraph 3 of Schedule 9 reads:

Statutory tenancy arising on operative date A person who is occupying a dwelling-house as his residence on the operative date shall become the statutory tenant of the dwelling-house if, on the assumption that this Act and the provisions of the Rent Act 1968 which are applied by Schedule 2 to this Act, including (where relevant) any amendments to those provisions, were in force at all material times before that date, he would be a statutory tenant of the dwelling-house on that date, and this Act shall thereafter apply to him, and by reference to him, on that assumption.

Section 4(2) provides: G

Subject to section 5 below, subsection (3), or, as the case may be, subsection (4) below shall have effect for determining what person (if any) is the statutory tenant of a dwelling-house at any time after the death of a person ("the original occupier") who was, immediately before his death, a protected occupier or statutory tenant of the dwelling-house in his own right.

This subsection defines the situation in which subsection (3) or subsection (4) may be invoked. There must be an original occupier from whom a right is said to be derived who was immediately before his death a protected occupier or statutory tenant. Thus it is his status that has to be established. Paragraph 3 of Schedule 9 lays down what status a H person occupying the dwelling-house on the operative date is to have. In considering this question we are to assume that the Act was in force at all material times before the operative date and then ask on that assumption if he would be a statutory tenant on the operative date. It is important to bear in mind that the paragraph requires the assumption to be made that the Act was in force for the purpose of determining whether the occupier would be a statutory tenant and not if anyone else would be or would have been a statutory tenant.

In the present case the defendant could only become a statutory tenant by virtue of the provisions of section 4. Section 4(1) converts a protected occupier into a statutory J tenant. For the reasons already stated, the defendant was never in my opinion a protected occupier and consequently she cannot rely on section 4(1) as directly governing her status. Indeed if she were a protected occupier she would not wish to do so. We pass on to section 4(2), which lays down the situation in which the subsequent subsections may be invoked. That situation is the death of the original occupier who was immediately before his death a protected occupier or a statutory tenant in his own right. If Mr Cooper was such a person, then the defendant could invoke section 4(3) as being a widow residing with Mr Cooper at his death. Consequently the material time for the purpose of section 4 K (3) is immediately before the date of Mr Cooper's death and thereafter. On the assumption that the Act was in force at the date of Mr Cooper's death, in my opinion section 4 would not be brought into play because Mr Cooper was not a protected occupier in his own right. Moreover, he was not in my opinion a statutory tenant in his own right because as I take the view that we cannot antedate the Act beyond Mr Cooper's death he never was a protected occupier.

In so far as paragraph 3 of Schedule 9 refers to "material times" in the plural, again, as with paragraph 2, this is necessary because section 4 refers to different events in its subsections. For the purpose of subsection (2) we start with L the death of a person, whereas subsection (1) relates to the termination of the licence or tenancy for various reasons.

I therefore take the view that the material times for the purpose of the Act will be the particular time in each case which in the relevant section or subsection is indicated as the time for the occurrence of the qualifying conditions which entitle the claimant to the particular status referred to. That being so, I do not think that it is possible for the defendant to go back to the time when her husband was working whole-time in agriculture.

I would therefore dismiss this appeal.

The appeal was allowed with costs in the Court of Appeal M (except the costs of a motion to expedite and dismiss the appeal) and costs in the county court (to date) on Scale 3. The county court order was set aside and the case remitted to the county court to try the further issue mentioned in the judgment of Megaw LJ. Leave to appeal to the House of Lords was refused.

A
COURT OF APPEAL

July 20 1978

(Before Lord Justice STEPHENSON, Lord Justice LAWTON and Lord Justice GEOFFREY LANE)

O'MALLEY v SEYMOUR

Estates Gazette June 16 1979

(1978) 250 EG 1083

Third Court of Appeal case on question whether transcation was a non-exclusive licence or a protected tenancy—A property owner can order his affairs so as to prevent his property being enmeshed in the Rent Acts, but present case distinguished from Somma v Hazlehurst and Aldrington
B **Garages v Fielder—Evidence of prior oral agreement in terms appropriate to tenancy—"Licence" did not reflect parties' true agreement—Tenancy disguised in trappings of a licence—County court judge justified in deciding that "licence" document was a sham**

This was an appeal by the plaintiff, John O'Malley, from the dismissal by Judge Dow at Clerkenwell County Court of his action against Colin Michael Seymour for possession of a ground-floor flat at 13 Ribblesdale Road, London N8. This was the third case in chronological order before the Court of Appeal raising the question whether a document constituted a
C non-exclusive licence excluding the application of the Rent Acts or whether it created a tenancy within the Acts. The fourth case, *Demuren v Seal Estates Ltd* has already been reported in ESTATES GAZETTE: see (1979) 249 EG 440.

Miss M T Catterson (instructed by Kingsford, Dorman & Co, agents for Merrill, Skerritt & Co, of Luton) appeared on behalf of the appellant; D Watkinson (instructed by Tyrer, Roxburgh & Dawson) represented the respondent.

Giving judgment, STEPHENSON LJ said: This is an appeal from a judgment of His Honour Judge Dow, given on November 4 1977 in the Clerkenwell County Court, dismissing the plaintiff-appellant's claim for possession of the ground-floor
D flat, 13 Ribblesdale Road, London N8. The whole issue in this case, as the judge said and as Miss Catterson has submitted on behalf of the unsuccessful plaintiff, is: on what terms did Mr Seymour, the defendant/respondent, enter into occupation of it? Did he go in under a short-term lease for six months? Or is he, as the plaintiff alleges, merely a licensee for a fixed term?

This is the third attempt which this court has had to consider recently by an owner of property to provide accommodation for somebody desperately in need of it without attracting the restrictions imposed on the rent he can charge and, more particularly, on his powers of eviction by the Rent Act 1974; and once again we have to consider whether the attempt has been
E successful or whether the judge was right in this case in holding that it failed. That is a problem which has perhaps been recently made clearer than it was at the time Judge Dow gave his judgment that the court ought to approach without any bias in favour of finding that the agreement was a tenancy and a genuine tenancy. What Buckley LJ said in the case of *Shell-Mex & BP Ltd v Manchester Garages Ltd* [1971] 1 WLR 612 at p 619 has to be borne in mind. He said:

> It may be that this is a device which has been adopted by the plaintiff company to avoid possible consequences of the Landlord and Tenant Act 1954, which would have affected a transaction being one of landlord and tenant; but, in my judgment, one cannot take
> F that into account in the process of construing such a document to find out what the true nature of the transaction is. One has first to find out what is the true nature of the transaction and then see how the Act operates upon that state of affairs if it bites at all. One should not approach the problem with a tendency to attempt to find a tenancy because unless there is a tenancy the case will escape the effects of the statute.

We have here the advantage of a note of the learned judge's judgment taken by Miss Catterson, who has presented the appellant's case. By an unfortunate oversight it has not been presented, or at any rate presented in time, to the respondent's solicitors and counsel for agreement, or to the learned judge for his approval. We agreed to look at it subject to Mr Watkinson's right, on behalf of the respondent, to challenge its G
correctness. He does not do so and I am not surprised. It is an admirably clear note; but we have to bear in mind that it has not been seen or approved by the judge when we come to look at the criticisms which are made of it by Miss Catterson. I do not wish it to be thought that in the ordinary way we would look at a judgment of a county court judge in the form of a note which he had not had an opportunity of revising. But I am quite satisfied, from the excellence of the note and what Mr Watkinson has said about it, having compared it with the note of Lord Gifford (who was then counsel for the respondent in the county court) and with a note taken by his solicitor, that we can safely act upon the note; but we must H
not, of course, attribute every word as it is set out in that note to the judge, because he might have wished to revise or correct it in minor particulars.

The questions which arise are, it seems to me: first, what on its true construction is the nature of the written agreement which the plaintiff relied and relies in his claim for possession, against the background of the evidence? Secondly, does the evidence prove and entitle the judge to hold that the document, on its proper construction, does not truly represent or reproduce the real transaction between the parties? Those are the questions which arise in all these cases. In this case the learned judge has, I think, construed the document as a tenancy J
rather than a licence because he said that "it in fact created a tenancy" (according to the note); but even if the document is at any rate capable of being construed as a licence, the judge has clearly found that it was a sham and did not embody the real transaction between the parties.

He approached the document in this way:

> I remind myself of the warnings issued over the years in cases dealing with the Rent Act legislation that courts must be astute to see that a coach and horses are not driven through the Rent Acts or, as Lord Denning has put it, that an "articulated vehicle" is not driven through them. Unless I am persuaded to construe the agreement in this case as a licence, it is quite plain that it is other- K
> wise a tenancy and that on the expiry of the six months specified in the agreement

which is exhibited

> Mr Seymour could hold over as a protected tenant. If, on the other hand, it is a licence, then the period has expired and the plaintiff is entitled to the order which he seeks.

In my judgment, that passage, if it accurately records the learned judge's concluded opinion, does indicate an approach too favourable to the defendant; but, for the reasons which I shall try to give, I do not think that the judge would have reached any different conclusion had he modified that language and that approach to accord with the judgments of this court L
in *Somma v Hazlehurst*, decided on March 3 of this year, and *Aldrington Garages v Fielder*, decided on July 7 of this year*.

He went on to say that it was "essential to look at the background to this matter." He stated that the plaintiff had "his main home in Eire where he has his family and farm. He has become the owner of more than one property in or around London, and he makes periodic trips to this country to look after the property he owns and to see if the necessary 'moneys' payable by the occupants has been received."

Then he called attention to other facts which were common ground. First of all, among the properties which the plaintiff M
owned was a property at 88 Cecile Park. Part of that property he advertised in, I think, the *Evening Standard* as to let—"a flat to let" in that property—and it was that advertisement which the defendant, Mr Seymour, saw; it was that advertisement which he answered, and it was as a result of answering that advertisement that he met the plaintiff at that other property. Now that property was let off in five or six bed-sitting rooms, and there was evidence before the learned judge of rent books in respect of two of those; rent books held by occupants

**Somma v Hazlehurst* was reported in (1978) 246 EG 311 and *Aldrington Garages v Fielder* at (1978) 247 EG 557.

who were not occupants under licence but were tenants. The learned judge found that the plaintiff was not a resident landlord and quite clearly knew very well that he was, as the learned judge put it, "saddled with the people in those bed-sitters." He rejected the plaintiff's evidence that he did not know that. He said if the plaintiff had been buying the property on that basis—that is to say the basis that the occupants were licensees—one would expect him to be furnished by the vendor's solicitors with the relevant licence agreements.

At the defendant's meeting with the plaintiff the plaintiff told the defendant that he did have another flat which might be suitable for the defendant. So both went off to that other flat, which was the ground-floor flat at 13 Ribblesdale Road of which the plaintiff has been refused possession. The defendant was told—and this is all common ground—that he would have to pay a deposit in relation to the condition of the flat and a month's rent in advance; and he would be allowed into occupation for a period of six months. So far, the background is one of negotiating for a suitable flat to let.

The learned judge then went on to deal with the written agreement which the plaintiff told him he got hold of from another property owner; an agreement which was clearly devised—and it is similar to other agreements which this court has looked at—to avoid the consequences of the 1974 Rent Act. That agreement, it was common ground, was presented to the defendant for signature on September 11 1976. But before that agreement was presented to him or signed, the evidence showed an oral agreement which is not in terms, as Miss Catterson rightly pointed out, referred to or found as an agreement by the learned judge. It is, however, in my judgment, the basis of his finding that the written agreement signed on September 11 1976 was a sham, and it emerges clearly from the evidence of the plaintiff in cross-examination. The judge's note of the plaintiff's evidence (which we have got) reads in this way:

> I told him I would want a month's rent in advance. He did pay this sum. He said he would return next day to give me the deposit money and a month's rent. He said he wanted to move in that week-end. It was a firm verbal agreement made on the evening. I took him to no 13. I expressed in no uncertain terms that the tenancy was for six months only.

The agreement itself is in these terms:

> This agreement is made the 11th day of September 1976 between John O'Malley (hereinafter called "the owner") of the one part and Colin Michael Seymour (hereinafter called "the licensee") of the other part,

and a curious feature, not referred to by the judge in the note which we have of his judgment, is that the agreement is signed in the top right-hand corner by Mr Seymour, the defendant, against the words which I have just read.

The agreement goes on:

> Whereby it is agreed as follows: (1) In consideration of the sum of £520 payable by the licensee to the owner as hereinafter mentioned, the owner grants to the licensee the right to use in common with others who have been granted or will be granted the like right the premises known as

and then filled in in manuscript

> ground-floor back flat, 13 Ribblesdale Road, Crouch End, London N8, together with the fixtures furniture furnishings and effects now in the said premises for six months from the date hereof.
>
> (2) The licensee agrees with the owner as follows: (i) to pay the said sum of £520 by equal instalments of £86.66 per month on the first day of each month, the first of such payments to be made on the signing hereof; (ii) not to damage or cause any damage to the walls or floors of the said premises or any part thereof and to preserve the fixtures furniture and furnishings and effects from being destroyed or damaged in any way and to replace all broken glass; (iii) upon the termination of this licence (whether by effluxion of time or otherwise) to pay to the owner a sum equal to £7.50 of the cost of washing and cleaning all counterpanes, blankets and curtains soiled, whether or not by the licensee, during the period of this licence; (iv) not to impede the use of the said premises or any part of it not nor [*sic*] upon the terms that each shares the cost of gas electricity and telephone services to impede the use of any services supplied to the said premises or any part of it by such other persons not exceeding three in number to whom the owner shall grant licence

I think the "not" is probably right, although it might more readily have been "nor."

> (v) not to assign this agreement nor to permit any other person, except as licensed by the owner, to sleep reside in or share occupation of the said premises or any part of it at any time; (vi) not to do or suffer to be done in or upon the said premises any act or thing which may be a nuisance damage or annoyance to the owner or to the occupiers of any adjoining property or which may vitiate any insurance on the premises against fire or otherwise or increase the ordinary premium thereon;
>
> (3) Provided as follows: (i) if the said instalments or any of them or any part thereof shall be in arrear or unpaid for at least seven days after the same shall have become due, or (ii) in the event of any breach by the licensee of the agreements herein contained, then the owner may terminate this agreement and the licensee's rights hereunder shall absolutely determine but without prejudice to any other remedies of the owner hereunder.
>
> (4) The owner agrees with the licensee to pay all general and water rates payable in respect of the premises but not any charges for the supply of gas or electricity or for the use of the telephone.
>
> (5) The licensee shall pay the sum of

and that is left blank

> which shall be retained by the owner until the termination of this agreement as a deposit to secure to the owner the performance by the licensee of the agreements on the part of the licensee herein contained, but without prejudice to the owner's rights under and in respect of this agreement. At the termination of this agreement and on the licensee vacating the said premises and subject to the proper performance by the licensee of his obligations herein contained, the owner will refund the said deposit to the licensee.
>
> (6) It is hereby agreed and declared that possession of the premises vests in and is retained by the owner subject to the right of the licensee to use of the premises given by this licence.

In fact a deposit was paid by the defendant by cheque dated September 9 for £50, and the learned judge also saw a receipt undated for £86, which it was accepted was paid before the defendant went into occupation.

The plaintiff sought to persuade the judge that before the agreement was signed he handed to the defendant a letter—a letter, I think, in the same form as was in evidence in the *Aldrington Garages* case and there subjected to some criticism. Nothing more need be said about that letter in this case, or what (if any) its effect could be on the transaction as a whole, because the defendant denied ever having received the document which the plaintiff swore he had given him and the learned judge accepted the defendant's denial.

The learned judge said of the defendant: "I am not dealing with an illiterate or an unsophisticated person. Mr Seymour has a good job and I found him an honest and straightforward witness." The defendant conceded that he had read through the agreement which I have read before he signed it, so there was no doubt about that; this agreement was signed by a sophisticated and literate person. The learned judge added that he did "not think for a moment that he" (the defendant) "appreciated the legal effect of the term 'licensee.' He was bothered about the limitation of the period to six months because he was reluctant to take the place with that limitation, but in the event he was so anxious for accommodation that he accepted it."

The learned judge then went on to consider the written agreement. He referred to the opening provision for the payment of £520 by instalments and said "the crux of the case is to be found immediately in clause 1 of the agreement . . .", that is in the opening words which stamp the agreement with the attempt, at any rate, to make the grant a grant of a right to use in common with others, a grant to a licensee to share with others, not a right of exclusive possession granted to a tenant. Of that the learned judge said: "I am convinced that that wording was designed to try and defeat one of the tests used to distinguish between a licence and a tenancy, that is to

A say whether there is exclusive possession. Mr O'Malley" (the plaintiff) "was honest enough to admit that he never contemplated for a moment foisting any other persons on the defendant. He knew that Mr Seymour wanted exclusive occupation for his own purposes. There was no suggestion ever made in the six months that Mr Seymour should accept any other person."

The learned judge went on: "Because he," that is the defendant, "signed a document referring to him as a licensee, the court is asked to infer that he does not have a tenancy under the Rent Act and that the plaintiff is entitled to the order he seeks." That might be criticised as an over-simplification of
B the plaintiff's case, but the judge went on: "The defendant says that this is not a genuine licence agreement but in effect a sham to avoid the Rent Acts, and that he is therefore entitled to remain."

The learned judge's opinion was that "Both parties here intended that Mr Seymour should occupy the flat for the payment of a periodic sum; Mr O'Malley intended that to be so and that Mr Seymour would not be disturbed in his occupation of the premises."

Then the learned judge went through the rest of the agreement part by part. One must remember, looking at his comments on the clauses, that they follow his initial comment
C on the first part of clause 1, which was that that was, as I understand his judgment, a device, and an unsuccessful device, to impart to the agreement between the parties a fundamental characteristic which it did not have. I do not think it necessary to refer to the learned judge's comments on the individual clauses. Miss Catterson does not quarrel substantially with any of them. Apart from the constant references to "licensee" and "licence," a label which may or may not be a true description of the nature of the agreement and the relationship between the two parties to it, there are clauses which are consistent, perhaps only consistent, with a licence; there are clauses which are consistent—perhaps more consistent, though none I think only
D consistent—with a tenancy; and there are clauses which are in my judgment, neutral and ambiguous. But I would accept Miss Catterson's submission that to turn this written agreement into a tenancy would require the court to rewrite substantial parts of it and that it is itself a licence and would by itself create a licence.

The learned judge went on to say that he had watched both the parties in the witness box and that he accepted the defendant's denial that he had ever seen this letter, as I have already stated. He came "unhesitatingly to the conclusion"—and I quote from the note—"that this is an attempt by the plaintiff to avoid the provisions of the Rent Acts. To set up
E this agreement as a licence is a sham; it in fact created a tenancy and I find that Mr Seymour is holding over as a protected tenant" and so he gave judgment for the defendant.

Miss Catterson, in what, in my judgment, was a model argument, has challenged that conclusion of the learned judge. She submits that the written agreement is appropriate to, indeed is, a licence, not a tenancy. She accepts that the learned judge was fully entitled to look at the background in order to see what the real transaction between the parties was. She does not challenge the right, and the duty, of the judge to find that the written agreement was a sham if that finding is justified on the primary facts which he finds, but she submits
F that his conclusion that this agreement was a sham is an inference which he ought not to have drawn from the facts he found.

She concedes that the learned judge was right in recording a fact (which is not in his note of the evidence) that the plaintiff did concede—and I had better repeat the actual words, "that he never contemplated for a moment foisting any other persons on the defendant. He knew that Mr Seymour wanted exclusive occupation for his own purposes. There was no suggestion ever made in the six months that Mr Seymour should accept any other person." But she says, first of all, that there are in this agreement clear, genuine provisions relating to sharing: in clause 1, which I have read; in clause 2(iv), G
which the judge called an obscure clause, about not impeding the use of the premises or services supplied to them by others; and in clause 6, agreeing and declaring that the possession of the premises was, among other things, retained by the owner.

She relies strongly on the fact found by the learned judge, and indeed admitted by the defendant, that he had read this agreement before he signed it, and the finding of the judge that he was not illiterate or unsophisticated. Though she cannot challenge the judge's finding that the defendant did not understand the legal consequences of the label "licence" and "licensee" the only part of the agreement which he was really interested in was the limitation of his occupation to six H
months. She said it was clear from his evidence that he did not regard himself as entitled under the agreement to stay on beyond the six months' period, and that evidence is quite inconsistent with his believing that he had a right to stay on given him by the law; he knew that that right was entirely dependent upon the will of the plaintiff, and that could only be true of a licence and not of a protected or regulated tenancy. "I read the document," said the defendant, "I noticed the limitation to six months' period. I told the plaintiff that I wanted accommodation for a period longer than six months. The plaintiff gave me to understand that the agreement could be renewed," and in cross-examination: "I do J
accept that I read through the whole of the agreement before I put my signature on it. The only query I raised was as to the six months' period. The plaintiff told me that it was his custom not to give an agreement for longer than six months" and then the plaintiff said he did not envisage any difficulty about renewing the agreement after six months. "I did notice the terms about people not exceeding three being nominated to share the flat with me. I thought I would be able to pay the rent myself and therefore no question would arise of bringing in other people to share in the payment of the rent."

Miss Catterson correctly submits that if there is no exclusive possession there cannot be a tenancy, although the converse K
is not true, and here, she says, are provisions in this agreement which show that possession is not to be exclusive; there is a right in the owner, the plaintiff, to share it with the defendant, and there is provision for others, not more than three, to share it. She submits that not only is that the position demonstrated by the written agreement, but that is what the judge should have found the oral agreement between the parties to have been when it was made firmly on the previous evening. She submits that the statement of both parties that there was no intention of sharing at the time the agreement was entered into is not really relevant, because an agreement may give you a right to share even if you do not intend to exercise it. L

The great difficulty in the way of her submissions is the finding of the learned judge that all these so-called rights to share, or references to sharing, were bogus, they were all part of the sham. If the "rights" really existed, if they were really agreed upon between the parties, they would take this ground-floor flat out of the Rent Act 1974 and enable the plaintiff to turn the defendant out at the end of six months. But the learned judge has found that that was not the agreement between the parties. The real transaction between the parties was the grant of a six months' tenancy with all the legal consequences that flow under the Rent Act 1974 from that fact. M

Miss Catterson summarised her contentions as well as she had made them. She said that the learned judge was wrong to give the weight he did to the evasion of the Rent Act. I have already indicated the extent of my agreement with that submission, but my opinion is that it did not alter the judgment which he would have come to had he had later decisions of this court before him.

Then she submits that the plaintiff knew that the defendant wanted exclusive occupation and would not contemplate putting anybody else in, but that is not inconsistent with his

A having an agreed right to do so. I have already dealt with that. But in connection with that submission it is relevant to point out, as was pointed out by my Lord, Geoffrey Lane LJ, in the course of the argument, that this case differs very materially from the two other cases, the *Somma* case and the *Aldrington Garages* case, not only that in one of those cases there was evidence about the possible use of that right which would have entitled the court to regard the provision of it as genuinely agreed and not a dead letter or a sham, but in each of those cases there was a contemporaneous agreement signed by a person wishing to share the same accommodation and granting exactly the same rights on the same terms, and those B two agreements had to be read together. So there was absent from those cases the striking feature of this case that there was, in fact, no one to share this flat with the defendant and no intention of there being anyone to share this flat with the defendant during the period of his occupation.

Finally, Miss Catterson submitted that the defendant having read this agreement before he signed it, the court can look at the agreement and discern in it the true nature of the relationship between the parties. Of course it must look at the agreement, but the question whether it discerns there the true relationship between the parties has, in my judgment, already been decided for us by the learned county court judge. There C was, in my opinion, clearly evidence on which he could have found that this agreement was a sham. I would go further and say on the evidence that was before him, if I had taken the same view of the witnesses as he did—as no doubt I should —I should have reached the same conclusion as he did, that it was a sham.

For those reasons, in spite of the excellent argument of Miss Catterson, I would affirm the judgment of the learned judge and dismiss the appeal.

Agreeing, LAWTON LJ said: A property owner is entitled to arrange his affairs so as not to get his property enmeshed in the Rent Acts. What he is not entitled to do is to arrange D his affairs in one way, which brings his property within the Rent Acts, and then to dress them up in another way so as to give the impression that it is outside the Rent Acts. The fact that he gets someone else to co-operate with him in that dressing-up does not affect the fact that he has done some dressing-up. What he has done is a sham.

The learned judge in this case decided that the plaintiff had been party to a sham. He had dressed up what was a letting for six months to look like a licence. If there was evidence to support such a finding, the learned judge was entitled to make it. In my judgment, the evidence was overwhelming that that was what the plaintiff had done. It was particularly convincing E evidence because most of it came out of his own mouth.

Mr Seymour met the plaintiff on September 9 1976. The meeting had come about because the plaintiff had inserted an advertisement in the *Evening Standard* that he had a flat to let. At the meeting the plaintiff and the defendant discussed the letting of a flat. In the end there was no letting of the flat which had been advertised, but the pair of them went to the flat which is the subject-matter of this appeal. The plaintiff, as my Lord has pointed out, never at any time suggested to the defendant that he was to have anything other than exclusive possession of the flat. In the course of the discussion with the defendant on September 9 1976 the plaintiff used the F language of letting. I read from the judge's notes of evidence: "I told him I would want a month's rent in advance. . . . He said he wanted to move in that week-end. It was a firm verbal agreement. . . . I expressed in no uncertain terms that the *tenancy* was for six months only." As a result of that meeting and before the agreement was produced and signed the defendant paid the plaintiff no less than £80 or thereabouts, being a month's rent in advance as asked for by the plaintiff, and had also paid the sum of £50 by way of deposit to protect the plaintiff from what he might well have called dilapidations. It was only after the money had been paid and after, in the plaintiff's own words, a firm verbal agreement had been made that this document dated September 11 came into existence. G That document did not reflect the agreement which had been made. Nevertheless he put it forward as an agreement which had been made.

In my judgment, the county court judge was entitled to find —indeed I would have found—that the licence agreement was a sham. I agree with my Lord that this appeal should be dismissed.

Also agreeing, GEOFFREY LANE LJ said: This appeal demonstrates how much each case of this sort must depend upon its own particular facts. The owner of the premises here was trying to put forward the same sort of arguments as did the owners in *Somma's* case, decided in March of this year, and H the *Aldrington Garages'* case, decided on July 7. Here, however, the basic, if not the only, problem which had to be decided was whether the written document which was proffered by the plaintiff and signed by the defendant purporting to record the agreement already reached between the parties was a genuine attempt to record the agreed terms or whether it was a sham. In other words, was it an attempt to disguise what was truly a tenancy in the trappings of a licence? The document itself, which was signed by the defendant, bore many similarities to the equivalent documents in the two cases which I have mentioned, but here there was, to say the very least of it, ample evidence that the document was indeed a sham. J The plaintiff's own answers in cross-examination to which my Lords have already referred really concluded the matter against him, because he effectively conceded that the prior oral agreement was firmly concluded and, secondly, that it was in the terms of a tenancy and not in the terms of a licence. The opposite contention was, on the evidence, scarcely arguable. There was plainly a tenancy here and not a licence. I would dismiss the appeal accordingly and would only add my tribute to the excellence of Miss Catterson's approach and argument in this appeal.

The appeal was dismissed with costs.

K

McHALE v DANEHAM AND OTHERS

(Before His Honour Judge EDWARDS in Bloomsbury and Marylebone County Court on January 8 1979)

Estates Gazette March 10 1979

(1979) 249 EG 969

Claim by owner for possession—"Temporary Holiday Accommodation Hire Agreement"—Whether agreement a "sham"—Licence or tenancy—Rent Act 1977 section 9—Possession order made

This was a claim by the plaintiff, Patrick McHale, who was L the owner of the property at 81 Shirland Road, Maida Vale, London W9, for possession of the first- and second-floor flat at that address which he had allowed to be occupied by seven people from abroad by virtue of an agreement dated June 25 1977. The plaintiff claimed that the agreement was a licence. Alternatively, that it was for a holiday letting within section 9 of the Rent Act 1977. The defendants, who were three of the original signatories to the agreement and their "guests," claimed that the agreement conferred on the original signatories a protected tenancy and not a licence, and in so far as the agreement purported to be for a holiday it was M a sham, as they were working in this country and had been at the time that the agreement was entered into, and that this fact was known at all material times to the plaintiff.

John H Fox (instructed by Jack Bernstein & Co) appeared for the plaintiff; Andrew Arden (instructed by the Paddington Law Centre) represented the defendants.

JUDGE EDWARDS, giving judgment, said that on or about June 9 1977, there was a meeting between the plaintiff Mr McHale and seven people, three of whom were among the present defendants. At this meeting it was arranged that the seven people could have the premises at 81 Shirland Road, Maida Vale, London, from June 25 1977.

On June 25 various people gathered at the premises and Mr McHale produced a written agreement dated June 25 1977. It is possible that one or two signed a day or two later, but nothing arose in connection with that. The agreement was for six months to December 25 1977.

The agreement was obviously central to the case. It was headed "Temporary Holiday Accommodation Hire Agreement" and it contained a number of clauses of the usual nature which were equally compatible with either a licence or a tenancy. However, one clause in particular, namely clause 12, recited that the agreement was for "temporary holiday accommodation only" and the attention of the signatories was particularly drawn to that clause. The agreement also contained clauses requiring access to the premises to be given to Mr McHale and requiring possession to be given up peacefully at the expiry of the term.

As regards the circumstances surrounding the signing of the agreement, what had happened was that Mr McHale had been introduced to seven people from abroad who he had been given to understand were on holiday in this country. There were two from South Africa, two from the United States and three from Australia. On June 25 all the parties gathered at the premises to be "let." The defendants said there was some discussion about the fact that the agreement was due to end on Christmas Day and also about their various jobs or occupations. There is certainly no doubt that they all read the agreement and signed it.

Because the occupants had nowhere else to go, the agreement was extended firstly for one month to January 23 1978 and finally for two months to March 24 1978. By that date four of the original signatories to the agreement had vacated the premises, but three were left together with three or four of their guests. They refused to go and as a result the plaintiff had brought possession proceedings against all the occupants.

It was argued by the counsel for the plaintiff first that the agreement was a licence and not a tenancy agreement. The learned judge was referred to *Somma v Hazelhurst* [1978] 1 WLR 1014 where Cumming-Bruce LJ had said, "We can see no reason why an ordinary landlord . . . should not be able to grant a licence to occupy an ordinary house." It was also argued on behalf of the plaintiff that the draftsman of the agreement had used the word "hire," rather than the word "tenancy," which was a much more common word in this context. Certainly the word "landlord" or "tenancy" did not appear in the agreement, but then neither did the word "licence." In His Honour's judgment, however, having looked at the wording of the agreement and the oral evidence, the agreement on a broad basis was much more like a tenancy agreement than a licence. His Honour felt that in general an agreement for occupation was usually a tenancy and that only in exceptional circumstances would such an agreement be a licence. The draftsman had not really got down to constructing a licence.

Counsel for the plaintiff's second point, which in His Honour's view was the real point in the case, was that this was a holiday letting within section 9 of the Rent Act 1977. That section read: "A tenancy is not a protected tenancy if the purpose of the tenancy is to confer on the tenant the right to occupy the dwelling-house for a holiday." Counsel for the defendants, on the other hand, argued that the agreement was a sham because all the ingredients of a sham were present, that is, the plaintiff wanted to let the premises, and he wanted to avoid the restrictions of the Rent Act, and counsel invited the court to find as a fact that the plaintiff well knew that the defendants were working in this country and could not therefore be on holiday. Furthermore, he argued, six, seven or nine months was too long for a holiday and Maida Vale was hardly a well-known holiday resort.

The plaintiff's counsel, on the other hand, submitted that even if the plaintiff knew that the defendants were working they could still be on holiday, because it was perfectly possible for a working holiday to be within section 9. The learned judge referred to *Snook v London and West Riding Investments Ltd* [1967] 1 All ER 518 at p 528 where Diplock LJ, as he then was, said that for there to be a "sham" there had to be a common intention to deceive. If, therefore, only one side were shamming, so to speak, then the agreement was not a sham.

The learned judge also referred to *Buchmann v May* [1978] 2 All ER 993, which was authority for the proposition that the onus of proof was on the defendants to avoid the provisions of the written agreement.

His Honour stated that quite clearly in cases where a sham was pleaded, the court had a duty to look at the whole background of the whole matter. However, he was disposed as far as possible to accept to the greatest extent that he could the evidence of all the witnesses. He did not think that there had been any course of cheating by Mr McHale, the plaintiff, who had been faced by a number of young people all of whom were from abroad and all of whom he thought were on holiday. Mr McHale did not really pursue the question of their working at all.

As to the question of whether there could be a "working holiday" His Honour was prepared to accept Mr Fox's submission in this respect notwithstanding the dictionary definition quoted in *Buchmann v May* that a holiday was a period of cessation from work or a period of recreation. His Honour could see no reason why a working holiday should not fall within the provisions of section 9 of the Rent Act. Although he did not place great reliance on the point, he thought that section 9 was enacted to encourage landlords to let their premises to holiday-makers. As to the extensions granted by Mr McHale, His Honour thought that they did not make any difference to the basic situation.

Not without hesitation and thought but firmly in the end, in His Honour's judgment the agreement was a holiday letting and did not confer a protected tenancy. Possession was ordered.

LANDLORD AND TENANT

RENT REVIEW

A **QUEEN'S BENCH DIVISION**

April 2 1979

(Before Mr Justice FORBES)

GREA REAL PROPERTY INVESTMENTS LTD v WILLIAMS

Estates Gazette May 19 1979

(1979) 250 EG 651

Landlord and tenant—Rent review clause in lease of office premises—Dispute referred to arbitration and consultative case stated by arbitrator on a point of law in course of arbitration—Proviso in lease that in determining the fair rack rental market value at date of review there should B be disregarded inter alia any effect on rental value of tenants' improvements—" Empty shell " before tenants' works—How to reflect inflation but not rental value of improvements—A judgment which valuers should study—Discussion of principles and different valuation methods—Whether to deduct annual equivalent of cost of improvements taken at review date from rental value of whole premises—Difficulties ensuing from difference between costs and values, assessment date and obsolescence—Whether to find from available evidence the rental value of improvements as a percentage of improved rental value at commencement of tenancy and C apply this percentage to improved rental value at review date—Judge's guidance to arbitrator

This was a case stated by the arbitrator under the Arbitration Act 1950 in the course of the arbitration (a consultative case) on matters arising from a review clause in a lease of parts of Dolphin House, North Street, Guildford, for office use. The tenants were GREA Real Property Investments Ltd. The landlord's interest was represented by Roy Granger Williams, the receiver appointed by the landlord's mortgagees, Crédit Lyonnais.

Ronald Bernstein QC and K Reynolds (instructed by Bischoff & Co) appeared on behalf of the tenants; Leolin Price D QC and G B W Nurse (instructed by Brecher & Co) represented the receiver.

Giving judgment, FORBES J said: This matter comes before me as a consultative case from the arbitrator of a dispute between the parties. It concerns the terms of a clause in a lease providing for a review of rent after seven years of the tenancy. The parties before the arbitrator have taken different views of the method by which the new rent should be calculated and both maintain that a point of law arises which the court should determine before the arbitrator proceeds any further.

The property in question is called Dolphin House and it E is a large modern edifice in North Street, Guildford. What appears to have happened is that the developers who were constructing the building ran out of resources before the building was finally completed. It is a typical modern steel and concrete building, in which the floors and roof are supported on a pattern of columns running from top to bottom of the building. Although the demise to the tenants, who are the respondents in this case, included certain other parts of the building, the substantial part of their demise relates to the third floor which is used as offices. It seems clear that when they took over that floor it consisted simply of an ex- F panse of bare concrete with a perimeter wall and a series of columns between floor and ceiling. The tenants carried out all the building works required to complete the floor so that it was fit for use, including such things as lavatories, cloakrooms and so on, in addition to the usual tenant's works such as office partitioning and matters of that kind.

In recognition of this the first six months of their tenancy was at a peppercorn rent and the clauses in the lease which deal with rent review have certain special provisions. The nominal claimant in this case is the receiver appointed by the landlord's mortgagees, Crédit Lyonnais.

I should examine the relevant provisions of the lease, which G is dated February 11 1971. The term is for 21 years from December 25 1969 and the reddendum provides first, as I have said, that the rent for the first six months should be a peppercorn. The rent for the remainder of the first seven years was £9,535. The rent for the remainder of the 21 years' term was to be " the rent as determined by the provisions of Clause 6 hereof." There were various additions to the rent concerned with insurance premiums and certain services provided by the landlord but these need not concern us. It will suffice to turn straight to clause 6:

PROVIDED ALWAYS and it is hereby agreed that at the expiration of the seventh and fourteenth years of the term hereby granted H (the time in each case being computed from the date of the commencement of the said term and being hereinafter referred to as " the date of review " the rent shall from and after the date of review be such sum as shall be agreed between the Landlords and the Tenants as representing the fair rack rental market value of the demised premises for the then residue of the term of years hereby granted as between a willing landlord and a willing tenant with vacant possession and in all respects on the terms and conditions of this Lease (other than the rent) and if the Landlords and the Tenants shall be unable to agree on the amount of such rent as aforesaid the same shall be decided by some competent person to be agreed to by the Landlords and the Tenants or in the event of failure so to agree by a person (hereinafter called " the Arbi- J trator ") to be named by the President for the time being of the Royal Institution of Chartered Surveyors who shall act as an arbitrator PROVIDED ALWAYS that the Landlords and the Tenants and such Arbitrator shall in agreeing or deciding the amount of the rent upon any such review disregard:

(a) any effect on the rental value of the demised premises of the fact that the Tenants have or their predecessors in title have been in occupation of the said premises.

(b) any goodwill attached to the said premises by reason of the carrying on thereon of the Tenants' business.

(c) any effect on the rental value of any improvement carried out by the Tenants or their predecessors in title otherwise than in pursuance of an obligation contained in this Lease and in K particular but without prejudice to the generality of the foregoing any such effect of any works of fitting out and finishing the demised premises to acceptable office standards carried out by the Tenants at the commencement of the term hereby created.

As I understand it, valuers would have no difficulty over the technical methods to be used for arriving at what was " the fair rack rental market value of the premises," having regard to the criteria set out in clause 6, were it not for the terms of the proviso which set out what has to be disregarded in arriving at the rent to be paid. It is because the valuers for landlord and tenant respectively have adopted wholly different valuation methods, having regard to the terms of this proviso, that the arbitrator has been persuaded to state this consultative case. It is difficult to decide a matter like this without knowing what the technical differences between the parties are, and both parties have helpfully put before me affidavits by the respective valuers explaining their differing approaches and conclusions.

It is a fundamental aspect of valuation that it proceeds by analogy. The valuer isolates those characteristics of the object to be valued which in his view affect value and then seeks another object of known or ascertainable value possessing some or all of those characteristics with which he may compare the object he is valuing. Where no directly comparable object exists the valuer must make allowances of one kind or another, interpolating or extrapolating from his given data. The less closely analogous the object chosen for comparison is the greater the allowances which have to be made and the greater the opportunity for error.

Now in this case no difficulty arises over subclauses (a) and (b) of the proviso to clause 6—the elimination of goodwill and of the effect of occupation by the present tenants. The only difficulty is over (c), the elimination of the value attributable to the tenants' improvements. These, it will be remembered, involved substantially the conversion of the tenants' part of the building from an empty shell to premises fit to be let to a commercial tenant. The occasions on which tenants carry out improvements of this character are sufficiently rare to present the valuer with this problem; it is virtually impossible to find a case where a building has been let in the state of an empty shell, so that directly analogous comparables are non-existent. There is, however, one fairly closely analogous case, namely the letting of this very building at the beginning of the current lease. The main difference between the situation then and the situation now is due to the passage of time and its effect upon rental values in the interim.

It is said that the determination of the proper method of valuation to adopt in this case involves a point of law. It has been pointed out to me that section 34 of the Landlord and Tenant Act 1954 contains provisions very similar to those of clause 6 in this lease:

> 34(1). The rent payable under a tenancy granted by order of the court under this Part of this Act shall be such as may be agreed between the landlord and the tenant or as, in default of such agreement, may be determined by the court to be that at which, having regard to the terms of the tenancy (other than those relating to rent), the holding might reasonably be expected to be let in the open market by a willing lessor, there being disregarded—
>
> (a) any effect on rent of the fact that the tenant has or his predecessors in title have been in occupation of the holding,
>
> (b) any goodwill attached to the holding by reason of the carrying on thereat of the business of the tenant (whether by him or by a predecessor of his in that business),
>
> (c) any effect on rent of any improvement carried out by the tenant or a predecessor in title of his otherwise than in pursuance of an obligation to his immediate landlord.

Paragraph (c) applies to tenants' improvements "carried out otherwise than in pursuance of an obligation to his immediate landlord." This section, of course, applies to determination of rents by the courts where a new lease of business premises is made under the Act. Its provisions are therefore applied very commonly in many cases up and down the country. I am told that there is no judicial pronouncement upon the valuation method proper to be adopted to reflect the words of the statute or the equally commonly applicable words to be found in leases of the kind before me. It is suggested by both counsel that in determining the present difference between the parties I should be mindful of the fact that any principle I may seek to lay down should be equally applicable to these other cases. Further, I am reminded that rent review clauses are now very common, owing to the incidence of inflation, and that some of the leases which contain them may be for much longer than the 21 years of this lease. Again it is suggested that any principle enunciated should be equally valid when applied to the last rent review of, say, a 50 years' lease as to the first review of the present one. I would not accept that my determination of what is legally appropriate to the interpretation of this contract between these parties should necessarily be swayed by consideration of similar wording in the statute. The provisions of the statute may well fall to be construed by what the courts consider to be the intention of Parliament; the provisions of a lease must be construed in the light of what the courts conceive to be the intention of the parties. This may be the same or it may not. Although it would be right to bear these other matters in mind, I shall stick to the lease. In doing so I shall of course remember that it provides for another review at the 14-year stage. The only question of law, however, seems to me to be the determination of the intention of the parties in relation to the process of agreeing the rent at the review stages.

It is now trite law to say that you cannot construe a contract *in vacuo*. "No contracts are made in a vacuum. There is always a setting in which they have to be placed. The nature of what is legitimate to have regard to is usually described as 'the surrounding circumstances,' but this phrase is imprecise: it can be illustrated but hardly defined. In a commercial contract it is certainly right that the court should know the commercial purpose of the contract and this in turn presupposes knowledge of the genesis of the transaction, the background, the context, the market in which the parties are operating." And again, "I think that all of their Lordships [Viscount Haldane LC, Lord Kinnear and Lord Dunedin in *Charrington & Co Ltd v Wooder* [1914] AC 71 at pp 77, 80 and 82 respectively] are saying, in different words, the same thing—what the court must do must be to place itself in thought in the same factual matrix as that in which the parties were." These are quotations from the opinion of Lord Wilberforce in *Reardon Smith Line Ltd v Yngvar Hansen-Tangen* [1976] 1 WLR 989 at pp 995-997B respectively.

I think the intention of the parties was clear. First, they both realised that in an inflationary period the rent appropriate at the beginning of the lease would almost certainly not represent a fair rent as between the parties at the end. Rather than make an estimate of the rent which, having regard to possible inflation, might be charged over the full 21 years, they chose to fix it only for the first seven and to provide for review at that date and at the 14-year date. Secondly, they realised that, because the tenant was paying for considerable works to be done which would ordinarily have been completed by the landlord, it would be fair that the landlord should not have the rental benefit of those works during the currency of the lease. For this reason the rental equivalent of the tenants' works was to be eliminated from the rent to be agreed at the reviews. In approaching any valuation here valuers must use some method of valuation which carries out the intention of the parties that the rent payable to the landlord should keep pace with inflation and that from such rent should be eliminated the rental equivalent (itself affected by inflation) of the tenants' works. Any such method must also proceed along lines which are logically sound and can be seen to be consonant with commonsense. In this connection it is important that in the exercise of drawing analogies and making comparisons like should be compared with like. You can do anything to an equation without altering its validity so long as you do it to both sides. If it is desired to compare, as here, the situation in 1969 with that in 1976 it is of the utmost importance that you should not assume *without evidence* that value and costs are synonymous terms, and any method of valuation which involves using these two concepts as if they were always interchangeable is logically suspect. It may be that seven years of inflation has had precisely the same effect in monetary terms on values as it has on costs, but it is much more likely that it has not. Further, a figure which is

A compounded of value and cost may, it seems to me, turn out to be as much a mathematical absurdity as a suggestion that three apples plus two oranges equals five bananas.

Although the matters primarily affecting landlord and tenant in agreeing the rent and rent review clauses were the incidence of inflation and the works carried out by the tenant, there will have been others of secondary but nevertheless of considerable importance. The tenant would have in mind in 1969 the cost of the works he was to carry out and of borrowing the capital to complete them, the length of the lease (including his assessment of the prospect of obtaining a new one under the 1954 Act), and the annual equivalent of his total cost spread over B the length of the lease. He would envisage spreading the cost over the whole length of the lease and not merely the first seven years because he would know that he would derive effective benefit from them for the whole period, since the rental value of his works would be eliminated from consideration at the rent reviews. He would also know, as would the landlord, that the benefit he would derive from his expenditure would cease at the termination of the lease, or any extension under the Act of 1954, and to this extent would be a wasting asset. Both parties would also consider the question of obsolescence and the incidence of the passage of time. In this case the works were really part of the building operations which should have been C carried out by the landlord and no one suggests that they would become obsolete and require replacement before the end of the term. But no doubt some classes of improvements in some long leases might cause the parties to contemplate the necessity for replacement before the termination of the lease. Even though there was no such possibility here both parties would be well aware that as time passed the improvements would be ageing and that at the review periods envisaged they would be respectively seven and 14 years old. The parties would also contemplate that in approaching any question of agreement at the reviews they would be advised by valuers, who would be careful to ensure that in any comparisons made, whether of D premises, circumstances or situations, like should always be compared with like.

The only principle of law which I think it necessary to enunciate is thus the simple one that any method of valuation must properly reflect the intention of the parties as expressed in the lease: that in arriving at that intention the court, and the valuer, must " place itself in the same factual matrix as the parties were "; and that this involves giving proper weight to the matters which as primary and as secondary considerations I have already set out.

The rival methods of valuation are set out shortly in the E case itself but have been considerably expanded by affidavits from Mr C H Toye, the landlord's valuer, and Mr E T Hartill, the valuer for the tenant. Briefly, the tenant's contention is that the correct way to disregard the effect on the rental value of the premises of the tenant's works is to value the premises, as at the review date, on the assumption that the tenant's works did not exist. As has already been pointed out, it is impossible to find examples of rents charged for comparable buildings in such a state, so that some other method must be sought to arrive at a valuation on this assumption. Mr Hartill accordingly starts with the rental value of the building as it is, improved, and at the review date. There is F no difficulty about this and comparable cases abound. He then goes on to envisage the hypothetical tenant calculating what it would cost *at the review date* to fit out those premises with the improvements as they exist (including the cost of servicing the capital borrowed for the purpose) and then determining the annual equivalent of that cost having regard to the length of the lease (including any prospect of a new one under the Act of 1954). Mr Hartill then deducts this annual equivalent from the assessed rental value of the whole premises to arrive at the rent due to the landlord from the review date. The method he chooses to find the costs of the works as at the review date is to take the historical cost of the works in 1969 and multiply that by a factor which is G the index of the increase of building costs generally between 1969 and 1976, a figure of 2.81. He also, incidentally, includes in his reduction from capital cost to annual equivalent a factor for a sinking fund in favour of the tenant.

Mr Price, for the landlord, attacks this method. He starts by arguing that, as the clause speaks of " the fair rack rental market value *of the demised premises*," a phrase which necessarily includes the improvements, there is no valid argument for adopting any method of valuation designed to find the appropriate rent for the " shell," that is, the building without the tenant's improvements. That, he says, would be to value not the demised premises but the premises before the carrying H out of the tenant's works. I cannot accept this argument in the broad terms in which it was originally stated. It does not seem to me to matter whether one starts with the value of the improved building and deducts the rental value of the improvements, or values the unimproved building direct, so long as one is properly taking into account inflation and the value of the improvements. The object of the valuation exercise is to find the rent payable to the landlord. If $A-B=C$ and the object is to find C it would be ludicrous to disregard the known value of C, if it were known, in order to go through the process of finding A and B and subtracting the one from the other. If there is evidence which enables a J valuer to value the landlord's share direct then it seems to me that he will have eliminated the effect on rental value of the tenant's share: to this extent Mr Price's argument is too wide. The difficulty may be that there is no such evidence.

To take an example: suppose three identical and adjacent four-storey Victorian houses in a good residential area of London. The owner of all three properties installs a lift in house A and lets it. The prospective tenant of house B wishes to leave the property as it is and endure the disadvantage of climbing up and down three flights of stairs. The prospective tenant of house C, however, would prefer a lift and, by agreement with the landlord, installs it himself. All three houses K are let for 21 years on leases with rent reviews at seven and 14 years, but tenant C's lease contains a proviso in the same terms as the lease in this case because of his expenditure on the lift. It would I think be an extraordinary thing if the valuer of property C were to be told that, at C's review, he must, whatever the circumstances, disregard the rent agreed at B's review because that was a rent of premises before the carrying out of tenant's improvements. Valuers work, as I have said, by analogous comparables, and a rent, freely agreed, for premises which are identical but for the tenant's improvements seems to me, other things being equal, to be the best evidence of the rental value of the improved premises " dis- L regarding the rental value " of those improvements. It is when other things are not equal or, as here, when no such comparable exists, that valuers may be driven to other shifts.

Mr Price's argument, however, seems to me to have more general validity if its terms are somewhat altered. To value premises by assuming that the tenant must pay for the improvements *at the time of the review* is to disregard the existence of the improvements and not necessarily their rental value. If, in the example I have given, you value house C by a method which involves deducting the annual equivalent of the cost of installing the lift at 1976 prices you may be " disregarding the rental value " of the improvements but if so M only by the accident that the rental values of premises with lifts have increased at the same rate as the cost of installing the lifts in them. This accident may in fact occur if the market in which these rental values operate is such that tenants regard lifts as virtual necessities. In such a situation any person considering the tenancy of a house which had no lift would regard it as essential that one should be installed and, provided he could achieve a satisfactory length of lease having regard to the expenditure required, and also the inclusion of safeguards such as those in the proviso to clause 6, might well install it himself. Mr Price is right to this extent that the problem facing the valuer is that the building he is valuing is in fact improved. The basic fact is therefore that the tenant is not going to have to pay for the improvement. If the improvement is not regarded as a necessity a tenant could no doubt be tempted to pay a higher rent for an improved property, but this would be according to his estimate of the desirability of the improvement. The more desirable, the nearer the acceptable increase in rent would approach the annual equivalent of the cost of installation. The measure of desirability of any class of improvement would depend on the views of the persons making up the market.

To subtract from the accepted rental value of an improved building the annual equivalent of the capital cost of the improvement *at the review date* (which is Mr Hartill's approach) thus may be a satisfactory method of disregarding the rental value of the improvement and of reflecting the incidence of inflation which the parties intended, but only if one or other of two conditions is satisfied. The first is that the rate of inflation has had the same effect on rental values as it has had on the cost of the improvements. The second is that, in the market which is being envisaged, the improvement would be regarded as a necessity or very nearly so.

But there is another factor which must be taken into account. At the start of the lease the tenant, having performed the calculations which produced the annual equivalent of the capital cost of the works, thought it justifiable to incur that cost because he would reap the benefit over the full period of the lease (plus any statutory extension). It does not in the least follow that a tenant faced with such expenditure at the first review date would regard it as economically sensible to pay the capital cost of a benefit which would only accrue to him over a shorter period; at the second review date the likelihood of such expenditure (particularly if inflated) being considered worthwhile may recede even further. The officious bystander may no longer be a fashionable interloper when considering implied terms in a contract but he may still provide a useful catalyst when determining the intention of the parties. I am quite certain that if he had intervened during the discussions which preceded the lease to say that, at the second review date, there would have to be envisaged a hypothetical tenant who might not regard it as sensible to install the improvements unless he obtained the premises rent free or even was paid by the landlord to occupy them, both parties would have said "But we both know that in fact no tenant will actually have to pay for any improvements: they already exist."

If, therefore, a method of valuation has to be adopted such as that put forward by Mr Hartill it will not only be necessary to demonstrate that one or other of the conditions I have already mentioned is fulfilled: there will also have to be some modification, if any is possible, to take account of the fact that what is being valued at each review date is the rental equivalent of a benefit to the tenant which the parties envisaged was to be incurred at the start of the lease and written off over the period of the lease and any extension of it. I might add that Mr Hartill's adoption of a sinking fund in deriving the annual equivalent of the capital cost seems here, if I have understood its purpose correctly, to be singularly inappropriate where the underlying fact is that the tenant will inevitably lose and the landlord gain the residuary value of the improvements at the end of the lease.

There is, too, the question of obsolescence. The improvements, the rental value of which is to be disregarded, at the first review will then be seven years old: at subsequent reviews they will, of course, be still older. Some effect should be given to this factor where appropriate as well as the others.

Now I have no evidence about the extent to which these improvements can be regarded as essential, nor have I any cogent evidence about the relative effects of inflation on site values, rental values, major building works or works of improvements such as the tenant here carried out. Nor do I know to what extent Mr Hartill's valuation can be regarded as giving effect to age or the fact that the benefit to the tenant should be regarded as a wasting asset. Such evidence will be necessary if any such exists, before it can be shown that Mr Hartill's method adequately reflects the intention of the parties, but this must be a matter for the arbitrator.

The landlord's valuer, Mr Toye, starts by assuming that in agreeing the rent for the first seven years the parties had in fact performed the exercise which is now being called for, namely they had, in fixing the rent for the demised (and therefore improved) premises disregarded the effect on that rent of the improvements. The rent they agreed was £9,535. By the use of readily available evidence of the 1969 values of comparable completed office buildings he deduces a market rent for Dolphin House, as improved, of £11,000. The difference between the two figures (£1,465) is 13.3 per cent of the rent deduced for the improved building. He then finds, again from comparable buildings, the 1976 rental value of the improved building to be £40,000. He applies the same percentage (13.3 per cent) as representing the value of the tenant's improvements (£5,320) and, subtracting this from the figure for the improved building, arrives at a final figure of £34,680. This method has the considerable merit that it is founded on an assumption which appears unassailable, namely that the rent initially agreed by the parties did in fact incorporate an agreed discount which the parties thought properly reflected the effect on rent of the tenant's works, and the additional attraction that it is both simple and relies on readily-available data, namely, the rents at the two relevant dates of comparable completed office buildings and the rent actually contained in the reddendum.

But while the calculation for 1969 would seem to rest on a firm base in relation both to the intention of the parties and to the dictates of commonsense I am not so sure about the 1976 calculation. This assumes that the proportion which the value of the tenant's works bears to the whole will remain constant whatever happens. It may be, however, that inflation has had, or would have in the future, different effects on site values, major construction works and works of improvement such as the tenant's here; these are the three components which make up the total value of the building. It may be also that, even if inflation has had different effects, the comparative magnitudes of the three component figures would mean that wide variations between the rate of inflation for one component on the one hand and the other two on the other would not seriously affect the validity of the final calculations. Again, I have no evidence of these matters which, in any case, should be for the arbitrator to evaluate and not for me.

The landlord's valuer, Mr Toye, before finally adopting the method I have already described canvassed another method. This involved considering the historical cost of both the tenant's works (ie the improvements actually carried out by the tenant) and the "landlord's works" which Mr Toye defined as the site and the remainder of the building works other than the tenant's works. The combination of tenant's works and landlord's works (as so defined) equals the totality of the improved building. He would then apply to any rental valuation of the improved building at the review date the percentage which the cost of the tenant's works bore to the costs of the whole improved building. He also expounded a refinement of this method involving a comparison of the capital value of the completed and improved building in 1969 with the historic cost of the tenant's works. It seems to me that both these methods suffer from the same defect which afflicts Mr Harthill's method. The landlord's works, as defined by Mr Toye, includes the value of the site as well as the landlord's building works. The percentage sum to which he is committed is thus based on a mixture of site value and building costs (itself a doubtful mathematical exercise) and its validity as an index must depend on the rate of inflation of site values and building costs remaining the same. It is only fair to add that if the proportion of the whole represented by the site value is very small there would be only a small error even if the rates of inflation varied widely. But there is still the difficulty that this approach also subsumes an equal rate of inflation for works of the character of the landlord's building operations (major steel and concrete framed building) and the tenant's (final fitting out for occupation). Mr Toye himself recognised that there were other defects in these methods in the circumstances surrounding this particular building and therefore rejected them in favour of the approach I earlier set out.

In summary, I conclude that the intention of the parties as expressed in the lease, having regard to the factual matrix in which the parties made their initial agreement, can be shortly set out as follows:

(1) in view of the possibility of inflation the rent must be reviewed at the seventh and 14th years in order to do justice between landlord and tenant;

(2) in view of the tenant's expenditure he was to be credited with the rental equivalent of the works the subject of that expenditure which might itself be affected by inflation; but this was to be the sole benefit which the tenant was to derive from his expenditure (apart from the initial rent-free period which can be disregarded at the review dates);

(3) the improvements themselves would be paid for once and for all at the start of the lease, and at the end of the lease (or any extension) enure for the benefit of the landlord, and in consequence the improvements so far as the tenant was concerned should be regarded as a wasting asset;

(4) the passage of time would mean that at the reviews the improvements would be no longer new; it would be seven- and 14-year-old improvements which fell to be valued;

(5) the valuers in making comparisons of any kind would be careful to compare like with like, bearing particularly in mind that the influence of inflation might differently affect values and costs.

As I perhaps have adequately indicated already, I have no facts which would enable me to decide whether any of the various methods of valuation suggested do in fact meet these requirements though I have considerable doubts about the extent to which certain methods can properly be said to be consonant with these intentions. I hope, however, that what I have said will enable the parties to address their minds to the question of what further evidence their valuers should adduce and the arbitrator to decide how far their valuations coincide with the intention of the parties as I have sought to express it.

It was ordered that the costs of this case should be reserved to be dealt with by the arbitrator. Leave to appeal to the Court of Appeal was given.

RATING

A **QUEEN'S BENCH DIVISIONAL COURT**

April 19 1978

(Before Lord WIDGERY CJ, Mr Justice BOREHAM and Mr Justice DRAKE)

COUNTY AND NIMBUS ESTATES LTD v LONDON BOROUGH OF EALING

Estates Gazette January 27 1979

(1978) 249 EG 339

General Rates Act 1967—Unoccupied property rate—Appeal against grant of distress warrant by magistrates—Whether altered valuation ineffective because valuation officer's proposals never reached owners—Owing to transfers of
B **ownership and moves notice of proposals sent to address which owners had left—Forwarding arrangements either not carried out or unsatisfactory—No objections having been received new rateable values entered on list—Effect of sections 87 and 67(6) and (7) of General Rate Act—Rating authority obliged to collect rates on basis of list in force—Valuation as altered in list conclusive evidence of value—Appeal dismissed**

This was an appeal by case stated from a decision of the Willesden magistrates granting a distress warrant for unpaid unoccupied property rate on Morris House, Cowley Road, London W3, a property owned by the appellants, County and
C Nimbus Estates Ltd.

Lord Colville of Culross QC (instructed by D J Freeman & Co) appeared on behalf of the appellants; C Fay (instructed by the Chief Solicitor, London Borough of Ealing) represented the respondents.

Giving the first judgment at the invitation of Lord Widgery CJ, DRAKE J said: This is an appeal by way of case stated from a decision of the Willesden Justices given on September 16 1976. On that date a complaint made by the London Borough of Ealing as the local rating authority alleging that the appellant company had failed to pay the unoccupied property rate in respect of premises known as Morris House,
D Cowley Road, London W3, was found to be good, and the justices granted a distress warrant in the sum of £59,124.73. The appellant company contended that the sum demanded was excessive because the true rateable value for the premises in question was lower than that shown in the valuation list on which the rating authority had based its demands.

In brief, in circumstances to which I will refer in more detail in a moment, the rateable value of the premises as shown in the valuation list had been amended with effect from December 21 1973 as a result of proposals made by the valuation officer on February 4 1974. The appellants say that the notice of such proposals was never validly served by the
E valuation officer in the manner required by the General Rate Act 1967 and accordingly the increased rateable valuation as shown in the valuation list was ineffective and invalid. They contend that the true and only valid valuation remained the lower one formerly in effect, and that they are liable only to pay rates calculated by reference to this lower figure.

The facts as found by the justices and recited in the case stated may be summarised as follows. In November 1973 the then owners of the property negotiated to sell the premises to a company called Nimbus Estates Ltd, which are to be distinguished from the appellant company, who are County and F
Nimbus Estates Ltd. The valuation officer for the area was informed by letters from the vendors' solicitors and from the proposed purchasers, Nimbus Estates Ltd that Nimbus Estates Ltd were buying the property. The letter from Nimbus Estates Ltd included in its heading the words "Reply to 7/7a Station Road, Finchley, N3 2SB." The sale was completed on December 21 1973, but at the direction of Nimbus Estates Ltd the property was in fact not purchased by them but was conveyed to another company called Freemanco (No 2) Ltd, which subsequently changed its name to the appellant company now named County and Nimbus Estates Ltd. Also subsequently to that conveyance Nimbus Estates Ltd changed G
its name to Nimbus Securities Ltd. No one informed the valuation officer of those matters, and so far as he was aware the purchaser had been Nimbus Estates Ltd, who had an address 7/7a, Station Road, Finchley, to which he had been requested to address any reply to the original letter received from that company.

The premises were then, subsequent to the conveyance, altered or enlarged or rebuilt, and the valuation officer on February 4 1974 made proposals to subdivide the premises from two rateable properties into four rateable properties. The effect of that as well was to increase quite considerably the total rateable value of the premises. He, the valuation H
officer, having assumed that the purchaser of the premises had been Nimbus Estates Ltd, proceeded to send his notice of proposals to the address 7/7a Station Road, Finchley, assuming no doubt that that was the registered office of Nimbus Estates Ltd. In fact that address was not the registered office of Nimbus Estates Ltd, and additionally Nimbus Estates Ltd had moved away from whatever activities they had carried on at any time at those premises and gone to some other address, and apparently the arrangements for forwarding mail were not fully carried out or satisfactory. The result of that was that the valuation officer's proposals never reached the appellants, who were in fact the owners of J
the property.

The valuation officer, not receiving any objections to his proposals for the obvious reason that the appellants had not received them, in those circumstances made his new valuations effective by directing the rating authority to enter them on the valuation list. The rating officer carried out that direction, and so the increased and new rateable values were duly entered on the list. At a later date when rate demands were received by the appellants on May 14 1975 based on the altered valuations, the appellants contended that the new valuations had not been made in the manner required by the K
General Rate Act 1967, the proposals not having been served on them or in the manner required by section 109 of the Act. Accordingly, the appellants say, the altered valuation was wholly ineffective.

The matter having come before the magistrates on a complaint by the rating authority asking that a distress warrant should be levied, the appellants then raised these arguments saying that because the proposals of the valuation officer had not been served in accordance with the Act the new valuation

A was wholly ineffective, and, therefore, they were not bound to pay the increase due under the altered valuation. The response of the rating authority to that was, and is in the arguments we have had addressed to us today in this court, in short, that the effect of section 87 of the General Rate Act 1967 and section 67(6) and (7) of that Act is to make the valuation as shown in the valuation list conclusive, and that the rating authority should not, and have no power to, go behind the valuation as shown in the list and to enquire into how the altered valuation got there and matters of that nature.

Section 87 of the Act reads as follows:

B The rating authority shall give effect to any directions which may from time to time be given to them by the valuation officer in pursuance of any provision of this Act authorising or requiring the valuation officer to cause or direct alterations to be made in a valuation list.

Section 67(6) reads as follows:

Subject to subsection (7) of this section, the valuation list in accordance with which, under section 2(4)(b) of this Act, any rate falls or fell to be made, as in force (or about to come into force) at the date of the making of the rate, shall be conclusive evidence for the purposes of the levying of that rate of the values of the several hereditaments included in the list.

C Subsection (7) deals with the case in which an alteration of the valuation list is made by providing as follows:

As respects any period during which, under this Act, an alteration of the valuation list referred to in subsection (6) of this section is for the time being to be treated as having had effect, the reference in the said subsection (6) to that list shall be construed as a reference to that list as so altered.

The effect of those sections, it is contended by the rating authority, and in my judgment rightly contended, is as follows. The rating authority are put under a mandatory duty by the provisions of this Act to collect rates on the basis of the
D valuation list in force, and, where an alteration is being made to that list, on the basis of the valuation as altered. Under subsections (6) and (7) of section 67 the valuation as altered shown in the valuation list is conclusive evidence as to its correctness.

Lord Colville on behalf of the appellants says that, while he must concede that section 87 is a section which prohibits in effect the rating authority from going behind the valuation, that does not debar the magistrates who are called upon to issue or consider the issuing of a distress warrant from going behind the valuation list to see whether the valuation has been properly entered as a result of valid steps taken within the
E procedure laid down by the Act. In my judgment, that is not a good argument. The magistrates are required to judge the position between the rating authority and the ratepayer, and Parliament has laid down by section 87 of the Act that there is a duty on the rating authority to give effect to directions which may from time to time be given to them by the valuation officer. The effect of that section, coupled with section 67, is, in my judgment, such as to bind the magistrates also not to go behind the valuation as shown in the valuation list and to give effect to it, and to issue a distress warrant where it is shown to them that the ratepayer has been properly summonsed to appear before the magistrates' court and the
F rate demanded is in accordance with the rate as shown on the valuation list.

In my judgment, the passage in *Ryde on Rating* 13th ed at p 861 correctly states the position when it says:

. . . if the rate is good on the face of it, and has been duly made and published, and there is jurisdiction to make a rate on the person charged, the duty of the justices becomes ministerial.

For these reasons, in my judgment, the justices were correct in this case in refusing to inquire into such matters as the validity of the steps taken by the valuation officer in serving the notice of proposals made by him, and were correct in granting the distress warrant in the same claim. For those G reasons I would dismiss this appeal.

Lord Widgery CJ and Boreham J agreed.

The appeal was dismissed with costs. Leave to appeal (if it were necessary) was refused.

COURT OF APPEAL

December 13 1978

(Before Lord Justice MEGAW, Lord Justice LAWTON and Lord Justice CUMMING-BRUCE)

WINDSOR SECURITIES LTD v LIVERPOOL CITY COUNCIL H

Estates Gazette April 7 1979 and April 14 1979

(1978) 250 EG 57 & 151

General Rate Act 1967 as amended by Local Government Act 1974—Unoccupied rate and penal rating surcharge—Important decision on principles governing relief—Discussion of meaning of "hardship"—Commercial premises purchased for development—Difficulties in securing surrender of all the tenancies—Project overtaken by economic troubles—Applications to city council for relief from unoccupied rate and surcharge on grounds of J hardship—Request by city treasurer for company's trading account and balance sheet not complied with—Applicants for relief should present a fully-documented case of hardship—Applications rejected by city council—No reasons given—Orders of certiorari and mandamus sought by applicants—Orders refused by court (except certiorari to quash one purported decision made without authority by city treasurer)—Submissions that the city council's rejection of applications was perverse, vitiated by error of law, a breach of rules of natural justice or defective for lack of reasons considered and dismissed

K This was an appeal to the Court of Appeal from the refusal of the Queen's Bench Divisional Court to grant orders of certiorari and mandamus on the application of Windsor Securities Ltd for the purpose of challenging decisions of Liverpool City Council rejecting applications for relief from the unoccupied rate and the rating surcharge assessed on the applicants' property, Peter's Buildings, Rumford Street, Liverpool.

Matthew Horton (instructed by Hextall, Erskine & Co, agents for Laces & Co, of Liverpool) appeared on behalf of the appellants, Windsor Securities Ltd; Ian Foster (instructed by the city solicitor, Liverpool) represented the respondent L city council.

Giving the first judgment at the invitation of Megaw LJ, Cumming-Bruce LJ said: By the first of these two applications dated May 2 1977 Windsor Securities Ltd seeks:

(i) an order of certiorari to quash a purported decision of the Liverpool City Council, acting through the treasurer, made on September 25 1975 whereby he purported to refuse an application by the applicant under paragraph 3 (A) of the First Schedule to the General Rate Act 1967 for relief on the ground of hardship from liability to pay the rating surcharge levied under section 17A of that Act, as amended, M in respect of Peter's Buildings, Rumford Street, Liverpool; and

(ii) an order of mandamus directed to the Liverpool City Council to reconsider the application and determine it according to law.

By the second application the same company seeks:

(i) an order of certiorari to quash a decision of the Liverpool City Council, acting through its General Services Committee, made on September 22 1976 whereby it refused an application by the applicant under the same provisions for relief on the ground of hardship from liability to pay both the unoccupied rate assessed pursuant to paragraph 1 of the First Schedule to the 1967 Act and rating surcharge in respect of the same buildings; and

(ii) an order of mandamus to reconsider the application and determine it according to law.

Before 1966 there was no liability for rates upon unoccupied property. By section 20 of the Local Government Act 1966, re-enacted in section 17 of the General Rate Act 1967, the law was changed. Thereafter local rating authorities were empowered to levy rates upon unoccupied property. The liability was imposed upon the owner. The assessment was to be 50 per cent of the assessment made on the hereditament when occupied, but by section 17 rating authorities were empowered to resolve whether to bring the liability to unoccupied rate into force in their areas and to determine the date from which such liability was to be imposed. Under the 1967 Act, once a rating authority had resolved to enforce liability to unoccupied rate, the only discretion that it had was that given under section 53, which provided: "A rating authority shall have power to reduce or remit the payment of any rate on account of the poverty of any person liable to pay it." The minister was given power by paragraph 3 of the First Schedule by regulations to provide that unoccupied rate should not be paid in respect of hereditaments of such descriptions as might be prescribed by the regulations or in such circumstances as might be prescribed. In 1967 the minister made regulations (SI 1967 No 954) exempting certain classes of owner after death, bankruptcy or upon winding-up of a company. They do not apply in these cases.

The Liverpool City Council had not resolved to implement the provisions of section 17 of the 1967 Act when the applicant embarked upon a commercial adventure which had as its purpose the purchase of Peter's Buildings, Rumford Street, in Liverpool, its demolition, and the construction on the site of a new block designed as office accommodation which would be put on the market and let. The applicant is described as a limited company incorporated on November 26 1971 for the purposes of property development. Peter's Buildings is described in a letter from the applicant to the chief executive of the council as comprising some 52 separate occupations in an old rambling office building which was wasteful of valuable city centre space. When it renewed its application in June 1976 and submitted a statement of case drafted by counsel it described it as follows:

2. The building is a large building comprising basement, ground and three upper floors. It was purpose-built for shop, office and other commercial use but by the time of purchase by Windsor it had become old-fashioned and dilapidated. However, the site occupied by the building is an important one and Windsor considered it ripe for redevelopment. It was for this purpose that Windsor decided to buy the building and it was on this basis that the purchase price of £375,000 for the property was calculated.

3. As is often the case in a purchase of this kind, vacant possession was not provided by the vendor. There were 58 tenancies expiring at various times up until November 30 1978. Windsor expected to be able to arrange with the tenants for the surrender of their leases, this also being standard commercial practice in a purchase of this kind.

In August 1973 the applicant agreed a price with its vendors. In October 1973 contracts were exchanged. Completion took place in March 1974. Meanwhile, on February 8 1974 the provisions of the Local Government Act 1974 came into force. They changed the law relating to unoccupied rate charge in three relevant respects:

(a) By section 15 power was conferred on rating authorities to assess the charge at different percentages of the occupied rate for different classes of property, up to 100 per cent.

(b) By section 16 a new levy called surcharge was imposed on unoccupied commercial buildings which had been empty for six months, the surcharge to be levied by doubling the normal rates for the first 12 months, to treble them for the second 12 months, to quadruple them for the third 12 months and so on progressively while the period of non-use lasted.

(c) By section 15 (5), which amended the First Schedule to the General Rate Act 1967 by introducing the new paragraph 3A, rating authorities were given power to reduce or remit both the unoccupied rate and the surcharge if they considered that the payment would cause hardship to the person liable for those rates.

On March 13 1974 the Liverpool City Council resolved to levy a rate of 100 per cent on all classes of empty property for two years from April 1 1974. Thus from April 1 the applicant, which had become the owner of Peter's Buildings by completion of the purchase in March 1974, became liable to pay both unoccupied rate at 100 per cent on any hereditament in the building which was unoccupied for six months after April 1 1974 and in addition a surcharge of a further 100 per cent on any hereditament which was unoccupied for a period of six months after February 8 1974.

The action taken by the applicant after acquiring the building is described in paragraphs 5 and 6 of its statement of case for relief to the council dated June 23 1976:

5. Having purchased the building Windsor have sought to implement its intentions as expeditiously as possible. On 29 July 1974 an application was made for outline planning permission and on 14 December 1974 permission was granted, thus demonstrating the City's support for the principle of redevelopment. Approval of details was sought in March 1975, but Windsor have had to wait until April of this year [1976] for the issue of the approval. No criticism is made of this delay in so far as Windsor see it as confirmation by the City of the importance they attach to having the site redeveloped in a way which will yield the greatest benefit.

6. Contemporaneously with the obtaining of planning permission Windsor have been using every endeavour to obtain vacant possession of the whole of the building since it has always been their wish to implement the planning permission immediately following its issue. Unfortunately unexpected difficulties have arisen in this respect and 11 tenants still remain in the building. This unfortunate situation is attributable to economic factors outside Windsor's control. In particular interest charges have risen; building costs have soared without any commensurate rise in rental values; and capital values have dropped severely. Windsor did not, and could not have been expected to, foresee these events at the time of their purchase. In consequence of that, however, Windsor has had to revise the economics of the redevelopment project. Accordingly their ability to pay compensation to their tenants in return for the surrender of their lease has been drastically restricted.

In 1975 the city council as rating authority served notice upon the applicant of assessment to unoccupied rate and rating surcharge in respect of such hereditaments in Peter's Buildings as qualified for assessment under the 1967 Act as amended by the 1974 Act. On February 18 1975 the applicant applied (the first application) for relief from the imposition of the surcharge on the ground of hardship in a letter addressed to the chief executive of the council. No application was then made for relief from the unoccupied rate charge imposed in respect of the same buildings.

It is not necessary to consider in detail the documents or correspondence in connection with the first application. The reason why that is unnecessary is that it is conceded before us on behalf of the respondent rating authority that the decision, made by them on the first application, adverse to the applicant, cannot be supported. The decision was made, or at least purported to be made, by the city treasurer, who had no lawful power to make it. The decision is attacked by the applicant also on other grounds. It is unnecessary to consider them. The decision, or purported decision, contained in the city treasurer's letter of September 25 1975, which by his letter of November 21 1975 he refused to modify, must

A be quashed. But since a second application was in fact made and considered, there would, in all the circumstances of this case, be no point in granting an order of mandamus requiring the rehearing of, and a fresh decision upon, the first application.

It is the second application which gives rise to the real question in this case.

On June 23 1976 in a letter to the city treasurer the applicant made an application (the second application) for remission of the unoccupied rate and a renewed application for remission of the rating surcharge demands in respect of various hereditaments in Peter's Buildings. The letter enclosed
B a formal application for remission of the rates so charged on the grounds of hardship. That document, headed "Owners' Statement of Case," presented the grounds on which the applicant claimed that it would suffer hardship if either unoccupied rate or rating surcharge were levied. It concluded, at paragraph 12: "In the premises Windsor request the city council to exercise its discretion in respect of this application for relief from unoccupied rate charge and rating surcharge liability on grounds of hardship." Paragraphs 2, 5 and 6 have been quoted earlier in this judgment. Paragraphs 7 and 8 read:

C 7. The City of Liverpool appears at the present time to be intent upon compounding Windsor's difficulties by requiring it to pay rates in respect of every hereditament within the building which has been unoccupied for longer than three months, and rating surcharge in respect of every hereditament in the building that has been out of use for more than six months. Windsor estimate that the total cost (including interest) may exceed £200,000, an increase of 53 per cent of the purchase price, if it has no success in getting vacant possession before November 1978.

8. Windsor contend that in these circumstances the payment of the unoccupied rate and rating surcharge would be a hardship. They feel that the consequence of meeting such a payment will
D be either to increase the rent which must be demanded or to compel the abandonment of the project for the foreseeable future. The abandonment of the project would clearly be a hardship to Windsor, as would the need to demand a higher rent since it would make the letting of space more difficult. If the higher rent cannot be obtained and the project is not abandoned hardship will result from the reduced return on Windsor's investment. The only way to seek to guard against such a reduced return will be to construct the building to a lower standard at less cost. That too would be a hardship since it will affect the attraction of the building in the letting market.

Paragraphs 9 to 11 develop the same theme further, arguing
E that "There is no way that Windsor can be certain of preventing the occurrence of this hardship" and explaining why, in the submission of the applicant, courses of action which might be suggested were not feasible or would frustrate the purpose of the redevelopment, which the applicant described as being "the common purpose of Windsor and the city council."

On July 1 the city treasurer replied, saying that the application would be submitted to the appropriate committees and asking for copies of the applicant company's final accounts for the past three years. There was no reply. On August 6 the city treasurer repeated his request for the
F accounts. On August 10 there was a temporising reply. The city treasurer tried again on August 18. On August 24 the applicant's solicitors gave their response to the treasurer's request for accounts. That letter should be read in full:

We thank you for your letter of the 18th instant and have to inform you that we have now received the following information from Windsor Securities Limited: "Windsor Securities Limited is a Development Company formed in 1971 for the purpose of carrying out development in the United Kingdom and elsewhere. To date the Company has not made any profits, as rental income has as yet been insufficient to cover interest charges. No Accounts have, therefore, been produced and it is regretted that it is, unfortunately, not possible for the Company to comply with the Council's request." G

Prior to receiving this information from our clients and not knowing that it would not be possible to comply with your request as regards the production of Accounts for the past three years, we had referred the request to counsel for his Advice; counsel's Advice was that if it had been possible to produce Accounts as requested, then it should be made clear that it is not accepted that the Accounts would be relevant evidence to the Application for relief on the grounds of hardship with regard to Peter's Building. Equally, of course, it is considered that our clients' inabiilty to produce such Accounts is not relevant to such Application.

The Committee will appreciate that the written case as H
submitted is a summary of the salient points of our clients' Application. Counsel has advised that, unless the Committee is disposed to grant the relief as applied for on the basis of the written case, then the rules of natural justice require that before a final decision is reached by the Committee, our clients be given the fullest opportunity to present their case and that they should be informed of certain matters. These matters include the sight of the Officers' Report to the Committee, what view the Committee takes of the Report, the written case and the Application generally, the reason for that view and the grounds on which it is based, what the Committee understands by "hardship" and what it considers to be relevant in identifying hardship. Counsel also advises that Rating Authorities are J
allowing and indeed inviting oral representations to be made on similar Applications and our clients would welcome the opportunity to be so represented in this case.

It is of course hoped that the Committee and the City Council will be disposed to grant our clients the relief as applied for on the basis of the written case but in view of the advice given by counsel we feel the matters mentioned above should be raised at this stage.

On September 6 the city treasurer replied:

I acknowledge receipt of your letter of 24 August, written in connection with your client's claim for hardship relief from the unoccupied rate and penal surcharge. The reason I requested K
the trading accounts and balance sheet of your clients was to enable the appropriate Committee of the City Council to fully examine all the relevant facts of the case. I would respectfully point out that it is for the Committee to decide what they consider relevant and not your counsel. The granting of hardship relief is entirely at the discretion of the Committee. A copy of your formal application, enclosed with your previous letter, will be submitted to the Committee, together with full details of the charges included in the City Treasurer's report. The Committee is due to meet on 22 September 1976, and I will write and advise you of their decision as soon as it is known.

Nothing further was heard from the applicant. On L
September 22 the application was submitted to the general services committee of the council, and the committee resolved that the application be declined. Their decision was communicated to the applicant's solicitors in a letter from the city treasurer dated October 4.

I refer to your letter of 23 June, with which you enclosed an application for remission of rates on the grounds of hardship which had been drafted by counsel. As requested by you, this application was submitted to the General Services Committee of the Council on 22 September, and the Committee resolved that the application be declined. In your letter of 24 August you requested that in the event of the Committee not granting relief, M
the reasons for the refusal and the Committee's interpretation of "hardship" should be communicated to you. I must advise you that the Committee gave the fullest consideration to your application, and it is not the practice to state their reasons for such refusal. Each application for relief is dealt with on its particular merits, and the action of the Committee is in every case discretionary.

It is that decision of the general services committee on September 22 which is the subject of the second application for orders of certiorari and mandamus. The grounds relied upon appear in paragraph 3 of the applicant's statement filed under Order 53 Rule 1(2) (as it then was). They may be summarised as follows:

(i) The council's refusal to grant any relief was a decision which no reasonable rating authority could have taken.

(ii) The council erred in law, the error appearing on the face of the record, " by misconstruing the purpose and ambit of its discretion to award relief on grounds of hardship."

(iii) This ground was not pursued before us.

(iv) the council failed to act fairly or to observe the rules of natural justice, in particular by failing to accede to the various requests made by the applicant in its letter of August 24 1976.

(v) The application for relief disclosed a *prima facie* case which was not challenged by the council. From the failure to give reasons for refusal, it should be inferred that the council had no good reason and that its refusal was a wilful failure to use the power to grant relief for the purpose for which it had been given by Parliament.

The applicant made both applications to the Divisional Court on November 1 1977. The applications were refused. On April 19 1977 it applied to another division of this court for leave to apply to the Court of Appeal, and leave was given.

We have not inquired in detail into the delay which has resulted in a period of 18 months passing before the hearing of these applications. No point is taken upon it by the respondents. We were told that part of the delay was due to the fact that the applicant attempted, unsuccessfully, to negotiate a remission or reduction, and that it made a third, unsuccessful, formal application for relief to the council which was dismissed early this year. We were told that this third application was stated to have been made without prejudice to the proceedings in this court in respect of the first purported decision of September 25 1975 or the decision of the general services committee of October 1976. After the refusal of the third application by the council, the applicant applied to the Divisional Court for leave to apply for orders of certiorari and mandamus in respect of that third decision. The applications were refused. On May 22 1978 the applicant sought leave from another division of the Court of Appeal for an extension of time and for leave to apply for orders of certiorari and mandamus. We were told that leave was refused, with an intimation that if there were good grounds for intervention of this court, it could be achieved in the current proceedings in respect of the decisions in 1975 and 1976.

Against that background, although at first sight the earlier decisions might be regarded as having, for practical purposes, been merged in or overtaken by the third decision (of 1978), I am satisfied that there is no objection *in limine* to these proceedings. This court can and should entertain them on their merits, and counsel for the respondents accepts that position.

The Applicant's case on the Second Application
The First Ground

Upon the first ground Mr Horton submits that upon the question whether the applicant had suffered hardship, no reasonable council could have come to any contrary view in respect both of unoccupied rate and of the rating surcharge. Upon this ground his submission can be summarised as follows:

(a) The facts relevant to the question of hardship or no are to be collected from the applicant's formal statement of case, against the background of their correspondence with the city treasurer leading to the decision in 1975. Hardship in this section means to suffer substantial detriment. The facts set out in the statement of case demonstrated that the applicant had been trying its utmost to obtain vacant possession of the whole building at the earliest possible date in order to demolish and to reconstruct in accordance with the detailed plans which the council had approved. It was not in any sense an example of an owner who was seeking to profit by continued vacancy of the hereditaments. On the contrary, its financial interest at all material times would only be promoted by achieving demolition of the whole building at the earliest possible moment. Only the unexpected change in the financial climate of the property market had up to the date of its application for relief frustrated its genuine attempts to negotiate with all its tenants vacant possession before November 30 1978 when all its leases determined.

(b) The applicant had made its plans and entered into contracts which committed it in October 1973, some five months before the Liverpool Council resolved to adopt the powers to levy unoccupied rate, which it could have done but had not done during the eight years that had elapsed since the power to levy such a rate was first conferred in 1966. And likewise it was committed to the purchase four months before the liability to rating surcharge reached the statute book.

(c) The mischief aimed at by the 1967 Act was the deliberate policy of some owners in keeping property unoccupied in order to benefit by a rising market. But though the applicant did deliberately negotiate the end of as many tenancies as it could, its purpose was one which any reasonable council, and Liverpool Council in particular, should have regarded as laudable. It was to obtain vacant possession of the whole building as early as possible in order to replace it as early as possible by a better building, which would when complete afford better office accommodation than the present building, and, incidentally, attract a far higher rating assessment for the benefit of the rating authority than the building they were trying to demolish. It was only prevented from proceeding to demolish soon after grant of planning permission by the unexpected combination of two factors out of their control: the financial factor flowing from the fall in capital property values, which gravely reduced the opportunities for borrowing, together with the rise in interest rates, and the recalcitrance of a small number of tenants in refusing to leave without exacting extortionate terms beyond the applicant's now reduced borrowing powers.

(d) Whatever epithet may be used, the object of the rating surcharge provisions in the 1974 Act was to provide a deterrent to those who were pursuing a policy of keeping commercial property empty in order to let or sell at a higher profit as a result of keeping property empty in a rising market. The surcharge is doubled every 12 months so that it will not be worth the owner's while to keep his commercial property unsold or without a tenant. Counsel relied upon the explanation of the purpose of the legislation given by Lord Widgery CJ in *Brent London Borough Council v Alfa Romeo (Great Britain) Ltd* (1977) (75 LGR 686, at p 691), where he said:

> One can usefully pause there for a moment just to remind oneself of the mischief at which this provision was aimed. It was concerned with such notorious cases as Centre Point, where large commercial buildings were left empty and, under the law as it then stood, not contributing to general rate funds at all, whilst their owners waited for a more suitable occasion on which to put them into use. The purpose of the legislation with which we are concerned is to control or prevent that kind of activity, and so one finds, not surprisingly, that an owner of a building which has been empty for a period may claim as a special defence to the surcharge that he has tried his best to let the building.

The applicant's dealings with Peter's Buildings since 1974 bear no relation to the kind of commercial decision that the rating surcharge was enacted to inhibit. The applicant loses a great deal of money with every month that passes before it can demolish and reconstruct. It could not sensibly grant new tenancies lasting beyond November 1978 and it is fear of

A allegations of harassment of remaining tenants that has prevented it from beginning to demolish piecemeal. So far from waiting for a more suitable occasion to put the building into use, the applicant has been trying to the limits of its financial capacity to demolish and reconstruct as soon as possible, and its intention has been frustrated by factors beyond its control.

(e) The cost of the unoccupied rate and rating surcharge with interest will add a liability of some £200,000 to the expenses of the project of purchase, demolition and reconstruction of a building which cost £375,000. If these levies have to be paid, the effect can only be to prejudice the B reconstruction project which the council has already approved. Either the applicant will have to build a cheaper and less commodious building, or it will have to abandon the project altogether, or proceed with it and try to recoup its losses by trying to charge higher rents than the market rents upon which it had based its calculations.

(f) The council have never suggested that the facts in the applicant's statement of case are untrue or materially incomplete. They have refused to indicate what facts or considerations if any are inconsistent with the applicant's case, have given no reasons for their decision, and filed no evidence in this court. This court should therefore proceed C on the basis that a *prima facie* case of hardship had been established in the sense that, when all the circumstances are regarded, the imposition of either unoccupied rate or rating surcharge must be seen as subjecting it to a substantial detriment it did not foresee, the liability for which has fallen on it as a consequence of factors which were outside its control and contrary to its interests.

(g) If the council found hardship, as they should have done, it is wrong to take the view on the proper construction of paragraph 3A of the First Schedule that the council has then a completely unfettered discretion whether to reduce or D remit the two charges.

In the treasurer's letter of September 6 he said that the granting of hardship relief is entirely at the discretion of the committee, and in communicating the decision of the committee he said that the action of the committee is in every case discretionary. The council have a discretion, but they exercise statutory powers. If they find hardship has been sustained, they have a duty to give effect to that finding by proceeding to consider by how much to reduce the rate or surcharge in question, or whether to remit it altogether. But no fact has been suggested at any stage by the treasurer or the council which could justify a refusal to exercise the E statutory discretion in favour of the applicant. Upon a proper construction of paragraph 3A the power, expressed in merely permissive language, is accompanied by a duty to exercise it in circumstances contemplated by the paragraph. Mr Horton points to the analogy of the licensing cases which establish that, when a licensing authority "may" grant a licence to an applicant with certain qualifications, the policy of the Act may be that licences are not to be withheld from qualified applicants unless there is a valid objection, and the cases on the construction of planning legislation, where the words "may grant permission" for a certain land use must in certain circumstances mean *shall* grant permission where F the context so implies.

The Second Ground

Mr Horton submits that the treasurer's letter dated September 6 1976 is to be regarded as part of the record, and that it discloses an error in law on its face by stating: "The granting of hardship relief is entirely at the discretion of the committee." This, he submits, is wrong. The discretion is a statutory discretion, and has to be exercised according to law. In particular the rating authority is not at liberty to have regard to irrelevant considerations or to disregard relevant considerations. It has to act fairly.

G *The Third Ground,* as I have already said, was not pursued before us.

The Fourth Ground

This ground is "that the Council's refusal to grant any relief is null and void and of no effect because in arriving at its decision the Council failed to observe the rules of natural justice in particular by failing to accede to the various requests made by the applicant in its letter of August 24 1976."

Mr Horton founds his case on this ground upon the cases in which, in the context of those cases, the court has indicated H the duties imposed upon an authority exercising discretionary powers. He relied upon the judgment of Lord Denning MR in *R v Gaming Board of Great Britain, ex parte Benaim and Khaida* [1970] 2 QB 417 at p 430, where he said, in relation to duties of the Gaming Board: "The statute says in terms that in determining whether to grant a certificate, the board 'shall have regard only' to the matters specified. It follows, I think, that the board have a duty to act fairly. They must give the applicant an opportunity of satisfying them of the matters specified in the section. They must let him know what their impressions are so that he can disabuse them." And at page 431/G he said: "I should have thought that J the board ought in every case to be able to give to the applicant sufficient indication of the objections raised against him such as to enable him to answer them. That is only fair."

So here Mr Horton submits that if on receiving the applicant's case, which, as he submits, any reasonable committee would regard as disclosing hardship, they were proposing to consider factors which negatived that inference, the committee should at least have indicated to the applicant what those factors were so that it could deal with them. Again, if they thought that there was a fact or opinion in the K statement which was not well founded, they should have given the applicant a chance to deal with it. He drew support from the passage in the judgment in *Maxwell v Department of Trade and Industry* [1974] QB 523, at p 534, where, in relation to the duty of inspectors making an investigation under section 165 of the Companies Act 1948, Lord Denning MR said, in respect of the duty to act fairly towards persons whose conduct is under scrutiny: "It is sufficient for the inspectors to put the points to the witnesses as and when they come in the first place."

Mr Horton adverted to the classification of cases made by Sir Robert Megarry V-C in *MacInnes v Onslow Fane* [1978] L 3 All ER 211 at p 218. Counsel submitted that the instant case could be regarded as a case in which the applicant, having presented what must be regarded as a *prima facie* case of hardship, was entitled to know what objection there was to relief.

The Fifth Ground

That the Council refused to give reasons for its decision that no relief should be given, and in the light of the fact that the applicant's case for relief disclosed a *prima facie* case which has not been challenged by the council, it should be inferred that its refusal of relief was a wilful failure by the council to use M for the purpose for which it was given the power to grant relief from liability to rates on grounds of hardship.

Mr Horton relied upon *Padfield v Ministry of Agriculture* [1968] AC 997, in particular upon the passage in Lord Reid's speech at pp 1032 and 1033: "But I do not agree that a decision cannot be questioned if no reasons are given. It is the minister's duty not to act so as to frustrate the policy and objects of the Act, and if it were to appear that that has been the effect of the minister's refusal, then it appears to me that the court must be entitled to act." And similarly he relied upon Lord Upjohn at pp 1061/G-1062/A.

The Respondent's Case on the Second Application

For the respondent Mr Foster accepted that the rating authority had not an unfettered discretion, in the sense that they were not entitled to be arbitrary or unfair. In respect of the applicant's first ground, he submitted that the case for the applicant as put before the council in the statement of case was not such that no reasonable council could reject it as showing hardship. He relied upon Lord Greene's judgment in *Associated Provincial Picture Houses Ltd v Wednesbury Corporation* [1948] 1 KB 223 at p 230, where he said: "It is true to say that if a decision on a competent matter is so unreasonable that no reasonable authority could ever come to it, then the courts can interfere. That, I think, is quite right; but to prove a case of that kind would require something overwhelming."

The rating authority having given no reasons, Mr Foster was asked by the court what he could suggest to show that the facts relied upon by the applicant did not amount to hardship. He ran into some difficulty. He submitted that this court had not the information available to the rating authority as to the city's need for money. But that could not be a relevant consideration. He said that one could conceive of worse cases, for example the case of an applicant who was a large employer of labour who could show that if the levies were paid they would go out of business and cause unemployment. But the fact that there may be worse cases of hardship cannot be relevant to the question whether this applicant suffered hardship. Then he submitted that if the financial prospects had become as bleak as indicated in its statement of case the company should on the facts disclosed therein have decided to abandon the project. It had embarked on the adventure on a commercial forecast which proved unsound. Then in the changed situation it insisted on pursuing an increasingly hopeless commercial policy and pursuant to that policy continued negotiating with tenants to get them out, in spite of changes in its financial situation. It must accept the legal liabilities involved in the pursuit of an impracticable commercial policy in which the unoccupied rate and the rating surcharge were liabilities known to them at latest in March 1974, the very month of completion. They were speculating with borrowed money. The speculation at an early stage was beginning to look doubtful. Yet they proceeded to empty 47 of the hereditaments without any ground of confidence that they could empty the whole building before November 1978. These are facts which the rating authority may reasonably have taken into account, and inferences which it may reasonably have drawn from the known facts. If they did, their decision would not be unreasonable whether they concluded that there was no hardship or that such hardship as was established by the statement of case was not such as should persuade the rating authority to remit or reduce the unoccupied rate or the progressive rating surcharge.

Conclusions

For my part the starting point is to scrutinise the facts set out in the statement of case submitted on behalf of the applicant, accepting such facts as therein are alleged as facts. This assumption must be made because Mr Foster for the rating authority concedes that the rating authority should be treated in the circumstances of this case as having accepted the truth of the facts alleged. When that statement of case is closely scrutinised, it becomes clear what the applicant did and did not establish, and the sequence of events is clarified.

(a) In paragraph 6 it is stated that the applicant had been using every endeavour to obtain vacant possession of the whole of the building contemporaneously with the obtaining of planning permission. From paragraph 5 it appears that the application for planning permission was made on July 29 1974 and outline planning permission was granted on December 14 1974. Application for detailed planning permission was made in March 1975 and approval given in April 1976. So it appears that the effort to persuade tenants to vacate began at the end of July 1974 and thereafter continued until 47 tenants had agreed to leave and had left. Thus the exercise of emptying the building began some six months after section 16 of the Local Government Act 1974 had come into force imposing on rating authorities a mandatory duty, subject to the hardship provision, to levy the rating surcharge in respect of commercial buildings which for a continuous period exceeding six months were not in use. The applicant does not suggest that it was unaware of the law in force, and as an established company with many years' experience of shop and office development could not do so. The resolution of the council bringing into force in their area liability to unoccupied rate was passed on March 13 1974, also some months before the applicant began to arrange that its property should be, as far as practicable, without tenants.

(b) It took the steps of arranging for 47 tenants to give up their tenancies before completing contractual arrangements with 11 tenants who, as appears from the case, refused to leave before the end of the term for which they held possession under their leases. So when it began to empty the building it was taking the chance that the building might not finally be vacant until some date later than November 1978.

(c) The reason why 11 tenants remained in possession is the drastic restriction in the applicant's ability to pay compensation. This was due to the combined effect of the rise in interest charges and building costs without a commensurate rise in rental values, and a fall in capital values of real property. The applicant could not have been expected to foresee those events at the time of the purchase, and in consequence has had to revise the economics of the redevelopment project. The hardship contemplated by paragraph 3A of the First Schedule is "hardship to the person liable." It was therefore wholly appropriate that the city treasurer should ask to see the accounts of the applicant for the last three years in order to enable the city council to have proper information as to the changed financial situation relied upon and in order to determine the effect of those changes when the applicant began to empty the offices, continued to empty more and more, and in order to enable them to appreciate the capacity of the applicant to pay the unoccupied rate and the rating surcharge without hardship, or alternatively to judge the degree of hardship that had been established. The response to the wholly reasonable request was surprising. It is found in the letter from the applicant's solicitors dated August 24 1976. Though the applicant sought relief on the ground of hardship due to changed trading and speculative conditions it refused to show any accounts to the committee. There are no figures in the statement of case showing what financial provision had been made for this development from the date of completion to the date when, after demolition and construction, a rental income would be obtained upon capital invested; or showing what revision of the initial provision was required to take account of the unforeseen factors. The applicant's solicitors described the statement of case as a summary of the salient points of their client's application, and asked to see the officers' report to the committee, what view the committee took of the report, the written case and the application generally, the reason for that view and the grounds on which it was based, what the committee understood by hardship and what they considered to be relevant in identifying hardship. In my view the applicant was not entitled at that stage to any of the information that it sought. If an experienced property company which has the benefit of professional advice decides to apply for rating relief on the ground of hardship to it, it is incumbent upon it to present its case for hardship docu-

A mented with such accounts and financial information as will enable the rating authority to arrive at an informed decision upon the material put before it. The city treasurer made it sufficiently clear in his letter of September 6 that its request would not be complied with and informed the solicitors of the date when the committee would consider the application. The applicant then decided not to supplement its case with any financial information, and cannot in my view be heard later to complain if the committee were not impressed with the case presented to them. Had the applicant agreed to provide the committee with its trading accounts, or had it presented to the committee a reasoned financial appreciation B of the project at Peter's Buildings showing the impact of the unexpected factors which gave rise to hardship, the committee might have decided to seek further information from them, indicating on what matters further information was required. But, faced as the committee were with a blank refusal by the applicant to support its case with any financial information, they cannot be criticised for failing to make a further attempt to extract relevant information, which only the applicant could give, before taking their decision.

Mr Horton has submitted that upon a reasonable consideration of the statement of case a rating authority could not reasonably decide that there was no hardship. In my view C he poses the wrong question. The duty of the rating authority upon an application for relief is to be collected from paragraph 3A of the First Schedule: "Without prejudice to section 53 of this Act, a rating authority shall have power to reduce or remit the payment of any rates payable in respect of a hereditament . . . if they consider that the payment would cause hardship to the person liable for those rates."

In my opinion the rating authority have to approach their task with an intention to answer one question rather than two separate questions. The single question is: In all the D circumstances has the applicant satisfied the council that the payment would cause it such hardship that, having regard to the purpose of section 17 or section 17A (as the case may be), we should reduce or remit the payment?

It is a single question because hardship is a matter of degree, and though payment of the rate or rating surcharge may inflict upon the owner substantial detriment the circumstances may be such that the rating authority may reasonably decide that they are not satisfied that they should reduce or remit the rate or surcharge. It does not seem to me to be sensible to think that Parliament imposed upon the rating authority a duty to break up their inquiry into two separate E compartments: First, has the applicant shown that in all the circumstances he has suffered something which can be described as "hardship"? Second, in all the circumstances should we reduce or remit the rate or surcharge having regard to that "hardship"?

The question whether payment would cause hardship to the applicant has to be resolved in the light of commonsense having regard to all the circumstances. Hardship is different from poverty, which is dealt with in section 53. Though "poverty" is not a precise expression, it usually involves extreme financial stringency such that the applicant has some F difficulty in meeting the conventional necessities of life. It is not an apt noun to apply to a limited company. Any levy of unoccupied rate or rating surcharge will reduce the trading income of a company owning commercial property, but not every such loss of income will in all the circumstances sensibly be considered hardship. The rating authority has accordingly to decide under paragraph 3A whether any hardship which payment would cause is such that, having regard to the purposes of section 17 or section 17A, the payment should reasonably be reduced or remitted. A commercial undertaking may run into financial difficulties on its trading account and be compelled to leave its property unoccupied for many months in circumstances in which G levy of unoccupied rate or rating surcharge would cause loss of profit. A property company may make a speculative purchase of property and empty the buildings in circumstances which attract the same levies causing the same loss of expected profit. It may be reasonable for the rating authority to reduce the charges in one case but not in the other. I conclude that on the limited information put before the council in the applicant's statement of case it would not have been unreasonable for the committee to reject Windsor's application for relief provided that it considered it in accordance with its duty.

H This brings me to the final question. The rating authority exercises a statutory power to decide whether to reduce or remit the charges assessed under section 17 or section 17A on the ground of hardship. Paragraph 3A confers a discretion. In the exercise of that discretion the authority has to behave fairly. It must direct itself properly in law. It must direct its attention to the matters which it is bound to consider. It must exclude from its consideration matters which are irrelevant to what it has to consider. If it does not obey those rules it may be said to be acting unreasonably. See *Associated Provincial Picture Houses Ltd v Wednesbury Corporation* [1948] 1 KB 223 at p 230. Likewise it must not so act as to J frustrate the object and purpose of the statute conferring the power: *Padfield v Minister of Agriculture* [1968] AC 997. So a statutory discretion cannot be accurately described as an unfettered discretion. It is nothing of the sort, and if the committee in this case had been shown to have approached their decision having directed themselves that they were entitled to act arbitrarily they would be shown to have misdirected themselves in law.

Counsel for the applicant submits that that inference should be drawn from the correspondence. It is sufficiently clear from his letters dated September 25 and November 21 1975 that the city treasurer then completely misunderstood K the object and purpose of paragraph 3A. His decision might have had to be quashed on this ground if it had not been necessary to quash it on the ground that he had no jurisdiction to usurp the functions of the rating authority.

When one comes to the next application, made in June 1976, the city treasurer by his letter of September 6 1976 tells the applicant that "the granting of hardship relief is entirely at the discretion of the committee," and that he is submitting the application to the council. The phrase "entirely at the discretion of the committee" suggests that the treasurer may still have been misdirecting himself upon the nature of the statutory discretion, as he had in the L previous year. But his error, if he was still in error, is not to be imputed to the committee unless it can be shown that it directed itself in the same way. So one looks at the decision letter of October 4 1976 communicated to the applicants by the treasurer on behalf of the committee. Speaking on their behalf he says: "Each application for relief is dealt with on its particular merits, and the action of the committee is in every case discretionary." This sentence is itself unexceptionable, unless the context of fact shows that as a matter of probability the words "in every case discretionary" should lead to the inference that the M committee had accepted the erroneous view of their functions held by the treasurer the year before. I think that it is going too far to draw the inference from that sentence in the decision letter that the committee fell into the same error as the treasurer had demonstrated in his correspondence in 1975.

As it has not been shown that the decision must have been unreasonable, there is no foundation for a submission that an inference adverse to the rating authority should be drawn from their decision to give no reasons for their determination.

A For these reasons I would dismiss the second application. As the committee are not shown to have acted unlawfully in arriving at their decision on the application made in 1976 it would not be right to make an order of mandamus directing them to hear and determine the application made in 1975.

I would propose that the order should therefore be: An order of certiorari will go to quash the purported decision dated September 25 1975 but no order of mandamus will issue directing the city council to hear and determine the applicant's application for relief dated February 18 1975.
B The applications for orders of certiorari and mandamus in respect of the city council's decision dated October 4 1976 are dismissed.

MEGAW LJ and LAWTON LJ delivered concurring judgments.

The court ordered that certiorari should go to quash the purported decision dated September 25 1975 (which was made by the city treasurer without the authority of the city council) but no order of mandamus to issue directing the city council to hear and determine the application which led to that decision. The applications for orders of certiorari and mandamus in respect of the city council's decision dated October 4 1976 were dismissed with costs.

C

COURT OF APPEAL

June 26 1978

(Before Lord Justice STEPHENSON, Lord Justice WALLER and Lord Justice CUMMING-BRUCE)

GRAYLAW INVESTMENTS LTD v IPSWICH BOROUGH COUNCIL

Estates Gazette June 16 1979

(1978) 250 EG 1079

D **General Rate Act 1967, section 17 and Schedule 1—Unoccupied hereditaments—Newly erected building deemed to become unoccupied on date when erection is completed—Service of completion notice by rating authority—Erection treated as completed on date of service of notice or such later date as is specified—Effect of paragraph 9 of Schedule 1—In this case erection of building was completed, including the work customarily done after substantial completion, such as partitioning and electric wiring, long before date of service of completion notice—County court judge in error in his construction of paragraph 9 and in purporting to add to the date specified in the completion notice a further**
E **six months which had in fact already expired—Rating authority's appeal allowed**

In this case the rating authority, Ipswich Borough Council, appealed against a decision of Judge Stinson at Ipswich County Court on a completion notice given by the authority to the present respondents, Graylaw Investments Ltd. The effect of the judge's decision was to extend the time before which the respondents' newly erected building was deemed to become unoccupied and therefore to attract rating liability under the provisions of the General Rate Act 1967, section 17 and Schedule 1.

R A W Sears QC, I L Croxford and T Davey (instructed
F by the solicitor, Ipswich Borough Council) appeared on behalf of the appellant authority; G R G Roots (instructed by Radcliffes & Co) represented the respondents.

Giving the first judgment at the invitation of Stephenson LJ, WALLER LJ said: This appeal relates to a new office building of high quality in Ipswich. Planning permission was granted for that building on January 30 1973. Work was started in September 1973. The building was structurally completed in late 1975 and finished, as found at the hearing before the county court judge, on March 25 1976. That was the date on which the letting agents wrote to the Ipswich Borough Council saying that the building was finished to a high standard and, in effect, offering it to the borough council as G
possible tenants. We have before us pictures of the building—a large modern building with very considerable floor area. One part of the building has been let, but at the date of the hearing before the county court judge by far the greater part of the building still remained unlet.

On March 7 1977 the local authority served upon the owners of the building, the respondents before this court, a completion notice, but gave roughly seven and a half weeks for the notice to take effect. The notice read:

> (1) You are the person entitled to the possession of the building newly/now being erected at: CROWN STREET, IPSWICH (comprising the ground floor, first floor, second floor, third floor and the H
> eastern wing of the fourth floor) in the Borough of Ipswich. (2) The Council ... being the Rating Authority for the said Borough is of the opinion that: (a) the work remaining to be done on the said building is such that the erection thereof can reasonably be expected to be completed within three months; and (b) the building when completed will be comprised in "a relevant hereditament" for the purposes of the General Rate Act 1967.

The notice is dated March 7 1977 and it finished:

> NOW THE COUNCIL HEREBY GIVES YOU NOTICE that the erection of the said building is to be treated for the purposes of Schedule 1 to the said Act of 1967 as completed on April 30 1977.

The respondents, Graylaw Investments Ltd, appealed, as J
they were entitled to do, to the county court judge and the case was heard on July 12 1977, when Judge Stinson allowed the appeal and gave the owners of the building six months from March 7 1977, the date of the notice, making it September 7 1977.

By the General Rate Act 1967 local authorities were given an optional power to rate the owners of unoccupied hereditaments after being unoccupied for a continuous period of three months. By a resolution passed in January 1975 the Ipswich Borough Council resolved that the provisions of Schedule 1 to the 1967 Act, giving such power, should apply to its area, so that from the time of the passage of that K
resolution it became possible for the local authority to rate the owner of an unoccupied building, provided they complied with the provisions of Schedule 1 to that Act.

I must read certain of the provisions of Schedule 1. Paragraph 1(1) says:

> Where, in the case of any rating area in which, by virtue of a resolution under section 17 of this Act, this Schedule is in operation, any relevant hereditament in that area ... is unoccupied for a continuous period exceeding three months, the owner shall, subject to the provisions of this Schedule, be rated in respect of that hereditament for any relevant period of vacancy; and the provisions of this Act shall apply accordingly as if the hereditament were occupied during that relevant period of vacancy by L
> the owner.

Under the heading, "Completion of newly erected or altered buildings," paragraph 7 reads:

> For the purposes of paragraph 1 of this Schedule, a newly erected building which is not occupied on the date determined under the subsequent provisions of this Schedule as the date on which the erection of the building is completed shall be deemed to become unoccupied on that date.

I hope it is a fair paraphrase to say that, if the provisions of the schedule are complied with, the building shall be deemed to become unoccupied on that date—that is the date M
on which the erection of the building is completed.

Paragraph 8(1)(b) reads:

> Where a rating authority are of opinion ... (b) that the work remaining to be done on a building within their area is such that the erection of the building can reasonably be expected to be completed within three months, and that the building is, or when completed will be, comprised in a relevant hereditament, the authority may serve on the owner of the building a notice (hereafter in this paragraph referred to as "a completion notice") stating that the erection of the building is to be treated for the purposes of this Schedule as contempleted on the date of service of the notice or on such later date as may be specified by the notice.

A By subparagraph (5) of that same paragraph 8:

> If a completion notice served in respect of a building is not withdrawn and no appeal in pursuance of subparagraph (4) of this paragraph is brought against the notice or such an appeal is abandoned or dismissed, the erection of the building shall be treated for the purposes of this Schedule as completed on the date specified by the notice; and if the notice is not withdrawn and such an appeal is brought and is not abandoned or dismissed, the erection of the building shall be treated for those purposes as completed on such date as the court shall determine.

Finally, I read paragraph 9 of the schedule, which is the important paragraph which we have to consider in this case:

> B In the case of a building to which work remains to be done of a kind which is customarily done to a building of the type in question after the erection of the building has been substantially completed, it shall be assumed for the purposes of paragraph 8 of this Schedule that the erection of the building has been or can reasonably be expected to be completed at the expiration of such period beginning with the date of its completion apart from the work as is reasonably required for carrying out the work.

It is accepted for the purposes of this case that work remained to be done of a kind which is customarily done to a building of this type. It is unnecessary to go into the full details of that work, but it included such work as the installation
C of partitioning, possibly the alteration of electrical wiring and so on. There is a picture in the brochure relating to the building which shows one floor—a very large floor—without any division at all, but carpeted, I think a witness said, for sales purposes. Work of a kind which is customarily done to a building, it is accepted, would include the installation of partitioning. In this case, the learned judge also allowed time for obtaining fire regulation and building regulation approval. There was a question of the length of time reasonably required for this work, particularly in regard to the application of these regulations, but the main question—indeed the only question which is left before us now—is
D when the time begins to run.

The learned judge proceeded on the basis that time began to run on the date of the service of the notice, and in some of the reported cases to which we have been referred the date of the service of the notice coincided with the date on which the building was substantially completed. In this particular case, however, the local authority—perhaps because it had encouraged the erection of this building—did not serve the notice until approximately 12 months after the date when the agents were saying that the building was finished to a high standard.

The question we have to consider is whether the judge
E was right in saying that the time began to run from the date of the notice, or whether time runs from the date of substantial completion. Mr Sears submits, on behalf of the local authority, that on a true reading of paragraph 9 the date from which time commences is the date of substantial completion. He submits that when reading the words of paragraph 9 "...shall be assumed for the purposes of paragraph 8 of this Schedule that the erection of the building has been or can reasonably be expected to be completed at the expiration of such period beginning with the date of its completion apart from the work as is reasonably required for carrying out the work," if you take the date of the service of the
F notice as the date from which time begins to run, that gives no effect to the words "beginning with the date of its completion apart from the work" in the passage to which I have just referred. If those words were omitted, the paragraph would read: "...at the expiration of such period as is reasonably required for carrying out the work," and it might well be arguable that that period would start with service of the notice. But when the words "beginning with the date of its completion apart from the work" are put in—and Mr Sears submits that in fact the paragraph should be read as if there were a comma at the beginning of that phrase and a comma after the word "work," which would make it clearer—those words are meaningless unless they make the G
date of substantial completion the starting date. In other words, "beginning with the date of its completion apart from the work" must mean, submits Mr Sears, apart from the work which remains to be done which is of a kind which is customarily done to a building of this type. That is to say, applying that to the facts of this case, beginning with the date of its completion apart from the work required to put in the partitioning and alter the electrical work. In other words, in this case, that would be either March 25 1976 or perhaps even earlier. But it is unnecessary to consider that because it is quite clear that the judge found that this was substantially completed on March 25 1976. H

I find that construction irresistible. Furthermore, it seems to me that it is logical. The date when the erection of a building has been or can reasonably be expected to be completed is a date which should not depend on notices. The date on which the erection of the building is to be treated for the purposes of this schedule as completed is the date of service of the notice, or on such later date as may be specified in the notice. Service of the notice is a condition precedent to liability to rates, and the date on which liability commences depends on the notice. The local authority, therefore, can do that which the Ipswich Borough Council did in this case, namely give a little more time to an owner of a building, J
without also having to have added to that more time an additional amount of time which in truth has already passed.

For these reasons I would allow this appeal and substitute the dates which were set out in the original notice.

Agreeing, CUMMING-BRUCE LJ said: Section 17(1) of the General Rate Act 1967 granted a new discretionary power to a local authority with respect to the rating of unoccupied property. Pursuant to that section the provisions of Schedule 1 to the Act govern the liability to rating. But it was against that background that the draftsman of the Schedule had to devise words apt, firstly, to enable the local authority to exercise its discretionary power and, secondly, to set out K
statutory assumptions which would effectually enable the local authority, where it so wished, to expose the owner to liability to rates where the building was actually completed or, in an appropriate case, where it had been substantially completed but some work remained to be done.

The way in which the draftsman solved the drafting problem is as follows. In paragraph 7, a building not occupied on the date determined as "the date on which the erection of the building is completed shall be deemed to become unoccupied on that date"—that is to say, there is a deeming provision which makes it necessary to discover the date on L
which the erection of the building is completed. In order to decide whether it is completed, one then goes to paragraph 8(1)(a) and (b), which provide both for actual completion and the situation where the work remaining to be done is such that the building can reasonably be expected to be completed within three months. In those situations the local authority is empowered to serve a notice upon the owner, and that notice is to state that the erection of the building is to be treated for the purposes of the schedule as completed on the date of the service of the notice or such later date as may therein be specified.

When one turns to paragraph 9, one finds that the drafts- M
man there has used words which give rise to a statutory fiction, so that in the situation covered by paragraph 9 one does not any longer look at the reality, but looks at the statutory fiction which the paragraph brings into existence. That statutory fiction, on the plain words of the paragraph, compels one to assume for the purposes of paragraph 8 that "the erection of the building has been or can reasonably be expected to be completed at the expiration of such period beginning with the date of its completion apart from the work as is reasonably required for carrying out the work."

It appears to me clear that when paragraph 9 was married

with paragraph 8, once the date of substantial completion has been found and the period which will be necessary in order to do the work that remains to be done as being work which is customarily done to a building, all the necessary ingredients of the statutory fiction have been found; then it is irrelevant that the date for notional completion so arrived at is earlier than the notice served upon the owner.

For those reasons, I am satisfied that the learned judge was in error and the appeal should be allowed.

Also agreeing, STEPHENSON LJ said: I agree with the construction of paragraph 9 which has been formulated by my Lords. Indeed, Mr Roots conceded that, if it means what it says, the relevant period must begin with the date of the building's completion apart from the work remaining to be done of a kind which is customarily done to the building. The eleven words which Mr Sears puts between commas, "beginning with the date of its completion apart from the work," have never been considered in any reported case and were not apparently adequately considered by the learned county court judge, from whose opinion we are differing.

In deference to him I add a few words of my own to the judgments which have already been given. An authority can state that the erection of a building is to be treated as completed at the date of the service of the completion notice or at some later date. That is plain from paragraph 8 of the First Schedule. It follows that the date when the authority form their opinion that its erection has been completed, or can reasonably be expected to be completed within three months, may be earlier than the date of the service of the notice and completion or the expiration of the three months may be earlier than the date of the service of the notice. It also follows that substantial completion may have taken place, and the period beginning with the date of completion apart from customary work remaining to be done to the building and reasonably required for carrying out the work may have started and expired, before the date of the service of the completion notice.

In this case the judge has found that substantial completion and what he happily called, "translation of the property from 'substantially completed' to 'completed' by doing work 'customarily done'," were much earlier than March 7 1977, the date when the completion notice was served, and that the period reasonably required was six months. The expiration of that period was, therefore, also in the past on the date of the service of the completion notice. It expired in September 1976. The authority was therefore entitled to serve the completion notice earlier than it in fact did, and was certainly entitled to determine the date on which the erection of this building was completed at any later date than September 1976, for instance, April 30 1977, the date which is specified in the completion notice. That accordingly is the date when this building is deemed to become unoccupied under paragraph 7, and consequently rateable under paragraph 1. The court, therefore, was not entitled to determine September 7 1977 under paragraph 8(5) of Schedule 1 by treating the six months as running from the completion notice, as was clearly confirmed to be its order in the explanatory letter which the court wrote on July 17 1977 to the solicitor to the Ipswich Borough Council.

For those reasons I agree that this appeal should be allowed and the authority's completion notice allowed to stand.

The appeal was allowed with costs.

REAL PROPERTY AND CONVEYANCING

A **COURT OF APPEAL**

October 16 1978

(Before Lord Justice BUCKLEY, Lord Justice ORR and Lord Justice GOFF)

ALLEN AND ANOTHER v GREENWOOD AND ANOTHER

Estates Gazette January 13 1979

(1978) 249 EG 139

Obstruction of light to greenhouse—Greenhouse a "building" within Prescription Act 1832—Access of light affected by neighbours' caravan and close-boarded fence alongside boundary adjacent to greenhouse—Evidence of disastrous
B **effect on tomatoes and pot plants—Extent of light to which dominant owner entitled—Specially high degree of light for normal use of greenhouse can be acquired by 20 years' user to the knowledge of servient owners—Prescriptive right can be acquired to something more than illumination, namely, to all the beneficial effects of light including that of the sun's rays on the process of plant growth—Prohibitory and mandatory injunctions granted to protect access of light to greenhouse**

This was an appeal by the plaintiffs, Hubert Allen and Marjorie Allen, his wife, against the refusal of Vice-Chancellor Blackett-Ord (of the County Palatine of Lancaster) to grant
C injunctions to restrain the defendants Tyrell Sandiford Greenwood and Lilian Alice Greenwood, his wife, from keeping a caravan and a close-boarded fence along the boundary of 15 Woodtop Avenue, Rochdale, Greater Manchester, in such a position as to cause a nuisance and obstruction of light to the plaintiffs' greenhouse in their garden at 13 Woodtop Avenue.

H E Francis QC and A W Simpson (instructed by Blackhurst, Parker & Yates, of Lancaster) appeared on behalf of the appellants; P V Baker and B C Maddocks (instructed by Brierley & Hudson, of Rochdale) represented the respondents.

Giving the first judgment at the invitation of Buckley LJ,
D GOFF LJ said: This is an appeal from an order of Vice-Chancellor Blackett-Ord dated October 6 1977, whereby he refused relief by way of prohibitive and mandatory injunctions in respect of the alleged obstruction of light to the appellants' greenhouse and dismissed the action.

The appellants have occupied 13 Woodtop Avenue, Rochdale, since September 1954, first as tenants, and since December 1974 as owners in fee simple. As the photographs show, it is a pleasant detached house with a spacious garden. The respondents are owners in fee simple of a comparable house and property, no 15 to the south, which they purchased
E in 1966 and have occupied ever since. The level of the appellants' garden is some 26 or 28 in lower than the respondents' and the respondents' land is supported by a low retaining wall commencing a little to the east of the greenhouse, with a low brick-built pillar. The appellants' greenhouse is of the following agreed dimensions. It is 16 ft long, 9 ft 6 in wide, 6 ft 9 in to the eaves, and then it has a pitched roof with the ridge 3 ft above that. It is built right up against the retaining wall. In fact that wall forms one of its walls. Two of the other three are built up to, or perhaps bonded into, the retaining wall to form a level support, and the glazed F
structure rises directly above these four walls. So on the south side the frame of the glass is actually resting upon, or secured to, the retaining wall itself. That wall belongs wholly to the appellants; it is not a party wall.

It is clear that the greenhouse was built soon after the erection of no 13 some time between 1939 and 1941, and user of it for the ordinary purposes of a greenhouse for at least 20 years next before action brought was duly proved. The appellants complain of an infringement of a prescriptive right to light to their greenhouse, founding their claim on section 3 of the Prescription Act 1832, which is in these terms:

> When the access and use of light to and for any dwellinghouse, G
> workshop, or other building shall have been actually enjoyed therewith for the full period of 20 years without interruption, the right thereto shall be deemed absolute and indefeasible, any local usage or custom to the contrary notwithstanding, unless it shall appear that the same was enjoyed by some consent or agreement expressly made or given for that purpose by deed or writing.

There was in this case no such consent. Prior to 1969 there were certain buildings or structures on the respondents' land, in addition of course to their house. There was an air-raid shelter a little to the west of the greenhouse, and a shed to the east of the greenhouse. After the war the air-raid shelter was put to use as a coal bunker. Mr H
Hubert Allen, one of the appellants, gave evidence at the trial. He was cross-examined, suggesting that these buildings or structures had obstructed the light to the greenhouse, but there was an open space of some 20 ft between the air-raid shelter and the garden shed, and it became clear that those structures did not cause any obstruction and that line was not pursued. In that year, 1969, the respondents caused the air-raid shelter, or coal bunker, and the garden shed to be removed. For a time they had a car-port to the west of their garage—that is, further away from the greenhouse—and that did not cause any obstruction. At this time they also had a J
wall built across their land from the pillar to the house, and laid out the part of their property to the east of that wall as a patio, leaving an open tarmac space in front of the greenhouse. So there was still no obstruction. Indeed, there was no trouble of any kind until May 1974, when the respondents applied for planning permission to erect a two-storey extension at the rear of no 15. Notice of that application was given to the appellants by the borough planning officer. Mr Allen inspected the plans and formed the view that such an extension would interfere with the access of light to the greenhouse and to his sitting-room. He, therefore, instructed solicitors, who wrote to Mr Greenwood on August 7 1974, K
saying:

> Mr Allen has inspected the deposited plans and finds that if the proposed extension is carried out it will cause a serious diminution of the access of light to our client's greenhouse and some diminution of light to his sitting-room. The buildings on our client's land have been erected long enough to have acquired a right to light and, therefore, we trust that you will not proceed with the extension in such a way as to infringe the rights of our clients. Mr Allen has felt it necessary to lodge a formal objection to your application.

A We trust that it will not be necessary for our clients to take any further steps to enforce their rights and we shall be glad to hear from you that the matter can be amicably disposed of.

There was no reply to that letter, but shortly afterwards the respondents started to park their caravan immediately alongside the greenhouse. Prior to that it had been parked to the west of the old car-port; that is, between the car-port and the road, where, of course, it was not an obstruction to the greenhouse.

On August 28 1974 the respondents put in posts, or supports, for a fence very close to the greenhouse. That fence, which is close-boarded and creosoted, was finished by
B September 1 from a point at the east end of the greenhouse nearly, but not quite, up to the west end. The fence was built to a height which, having regard to the difference in level of the two properties, brought it about 18 in above the eaves. The appellants' solicitors wrote a further letter on September 6 1974 in the following terms:

Further to our letter of the 7th August 1974 we are now informed that you have erected a boarded fence on the boundary of your property adjacent to that of our client's the top of which in part by reason of the difference of levels extends 9 ft above the level upon which our client's greenhouse stands and is at a distance of approximately 5 in from it.

C This fence naturally interferes very seriously indeed with the access of light to the greenhouse.

We have already pointed out to you that our client has acquired a right to light by prescription in view of the fact that the greenhouse has occupied its present position for a period considerably in excess of 20 years.

Unless therefore the fence in its present form is removed our client will institute proceedings for an order to enforce its removal without delay.

We must therefore request you to advise us within the next 7 days that this will be done.

D On receipt of that letter the fence, so far as it was immediately opposite the greenhouse, was dismantled, but the caravan was then again placed alongside the greenhouse. Between October 1974 and February 5 1975 the fence was restored and extended right to the west end of the greenhouse and approximately 5½ ft beyond. When first restored it came only up to eave level, but it was later raised to 18 in above, when the appellants' gardener placed boxes of seeds and young plants at about eave level to get as near to the light as possible to prevent them from getting drawn up.

The position, therefore, when the action commenced was, and it still is, that there is this fence from the east end of the
E greenhouse to a point some 5 ft 6 in beyond it on the west, only about 6 in away from the glass, with the caravan parked close up to the fence.

It was clear from Mr Allen's evidence that the greenhouse had been used at all material times in the ordinary normal way in which such an appurtenance would be used in a private garden. He was asked: "What use—since 1954 when you went into occupation of the house—have you made of the greenhouse?" and he answered: "I have invariably grown tomatoes during the summer. We have raised our own plants from seed." Then the learned Vice-Chancellor asked: "Do you mean tomato plants?" and Mr Allen answered:
F "Tomato plants. Sometimes we have bought tomato plants; sometimes we have raised them from seed. Then we have geraniums, and we raise cuttings and so forth. Indeed, we are accustomed also to grow pot plants for the purpose not only of being able to see them in the greenhouse, but also when required to be able to take them inside the house. My wife is fond of plants and flowers inside the house, and I am myself for that matter. Of course, the greenhouse has served a useful purpose in that sense." Then counsel went on: "You have mentioned so far tomatoes, geraniums and house plants? (A) Yes. (Q) Do you raise any plants other than tomatoes from seeds? (A) Oh, yes, a whole range of biennial-type bedding plants that one raises from seed in the spring. (Q) G
What sort of plants? (A) Stocks, antirrhinums, various kinds of marigold, zinnias, violets, pansy—the red ones that are rather tender; that sort of plant, you see."

He also gave evidence of the disastrous effect of the obstructions and he said:

Some things do not like too much light, I can put ferns there, but it is no use trying to grow flowers there, so, in effect, I am in difficulty now cultivating flowers in pots in that greenhouse, which I was able to do earlier, prior to the existence of the fence. It used to be nice to walk into the greenhouse and see flowers, and also it used to supply the house with flowering plants.

In reply to a question from Mr Francis: "What about H
seedlings?" Mr Allen said:

We are able to grow some, but we are still somewhat deprived, because we cannot; we used to erect a temporary bench at the southerly side to put the seed trays on before they germinated, and just for a short time after germination. There is no point in that now. Virtually speaking, the southern half of my greenhouse is useless for what were previously its normal functions. In respect of my seedlings, when I grow them I have to bring them towards the centre of the greenhouse more than would have been necessary. The result is that there is a greater depth between the seed pans and the glazing, and there is a greater tendency for them to be drawn, and be less sturdy plants.

and then the learned Vice-Chancellor summed up his evidence J
by saying:

What it boils down to is that you say that half of your greenhouse is really very little use.

to which Mr Allen replied:

Yes.

He was cross-examined, but the cross-examination did not shake him, and the learned Vice-Chancellor found as follows:

The result, of course, is totally to exclude all sunlight from the south except to a small extent when the sun is high, and K
roughly half of the greenhouse gets no sun at all . . . I accept the evidence of Mr Allen, which was very fairly given, that the use of the greenhouse for the purposes for which he was expected to use it, namely growing tomatoes and pot plants, has been seriously interfered with.

He further found:

. . . there is, of course, plenty of light left in the greenhouse for actually carrying out any operations. If you want to pick the tomatoes, or pot some plants, or indeed if you want to get away from the family and read a book, there is plenty of light for those purposes in the greenhouse.

The learned Vice-Chancellor further held: L

Here, of course, the structure was obviously a greenhouse, but there is no evidence that the owners of the servient tenement, the successive owners, knew the precise use which was being made of it. Some plants require more light than others. . . . There is ample light left in the greenhouse, I find, for everything except the special purpose of growing certain plants.

It is to be observed that he did not say that there was anything special about the plants; nor indeed was there. So the *ratio decidendi* of the learned Vice-Chancellor's decision is that a greenhouse is a building which by its nature requires special light and there is no nuisance, although that purpose M
be defeated, so long as there is enough light to work in the place, though the work be rendered useless or seriously less effective, so long as there is sufficient light to use it as if it were a living-room in a house, and not a greenhouse—at all events, unless the servient owners have some particularly detailed knowledge of the user, which the Vice-Chancellor described as "precise."

Such being the facts, the first question which arises is whether a greenhouse is a "building" within the meaning of section 3 of the Prescription Act 1832, and *Clifford v Holt* [1899] 1 Ch 698 is an express decision that it is. The respondents reserved the right to argue in this court that it was not, but they have not done so and I proceed on the basis that it is. Also, a greenhouse is not to be regarded simply as a garden under glass, but as a building with apertures, namely, the glass roof and sides: see *Gale on Easements*, 14th ed p 239.

Accordingly, the appellants submit that they had, by the time they commenced their action, acquired a right of light to the greenhouse, and they say "The law will protect the dominant owner in the enjoyment of so much light as, according to the ordinary notions of mankind, he reasonably requires for all ordinary purposes for which the building is adapted," and further that the evidence shows, as it does, that the user throughout the 20 years has been the normal and ordinary use of a greenhouse in a private garden. Therefore, they submit, they are entitled to such light as is reasonably required for the continuation of that use, as being light required for the ordinary use of that type of building. Alternatively, they say that as they have in fact so used the greenhouse for the full period of 20 years, they are entitled to such light as is required for the normal use of the greenhouse, even though that use should be regarded as one which calls for a specially high degree of light.

The respondents, on the other hand, say that the appellants are, at the least, claiming a specially high degree of light, which they cannot have because, it is submitted, one can only prescribe for light which is required for ordinary residence, or ordinary business, in the tenement in question; and they say that ordinary business means a business requiring an ordinary amount of light, not necessarily to be measured by the business in fact being carried on there or the business for which the tenement is adapted. But their objection goes much deeper than that, for they say that in any event one can only prescribe a right to light, whether ordinary or special in degree, for purposes of illuminations, not a right to the direct rays of the sun, or to heat, or to other beneficial properties from the sun's rays. In effect they submit that the appellants are out of court because of the Vice-Chancellor's finding, which I have read, as to the amount of light remaining in the greenhouse. They argue that the appellants are not claiming light for human purposes or activities, but for its beneficial effect upon the plants or, looking at it another way, for the purpose of carrying out some chemical process. By analogy with a factory or workshop they submit that there is no actionable wrong if there be enough light to enable the workman to see what he is doing, even if he is carrying out some process which itself requires light, and there is insufficient to make it work.

The starting point for the resolution of these contending arguments must be in the case of *Colls v Home & Colonial Stores Ltd* [1904] AC, 179, which does, as it seems to me, establish the basic principle that the measure of the light to which right is acquired, of which it has to be seen whether there is such diminution as to cause a nuisance, is the light required for the beneficial use of the building for any ordinary purpose for which it is adapted. I think this emerges clearly from the speech of Lord Davey, with which Lord Robertson agreed, and from that of Lord Lindley. Lord Davey, at p 198, approved a passage from the judgment of James LJ in *Kelk v Pearson* (1871) 6 Ch App 809, which is in these terms:

> I am of opinion that the statute has in no degree whatever altered the pre-existing law as to the nature and extent of this right. The nature and extent of the right before that statute was to have that amount of light through the windows of a house which was sufficient according to the ordinary notions of mankind, for the comfortable use and enjoyment of that house as a dwelling-house.

and now come the important words—

> or for the beneficial use and occupation of the house if it were a warehouse, shop or other place of business. That was the extent of the easement, a right to prevent your neighbour from building on his land so as to obstruct the access of sufficient light and air to such an extent as to render the house substantially less comfortable and convenient.

Lord Lindley, at p 208, approved a similar passage in the case of *City of London Brewery Co v Tennant* (1873) 9 Ch App 212, and Lord Lindley said:

> That doctrine, as stated in (the case I have just mentioned) is that generally speaking an owner of ancient lights is entitled to sufficient light according to the ordinary notions of mankind for the comfortable use and enjoyment of his house as a dwelling-house, if it is a dwelling-house, or for the beneficial use and occupation of the house if it is a warehouse, a shop, or other place of business. The expressions "the ordinary notions of mankind," "comfortable use and enjoyment," and "beneficial use and occupation" introduce elements of uncertainty; but similar uncertainty has always existed and exists still in all cases of nuisance, and in this country an obstruction of light has commonly been regarded as a nuisance, although the right to light has been regarded as a peculiar kind of easement.

Those passages do, in my judgment, tie the measure of the light to the nature of the building and the purposes for which it is normally adapted.

Mr Baker relied considerably on a dictum of Bray J in *Ambler v Gordon* [1905] 1 KB 417, at p 422, where he said: "I think that the word 'ordinary' is used solely with reference to light, and an ordinary user or ordinary business means a user or business which in fact requires only an ordinary amount of light," and he submits that there is only one standard, which does not vary with the type of building. That case, of course, is not binding upon us, and I confess for my own part that I do not wholly understand the conception of "an ordinary amount of light" in the abstract. It seems that what is ordinary must depend upon the nature of the building and to what it is ordinarily adapted. If, therefore, the building be, as it is in this case, a greenhouse, the normal use of which requires a high degree of light, then it seems to me that the degree is ordinary light. Therefore, subject to the respondents' argument, to which I shall refer as the overriding argument, and which I shall consider later in this judgment, that the light of which the appellants have been deprived is not, on the findings, required for illumination but for the process of raising and growing plants, I would hold that in the case of a greenhouse, light required for its normal use is ordinary and is, therefore, acquired under the Act by 20 years' enjoyment.

But lest I be wrong on that, I turn to consider, still subject to the overriding argument, whether assuming it is a specially high degree of light, the right to it is acquired by 20 years' user to the knowledge of the servient owners. The learned Vice-Chancellor said that the case of *Ambler v Gordon* is "a clear decision that no claim of right by prescription for an extraordinary amount of light lies when the servient owner is unaware of the use being made of the dominant tenement." That I accept and apply, but the learned Vice-Chancellor went on, as I have already read, to say that there was no evidence that the owners of the servient tenement knew the precise use which was being made of it. In my judgment, however, with all respect, that was an error. It is an irresistible inference from the photographs that the respondents and their predecessors, owners successively of no 15, must have been fully aware at all times of the way in which the greenhouse was being used. As Buckley LJ said in argument, the contrary is inconceivable. There is no suggestion in this case that the greenhouse is now, or ever has been, used for any exotic purpose. The problem, therefore, is whether a right to a specially high degree of light can be acquired by known enjoyment of that specially high degree for the full period of 20 years. Strangely enough, there is no decision on this question, although there are conflicting dicta.

A The first is that of Vice-Chancellor Malins in *Lanfranchi v Mackenzie* (1867) LR 4 Eq 421, where he said at p 430:

> I intend to decide this case on broad general principles, and my view of the law is this, that if there be a particular user, and the quantity of light claimed for that is such as would not belong to the ordinary occupations of life, a person who claims that extraordinary quantity of light cannot establish his right to it unless he can show that he has been in the enjoyment of it for 20 years. If a man cannot establish a right within 20 years to an ordinary quantity of light, how can he establish in a less period the right to an extraordinary quantity? All he can establish is the right to the quantity of light he would be entitled to for ordinary purposes. If he has been in the enjoyment of an extraordinary user for 20 B years, that would establish the right against all persons who had reasonable knowledge of it.

Lord Davey, however, threw some doubt upon that in *Colls'* case at pp 203 and 204, where he said:

> If the plaintiffs had intended to claim and rely on a special easement of that description, it was for them to state their claim and prove the facts to support it. It is unnecessary to say, therefore, whether such a claim would be good in law. Vice-Chancellor Malins thought it could be sustained if the special user was had with the knowledge of the owner of the servient tenement. I will only say that I see some difficulties in the way, and reserve my opinion.

C In *Ambler v Gordon* [1905] 1 KB 417, Bray J went further and specifically disagreed with Vice-Chancellor Malins, saying at p 424: "I am, however, prepared to go further, and to hold that even 20 years' enjoyment to the knowledge of the servient tenement will not give a larger right."

With all respect to Lord Davey, who after all merely reserved his opinion, and to Bray J, I would adopt Vice-Chancellor Malins' dictum. It is clear that a right to a greater degree of light than such as is normally obtained by prescription could be the subject of a valid grant, and in my judgment, therefore, it is capable of being acquired by pre- D scription. That being so, provided it is enjoyed for the full period of 20 years to the knowledge of the servient owners, I fail to see any ground upon which it should be held not to have been acquired by prescription.

Of course, where the operation which needs special light is carried on indoors it may be very difficult in fact to prove sufficiently precise knowledge, but here the user was completely obvious. The learned Vice-Chancellor, Blackett-Ord, decided this point against the appellants, as I have already observed, on the ground that there was no evidence that the owners of the servient tenement knew the precise use which was being made of the greenhouse, but with all respect to E him, in my judgment the evidence was amply sufficient to prove knowledge, and sufficient knowledge.

In my judgment, therefore, the crux of this case at the end of the day is the overriding argument, which I must now consider. The respondents argue on this as follows:

(1) In *Colls'* case the House of Lords was seeking to limit, or restrict, the extent of the right to light, so as to prevent undue restrictions on the development or improvement of surrounding land or buildings, and the court should be very chary of any extension of the right;

F (2) That although the standards prescribed by the speeches in *Colls'* case are expressed in terms susceptible of a wider interpretation, in their context they must be taken as referring to illumination only;

(3) That in all cases, at least since *Colls'*, the right to light has been tested or measured in terms of illumination only. They refer, for example, to Mr Waldram's calculations and the theory of the "grumble point" (see *Charles Semon & Co v Bradford Corporation* [1922] 2 Ch 737 at pp 746 and 747) and to the case of *Hortons Estates Ltd v James Beattie Ltd* [1927] 1 Ch 75, where the question was whether the extent of the right to light should vary according to locality, and Russell J said, at p 78: ". . . the human eye requires as much G light for comfortable reading and sewing in Darlington Street, Wolverhampton, as in Mayfair."

Mr Maddocks on their behalf, in his supporting argument, referred also to *Warren v Brown* [1900] 2 QB 722 at p 725, where the test was stated to be "all ordinary purposes of inhabitancy or business," and to the test applied by this court in *Ough v King* [1967] 1 WLR 1547, "ordinary notions of contemporary mankind." These, however, I think, are at best neutral and possibly tell the other way, since a greenhouse is perfectly normal and ordinary in private gardens. So far as the last case is concerned, however, Mr Maddocks H relied upon the fact that this court approved of the county court judge having had a view, which again, he suggests, points to illumination as the test, though that I take leave to doubt.

(4) That in no case since *Colls* has the right to light been established, save on the basis of what is required for illumination.

That is true, but in *Lazarus v Artistic Photographic Co* [1897] 2 Ch 214 Kekewich J expressly extended the right to light for photography, which is not simply illumination but extra light required to effect a chemical process. That case J was wrongly decided, because he held that such a right could be acquired though the special light required for the purpose had been enjoyed for part only of the 20 years, but nevertheless it has, I think, some value as a negation of the respondents' argument. Moreover, in *Colls'* case itself, at p 203, Lord Davey instanced a photographic studio. True, he was there saying that one could not increase the burden on the servient tenement by changing over to such user within the 20 years, but at least he clearly envisaged a claim to light for such a purpose as a possibility.

(5) A distinction must be drawn between the heat and other properties of the sun and the light which emanates K from it, and the respondents say that, having regard to the Vice-Chancellor's findings, the only complaint that the appellants can have is loss of heat or radiant properties, and they postulate the example of a swimming pool, part of which is fortuitously warmed by sunlight coming through a south window. They say, and I have no doubt rightly, that the owners could have no cause of action against one who, while leaving fully adequate light for the complete enjoyment of the swimming pool, so shaded the sun as to deprive it of this chance warmth. That, I think, is a very different case from the present. L

(6) That in reality or in substance the injury here is not deprivation of light, but of heat or other energising properties of the sun and it is the plant life and not the human beings who are deprived.

I do not think this last point is in any case wholly accurate, as plants need light as well as heat, but it seems to me, with all respect to the learned Vice-Chancellor and to counsel, to lead to an absurd conclusion. It cannot, I think, be right to say that there is no nuisance because one can see to go in and out of a greenhouse and to pot plants which will not M flourish, and to pick fruit which cannot properly be developed and ripened, still less because one can see to read a book.

The appellants answer all this simply by submitting that they are entitled, by virtue of their respective right to light, to all the benefits of the light, including the rays of the sun. Warmth, they say, is an inseparable product of daylight, and they stress the absurd conclusion which I have already mentioned, to which the contrary argument inevitably leads. This reply commends itself to me, and I adopt it. So the overriding argument, in my judgment, does not prevail, and

A for the reasons I have already given the appellants are right, both on their primary and their alternative case, and I would allow this appeal.

Subject to any observations of my brethren or of counsel, I would grant an injunction on the following lines: restraining the defendants by themselves, their servants, contractors, workmen or otherwise from continuing to keep the caravan and fence in such a position on the defendants' property as to obstruct or diminish the access of light to the southerly and south-easterly walls and glass roof of the said greenhouse to such an extent as to cause a nuisance. Secondly, a mandatory order that the defendants do forthwith remove
B the said caravan and fence from such a position as so to obstruct or diminish the access of light to the said southerly and south-easterly glass walls and glass roof of the said greenhouse.

I desire, however, to add one important safeguarding proviso to this judgment. On other facts, particularly where one has solar heating (although that may not arise for some years) it may be possible and right to separate the heat, or some other property of the sun, from its light, and in such a case a different result might be reached. I leave that entirely open for decision when it arises. My judgment in this case is based upon the fact that this was a perfectly ordinary green-
C house, being used in a perfectly normal and ordinary manner, which user has, by the respondents' acts, been rendered substantially less beneficial than it was throughout the period of upwards of 20 years before action brought, and if necessary upon the fact that all this was known to the respondents and their predecessors for the whole of the relevant time.

ORR LJ: agreed.

BUCKLEY LJ delivered a concurring judgment, in the course of which he said:

D In the case of an easement of way, the extent of the use to which a claim can be established, whether on foot only or with animals or vehicles, depends, whether under the doctrine of lost grant or prescription, on the nature of the user which has been enjoyed as of right for a sufficient period. By analogous reasoning it seems to me that the same principle should apply to an easement of light. If in any case it could be shown that a use of the dominant tenement for a period and in circumstances justifying the implication of a lost grant has been such as to make an exceptional amount of light necessary for the particular use to which the tenement has been put, I can see no reason why a lost grant of
E such an exceptional amount of light should not be presumed. If in any case it can be shown that there has been such a use of the dominant tenement for 20 years or upwards before action brought, I see no reason why a prescriptive right should not be obtained in respect of the dominant tenement to such an exceptional amount of light as may be necessary to the satisfactory enjoyment of that use; but the use, to demonstrate a claim as of right to the exceptional amount of light, must be one which, according to ordinary notions, reasonably requires such an exceptional amount of light and must be known at all material times to the occupier of the servient tenement. Whether the use must be continuous
F throughout the period is not a question which arises in the present case, for it is not disputed that in this case the greenhouse has been used by the occupiers of no 13 as an ordinary domestic greenhouse throughout the 20-year period.

Accordingly, I agree with Goff LJ that if the plaintiffs' claim is not good on the basis of the ordinary use of light for a building of the character of a greenhouse, it succeeds as a claim to an extraordinary amount of light enjoyed as of right to the knowledge of the occupiers from time to time of no 15, throughout the period of upwards of 20 years before action brought.

The court made an order for an injunction in the terms G
proposed by Goff LJ. The appeal was allowed with costs in the Court of Appeal and below. Leave of appeal to the House of Lords was refused.

COURT OF APPEAL

December 8 1978

(Before Lord DENNING MR and Lord Justice BRANDON)

LEWISHAM LONDON BOROUGH v LESLIE & CO LTD

Estates Gazette June 30 1979

(1978) 250 EG 1289 H

**Limitation Act 1939—"Concealed fraud"—Tower blocks erected by contractors employed by London County Council —Serious defects in construction discovered 10 years later —Action by LCC's successor in title against contractors based on allegations of breach of contract and negligence —Limitation Act 1939 pleaded by contractors—Reply by authority that time had been prevented from running under the statute by the "concealed fraud" provisions in section 26—Claim struck out by the official referee on the ground that the "concealed fraud" provisions did not apply where defects could or should have been discovered by the LCC's
own staff of architects and supervisors—Official referee** J
wrong to have struck out plaintiff authority's claim—Triable issue shown, although perhaps a case for a preliminary issue on application of the statute

This was an interlocutory appeal from a decision of Judge Fay QC, the Official Referee, striking out a statement of claim in an action by the London Borough of Lewisham against Leslie & Co Ltd, contractors. The action related to faults discovered in four tower blocks built on the Dacres Road site at Lewisham. The facts are given in the judgment of Lord Denning MR.

C Beaumont (instructed by Robert A Joy, Lewisham Town K
Hall) appeared on behalf of the appellants; A May (instructed by Masons) represented the respondents.

Giving judgment, LORD DENNING MR said: In 1960 (as long ago as that) the London County Council (as it then was) determined to erect four new tower blocks on a site called the Dacres Road Site at Lewisham. They employed as contractors to build these tower blocks a firm called Leslie & Co Ltd. The tower blocks were duly built and finished in about 1961.

We do not hear of any trouble with regard to them until 1971—10 years later. By this time the property belonged to the London Borough of Lewisham, who were the successors
in title to the London County Council. Then a fireman, who L
went up to the top of one of the buildings to make checks, noticed that there was a gap of about an inch or two between the brick and the concrete. This rather worried him because it showed that there might be something wrong with this construction—that the walls were not vertical or properly in line. He reported it. Investigations were made. A great many faults were found in these tower blocks. It was a dangerous situation. The walls were not upright. They were an inch or two out in different places. Quite a number of other faults were found. I will only take one instance. These tower blocks have a concrete wall inside and a brick cladding on the outside. These
two ought to have been tied together by ties called Abbey M
ties. When investigations were made, it was found that slots had been made ready to receive the ties, but the ties had not been inserted. To take one example: 165 slots were all ready for the ties to be put in, but there were only 38 ties inserted altogether. So less than a quarter of the ties had been put in. I will not go into the other defects which were found, but the upshot of it was that extensive and expensive repairs had to be done by the borough of Lewisham to these tower blocks to make them habitable. The occupants had to go out from time to time while the work was being done. It cost about £150,000 to put all these defects right.

The London Borough of Lewisham—now the owners of the property—brought an action against the original contractors—Leslie & Co Ltd. They brought it on July 26 1976. Their claim was that the work had not been properly done. There were all these defects which had been concealed, so that no one knew about them for 10 years: and they claimed damages for breach of contract and damages for negligence. In answer to that claim the contractors—Leslie & Co Ltd—put in a defence. They denied that there had been any defects. Furthermore, they expressly relied on the Limitation Act 1939. It provides, in the case of a contract, a limitation period of six years from the date of the breach of contract. They said in this particular case that the tower blocks were completed in 1961. Six years passed, and more. Indeed, 15 years passed before the writ was issued. They said that the action was statute barred—it was barred by the lapse of time. In answer to that, the Lewisham Council put in a reply saying that their right of action had been concealed by the fraud of the contractors, their servants or agents, within the terms of section 26 of the Limitation Act 1939. It says that, if the right of action is concealed by the fraud of the defendant or his agents, time does not run until the plaintiff has discovered the fraud or could with reasonable diligence have discovered it.

I know it sounds terrible to talk about "concealed fraud," but it does not mean what it says. It applies in these building contract cases where bad work has been done and has been hidden from view. In such cases time does not run during the time when it was hidden from view. It starts from the time when the cracks or other signs of danger appear, that is, from the time when the owner discovers it, or could with reasonable diligence have discovered it.

The two most recent cases in this court are *Applegate v Moss* [1971] 1 QB 406; and *King v Victor Parsons & Co* [1973] 1 WLR 29. I would just like to quote a few words of what I said at p 33 in the *King* case:

> In order to show that he "concealed" the right of action "by fraud," it is not necessary to show that he took active steps to conceal his wrong doing or breach of contract. It is sufficient that he *knowingly* committed it and did not tell the owner anything about it. He did the wrong or committed the breach secretly. By saying nothing he keeps it secret. He conceals the right of action. He conceals it by "fraud" as those words have been interpreted in the cases. To this word "knowingly" there must be added "recklessly" . . . It may be that he has no dishonest motive: but that does not matter. He has kept the plaintiff out of knowledge of his right of action: and that is enough.

The Lewisham Council relied upon that doctrine. They said that in those circumstances these defects were hidden and concealed from their predecessors and them for all those years: and, accordingly, that time did not run against them.

The pleadings seem to me to show a perfectly good triable dispute which should go for trial to this extent—at all events—to see whether the Statute of Limitations was a bar, or not. But, instead of asking for it to be tried, even as a preliminary issue, the contractors took out a summons in May of this year asking for the statement of claim to be struck out on the ground that there were not any adequate particulars of concealed fraud. They went before Judge Edgar Fay, the Official Referee, and asked for it to be struck out under the inherent jurisdiction of the court. Judge Edgar Fay did strike it out. He seems to have recognised that, in the case of a small house, if the contractor builds it with a lot of defects, such as bad foundations, and hides them, the owner of the house can sue the contractor without being defeated by the Statute of Limitations. But Judge Edgar Fay drew a distinction between small houses of that kind and these great tower blocks. He pointed out—and this is no doubt true—that the county council had their full establishment of architects and supervisors on the spot all the time; and, in the circumstances, the judge thought that any defects could or should have been discovered by the council's own people—if they had used reasonable diligence.. He said:

> I am content to assume that the parties were of equal weight and that neither was able to take advantage of the other. Here was a large architects' department on the one hand, and what I assume to be a substantial builder on the other. In these circumstances, where there was full supervision . . . there is nothing unconscionable in placing reliance on the Statute of Limitations. There was no special relationship, and no element of trust . . . I cannot find in this case any circumstance that allows room for the operation of the doctrine of concealed fraud . . . I find that the defendant has established a clear and unanswerable case, and I propose to make the order asked for in the summons.

I am afraid that I cannot agree with the judge. It seems to me that there is no difference in principle between a small house and a small building-owner and a great tower block and a great council, even though it does employ architects and supervisors. It only requires a little imagination to think of circumstances in which workmen may do their work badly, leaving defects—which the architect or supervisor would not discover, even by using reasonable diligence. In this particular case the contractors omitted all these ties which should have tied the walls together, and then they put the brickwork on top, making it impossible for the defects to be discovered afterwards. It is all very well to talk about the council having its own architects and supervisors. But these gentlemen may have been misled by the contractors. The bad work may have been done when they were away, or their backs were turned for some good reason. The architects and supervisors may even have turned a blind eye to the defects—for some reward paid by the contractors. I do not suggest that it was in this case—but there are still people about who have an "itching palm," as Shakespeare described it. In the circumstances, it seems to me that this is essentially a question depending on the particular facts of the case as to whether or not there is what is called "concealed fraud," such that time does not run.

In this case I cannot consider it a matter for striking out the claim. On the other hand, it might well be a proper case for a preliminary issue to be ordered whereby the official referee can inquire into these preliminary matters and see whether or not the Statute of Limitations is an answer, or not.

Before parting from the case, I would like to say that the case was pleaded not only in breach of contract but also in negligence. We are told that by some arrangement—not yet on the pleadings—the charge of negligence has not yet been proceeded with. It may be (I do not know whether it can be) suggested that the plaintiffs might wish to resurrect it. Whether they can do that or not I cannot say. I would only suggest that in the light of recent authorities, where a contracting party is under a duty of care, the other side may sue, not only in contract, but also in tort for negligence: see *Esso Petroleum Co Ltd v Mardon* [1976] QB 801, *Midland Bank Trust Co Ltd v Hett Stubbs & Kemp* [1978] 3 WLR 167. In the case of the tort of negligence, the time does not begin to run until damage is sustained: that is, in the case of a defective building, when the cracks or other signs of damage appear, or the defects could with reasonable diligence have been discovered: see *Sparham Souter v Town & Country Developments (Essex) Ltd* [1976] QB 858, approved by the House of Lords in *Anns v Merton London Borough Council* [1978] AC 728. At all events, if there is a claim for negligence, all these points regarding the Statute of Limitations have to be considered in a different context altogether. I only mention that point to show that it should not be overlooked or forgotten. But, treating this case simply as a claim for damages for breach of contract, it seems to me it is not a case which can be struck out at this stage. At most it should be a case for a preliminary issue to see whether the Statute of Limitations applies or not.

So I would allow the appeal, accordingly, and an application can be made, if desired, for the preliminary issue to be heard.

BRANDON LJ agreed.

The appeal was allowed with costs here and below. An application for leave to appeal to the House of Lords was refused.

HARLOW DEVELOPMENT CORPORATION v MYERS

(Before His Honour Judge D E HILL-SMITH in Harlow County Court on October 18 1978)

Estates Gazette March 31 1979

(1978) 249 EG 1283

Restrictive covenant—" Open fronts " policy in housing design —Whether covenant enforceable by mandatory injunction —Whether breaches had been permitted in other instances— Whether covenant capriciously enforced—Whether court had any discretion to refuse mandatory order—Old and modern decisions compared—Mandatory injunction granted for removal of offending wall

This action was brought by Harlow Development Corporation for a mandatory injunction requiring the defendants, Mr and Mrs Myers, to remove a wall erected by Mr Myers outside their house at 4 Burnett Park, Harlow, Essex. It was not denied at the trial that the wall constituted a breach of a covenant not to erect any buildings or other structures on the property without approval. The question was as to the appropriate remedy, and in particular whether it was a proper case for a mandatory injunction.

D Hands (instructed by the solicitor to the corporation) for the plaintiffs; R D Naish (instructed by Attwater & Liell, of Harlow) for the defendants.

Giving judgment, JUDGE HILL-SMITH said the master plan for Harlow New Town, prepared in 1947 by Sir Frederick Gibberd, incorporated the principle of open, or common, fronts, in order to secure overall landscaped design.

The policy was reaffirmed in 1959 by a resolution of the corporation which provided that individual exceptions be permitted only where the Planning Board was satisfied that special circumstances made open fronts impracticable, and that every encouragement be given to tenants to cultivate flower beds.

The house in question was in Burnett Park, on the south-west periphery of the designated area. The defendants purchased a plot there in March 1973 (a policy of selling off plots to individual purchasers having evolved) and engaged a builder to erect a house on it. A guide to developers, issued by the corporation and dated 1967, set out in terms the open-front policy, and Mr and Mrs Myers were required to covenant not to erect any buildings or other structures on the property other than those previously approved by the transferor or the corporation.

The cost of building proved higher than expected, and in 1976 Mr Myers persuaded the corporation to let him sell off part of his plot. They did so on condition that he paid them half the proceedings of sale.

Mr Myers levelled the front part of the land surrounding his plot, but had indifferent success at cultivating it. It remained open and barren, and invited trespassers, including youths who "came like butterflies to the candlelight shed by a young lady at number 6." Dogs fouled the front. People crossed the land to get at blackberries, and also as a short cut into a cul-de-sac. So, in 1977 Mr Myers built a wall along the frontage and into the adjoining cul-de-sac to link up—or nearly so—with a hedge. He omitted to write first to the Development Corporation. He said he had forgotten about the covenant. When he was shown office copies of the transfer containing the covenant, he politely asked for the corporation's indulgence, saying that the wall improved the appearance of the property; but the corporation were adamant that the open-front policy must be adhered to, and now sought a mandatory injunction for removal of the wall.

A claim for damages was abandoned. Mr Myers did not now deny that the wall constituted a breach of covenant; but he said that it was built to abate nuisance by trespassers and animals, and in ignorance of the covenant. He said that, since the corporation had permitted others to build walls elsewhere within the new town, it would not be fair to order him to pull down his, and nominal damages would be more appropriate.

If the corporation had permitted indiscriminate breaches of the open-fronts policy, that would clearly go to the exercise of the court's discretion whether to grant an injunction. A capricious application of the policy, just like excessive delay in bringing the case, would disentitle them to relief. But His Honour was satisfied that, in the vast majority of cases referred to, the infringements were not authorised by the corporation. What, if any, steps were taken when they were discovered, he did not know. The grant of permission to the occupant of No 1 to build a wall seemed to fall perilously near to the capricious, but that wall had some of the characteristics of a retaining wall, and was set back some distance from the pavement line, albeit that that distance diminished the further from the house line one travelled. His Honour was satisfied that the open-fronts policy was broadly accepted in Burnett Park.

In *Doherty v Allman* (1878) 3 App Cas 709, Lord Cairns LC said (p 720) that if parties had, with their eyes open and for valuable consideration, entered into a negative covenant, the court would enforce such a negative bargain, and there would be "no discretion to exercise." Lord Blackburn (p 730) used words to the same effect, but added that circumstances might change, so that it would not be reasonable to enforce the covenant, and cited *Duke of Bedford v Trustees of the British Museum* (1822) 2 My & K 552, per Lord Eldon. In *Lord Manners v Johnson* (1875) 1 Ch D 673, Hall V-C said at p 679 that the covenantee obtained the covenant free from being embarrassed by the question whether any, and if so what, injury was caused by any breach of a covenant. In *Achilli v Tovell* [1927] 2 Ch 243, Astbury J at p 247 said he had no jurisdiction or discretion to force the plaintiff covenantee against her will to accept damages instead of an injunction.

Recent cases had tended to soften the rigour of earlier decisions. In *Shepherd Homes Ltd v Sandham* [1971] Ch 340, Megarry J (as he then was) said even a blameless plaintiff could not as of right enforce a negative covenant by mandatory injunction. The basic concept was that of producing a "fair result," which involved the exercise of a judicial discretion.

That statement as to the "basic concept" was found to be of "doubtful usefulness" by the Court of Appeal in *Charrington v Simons & Co Ltd* [1971] 1 WLR 598 at p 603, but Brightman J, faced with the prospect of ordering the demolition of houses in order to preserve restrictive covenants entered into by the defendant's predecessor in title, awarded damages at 5 per cent of the profit made by the estate developer who was in knowing defiance of the covenant (*Wrotham Park Estate Co Ltd v Parkside Homes Ltd* [1974] 1 WLR 798).

In the present case the plaintiffs had suffered no loss, and the defendants had made no profit. Nominal damages were not appropriate, and the normal remedy would be a mandatory injunction.

The defendants said there had been a change of circumstances between the time of the imposition of the covenant and that of the breach; but His Honour found there had been no change such as envisaged by Lord Eldon in the *British Museum* case. The defendants entered into the covenants which were exacted for the benefit not of the corporation but of others living in the locality and in pursuance of what His Honour was satisfied was a consistent policy.

His Honour made the order sought by the plaintiffs; by consent, he gave the defendants three months to comply.

A **ABBEY NATIONAL BUILDING SOCIETY v MORRIS**

(Before His Honour Judge BARR at Uxbridge County Court on October 20 1978)

Estates Gazette April 7 1979

(1978) 250 EG 66

Mortgagees obtain unfettered order for possession—Mortgagor goes back into possession without consent—Mortgagees seek writ of restitution—County Courts Act 1959 section 74—Whether writ of restitution appropriate in county court—Writ granted notwithstanding note in Green Book—Whether appropriate case also for committal for contempt

B This was an appeal from a registrar's decision in an *ex parte* application by the plaintiff building society for a writ of restitution. The application was supported by an affidavit which set out the following facts:

The plaintiffs had on September 21 1977 obtained an unfettered order for possession by reason of the default of the defendant, John Francis Morris, of his obligations under a mortgage. Possession was delivered up to the plaintiffs under a warrant for possession. The premises were thereupon secured against re-entry. The defendant went back into possession without consent and failed to comply with written demands to restore possession to the plaintiffs.

C The affidavit further recited that a warrant for possession had been issued and executed pursuant to an order of a divorce court dated April 25 1977, in proceedings brought by the defendant's wife, and that in that case, also, after the warrant had been executed, the defendant had re-entered the premises and changed the locks.

The registrar refused the plaintiffs' application on the ground that "a writ of restitution is not applicable in the county court." D

The plaintiffs appealed against the registrar's decision and also applied for the committal of the defendant to prison for contempt of court in failing to comply with the order for possession (notices on Forms 140 and 194 having been served upon him).

Roger de Haar (instructed by Tucker, Turner & Co) appeared for the plaintiffs; the defendant did not appear and was not represented.

His Honour JUDGE BARR held that, notwithstanding the note on p 83 of the 1978 *County Court Practice*, he construed section 74 of the County Courts Act 1959 (on the plain words of subsections (1) and (2), and particularly paragraph (a) of subsection (1)) as giving him jurisdiction to issue a writ of restitution. E

The note to which His Honour referred is headed "Restitution," and reads, *inter alia*:

It is doubtful, however, whether an order for restitution is a form of relief covered by the above section [section 74].

Counsel cited in argument *Alliance Building Society v Austen* [1951] 2 All ER 1068, where Roxburgh J held that although the ordinary remedy [in the High Court] in such a case was an order of restitution, the court had jurisdiction to commit for contempt in such circumstances. His Honour was satisfied that the defendant was in contempt, and that it was an appropriate case to commit the defendant to prison for seven days. F

Order accordingly.

TOWN AND COUNTRY PLANNING

QUEEN'S BENCH DIVISION

November 1 1978

(Before Sir Douglas FRANK QC, sitting as a deputy judge of the division)

RICHMOND UPON THAMES BOROUGH COUNCIL v SECRETARY OF STATE FOR THE ENVIRONMENT

Estates Gazette January 20 1979

(1978) 249 EG 244

Town and Country Planning Act 1971—Application by local council to quash decision of Secretary of State refusing consent to demolition of building in conservation area—Council had given themselves permission to develop an area which included the building in question—Secretary of State entitled to consider merits of proposed redevelopment despite permission being deemed to have been granted by him—No ground for criticism that Secretary of State had failed to consider evidence at inquiry as to costs of preservation—Application dismissed

This was an application by the Richmond upon Thames Borough Council for an order quashing a decision by the Secretary of State for the Environment in which he had refused consent for the demolition of an old mission hall by the Thames at Twickenham which was included in a conservation area. The council had granted themselves planning permission for the redevelopment, such permission being deemed to have been granted by the Secretary of State by virtue of the Town and Country Planning General Regulations 1974.

D Hands (instructed by A W B Goode, municipal offices, Twickenham) appeared on behalf of the applicant council; H K Woolf (instructed by the Treasury Solicitor) represented the Secretary of State.

Giving judgment, SIR DOUGLAS FRANK QC said: This is an application to quash a decision by the Secretary of State refusing consent for the demolition of a building in a conservation area. The building concerned is an old mission hall by the Thames at Twickenham. The council had given themselves planning permission for the redevelopment of the site with a yard for yacht storage and a two- and three-storey building to be used for housing old people. By the Town and Country Planning General Regulations 1974 that permission was deemed to have been granted by the Secretary of State. The main ground for refusing consent for the demolition of the building was that the proposed new building would intrude to an unacceptable degree in the conservation area as such. The Secretary of State's decision is attacked on two grounds and I shall deal with each in turn.

First, it is said that the Secretary of State was not entitled to take into account the merits of the applicant's proposed development, for which planning permission exists. Mr Hands for the council submitted that as the Secretary of State was deemed to have granted permission for the new development, it was not open to him to give further consideration to the scheme already permitted. He drew attention to section 277 A of the Act of 1971 as amended by the Town and Country Amenities Act 1974 which provides, *inter alia*, that section 55 shall have effect in relation to buildings in a conservation area and that the effect of that is to apply the listed building procedure. He then referred to a decision of the Secretary of State and also to the case of *Kent Messenger Ltd v Secretary of State* (1976) 241 EG 25 for the proposition that in dealing with an application for listed building consent or demolition of a listed building the Secretary of State must have regard to the desirability of preserving the building, and the quality of the proposed replacement building is not material. I should, however, point out that as yet there is no decision by the court that that proposition correctly states the law. However, Mr Woolf submitted that, whatever may be the law in respect of a listed building consent, different considerations may apply to the demolition of a building in a conservation area. He referred particularly to subsection (5) of section 277 which provides:

> Where any area is for the time being designated as a conservation area, special attention shall be paid to the desirability of preserving or enhancing its character or appearance in the exercise, with respect to any buildings or other land in that area, of any powers under this Act, Part I of the Historic Buildings and Ancient Monuments Act 1953 or the Local Authorities (Historic Buildings) Act 1962.

He submitted that the function described in that subsection could not be performed without seeing what was to be substituted and how it would fit into the conservation area.

I agree with Mr Woolf and I further tested the matter in this way. I asked Mr Hands whether his argument would hold good in the ordinary case, that is to say, where the applicant was a private developer and permission had been given by the local planning authority for the redevelopment. I then asked, would the Secretary of State be prevented from taking the permission into account? Mr Hands said that that was a different case because in the instant case the permission is one deemed to have been given by the Secretary of State. I cannot think that the provision in the General Regulations stating that a permission is deemed to have been granted by the Secretary of State can produce as a side wind a prohibition against his considering such a deemed permission merely because the local planning authority happened to be the developers. I further think that the council's submission cannot be reconciled with the intention of Parliament that conservation areas shall be conserved.

Mr Hands' second submission has rather more substance. It is that the respondent failed to give any, or any proper, consideration to the evidence given at the inquiry about the cost of preserving the buildings. I think there can be no doubt that the question of cost is relevant. Certainly the inspector recorded the evidence given in respect of it and it was apparently unchallenged as to amount, but it was said in argument that the cost was not an unreasonable sum. In his conclusions the inspector said: "It is evident that a substantial amount of work would need to be done to bring the hall and cottage back into use" and he went on to make some comments on the condition of the property. It seems to me necessarily inherent in that conclusion that it is a finding of fact that money would have to be spent in carrying out a

A substantial amount of work. Although I think it would have been better had the inspector dealt more explicitly with the question of costs, I am satisfied that he took it into account and that he has sufficiently stated that he did so. Mr Hands also said, although it is not in his notice of motion, that even if cost was dealt with, the utility of a repaired building was not considered. In fact the inspector said: "I accept that Richmond Council would have difficulty in putting the buildings to use but I was not satisfied that the efforts which had been made to find a new use or user were exhaustive." In my view that is a sufficient answer to this submission.

B In view of the foregoing this appeal will be dismissed.

The appeal was dismissed with costs.

QUEEN'S BENCH DIVISION

November 3 1978

(Before Sir Douglas FRANK QC, sitting as a deputy judge of the division)

WALTERS v SECRETARY OF STATE FOR WALES AND ANOTHER

Estates Gazette January 20 1979

C (1978) 249 EG 245

Town and Country Planning Act 1971—Refusal of outline planning permission for housing development—Appeal against inspector's decision—Inspector would have been wrong if he had justified refusal on the ground that the proposed development would be uneconomic, but he did not decide on this ground—Inspector was considering possibility that, owing to severe design constraints, a scheme could not be devised for which planning permission ought to be given—Inspector also not in error because he decided that he could not make a final judgment on access proposals without details of the form of the rest of the development
D **—Appeal dismissed**

This was an appeal by David William Walters against the decision of an inspector appointed by the Secretary of State for Wales to determine an appeal from the refusal of Swansea City Council to grant outline planning permission for the erection of five houses with garages on land at Uplands, Swansea. The inspector upheld the refusal of the city council to grant permission.

G Bartlett (instructed by Collyer-Bristow & Co, agents for Collins, Woods & Vaughan Jones, of Swansea) appeared on behalf of the applicant; H K Woolf (instructed by the Treasury E Solicitor) represented the Secretary of State, the first respondent. The second respondent, Swansea City Council, was not represented and took no part in the proceedings.

Giving judgment, SIR DOUGLAS FRANK QC said: This is an appeal under section 245 of the Town and Country Planning Act 1971 against a decision by M I Montague-Smith BArch MCD RIBA MRTPI, an inspector appointed by the first respondent to determine an appeal against a refusal by the second respondents to grant planning permission. The planning permission applied for was outline permission for the erection of five detached houses with garages on land at Uplands, Swansea. The inspector decided to dismiss the appeal F on grounds which are summarised in his decision letter in these terms:

> 13. In conclusion, while it may be possible to design a scheme of acceptable quality and appearance, it is not certain that this could be achieved within the limits imposed by road safety requirements, economic factors, and market demand. This is not a case, I feel, where outline approval can be granted with matters of access, layout, and external design left as reserved matters to be decided later. This information is needed at the outset before a firm decision can be made on whether the site is in fact capable of being developed to an acceptable standard. In these circumstances I do not feel that I should grant outline planning permission in the absence of satisfactory details.

The inspector's decision is attacked on two grounds and I shall deal with them in turn. G The first is that "in concluding that it was not certain that a scheme of acceptable quality and appearance could be designed he had regard to considerations which were irrelevant, namely, economic factors and market demand." Mr Bartlett for the appellant submitted that although there may be cases where financial considerations are relevant, in general they are not in themselves relevant and it cannot be relevant to consider whether the development would be economic. He said in effect that it was for the developer to decide whether to carry out the development and that it was not the purpose of planning control to ensure that development was carried out. He referred to *J Murphy* H *& Sons Ltd v Secretary of State for the Environment* [1973] 2 All ER 26 where Ackner J held that a planning authority was concerned with the question how the land is to be used and not whether the proposed development would be a wise commercial venture, and so the Secretary of State on appeal was not entitled to have regard to the cost of developing a site in determining whether planning permission should be granted. He recognised that Forbes J said in the subsequent case of *Sovmots Investments Ltd v Secretary of State for the Environment* [1977] 1 QB 414 at p 424:

> If Ackner J was intending to say that costs can never be a relevant consideration either in a planning appeal or on a com- J pulsory purchase order . . . then I find myself unable to agree with him.

However, in a later case of *Hambledon and Chiddingfold Parish Councils v Secretary of State for the Environment* (1976) JPL 502, Ackner J said:

> Having had the benefit of reading Forbes J's judgment, my present impression is that I may have stated the general proposition, which I still think is correct, too widely.

Mr Woolf first drew attention to para 2 of the decision letter, in which the inspector said that the determining issue was whether it was possible to decide the likely effect of the K proposal on the surrounding area in the absence of any details showing means of access, site, design and external appearance of the proposed development. Thus, it follows from that paragraph read in conjunction with para 13 that the reasoning underlying the inspector's decision was that this case could not be decided on a bare outline application in the absence of detailed plans showing how the development would be carried out. He therefore left open the question whether if a full application were made permission should be granted. Dealing specifically with economic matters and market demand, he said it was proper for the Secretary of L State to take into consideration whether there was any likelihood of the development being carried out. That is particularly relevant with an outline permission so that it can be seen whether difficulties can reasonably be overcome or, on the other hand, if the expenditure would be out of proportion. He submitted that if a development was unlikely to be carried out for economic reasons, it would be proper to refuse permission in order to avoid an accumulation of planning permissions. He referred to one as yet unreported case, *Brighton Borough Council v Secretary of State for the Environment* (June 29 1978), in which I held that it was not improper to M take into account the financial consequences of granting planning permission because the money realised would be used for the maintenance of a listed building.

I have no doubt that in certain circumstances it is proper to take into account the financial consequences of a planning permission, as in the *Brighton* case and in *Niarchos (London) Ltd v Secretary of State* (1978) 35 P&CR 259. However, in neither of those cases was the question the same as it was assumed to be in this case, namely, whether it would be feasibly economical for the development to be carried out, nor did they remotely touch on that point. With respect to

A Mr Woolf, I think the fallacy of his argument was demonstrated when he said it would make a difference if the applicant were an eccentric millionaire indifferent to the financial consequences of carrying out the permitted development. In my judgment, the grant of planning permission cannot depend on the resources and intentions of the applicant. It is for the developer to make the economic decision whether to carry out the development and not the local planning authority. The examples given by Forbes J in the *Sovmots* case related to inquiries other than those concerned with the grant of planning permission. The prevention of an accumulation of sterile planning permissions which Mr Woolf gave B as a justification for an economic inquiry is taken care of by the time conditions imposed by the statute.

If the inspector's letter is to be construed as conveying that permission would be refused if it were shown that the development could only be carried out at a loss because of the design constraints and the market demand, then I would hold that he was wrong. However, I have come to the conclusion that that is not the right construction to put on his letter. In my judgment, reading para 13 of the letter in the context of the letter as a whole, the inspector was saying that it was conceivable that because of the severe design constraints which he had already described there was a possibility that C it would be impossible to devise a scheme for which planning permission ought to be given. I do not think that he was intending to convey that if a satisfactory scheme was submitted, that permission should be refused on the ground that it would be uneconomic. I consider that the statement challenged by the appellants was no more than an observation and was not germane to the decision. Accordingly, I hold that the appeal fails on the first ground.

I turn now to the second ground in the notice of motion, which is that the inspector erred in law in that " he concluded that without any access details it was not possible to judge D whether road-safety requirements could be met without sacrificing important planning considerations " whereas plans showing access details were tendered to him in evidence. In paras 11 and 12 of his letter the inspector had dealt with the need to provide suitable access and the possible consequences of so doing, such as opening up views of the site, and he concluded that " without any access details it is not possible to judge whether road-safety requirements could be met without sacrificing important planning considerations." There were in fact produced to the inspector three plans showing, among other things, possible means of access to the site as developed. These plans did not form part of the application but had been E part of applications previously made to the local planning authority and refused. Mr Bartlett submitted that these plans constituted evidence specifically adduced to deal with the access point and that in ignoring them in his decision the inspector had failed to have regard to a material consideration of some importance. Alternatively, if he had had regard to the plans, then his expressed reasons were inadequate in that they did not deal with material of importance, and he referred to *Hope v Secretary of State for the Environment* (1976) 31 P&CR 120, and *Ellis v Secretary of State for the Environment* (1976) 31 P&CR 130. I do not think it follows that because the inspector did not mention those plans that F he did not have regard to them. It is not incumbent on an inspector to set out the evidence on which he relies. Moreover, having heard that the plans were produced, it seems to me probable that the inspector not only had regard to them but favourably regarded them, which probably contributed to his conclusion that " reasonable visibility could be achieved." However, it is manifest that it is impossible to make a final judgment on access details in ignorance of the form of the other development for the approved houses the disposition of which of necessity had a bearing on the access details. Further, in my judgment the decision letter, particularly in para 12, adequately sets out the reasons for requiring access details.

Under the circumstances this appeal must be dismissed. G

The appeal was dismissed with costs.

QUEEN'S BENCH DIVISION

June 29 1978

(Before Sir DOUGLAS FRANK QC, sitting as a deputy judge of the division)

BRIGHTON BOROUGH COUNCIL v SECRETARY OF STATE FOR THE ENVIRONMENT AND ANOTHER

Estates Gazette February 24 1979

(1978) 249 EG 747 H

Town and Country Planning Act 1971—Local authority challenge decision of inspector who allowed appeal from refusal of planning permission to trustees of private school —Proposals to develop unused part of playing fields by building houses and flats—Whether inspector in error in taking account of finance needed to restore and preserve the school, a listed building in a conservation area—Planning benefit likely to accrue from money derived from development—Inspector entitled to take account of these considerations—Other criticisms based on alleged lack of evidence and disregard of development plan also rejected —Inspector's decision upheld J

This was an application by Brighton Borough Council to quash the decision of an inspector appointed by the Secretary of State to determine an appeal by the trustees of St Aubyn's, a private preparatory school in Rottingdean, against the council's refusal of planning permission for the development of an unused part of the school's playing fields. The inspector had allowed the appeal.

R Seabrook (instructed by Sharpe, Pritchard & Co) appeared on behalf of the applicant council; C Fay (instructed by the Treasury Solicitor) represented the first respondent, the Secretary of State; D Keene (instructed by Malcolm, K Wilson & Cobby, of Worthing) represented the second respondents, St Aubyn's School Trust Ltd.

Giving judgment, Sir Douglas Frank QC said: St Aubyn's is a private preparatory school at Rottingdean, which is within the area of the applicant borough council. The school building is in a conservation area and is a grade II listed building. Adjoining it are the school playing fields, which are not in the conservation area; the southern part of these are unused and that is the part farthest from the conservation area.

On November 22 1976 the school trustees applied for L planning permission to develop the unused playing fields area, which I shall now call the appeal site, by building houses and flats on it. On February 1 1977 the borough council refused permission and they did so on three grounds. The first ground was that the development would be contrary to the development plan. The second one was on amenity grounds, with particular regard to visual amenity and on the ground that the development would detract from the character and identity of the village. The third ground was that the development would constitute over-development or would be out of character. The trustees appealed against that refusal and the Secretary of State appointed one of his inspectors to hold M a public inquiry and delegated to him the duty of deciding the appeal. The inspector whose concern it was was G Borough-Copley OBE BSc(Eng) FICE. I mention his qualification because it is not altogether irrelevant.

The appeal was heard on July 12, 13 and 14 1977 and apart from the trust and the borough council at the inquiry there also appeared the local preservation society and two housing associations, each of which said they would like a part of the appeal site for development of houses for old people. On August 30 the inspector made his decision and he allowed the appeal. In the second paragraph of his report he said:

From my inspection of the site and the surroundings and from the evidence given at the inquiry, I am of the opinion that the determining issues are firstly, whether the visual and physical importance of the appeal site as opposed to the playing fields in relation to the various parts of Rottingdean is such that it should remain undisturbed; and secondly consideration of other special factors.

He then went on to deal in some detail with the first issue. He concluded that the appeal site, as distinct from the playing fields, was not of such visual and physical importance that development should be excluded. Nothing really turns on that finding, but it is to be noted in passing that that was the first issue and I think it was really the main ground upon which the development was disapproved of by the borough council.

Then he turned to the special factors of which he said there were four, the first of which was the upkeep of the school buildings and playing fields; he referred to a structural survey and requirements made by the fire department and stated that action was being urged by the Ministry of Education. He then said: ". . . the most cursory inspection indicates the magnitude of the works required on maintenance and refurbishment." I observe in passing that in coming to that conclusion the inspector not only had evidence of a surveyor's report and other evidence to which I shall refer later but he was able to assess that evidence on his inspection for which he was qualified by virtue of his professional experience and qualifications. He then stated:

In the ordinary way finance would not be a planning consideration. However, in this particular instance because of the importance and prominence of the school buildings within the conservation area and so that the school can continue to be successful such that its presence, which does much for Rottingdean, can be assured, it is proper to take account of the advice given to your clients so as to ensure continuity.

The first two grounds of the appeal to this court are concerned with that part of the inspector's decision. It is said, in short, that the inspector was wrong in law in taking into account the financial state of the trust and the cost of works to be carried out or which ought to be carried out to the school. Mr Seabrook said that the inspector was only entitled to take into account planning considerations, which considerations must relate to the use of land and that there was no place for consideration of the applicant's resources. He said that what the applicant proposed to do with the proceeds of the development was not relevant; that had nothing to do with the use of the land under consideration and there was no obligation to use the money for particular purposes.

Now, as a generality, it may be true that financial considerations are not relevant to planning. Thus, it is not for the Secretary of State to consider whether or not it would be economic for the applicant to carry out the development. That I think is what is ordinarily meant by planning considerations and that was what was under consideration in the case of *J Murphy & Sons Ltd v Secretary of State for the Environment* [1973] 1 WLR 560. I think that it is that sort of consideration which was in the mind of Ackner J in another case and broadly to what Forbes J was referring in *Sovmots Ltd v Secretary of State for the Environment* [1977] QB 414. However, both Forbes J and Ackner J, I think, made it clear that they had no intention of saying that financial considerations can never be relevant, and indeed in a more recent case, *Niarchos (London) Ltd v Secretary of State for the Environment* (1977) 76 LGR 480, I held that the Secretary of State was wrong on the facts of that case in not taking into account the financial consequences of his decision. An analysis of the *Niarchos* case shows that what the planning authorities tried to do was to coerce the landowner into converting his premises from offices to residential use and accordingly permission for office use to continue was refused. The planning authority and the Secretary of State decided that they were not bound to take into account the economic aspects of the conversion into houses and I held that that was wrong.

The instant case seems to be in parallel in so far as the inspector was saying that, in deciding whether to grant planning permission, he should take into account the financial implications of works required for planning purposes, namely the refurbishment, the restoration, fireproofing and the various works which were necessary to the grade II listed buildings in a conservation area. Now it is clear that the restoration and maintenance of that building was a planning matter and an important planning matter having regard to its listing, situation and its prominence in the conservation area. Once that situation has come about it seems to me that the Secretary of State is entitled—and that is all I have to decide—to take into account the possibility that by granting permission there will thereby result a planning benefit in the maintenance and refurbishment of an important listed building. Under the circumstances, therefore, I hold that the inspector did not wrongly take into account the financial state of the trust and the cost of the maintenance, refurbishment and upkeep of the structure of the scheme.

I now turn to the next ground of challenge: it was that there was no evidence before the inspector from which he could properly or reasonably infer or find, as he implicitly found, that the existence of the school would be prejudiced were he to dismiss the appeal. It is right to bear in mind that the rules of evidence do not apply to planning inquiries. What the inspector seems largely to have relied upon is the evidence which had been given by the chairman of the trust that he had been advised informally by an architect that the cost of all the necessary work would be in the order of £100,000. But he also said:

If these proposals are approved we should be able to provide sufficient capital to bring the school up to date and so safeguard its future in Rottingdean. Only thus can we ensure the preservation of the major part of our property as open space playing fields. If our proposals are rejected we can see no way of raising the capital essential to the future of the school in Rottingdean, and a major doubt must be cast over its continuance on its present site.

Neither the borough council nor any other party at the inquiry led as to the work necessary to the building or to the cost of it. I therefore decide this point on the very short ground that the inspector was entitled, if not bound, to accept such evidence as there was before him, however tenuous it may have been, and the fact that it would be inadmissible in an ordinary court is irrelevant. Hence ground two fails.

The third ground was that there was no evidence from which the inspector could infer that there was appreciable support from two local housing associations who wanted to develop the appeal site. The fact is that there was evidence given by two housing associations, and one of them produced an architect's plan as a feasibility study for the provision of houses for old people. Although they did not show housing on the whole site, and indeed a large part of the site is shown as possible public gardens, I understand that it would not be right or feasible to put buildings on other parts of the site. Be that as it may, there was very appreciable support from two housing associations who want to develop the site, and adequate evidence before the inspector to support these findings. Whether they want to develop the whole or not is of no significance, because the inspector did not say that they wanted to develop the whole, although, as I understand the evidence, I think that certainly one of them would like to do so.

Now I turn to the last ground of challenge to the decision, which is that the inspector failed to have regard to the provisions of the development plan, which shows the land allocated as an independent school. Mr Seabrook readily acknowledged that this was not his best point; indeed I

A think that there is nothing in it at all, because the whole inquiry was about whether the site should remain for the purposes and use shown in the development plan, that is as an independent school. The inspector was necessarily having regard to the development plan position. Under all those circumstances this application must be dismissed.

The application was dismissed, the applicant council to pay the costs of both respondents.

QUEEN'S BENCH DIVISIONAL COURT

December 12 1978

B (Before Lord WIDGERY CJ, Mr Justice KILNER-BROWN and Mr Justice ROBERT GOFF)

BROXBOURNE BOROUGH COUNCIL v SECRETARY OF STATE FOR THE ENVIRONMENT AND OTHERS

Estates Gazette March 10 1979

(1978) 249 EG 959

Town and Country Planning Act 1971—Local planning authority challenge decision of Secretary of State to quash enforcement notices—Change in character of use of land from predominantly rural use for storing, sorting and grading timber in the round to industrial timber yard with C **stacks of planks—Present use, however, not materially different from that specified in established use certificate—Effect of certificate—Function under section 94 of Act to render specified use, so long as it persists, immune from enforcement action—Secretary of State acted correctly in rejecting inspector's recommendations, based on investigation of actual use before issue of certificate, that the enforcement notices should, with an exception, be upheld—Wrong to go behind the established use certificate—Appeal and motion to quash dismissed**

These proceedings consisted of an appeal under section D 246 and an application to quash under section 245 of the Town and Country Planning Act 1971. Both were designed to challenge decisions of the Secretary of State concerning enforcement notices served by the appellants, Broxbourne Borough Council, in relation to Doverfield, Goff's Lane, in Hertfordshire, a piece of land of about 3½ acres. The Secretary of State was the first respondent, St Ouen Diamond & Fine Art Co Ltd, the freeholders, were the second respondents and Sir William Burnett & Co (Timber) Ltd, leaseholders, were the third respondents.

C Fay (instructed by P H Sturrock, solicitor to Broxbourne Borough Council) appeared on behalf of the appellants; Harry E Woolf (instructed by the Treasury Solicitor) represented the first respondent; the second respondents were not represented and took no part in the proceedings; and Viscount Colville of Culross QC (instructed by Braby & Waller) represented the third respondents.

Giving the first judgment at the invitation of Lord Widgery CJ, ROBERT GOFF J said: There are before the court two matters brought by the Broxbourne Borough Council, each arising out of the same decisions by the Secretary of State for the Department of the Environment. The first is an appeal under section 246 of the Town and Country Planning Act 1971 and the second is an originating F motion under section 245 of the same Act. Both arise out of decisions by the minister on appeals to him against enforcement notices served by the appellant council relating to the use of certain land in Hertfordshire.

The land in question is an "L"-shaped piece of land called Doverfield, Goff's Lane, Goff's Oak in the county of Hertfordshire. The site is about 3½ acres in area. In the north-east corner there is an access road about 200 ft long leading to Goff's Lane. The northern side of the site abuts on to the back gardens of houses on Goff's Lane and is separated from them by a close-boarded fence. To the east of the site is a caravan site, to the south of which there is open land. The remainder of the site is surrounded by G allotments, woodland and farmland.

The history of the site was the subject of detailed findings by the inspector following a local inquiry which he held in April and May 1976 following the appeals against the Enforcement notices. In his decision letter the minister accepted the facts found by the inspector, and it is enough that I should refer to the summary of those facts in the decision letter itself. It is there found that until 1960 the site was owned by a Mr Skeggs, who used it for the storage and cutting of trees, logs and fencing posts. Part of the site was also used by a Mr Fletcher, a tree-felling and tree surgery specialist, for the storage of planks and logs and for cross- H cutting and resawing timber, and part by a Mr Parks who was concerned with tree surgery and site clearance and cut and trimmed trees for firewood. In addition, Mr Clothier, a haulage contractor, used the site from 1958 as a yard and store for logs brought in from the forest. There is then reference to use by Mr Symonds, who took over Mr Skeggs' business in 1960, and to its continued use until 1975, showing that the use was primarily for the storage, sorting and grading of timber in the round.

On October 6 1972 there occurred an event of crucial importance in the case. On that date an established use certificate was issued in respect of the site in the following J terms:

> It is hereby certified that the use of the above land and building for the storage, sawing, resawing and disposal of timber in the round and the storage, maintenance, repair and overhaul of vehicles and plant incidential thereto, was on the 3rd day of October 1972 established within the meaning of paragraph (a) of section 94(1) of the Town and Country Planning Act 1971.

Thereafter, as is found (I quote from the minister's decision letter): "Mr Fletcher, Mr Parks and Mr Clothier remained on the site until 1975. When a workshop was erected in 1967 K Mr Clothier moved his depot to the site. He worked as a subcontractor to Mr Symonds and also on his own account. All the occupants of the site had vehicles, plant and other equipment there."

The next crucial event occurred in 1974 when the site was sold to a company called St Ouen Diamond & Fine Art Co Ltd. In June of 1975 that company leased the site to Sir William Burnett & Co (Timber) Ltd, and subsequently it has been used by Burnett as a bulk storage depot for timber planks. No timber in the round is handled there, but occasionally sawing is done; the depot is used as a buffer stock area to supply Burnett's timber mill at Cuffley and is used L for the storage of planks in stacks for that purpose. It is clear that in point of fact a material change did take place in 1975 following the change of ownership. Substantial concrete roadways were completed by July 25 1975. These consisted of a concrete road running south-west from the site entrance with four concrete spur roads adjoining the road at right-angles on the western side. The area between the spurs has been covered with hardcore. These concrete roadways are now used by lorries transporting the timber planks to and from the site, and the timber is now stacked on the hardcore area. In addition to the roadways two buildings have been constructed. One is a wooden sectional building M of modest dimensions, and the other is a breeze-block extension which was added on to the back of the existing workshop at the south-eastern corner of the site.

As a result of these new activities the inspector held in findings which were accepted by the Secretary of State that:

> the character of the use has changed quite dramatically. The former use had rural connotations with logs being stored on what was predominantly a green field site stacked to a fairly low level. Now the site has the appearance of an industrial timber yard with high rectangular stacks of timber concentrated in a small area of the site serviced by concrete roadways.

A This change led to complaints by the local residents. Following those complaints the appellant council served four enforcement notices dated respectively July 28, July 31, and September 1 1975 and January 13 1976. These were served on both the second respondents (the freeholders) and the third respondents (the leaseholders). They have been referred to as notices A, B, C and D. Notice A was concerned with the construction of the concrete roadways and required them to be broken up and the land to be restored to its condition before the development took place within two calendar months. Notice B, which is by far the most important notice in the case, is concerned with use of the land for the purpose B of the stacking and storage of planks of timber. It required the respondents to discontinue that use within one month. Notice C was concerned with the erection of the wooden shed and required its removal within one month. Notice D was concerned with the breeze-block extension to the workshop and also required its removal.

The second and third respondents appealed against these enforcement notices, and the appeals against notices A, C and D were on the grounds specified in subsections (a) and (g) of section 88(1) of the Act. The appeals against notice B were on the grounds specified in subsections (a), (b), (d), (f) C and (g) of that subsection.

As I have said, a local inquiry was held. The inspector investigated in great detail the actual uses of the site before the events of 1975. He reported to the minister on June 25 1976 and the inspector considered first the appeal against notice B. His conclusion was that the change of use was so very marked as to constitute a material change in the use of the site after January 1 1964. He accordingly recommended that the appeal under section 88(1)(d) should fail. So far as the appeal in respect of notice B under section 88(1)(a) was concerned, he considered the present use was completely out of place in the approved green belt, and, having considered D also the weight of traffic and the inadequacy of the access road, his recommendation was that the application for planning permission should be refused. On the assumption that those recommendations were accepted, he also recommended that the appeals in respect of notices A and C should likewise be dismissed, but he recommended that the appeal in respect of notice D should in any event succeed on the grounds that the extension, which was to cover an oil tank, was there on the grounds of safety and was inoffensive.

The Secretary of State, having considered the inspector's report, decided to quash all four notices. He decided that the appeal in respect of notice B should succeed on ground (d) in E section 88(1)—ie that it was protected by time. But he declined to grant planning permission in respect of that development. He also allowed the appeals in respect of notices A and C as well as notice D, in each case granting planning permission for the retention of the development.

The appellants now appeal against the Secretary of State's decision in respect of notices A, B and C, but not D, and it is common ground that, if the appeal on notice B is successful, then they will also succeed in respect of notices A and C. Conversely, if they fail on notice B, they will fail on the other two. Therefore, notice B is the crucial notice.

F Before dealing with the merits of the appeals, there is one procedural point I have to mention, and that is that the appellants appeal both under section 246 of the Act and by originating motion under section 245. That is because they are asking for the planning permission granted by the Secretary of State in relation to notices A and C under section 88(5)(a) of the Act to be quashed, and such a matter is expressly referred to in section 245(3) by reference to section 242(3)(f) of the Act. By an order made on May 16 1978 this court ordered that the originating motion should be heard immediately after the appeal.

I now come to the substance of the appeal. In order to understand the point at issue it is necessary to refer to G section 94 of the Act of 1971, which is concerned with certification of established uses, and in particular subsection (7) of that section, which provides as follows:

> An established use certificate shall, as respects any matters stated therein, be conclusive for the purpose of an appeal to the Secretary of State against an enforcement notice served in respect of any land to which the certificate relates, but only where the notice is served after the date of the application on which the certificate was granted.

The crucial passage in the Secretary of State's decision letter is contained in paragraph 12 of that letter, which reads as follows. I omit the first sentence. The paragraph then H continues:

> It is also agreed that the established use certificate is conclusive for the particular use specified in it and that the main issue before the Secretary of State is whether the present use is so different in degree and character from the use specified as to constitute a material change of use. Section 94(7) of the 1971 Act states that the established use certificate shall "be conclusive for the purpose of an appeal . . . against an enforcement notice" as respects any matters stated therein. The view is taken that in the light of this provision the Secretary of State must accept that the use of the site before January 1 1964 was as described in the established use certificate and it is not open to him to consider the nature of the use in respect of which J the certificate was sought. (Had he been able to reopen the matter, however, he would have taken account of the fact that the former use of the site had rural connotations, being described as a forestry contractor's yard and depot in the application for the certificate.) It is concluded that the present use of the land for the purpose of the stacking and storage of planks of timber is not so different from that described in the established use certificate as to constitute a material change of use of the land. The use described was established before 1964 and the appeals succeed on ground (d). The notice will be quashed accordingly.

It is to be observed that the Secretary of State's approach K to the matter was, therefore, very different from that contained in the recommendations of the inspector. The inspector's report contained elaborate findings of fact relating to the use of the land before 1975. However, it being accepted that there was no change of use between the established use certificate in 1972 and the events of 1975, the Secretary of State approached the matter on an entirely different basis. He simply had regard to the use as specified in the established use certificate and the present use of the land, and asked himself the question whether the present use of the land for the purposes of stacking the storage of planks of timber was so different from the use specified in the certificate as to constitute a material change of use of the land. L That question he answered in the negative. The evidence as to the actual use of the land before 1975 he, therefore, treated as wholly irrelevant.

Mr Fay for the appellant council submitted that in so approaching the matter the Secretary of State erred in law. He submitted that the proper course for the Secretary of State was to compare the existing use with the last preceding use as it in fact existed in order to decide whether there had been in fact a material change in the use of the land. Section 94(7) constituted, he submitted, no bar to such an inquiry because that subsection provided only that an established M use certificate should be conclusive as respects any matter stated therein, and the certificate in question was wholly silent as to the scope and intensity of the use at the time it was issued.

In my judgment, this criticism of the Secretary of State's approach is not well founded. The purpose of an established use certificate is clear. It does not render a use lawful. To that extent it is unlike a grant of planning permission. Therefore, if, for example, the use specified in an established use certificate is abandoned, it cannot lawfully be resumed.

A Its function is to render the specified use, as long as it persists, immune from an enforcement notice. It, therefore, precludes the necessity of investigating events which may have occurred many years before as to what was the established use as at the date of issue of the certificate. If Mr Fay's submission were to be correct it would deprive an established use certificate of all efficacy. In each case it would be necessary to investigate the actual use at the date of the issue of the certificate in order to ascertain whether there had in fact been a material change of use, a fresh investigation which it was the object of the established use certification procedure to obviate.

B Nor, in my judgment, does it assist Mr Fay to argue that the certificate specifies no particular level or intensity of activity and, therefore, to submit that the level or intensity of activity was not a matter stated in the certificate and so was a matter on which the certificate was not conclusive. The short answer on this point is that the use was a matter stated in the certificate. Since no limit was placed upon either the part of the site to be so used or the intensity of the use, the use so specified was without limit as to space within the site or intensity. Mr Fay cannot, therefore, now complain that the Secretary of State has erred in law in
C holding that there has been no material change of use by reason of the area of the site now employed or the present intensity of the use.

It follows, in my judgment, that the Secretary of State applied the correct test in the present case. It is clear from the decision letter that the Secretary of State was somewhat unhappy as to the conclusion he felt bound to reach, having regard to the findings of fact in the inspector's report, which appear to indicate that the use by the appellant council's predecessor in title specified in the established use certificate was somewhat wider than the use actually enjoyed at the time of the issue of the certificate. Indeed, as the argument
D proceeded, it became only too clear that what Mr Fay was seeking to do was to invite the court to read the certificate in a qualified manner. But the duty of the minister and of this court is to apply the correct legal principles, and, if it were not to do so and were to accept Mr Fay's submissions, the result would be that established use certificates would be deprived of their proper force; their purpose would be undermined, and indeed purchasers of land in respect of which there existed established use certificates might be misled into paying too high a price for the land.

For these reasons, in my judgment, the appeal under section 246 should be dismissed. It must, therefore, follow
E that the originating motion under section 245 should likewise be dismissed. But the case has a moral. It demonstrates that planning authorities should exercise great care concerning the terms of established use certificates which they issue. If a certificate is not drawn with care and expressly limited to the precise use in question, then its issue can lead to the consequence that the authority may, through its own act, find itself thereafter precluded from preventing a use for which planning permission would not have been granted simply because the certificate had been issued in terms wider than were necessary.

F Agreeing KILNER BROWN J said: I would merely wish to add a word or two with reference to what my Lord has described as "the moral" to be drawn from this case. In my judgment, it does underline the duty of a planning authority to consider with care whether or not the proposed terms of the certificate sufficiently particularise the existing use. The object of the exercise contemplated by section 94 is not only to avoid an examination of the history and existing fact but also to identify both the nature and the extent of the use. This demonstrates that it is clearly important to be careful not only in order to avoid the conclusive effect of a loosely and widely phrased certificate but also, as all counsel have agreed and as my Lord has pointed out, for the G
assistance of a prospective purchaser.

For the reasons given by my Lord, with which I agree, I concur.

LORD WIDGERY CJ also agreed.

The appeal under section 246 and the originating motion under section 245 of the Town and Country Planning Act 1971 were both dismissed. The court awarded one set of costs only, to be paid to the first respondent, the Secretary of State.

H

QUEEN'S BENCH DIVISIONAL COURT

January 23 1979

(Before Lord WIDGERY CJ, Lord Justice BRIDGE and Mr Justice CAULFIELD)

EMMA HOTELS LTD v SECRETARY OF STATE FOR THE ENVIRONMENT AND ANOTHER

Estates Gazette April 14 1979

(1979) 250 EG 157

Town and Country Planning Act 1971—Enforcement notice —Appeal from Secretary of State who had upheld notice —Larger area of hotel used as non-residents' lounge and bar than that for which planning permission had been given—Whether material change of use because of change in the proportion of elements making up composite use of single planning unit—Class XI in Schedule to Town and Country Planning (Use Classes) Order 1972—Enlargement of small permitted area of non-residents' bar to include de facto use of much larger lounge bar area not sufficient to take premises outside Class XI—Appeal against Secretary of State's decision allowed J

This was an appeal under section 246 of the Town and Country Planning Act 1971 against the decision of the Secretary of State, who had dismissed an appeal under section 88 of the Act against an enforcement notice requiring the use of part of premises known as the Boston Hall Hotel, The Leas, Westcliff-on-Sea, as a lounge and bar, open to non-residents to be discontinued. K

C Cochrane (instructed by Kingsford, Dorman & Co, agents for Drysdales & Jones, of Southend-on-Sea) appeared on behalf of the appellants; A Collins (instructed by the Treasury Solicitor) appeared on behalf of the first respondent, the Secretary of State; the second respondents, Southend-on-Sea Borough Council, were not represented and took no part in the proceedings. L

Giving the first judgment at the invitation of Lord Widgery CJ, BRIDGE LJ said: This is an appeal against a decision of the Secretary of State for the Environment given by letter dated March 3 1977 dismissing an appeal under section 88 of the Town and Country Planning Act 1971 against an enforcement notice requiring the cessation of a use of a certain part of premises known as Boston Hall Hotel.

The history of the matter is this. The planning unit with which we are concerned comprises what were at one time three private dwelling-houses, 25, 26 and 27 The Leas on the seafront at Westcliff-on-Sea. In 1969 permission was M
granted for the conversion of two of those dwelling-houses into an hotel (nos 25 and 26) and that permission was duly acted on. Then, early in 1974, permission was sought and obtained to incorporate no 27 into the hotel complex, and permission was granted subject to conditions not presently relevant to convert those premises, which were then used as flats, into six bedrooms, a lounge and a bar. Shortly thereafter, permission was sought by the present appellants to change the use of the bar, which by then had been provided or was to be provided in the newly incorporated no 27, to a use as a non-residents' bar.

A Mr Cochrane on behalf of the appellants submits that that was a planning permission which was never required, and of course the appellants are not to be prejudiced in any way by the fact that they may have sought a permission if, on the true view, they did not require any such planning permission in order to do that which they wished to do. The express planning permission granted in March 1974 for the use of a non-residents' bar in the new part of the hotel extended only to a very small area on the ground floor. But in due course the appellants came to the conclusion that they wanted to use a very much larger area, including a substantial lounge adjacent to what I will call the permitted non-
B residents' bar, also as a lounge bar which should be open to non-residents. They applied to the licensing justices for the necessary licence to enable them to carry out their wishes. They were refused by the licensing justices, but in due course a licence was granted to them by the Crown Court on appeal in February 1976.

It was the implementation of what had been permitted from the point of view of the licensing law which evoked the service of the enforcement notice in this case which requires the appellants to cease using the pink land (the pink land is indicated on the plan as referring to the large
C lounge bar area which *de facto* is being used as a lounge and bar open to non-residents) for the sale of intoxicating liquor to, and for the consumption of intoxicating liquor by, the general public.

The Secretary of State, having received a report from an inspector who held a public inquiry into this matter, gave his decision in the letter referred to in the following terms:

> The evidence and facts found by the inspector, which are accepted, show that planning permission for the use of nos 25 and 26 The Leas as an hotel was granted in 1969. Permission to convert no 27 into six bedrooms, lounge and bar as an extension to nos 25 and 26 was granted on February 11 1974.
> D The whole building nos 25, 26 and 27 The Leas is now known as the Boston Hall Hotel, but no 27 is also known as Oak Hall. On March 28 1974 planning permission was given for a part of the ground floor of Oak Hall to be used as a non-residents' bar. The larger area which is the subject of the enforcement notice is now in use together with the permitted area as a non-residents' bar and lounge.

Then the Secretary of State refers to the submission on behalf of the appellants made under paragraph (b) of section 88(1), and continues in paragraph 7: "The view is taken that the planning unit is the entire unit of occupation, that is nos 25, 26 and 27 The Leas." Pausing there, that is a
E conclusion of the Secretary of State with which no one quarrels and a conclusion which was clearly right.

The letter continues:

> It is considered that these premises are in a composite use for the purposes of an hotel and a non-residents' lounge and bar. When a private hotel extends facilities to non-residents a clientele from a much larger area may be attracted and the increased number of people and vehicles will affect the character of the use of the premises and to some extent the amenities of the area. A small non-residents' bar within an hotel may not make much impact on the area or amount to a material change of use of the premises as a whole. However, it is concluded that the
> F introduction of a non-residents' bar and lounge on the scale of Oak Hall, offering facilities normally available in a public house, with the consequent increase in customers, traffic, noise and disturbance of the neighbourhood changed the use of the premises from a private hotel to a composite use for the purposes of the hotel and a fully licensed non-residents' bar and lounge and was sufficient to constitute a material change in the use of the planning unit as a whole.

Accordingly, the Secretary of State concluded that planning permission was required, and went on to consider the issue which was before him on the planning merits and to decide that against the present appellants.

At an early stage in the argument, Mr Cochrane for the G
appellants conceded, and in my judgment rightly conceded, that, viewed independently of any provision in the Use Classes Order, to which I shall have to turn in a minute, it could properly be said that use of premises as a licensed public house was a materially different use from use of premises as a private residential hotel. That concession having been made, it would certainly seem to me that, if one had to decide this appeal independently of the Use Classes Order, one would be bound to say that there was material on which the Secretary of State could properly conclude that here there was a composite use of a single planning unit in relation to which a change in the proportion of the elements making H
up the composite use had taken place, which was capable of affording material on which the Secretary of State could conclude as a matter of fact and degree that a material change of use of the whole had taken place.

However, the matter does not rest there, for Mr Cochrane relies on the well-known provisions of article 3 of the Town and Country Planning (Use Classes) Order 1972, which it will be recalled provides:

> Where a building or other land is used for a purpose of any class specified in the Schedule to this order, the use of such building or other land for any other purpose of the same class J
> shall not be deemed for the purposes of the Act to involve development of the land.

The class in question here is Class XI, which refers to "use as a boarding or guest house, or an hotel providing sleeping accommodation."

Before the extension of the non-residents' licensed bar area involved in what is now alleged to have been a material change of use for the purposes of planning law took place, there cannot be the slightest doubt that the whole unit, nos 25, 26 and 27, constituted an hotel providing sleeping accommodation, and was within that class. The question K
I ask myself is whether the fact that the permitted small area included in no 27 for use as a non-residents' bar has been enlarged to take in the area to which the enforcement notice relates for some reason takes the entire premises outside the ambit of Class XI.

For my part, I am quite unable to see that it does. Mr Collins, who has said everything that could be said on behalf of the respondent Secretary of State, has drawn attention to the fact that in the definition of a "shop" there are specifically excluded from the definition, *inter alia*, hotels and premises licensed for the sale of intoxicating liquor, L
which are there dealt with as separate entities. That confirms the provisional view I expressed that they are indeed different uses. But it does not follow from that, in my judgment, that an hotel providing sleeping accommodation in any way ceases to come within Class XI if it includes as one of the incidents of the hotel use use of part of the hotel premises as a bar open to non-residents. Both bars and dining-rooms are common features of hotels providing sleeping accommodation. I suppose the vast majority of hotels up and down the country include such facilities, and in many, many cases those facilities are open to the general public and not confined to use by persons who are residing in the hotels. M

For those reasons I have come to the conclusion that the Secretary of State, who does not appear to have addressed his mind in his decision letter to the issue raised under Class XI of the Use Classes Order, came to a conclusion which cannot be supported in law, and I would allow the appeal and remit the matter to the Secretary of State to be dealt with in accordance with the judgment of this court.

LORD WIDGERY CJ and CAULFIELD J agreed.

The appeal was allowed with costs against the Secretary of State.

QUEEN'S BENCH DIVISION

May 10 1978

(Before Mr Justice FORBES)

GEORGE WIMPEY & CO LTD v SECRETARY OF STATE FOR THE ENVIRONMENT AND ANOTHER

Estates Gazette April 21 1979

(1978) 250 EG 241

Town and Country Planning Act 1971—Application to quash decision of Secretary of State who dismissed appeal from council's refusal of planning permission for housing development—Refusal on ground that proposals were premature in relation to available communal facilities and sewerage arrangements—" Severe strain on infrastructure " —Alleged failure of Secretary of State to give adequate reasons and to take account of relevant matters—Criticisms rejected—No need for Secretary of State to say in terms that he accepted inspector's findings of fact—No failure to take account of water authority's duty to provide sewerage under section 16 of Water Act 1973—No failure to consider an alternative claim by applicants

This was an application under section 245 of the Town and Country Planning Act 1971 to quash the Secretary of State's decision upholding the refusal of Maidstone District Council to give planning permission to the applicants, George Wimpey & Co Ltd, to carry out housing development on land near Staplehurst, Kent. The facts and grounds for the refusal are fully set out in the judgment of Forbes J.

A B Dawson QC and R J Furber (instructed by P Callcutt, Legal Department, George Wimpey & Co Ltd) appeared on behalf of the applicants; Harry Woolf (instructed by the Treasury Solicitor) represented the first respondent, the Secretary of State; the second respondents, Maidstone District Council, were not represented and took no part in the proceedings.

Giving judgment, FORBES J said: In this case Mr Dawson moves to quash a decision of the Secretary of State for the Environment in relation to some land in Kent. The land in question is some 6.214 hectares, which I am told is 14.30 acres, of land near Staplehurst in that county, and it is owned by the applicants a well-known company, George Wimpey and Co Ltd. I should add perhaps that as far as I understand from the case the site lies adjacent to some other development which Messrs Wimpeys have completed or are completing. There was an application for planning permission on February 4 1976, and that was refused by the local planning authority on May 5. There was an appeal on May 27 and the inspector held an inquiry on October 12 1976; his report is dated November 15 1976; thereafter the Secretary of State gave his decision on January 31 1977. The inspector recommended dismissal of the appeal and the Secretary of State upheld the inspector's recommendation.

The land is what is known as " white " land, that is land shown in a development plan as an area where existing uses are intended to remain for the most part undisturbed. But there was a special situation in relation to this particular land brought about largely by a circular, no 122 of 1973 from the Department of the Environment, which dealt with land in what were called growth areas as defined in strategic plans. As I understand it, strategic plans are produced for large areas or regions, in this case one for the south-east of England, and within it there may be areas shown in those strategic plans as what are called growth areas. This circular is dedicated among other things to indicating what are the general presumptions to be applied by planning authorities in relation to such areas. The particular part with which we are concerned is annex A to the circular and this harks back to begin with to another circular, 102 of 1972, which had dealt with the need to give clearance for substantial additional areas of building land. The annex goes on in para 2 to indicate that in general the fact that land is shown as " white " land in the development plan cannot carry with it an automatic presumption against housing development. It then continues in para 3 in these terms:

> Different considerations apply to the treatment of " white " land within and outside growth areas as defined in strategic plans approved by the government. In either case the development of " white " land should be orderly and not haphazard, and may, where appropriate, reflect regional strategies as they emerge and those structures and local plans which are fully formulated.

Para 4:

> In growth areas planning applications must be considered against the background of a strong general presumption in favour of housing. But as a growth area is a general location only, this strong general presumption applies only to those parts of the area that clearly would fall to be considered for inclusion in any specific growth area plan. . . . For those parts the presumption should only be overriden if there are exceptionally compelling planning objections to a development which, in the circumstances of the case, outweigh the need to make the land available for housing.

Para 5 starts:

> Outside growth areas, white land can be regarded as suitable for housing, provided its development represents a natural extension of existing development and is compatible with its size, character, location and setting.

It goes on, that paragraph, to say:

> and provided there are no further compelling planning objections (which would of course include those mentioned in para 4 above as well as valid objections on the grounds of inadequate infrastructure . . . the absence of any demonstrable need for the development proposed should not by itself be treated as a planning objection. The question of need is relevant only to the extent that other objections arise and it has to be decided whether the evidence of need outweighs them.

I do not think I need read paras 14, 15 and 16 which deal with infrastructure. They do indicate that problems may arise where development is proposed where the necessary infrastructure is not yet developed. And infrastructure quite clearly includes the question of sewering an area which is to be developed. It includes other things such as roads, schools and presumably other matters like welfare services. Indeed it mentions amenities and recreation facilities such, no doubt, as open spaces and village halls.

The land which is the subject of this motion is included within a growth area in a strategic plan as I understand it, because it is shown on the strategic plan as a medium-growth area. By looking at another map it appears to indicate that it is in area no 2, planning area no 2. The point about this is that there was an argument at the inquiry on the question of whether or not the land in question fell to be considered under para 4 or para 5 of that circular. As will be apparent in a moment, the applicants' case at the inquiry proceeded on a dual or alternative basis—one argument if the land was shown to be under para 4 and another argument if it was shown to be under para 5. It should be noted that the only actual mention of infrastructure in this circular or the annex to it is in regard to the matters in para 5. That was one point—there was a question as to whether the land was within the one paragraph or the other, and the deployment of the various arguments relating to those paragraphs therefore proceeded on that alternative basis.

The real difficulty in this case, and this seems to have been accepted by both the planning authority and the applicants, was that the addition of this extra housing would in fact cause difficulties in the infrastructure. I do not like the term, but it is perhaps a convenient one to use. It was said among other things that the primary school there was overcrowded, the nursery school was booked up to 1979, the village hall was inadequate. It was suggested that there were not enough shops because there were no additional ones to cope with an increase of something just under an additional 50 per cent in the population. It was suggested that there was insufficient open space. And all these matters were urged at the inquiry by the planning authority, and a large number of them were accepted, at any rate factually, as I understand it, by the applicants, as being matters which indicated that permission for residential development of this land was premature until the necessary matters of infrastructure had been put right. In particular one of those was the question of the sewerage. There is no doubt at all that it was accepted by the appellants at the inquiry that the existing sewerage system was in fact inadequate. I need not go into the reasons for it or the exact extent of the inadequacy, but it is quite clear that fresh sewerage facilities would have to be provided if the appellants' land were to be developed for housing.

The inspector reported on this and as usual he set out a series of findings of fact. Some of them indicate quite clearly that he found as a fact that there were inadequacies in the infrastructure. He said the village hall was inadequate, the primary school was overcrowded, the nursery school was booked up to 1979, and the appellants accept that the existing sewerage system from the Wealden Estate to the sewerage disposal works was inadequate. So in all these matters the inspector found inadequacies as a fact. In relation to shops and open spaces he did not go so far as to find inadequacy as a fact. He merely stated as part of the evidence: "In 1973 when the population was 3,500 there were 27 shops and the number has not been increased." The population has in fact increased to 5,000. So he did not pass upon the inadequacy of the shops there. So far as open space is concerned he said: "The village has 14 acres of public open space and 16.2 acres of private open space." He did not in fact pass upon the question of whether that was or was not adequate. Those really are the important findings of fact.

He then went on in his conclusions:

> Bearing in mind the foregoing facts it seems to me that new housing at Staplehurst may run too far ahead of the provision of necessary community facilities such as schools, open spaces and the like. It is therefore essential that large-scale new housing should be phased in order to allow these facilities to catch up. In the circumstances postponement of development of the appeal site to the phases programmed for after 1981 is the right course. The appellants' argument regarding the suitability of the site for housing and the general need to stimulate private house building is not sufficient to outweigh this major consideration, particularly as there is land at Staplehurst already with planning permission and other land allocated for development in the period up to 1981. The proposal is also premature in relation to sewerage arrangements. With the present facilities there would be additional surcharging at a pumping station and of a sewer from the rising main, leading to contamination of open watercourses. Permission for the development should not be granted now when a satisfactory resolution of the sewerage difficulties is not imminent.

The Secretary of State's letter is a very short one. After a formal opening paragraph it sets out those two paragraphs of the inspector's conclusions which I have just read and then goes on in these terms.

> Although the appeal site is located in a growth area in the strategic plan for the south-east, this is a general location only and the site does not fall to be considered for inclusion in any specific growth area plan. The draft informal district plan for Staplehurst village shows the site as being suitable for development at a later date. However, on the basis of the evidence submitted, it is considered that the immediate release of this land for development would place a severe strain on the infrastructure and social services available in the village and this serious objection is not outweighed by any overriding need for the development proposed. The Secretary of State accordingly agreed with the inspector's conclusions and accepts his recommendation. Therefore he hereby dismisses the appeal.

Out of this Mr Dawson grounds his application on two broad submissions, that is two submissions as to the areas in which the Secretary of State has, he says, gone wrong. They are: (1) that the decision was not within the powers of the Act; and (2) that the relevant requirements have not been complied with and the interests of the applicants have been substantially prejudiced. He then goes on to particulars. I need not I think rehearse the law applicable to this particular jurisdiction. So far as is relevant to this case the Secretary of State acts outside the powers of the Act if he takes into account any matter which is not relevant or if he fails to take into account any matter which is relevant. I need not I think go into any further details on the principles which govern what is or is not within the powers of the Act. So far as the second limb of Mr Dawson's main general submissions are concerned, those arise under the Inquiries Procedure Rules and are concerned with rules 12 and 13. So far as is relevant rule 12 requires the appointed person, that is the person we normally refer to as the inspector, to make a report in writing to the Secretary of State which shall include the appointed person's findings of fact and his recommendations or his reasons for not making a recommendation. And where the Secretary of State differs from the appointed person on a finding of fact and by reason thereof is disposed to disagree with a recommendation made by the inspector, he has at least to give an opportunity to the parties to make representations about the matter in relation to which he differs from the appointed person on findings of fact. He can of course if he wishes, and in certain circumstances I believe has to, reopen the inquiry.

So far as rule 13 is concerned, that requires the Secretary of State to notify his decision in writing to the parties. But, and this is important, it also requires him to give his reasons. There is ample authority: *Re Poyser and Mills' Arbitration* [1964] 2 QB 467 is typical, in which it is said, and I do not quote exact words, that the reasons given must be clear and intelligible reasons and must grapple with the substantial issues involved at the inquiry.

The particular matters about which Mr Dawson complains are three. He says in his particulars that the Secretary of State first of all failed to give any comprehensible reasons for his decision because: "He failed to state whether he accepted all or any of the appointed person's findings of fact." His second particular is that the Secretary of State "took into account alleged inadequacies of the present sewerage system but failed to take into account the duty of the water authority under section 16, Water Act 1973." That is put, as it were, as a subparticular of a failure to give proper and comprehensible reasons. I think in fact it has been argued not under that heading but under the heading that it was outside the powers of the Secretary of State in the sense that he took into account irrelevant matters or failed to take into account relevant matters. Then the third particular complaint that Mr Dawson makes is that: "The appointed person failed to make a proper or sufficient report in that he failed to report on the case presented by the applicants on the alternative basis that the land was not in a growth area."

If I may take those points in that order, it is quite true that the Secretary of State did not state whether he accepted all or any of the appointed person's findings of fact. But it should be noted that the findings of fact are effectively matters which were virtually unchallenged at the inquiry. The most important point, the sewerage one, was accepted by the appellants and the inspector so found. But the short conclusions of the inspector are introduced by the words "bearing in mind the foregoing facts," and the Secretary of State says that he accepted the conclusion of the inspector. Mr Woolf argues that it is quite clear from the report read as a whole and the Secretary of State's letter read as a whole that the only reason for the conclusions were the facts which the inspector had found in the previous paragraph. I do not think it can be said that inevitably and in every case the

A Secretary of State has got to say whether or not he accepts his inspector's findings of fact. There must be many cases, and in my view this is one, where it is so apparent from reading the report and the decision letter that the Secretary of State is accepting the inspector's findings of fact, that it is quite unnecessary for the Secretary of State to put that down in writing. It seems to me that broadly one has to remember that, when dealing with cases of this kind and looking at a decision letter of the Secretary of State, a court has to bear in mind that it has got to approach this matter in a commonsense way. There is ample authority for this. You do not construe the inspector's report or the Secretary B of State's letter in the same way that one would approach the construction of a statute or a contract. You do not approach it in the analytical context, as it were, of the schoolman. You look at it in commonsense. You also look at it, it seems to me, against the background that the Secretary of State's letter, and for that matter the inspector's report, is being addressed to persons who know all about the matter. They are the parties who have been present at the inquiry, and they know what the arguments have been, what arguments have been deployed, and so on, on all sides. So the letter and to a certain extent the report are for consumption by a well-tutored audience. Bearing those matters in mind C it seems to me that it is impossible to say that in a case such as this it is incumbent upon the Secretary of State to put in a few words relating to the findings of fact. The Secretary of State said he agreed with the inspector's conclusions. What is being suggested is that he ought to have added " his findings of fact and his conclusions." I do not think this is necessary in every case as I have indicated, and I do not think it is necessary in this case. That is the substance of the complaint that the Secretary of State failed to state whether he accepted all or any of the appointed person's findings of fact. I do not think I can uphold Mr Dawson's submissions on that point.

D The second point is that he took into account alleged inadequacies of the present sewerage system but failed to take into account the duty of the water authority under section 16 of the Water Act 1973. If I may very briefly deal with the background. In 1973 the Water Act of that year made a profound difference to the position about responsibilities for sewerage. Up to 1974, when the Act came into force, responsibility for sewerage was that of district councils, or whatever their successors were, under the Public Health Act 1936. That Act by section 14 cast a duty upon a local council adequately, or effectually I think was the word, to E drain its district. It provided by section 34 that a private person could connect his property to a public sewer. By the Water Act 1973 the responsibility for sewerage was transferred from local councils to water authorities. Profound changes were made, one of the more important of which was that from then on the cost of sewerage was found not by the ratepayers of the local authority but by the consumers because the water authority charged a sewerage rate on the consumer as part of the water rate. Section 14 of the 1973 Act, although in some respects in different phraseology, really makes no difference to the general duty of effectually sewering a district except of course that it casts the duty on a water authority F instead of the local district council. But section 16 of that Act provides in effect that owners of premises may require the water authority to provide a public sewer for the drainage of new buildings which they propose to erect. And there are certain consequences when they do which can broadly be put this way, that the owner of the premises has in effect to pay for the cost of providing the new public sewer. But the difference is this, that whereas under the old provisions the only right of a private owner was to connect with an existing sewer, this section casts the duty on the authority to provide a new one, subject, as I say, broadly to the private person being ready to defray the cost.

Mr Dawson says that that profound difference has made G also a difference in relation to the whole question of prematurity due to inadequacies of the infrastructure. Because, so he says, where the inadequacies of the infrastructure are concerned with sewerage it is no longer possible to say that a development is premature because of lack of sewerage facilities. It cannot be premature for that reason because the private developer is enabled to require the water authority to provide the necessary new sewerage to drain the new premises to be built. It should, however, be noted that power to require the new sewer appears to me, by the terms of subsection (3) of section 16, to arise only when the private sewer which it is desired to drain into the new public sewer has already been H provided by the private owner. Of course you cannot do that without planning permission in the first place. You cannot construct a private sewer without planning permission. So the key to section 16 appears to me to be this, if looked at in realistic terms; that before any private individual can make any use of the powers he has now acquired under section 16 to require the provision of a new sewer, he must have planning permission for and have actually built the private sewer which he wishes to drain into the new public sewer.

As I say, Mr Dawson's argument is that this new power has, as it were, put paid to any argument on prematurity in relation to sewerage. And of course if that is so it puts paid J to the provisions of circulars 102 of 1972 and 122 of 1973 which proceed on the basis that certain things should be done or should not be done if there are inadequate sewerage arrangements available. The difficulty about Mr Dawson's argument seems to me to be this. Even if he is right, and I am at the moment perhaps not disposed to think that he is, there is absolutely no indication at all that this point was in fact put to the inspector. It forms no part as I see it, because one can look at the report by the inspector of what the appellants' case was; it forms no part of the appellants' case as reported by the inspector. I have also seen, because Mr Woolf had no objection to its going in, the proof of the only K witness, a written proof that is, of the only witness called by the appellants at the inquiry, a Mr Hopkins. Of course it is clear that he must have expanded that written proof orally at the inquiry, but again there is nothing in that proof which indicates that this point was being taken at the inquiry, and indeed without going into it it is clear to me that there are contra-indications in the inspector's report. Mr Dawson therefore is faced with the point that what he is really complaining about is that the Secretary of State failed to take into account a matter which was never put before his inspector, and on which therefore no report was made by L the inspector to the Secretary of State. Mr Dawson is then forced back to this stance: well, he says, it is a matter of law which the Secretary of State ought to have taken into account *proprio motu*, as it were, because it is background law which he ought to know about.

In the first place I am not at all sure that it is a legitimate legal argument to say—it is partly law and partly fact and partly law and partly opinion—but I do not think it is legitimate to say that because this change has been made by section 16 it follows that all arguments about prematurity have gone. It surely must remain in realistic terms that it is important to consider the orderly phasing of residential and M other development; because even though water authorities may be under the duty which has been referred to under section 16, if everybody in the country who had a piece of land that he wanted to develop was able to say: I am entitled to develop despite the lack of sewerage facilities because I can always require the water authority to lay a new sewer under section 16—if in fact they all were hammering on the doors of all the water authorities to lay such sewers all at once —it is quite clear that the civil engineering resources of this country, quite apart perhaps from its financial resources, would be grossly overstretched. It seems to me it must remain within the Secretary of State's competence to say: no, that is a disorderly way of proceeding. He must still be able to say, despite this possible requirement, that a developer should not have planning permission until something can be done about it. Mr Dawson's point however, it seems to me, is simply that whether that is so or not the Secretary of State did in fact fail to take into account this duty. There is nothing actually in the Secretary of State's decision letter which indicates whether he took it into account or not. He may have taken it into account and he may have come to the conclusion which I have just indicated. He may not have taken it into account. He may have regarded it as quite irrelevant. But whichever he did it is a matter for the applicant in a case such as this to satisfy the court that he did in fact fail to take this matter into account. I remain entirely unsatisfied about that point.

Mr Dawson suggested as an alternative that if he had taken the point into account, the decision letter disclosed no trace of that, and in those circumstances it was a failure to comply with rule 13. That may be a very nice argument, but it does not fall to be considered because it is not set out in the notice of motion. It seems to me that in those circumstances I am entitled to take the view that it is not a tenable proposition. In any event, as I say, it is really rather hard upon the Secretary of State that he should be taken to task for failing to take into account an argument which his inspector was never apprised about and which in consequence was never laid before the Secretary of State himself.

The last point taken by Mr Dawson is that the inspector failed to report on the case presented by the applicants on the alternative basis that the land was not in a growth area. This throws one back to paras 4 and 5 of the annex to the circular I have mentioned. I have already indicated in the course of argument what I think the circular means, and I have not the smallest doubt that the only logical way of construing it is that para 4 refers to land inside growth areas in strategic plans and para 5 refers to land outside growth areas in strategic plans. But that, Mr Woolf assures me, is not what the Secretary of State intended. What he intended to say in para 5 was that that paragraph applied to areas within growth areas in strategic plans which would not clearly fall to be considered for inclusion in a specific growth area plan. By some process of divination wholly unassisted by logic Mr Hopkins apparently arrived at the same conclusion. He put forward two alternative bases for argument to the inspector. The first was: let us suppose, he said, that the land was within para 4. Then the criteria which are set up by para 4 are so and so and he set them out. There are six of them and I do not think I need deal with them at all. He correctly set out the criteria which are referred to in para 4, and he went on in this way: "Having regard to their own allocation of the site for housing the planning authority obviously raise none of these objections to the proposed development." That was all he said about the para 4 hypothesis. He then went on to say: "all right, if it is not para 4, then it must be para 5." And again he set out the third criteria relevant to para 5, the third of which is that the development was not subject to valid objections on the ground of inadequate infrastructure. In dealing with that he went on, as far as one can see, to deal with the sewerage problem. It was accepted by the applicants that sewerage problems existed. The inspector nowhere refers to this alternative way of putting it under either para 4 or para 5. But I am quite satisfied that the whole of his report proceeds on the basis that this was a para 5 case, not perhaps surprisingly, because really it is very clear from the terms of Mr Hopkins' own proof that he was shrugging off the para 4 case and dealing substantially with the para 5 one. The inspector, it seems to me, dealt with it on the basis of the infrastructure argument as if it were a para 5 case, not perhaps surprisingly, as the infrastructure argument is only relevant under para 5, as I have already pointed out. The Secretary of State in his decision letter specifically says: "Although the appeal site is located in a growth area in the strategic plan . . . this is a general location only and the site does not fall to be considered for inclusion in any specific growth area plan," ie it is a para 5 case. That in fact has been what the planning authority had said at the inquiry. I find no difficulty in coming to the conclusion that the whole of the argument at the inquiry really proceeded on the basis that while the para 4 case was being made, not much if any reliance was being placed on it, and the real issue at the inquiry arose on the supposition that this was a para 5 case.

The notice of motion makes it clear that it is the failure to report on the para 5 basis that is being complained of. I can find no justification for that complaint. There is no alternative claim in the notice of motion that alternatively the inspector failed to report on the para 4 case. If there had been, then perhaps one would have had to have looked into the matter to see whether in fact there was any significant prejudice to the applicants in that failure. Of course in a case such as this where it is quite clear that that case was merely, as it were, stated and then shrugged off, there may well have been little if any prejudice to the applicants in the inspector not reporting on a matter which was so clearly not regarded as of great importance by the applicants themselves. But there is no such alternative claim in the notice of motion, and in my view this point also fails.

The result is that, entertaining as I find Mr Dawson's arguments as always, I fear I must find against him on this matter and this motion must fail.

The application was dismissed with costs.

QUEEN'S BENCH DIVISION

February 8 1979

(Before Sir Douglas FRANK QC, sitting as a deputy judge of the division)

GEORGE WIMPEY & CO LTD v NEW FOREST DISTRICT COUNCIL AND ANOTHER

Estates Gazette April 21 1979
(1979) 250 EG 249

Town and Country Planning Act 1971—Appeal under section 245 from Secretary of State's decision upholding planning authority's dismissal of builders' application for permission for residential development—Difficulty about access by construction vehicles to development site without detrimental effects of such traffic through existing housing estate roads—Secretary of State justified in holding that access agreements over other land did not give applicants "control" of such land so as to allow authority to impose a condition restricting construction traffic to the use of that land—Effect of section 30(1)(a) considered—Secretary of State at fault, however, in not applying his mind to the alternative of a negative condition under section 29 excluding access for construction traffic through specified housing estate roads—Decision quashed on latter ground

In this appeal George Wimpey & Co Ltd challenged a decision of the Secretary of State for the Environment in which he upheld the refusal of New Forest District Council to grant planning permission for residential development at Totton, Hampshire.

Iain Glidewell QC and R J Furber (instructed by P J Ward) appeared on behalf of the appellants; S D Brown and R Wilmot-Smith (instructed by the Treasury Solicitor) appeared on behalf of the second respondent, the Secretary of State; the first respondent, New Forest District Council, were not represented and took no part in the proceedings.

Giving judgment, SIR DOUGLAS FRANK QC said: This is

A an appeal under section 245 of the Town and Country Planning Act 1971 against the dismissal by the second respondent of an appeal against the refusal of planning permission by the first respondents for the residential development of 10.8 acres of land at Totton, Hampshire.

The existing access to the land concerned is by roads serving housing estates and it was because of the likely detrimental effect of construction vehicles necessary for the development of 10.8 acres that the appeal was dismissed. However, the applicants had entered into two agreements for the purpose of providing a means of access for construction traffic avoiding existing development. The first agreement B was with the owners of land (called the blue land) adjoining the western boundary of the applicants' land whereunder the owners of the blue land granted a licence to the applicants for four years from October 6 1977 for a right of way for vehicles along a strip of land referred to as the yellow land. The agreement was expressed to terminate upon the exercise of an option held by Laing Homes Ltd (referred to as "Laing") to purchase the blue land. By an agreement dated November 15 1977 Laing agreed that when they exercised their option the applicants would have a right of way on the yellow land for a period of five years from the date of the agreement. However, the agreement was subject to this C important proviso:

> Provided Always that this Agreement shall be terminable upon not less than one month's prior written notice by Laing expiring at any time if the use of the right of way by Wimpey conflicts with inconveniences or interferes with Laing's development of the blue land subject however to Laing making available if possible an alternative right of way over the blue land if this can be done without prejudicing, inconveniencing or interfering with Laing's development of the blue land.

The Secretary of State's decision was given in a letter dated April 21 1978. It is necessary only to refer to one D paragraph and that is as follows:

> 4. It is not thought that the agreements that the company have entered into which would enable construction traffic to obtain access to the site from the A336 across adjoining land are such that this land can be considered to be "under the control of the applicant" for the purposes of section 30(1)(a) of the Act. As it is not possible to deal satisfactorily with the question of access by construction traffic it is concluded that the appeal should be dismissed; but for this difficulty the Secretary of State would have been minded to allow the appeal. The Secretary of State hereby dismisses the appeal.

When the matter came before me, the only ground to be E argued was that "the second respondent misdirected himself in law in holding that two agreements of which the applicants had the benefit did not put under the control of the applicants the land the subject of the said agreements so as to enable the second respondent to impose a condition regulating the development or use of this land." The determination of this question depends on the provisions of section 30 of the Town and Country Planning Act 1971 and, in particular, these words:

> (1) Without prejudice to the generality of section 29(1) of this Act, conditions may be imposed on the grant of planning permission thereunder—
>
> F (a) for regulating the development or use of any land under the control of the applicant (whether or not it is land in respect of which the application was made) or requiring the carrying out of works on any such land, so far as appears to the local planning authority to be expedient for the purposes of or in connection with the development authorised by the permission; . . . .

Mr Glidewell submitted that the phrase "under the control of the applicant" does not involve having an estate or interest in land. He said that the land is under the control of the applicant if he has a right in law to do or cause to be done whatever the condition requires. He further said that G the relevant time for having that control is at the time when the planning permission is granted and it is irrelevant if subsequent events may render the development impossible. In answer to a further contention that the licence could not enure for the benefit of a purchaser because it was personal to the applicant, he said that if the purchaser was unable to obtain a licence he would be unable to carry out the development; that possibility was not a reason for refusing planning permission.

Mr Brown, for the Secretary of State, accepted that the word "control" must be given its ordinary meaning and that is the right to regulate the use of the land. He said that the H issue was the period of time when that right must be vested in the applicants. He submitted that at the time when the application is determined the applicants must be able to demonstrate that they have the necessary right for such period as appears to the local planning authority (or the Secretary of State on appeal) to be expedient. He argued that the contrary submission of Mr Glidewell would nullify the purpose of the subsection and he pointed out that the licence here was personal to the applicants, was potentially for one month only and in any event was limited in time.

One of the difficulties I find is in determining the purpose of limiting the power to impose conditions under this sub- J section to land under the control of the applicant. I can find no guidance in the statute and there is no other case in which the question has arisen. It seems to me that without the words the subsection would be equally effective, for if the applicant were unable to obtain the necessary control then he would not carry out the development.

One possible explanation is that the words were inserted so that conditions could not be imposed which affect land other than the applicants' to which the owner of that other land had not had an opportunity of objecting. However, that seems unlikely because the imposition of a condition could not in K any case bind the owner of that other land. It is common ground that control does not necessarily involve having an estate or interest in the land but only such right as is required to implement the condition. Accordingly, there is no dispute as to the meaning of the word "control" and it seems to me that it then becomes a question of fact and degree for the Secretary of State whether the control is of a degree and kind sufficient to satisfy him that the condition will be complied with. I think that some support for that view is contained in the words "so far as appears to the local planning authority to be expedient for the purposes of or in connection with the development authorised by the permission." It seems to me L logically to follow that the character of the applicants' control should be such as appears to be expedient for the purpose of implementing the condition. In other words, it is a matter of fact and degree for the local planning authority or the Secretary of State as the case may be. There is ample evidence for the Secretary of State to have come to the conclusion that the agreements are not such that the land can be considered to be under the control of the applicant for the purposes of section 30(1)(a) of the Act, which in this case is to impose a condition to ensure that construction traffic uses the yellow land.

That would have been the end of the matter. But a second M point arises: that the Secretary of State had not applied his mind to the question whether he could have achieved his object by a negative condition under section 29. Mr Glidewell, having considered the point, with consent added a second ground of appeal namely:

> As a matter of law the second respondent could have granted planning permission subject to a condition that there should be no access for construction traffic to and from the appeal site over Malwood Gardens and Barnsfield Crescent and he failed to consider whether such a condition would achieve the desired effect.

A Mr Brown conceded that the new ground was valid and, accordingly, I quash the decision. The effect of my decision is that the Secretary of State shall now consider whether such a condition would achieve the desired effect.

COURT OF APPEAL

April 11 1979

(Before Lord DENNING MR, Lord Justice ORMROD and Lord Justice BROWNE)

B ATTORNEY-GENERAL AT THE RELATION OF CO-OPERATIVE RETAIL SERVICES LTD v TAFF-ELY BOROUGH COUNCIL AND OTHERS

Estates Gazette May 26 1979

(1979) 250 EG 757

Town and country planning—Rival applications for planning permission for proposed hypermarket on sites not far from each other—Whether applications should, as involving a "county matter," have been reserved for the county council —Clerk to borough council without authority purported to issue planning permission to one of the applicants—
C **Effect of such action—"Ratification" by borough council —Purported grant of permission a complete nullity—Permission of no effect in hands of bona fide holder—Slough Estates Ltd v Slough Borough Council distinguished—Action incapable of ratification—Various points of planning law discussed—Decision of Sir Douglas Frank in favour of borough council and recipient of permission reversed**

This was an appeal from a decision of Sir Douglas Frank QC, sitting as a deputy judge of the Queen's Bench Division, in an action by Co-operative Retail Services Ltd against the Taff-Ely Borough Council and Tesco Stores Ltd, claiming that a purported planning permission for a hypermarket in favour
D of Tesco Stores Ltd was null and void. Sir Douglas Frank gave judgment in favour of the defendant borough council and Tesco Stores Ltd. The site to which the purported permission related was on land at Gellihirion Farm, Upper Boat, Treforest, Mid-Glamorgan. The proceedings were begun by Co-operative Retail Services Ltd as plaintiffs. Subsequently the plaintiffs obtained the Attorney-General's fiat and the action became a relator action.

L Read QC and J C Harper (instructed by Gellatlys) appeared on behalf of the appellants; G Eyre QC and V Pugh (instructed by G Hockin, clerk of the borough council) represented the Borough Council of Taff-Ely, respondents;
E L J Davies QC, G Langley and W Featherby (instructed by Berwin Leighton) represented Tesco Stores Ltd, respondents; Harry Woolf and D Latham (instructed by the Treasury Solicitor) appeared with the leave of the court on behalf of the Department of the Environment. Sir Robert McAlpine & Sons (South Wales) Ltd and Mid-Glamorgan County Council had been defendants at one time, but the action had been discontinued against them and they took no part in the appeal proceedings.

Giving judgment, LORD DENNING MR said: Superstores are coming into fashion. They are stores built on the outskirts of a town. One storey with many shops surrounded by
F a capacious car park. The shoppers drive out to them and do all their shopping in the one place.

This case concerns two planning applications for a superstore. Two rival concerns proposed to set up a superstore in the outskirts of Pontypridd in the Borough of Taff-Ely. Each of the two concerns was looking for a suitable site. Only one superstore would be allowed. So each wanted to get its own planning permission and so outdo the other. One concern was the Co-operative Retail Services Ltd. They had their eyes on a piece of rough grazing land at Heol-y-Bwnsy. On October 17 1975 Co-operative Retail Services applied for outline permission to develop it as a "trading centre for retail outlet with parking facilities." I will call it the "Co-op site." G
The other concern was Sir Robert McAlpine Ltd with Tesco in mind. They had their eyes on a piece of derelict land at Gellihirion. It was about half a mile away from the Co-op site. On March 16 1976 (five months behind their rivals) McAlpine applied for outline permission to develop it as a "superstore, garden centre and petrol filling station." I will call it the "Tesco site." But it must be remembered that Tesco Ltd were not the owners, nor were Sir Robert McAlpine, but only an unknown company called Maidsfield Properties Ltd of Cardiff.

Up till the present day Co-op have lost and Tesco have won. The Borough Council of Taff-Ely (which is a district council) have refused the Co-op application and have granted H
the Tesco application. A formal grant was issued on November 4 1976. But that grant in favour of Tesco has been challenged. The question for us is whether it was valid or not.

Outline of planning control

Much of the difficulty in the case arises because in 1975 there was a complete change in the structure of planning control. Previously the local planning authority had been the county council. They used to delegate some matters to the local districts, but still the county was the planning authority: see section 1(1) of the Town and Country Planning Act 1971. But when local government was reorganised the planning structure was reorganised too. Each "county" was a large J
area comprising several "districts." The county council (here the Glamorganshire County Council) became the planning authority for the county: and each district council (here the Borough Council of Taff-Ely) became the planning authority for its district: see section 182 of the Local Government Act 1972. This was obviously capable of producing many problems, because many applications would concern both the district and the county. So the Act contained elaborate provisions so as to ascertain what was a "district matter" and what was a "county matter." These were set out in Schedule 16, paragraphs 15 to 33, to the Local Government Act 1972: and in the Town and Country Planning General Development K
Order 1973 (1973 SI No 31) and in directions given by the minister in the Town and Country Planning (Development Plans) Direction 1975: these came into operation on October 11 1975. That was only shortly before the application for the superstore at Pontypridd. So all the local authorities found themselves wading through schedules, orders and directions. I sympathise much with them. I find myself in the same position as that described by Harman LJ in his classic passage in *Davy v Leeds Corporation* [1964] 1 WLR 1218 at pp 1224-5:

> To reach a conclusion on this matter involved the court in wading L
> through a monstrous legislative morass, staggering from stone to stone and ignoring the marsh gas exhaling from the forest of schedules lining the way on each side. I regarded it at one time, I must confess, as a slough of despond through which the court would never drag its feet, but I have, by leaping from tussock to tussock as best I might, eventually, pale and exhausted, reached the other side.

In an endeavour to summarise the position, I would say this. In the first instance every application for planning permission has to be made to the district planning authority; but they have to send a copy of it straight on to the county planning authority. If it "appears" to the district planning authority to M
relate to a "county matter," then the district planning authority cannot grant permission themselves but must refer it to the county planning authority: see paragraph 15(2) of Schedule 16 to the 1972 Act. Even if it appears to the district planning authority to relate to a district matter and not a county matter, nevertheless the county planning authority can step in and direct the district planning authority as to how they are to determine it: see paragraph 19 of Schedule 16.

The criteria for determining whether or not a matter is a "county matter" are to be found in paragraph 32. These are very difficult for anyone to understand. For one thing it is

A necessary to look at many documents which are in the keeping of the county planning authority, such as the structure plan for the area, the development plan, and statement of planning policy. In order to help the planning officers, the Mid-Glamorgan County Council had issued a development control scheme in which they advised that applications should be divided into three basic catagories—1, 2 and 3—and gave advice as to what should happen to each category.

There were good grounds in this case for thinking that an application for a superstore would be a "county matter": for it might conflict with the fundamental provisions of the development plan (paragraph 32(d)(ii) of Schedule 16) and it might be
B inconsistent with the statement of policy adopted by the county planning authority (paragraph 32(d)(iv)).

In March 1976, when the Tesco application came before the planning officer of the district council, he took the view that it was a "county matter." He placed it provisionally in a special category 2. That meant that it should not be granted except by agreement with the county council. In particular, if there was a difference between the planning officers, there should be a subcommittee of county and district members to consider the matter further.

Facts of this case

On June 7 1976 the planning committee of the district council
C considered the application for the Tesco site. They received the recommendation of the district council's planning officer in which he said that there was only one objector. It was the county planning officer, who considered it premature and unnecessary in view of similar development at Caerphilly. The district planning officer recommended that planning permission in outline be granted. He added significantly:

> If this recommendation is accepted it will be necessary to refer the matter to a joint meeting with Mid-Glamorgan County Council in accordance with the terms of the agreed development control scheme.

The planning committee appointed a subcommittee to
D consider the matter on the sites. On June 15 1976 the subcommittee met on the sites. They considered the planning officer's reports. They made their own report in these words:

> After inspecting the sites and following a discussion it was RECOMMENDED as follows:
>
> (1) That the application *(for the Co-op site)* be not granted for the reasons detailed in the planning officer's report;
>
> (2) That the application *(for the Tesco site)* be granted in outline subject to the conditions detailed in the report;
>
> and that the planning officer take the necessary action.

On July 6 1976 the district council met and made this formal minute: The report of the subcommittee "was considered and
E adopted."

Now as I read those minutes, there was *not* then a decision by the district council to grant permission for the Tesco site. The district council only recommended that the planning officer take the necessary action, that is, to refer the matter to a joint meeting with the Mid-Glamorgan County Council in accordance with the agreed development control scheme. That interpretation is borne out by a letter of July 12 1976 which the district planning officer wrote to the county planning officer:

> In accordance with the terms of the agreed development control scheme it will now be necessary to convene a joint meeting between
> F our respective authorities in view of your recommendation to reject the application.

Soon afterwards, on July 30 1976, the owner of the Tesco site agreed to sell it for £500,000 to Tesco, who paid a deposit of £50,000, but reserved liberty to rescind it if no planning permission was issued on or before March 25 1977. The joint meeting of the two authorities was arranged for Monday September 27 1976. The county council representatives all came, but only one of the district council representatives. So there was no quorum and no business could be transacted. It was agreed that consideration of the matter be deferred pending the receipt of a further report.

G *The bombshell*

Now there came a bombshell into the offices of the district council. On October 22 1976 the solicitors for the owner of the Tesco site asserted that the district council had on July 6 1976 resolved to grant planning permission for the Tesco site, and had wrongfully failed to issue the appropriate grant of permission. This was a vital matter for the owner of the Tesco site for two reasons. (1) He wanted to get his £500,000 for the land; and (2) he wanted the grant to be dated July 6 1976. This was because after August 1 1976 there was a tax in the shape of a land development charge payable. The solicitors said they had taken legal advice. They threatened to bring proceedings
H for mandamus and also to refer the matter to the local authority ombudsman; and also to claim damages in excess of £250,000. But they did not make it clear for whom they were acting. The clerk to the district council thought they were acting for Tesco, whereas in fact they were acting for the owner of the site.

The shaken clerk

The clerk to the district council was shaken to the core by this threat. Now he knew that the county council had arranged to consider the Tesco application on November 12 1976. He knew they were likely to refuse it. He thought that if they did, his district council might be liable in damages. So he took a great deal on himself. On November 4 1976 he himself issued
J the formal permission for development. He had received no authority from the council to issue it. He took it on himself. He got a printed form headed "Permission for Development." He filled in all the details, signed it, and back-dated it to July 6 1976. It ran:

> Taff-Ely Borough Council as the local planning authority hereby grant permission for the proposed development. . . .
>
> Dated the 6th day of July 1976.
> (Signed) G HOCKIN

He sent it round by hand that very morning to the solicitors for the Tesco application, who wrote to their client in these
K words:

> *Tesco Stores Limited—*
> *Land at Pontypridd*
>
> With a *sense of triumph* I produce herewith in pursuance of the contract a copy of the planning permission. . . .

It was indeed a triumph. He would get from Tesco £500,000 for the owner of the site: and he had got a grant dated July 6 1976, which saved any land development charge.

On the next day the clerk to the district council wrote this illuminating report to the finance committee of the district council. He gave his reasons frankly:

> L It was necessary to act promptly to avoid the possibility of the county council taking advantage of section 12 of the 16th Schedule to the 1972 Act giving their statutory declaration to the district planning authority as to how the district planning authority are to determine any particular application because, if that notice had been served on the district council, the district council's hands would have been tied irrevocably and the bill would have fallen as a burden on the ratepayers of this Borough.

The grant of permission was duly shown to Tesco. They were satisfied with it: and on the faith of it, on December 6 1976 they bought the site from the owner and paid £500,000 for it. The newspapers in South Wales got to know about all
M this. There was a tremendous hubbub in the press. The district council attempted to put this right. The clerk was no doubt anxious about his position in issuing the planning permission without specific authority. So he made a long report to the full district council on December 14 1976. They then resolved:

> That the reports of the officers be accepted; that the action taken by the clerk in issuing the planning consent be confirmed, and that it be recorded that there is no reflection on the integrity and diligence of the officers of the authority.

Sir Douglas Frank QC made a stern comment. He said:

> Although the clerk's integrity is not in question, I find it impossible to justify the deliberate flouting of the agreement with the county

A council and the district council's subsequent condonement of his actions. The clerk to the county council was right to be incensed.

The Co-op take action

The Co-operative Retail Services Ltd were much aggrieved by all that had happened. Their own application had been turned down by the district council on July 6 1976. They had appealed to the minister. But there was room for only one superstore near Pontypridd. Their own appeal would inevitably be rejected if the Tesco grant were valid. So on December 30 1976 the Co-op issued a writ in their own name claiming that the grant to Tesco was invalid and of no effect. Soon afterwards they realised that it was a matter of public
B concern. So they related the facts to the Attorney-General: and he has granted his fiat for relator proceedings. The trial judge, Sir Douglas Frank QC, has held that the grant to Tesco was valid. There is now an appeal to this court.

Was the "grant" valid?

To my mind the purported "grant" back-dated to July 6 1976 was utterly invalid. It would have been valid if the district council on July 6 1976 had resolved on that day to grant permission themselves for the Tesco site. But, reading all the minutes, the district council did not grant permission on that day. It appears from the report of the planning officer that this case came within category 2—and that no grant should be made
C except after consultation with the county council. Arrangements were made accordingly for a joint meeting on September 27 1976. That was fruitless for want of a quorum. So the next step was to wait till November 12 1976 when the county council would consider the matter. When the clerk to the district council on November 4 1976 issued the grant of permission, he did so without any authority. He did it because he was shattered by the bombshell. He thought that the district council might be liable in a huge sum. So he deliberately issued the permission before the county council could consider the matter. That was an entirely mistaken step by him.

D *Must it be taken at its face value?*

It was argued that, despite its invalidity, the printed grant of permission signed by the clerk was good in the hands of any bona fide holder of it. Tesco (who bought the land from the owner) were bona fide holders of the grant and should be entitled to rely on it. Reliance was placed on my words in *Slough Estates Ltd v Slough Borough Council* [1969] 2 Ch 305 at p 315:

> The grant of planning permission has to be in writing . . . and it runs with the land. The grant is not made when the county council resolve to give permission. It is only made when their clerk, on their authority, issues the permission to the applicant. Seeing
> E that it has to be in writing, you can only look to the permission itself and the documents incorporated in it.

And on the words of Lord Reid in the House of Lords [1971] AC 958 at p 962 when he said that a written grant of planning permission "is available to purchasers from the person who originally obtained the permission. They may have no means of discovering what facts were known to the planning authority." Those words are truly applicable to a grant of planning permission which has been authorised by the planning authority. But they have no application to a grant which has never been authorised: and is issued by the clerk of his own head. It is then no better than a grant which was issued by a mistake—
F issuing a grant instead of a refusal, as in *Norfolk County Council v Secretary of State for the Environment* [1973] 1 WLR 1400; or when the signature of the clerk has been forged or written by a subordinate without any authority whatever. In all these cases the purported grant is *ultra vires* and void and of no legal effect whatever. It is true that until the reason for its invalidity is discovered—and pronounced upon by the courts —people may in good faith have acted on it. The courts may in their discretion allow these actions to stand as between innocent third persons. But that does not serve to validate the invalidity or to resurrect the nullity: see the discussion by Professor Wade in his book on *Administrative Law*, 4th ed, pp 296-301.

Applied to this case the grant of planning permission was G
ultra vires and void. So Tesco cannot rely on it. They cannot build a superstore on the faith of it. But the conveyance of the site by the owner may have been good: unless it can be avoided on the ground of mutual mistake—or some other ground; or damages obtained from someone or other.

The supposed ratification

It was suggested that the resolution of the district council on December 14 1976 amounted to a ratification of the clerk's action on November 4 1976. This cannot be right. It only "confirmed" the clerk's action. In any case, however, an act which is *ultra vires* and void cannot be ratified: see *Ashbury Railway Carriage and Iron Co v Riche* (1875) LR 7 H
HL 653. The district council could not possibly retrospectively validate the permission—so as to make it speak as of July 6 1976.

Paragraph 51

I ought to say a word about paragraph 51 of Schedule 16 to the Local Government Act 1972. That is inserted so as to cope with cases where the two planning authorities have got mixed up—as to which of them should grant planning permission. In the complexities of the statute, a district council may grant planning permission when it should have been the county council or vice versa. Paragraph 51 provides that the validity of it shall not be challenged on the ground that it J
should have been done by the other. It has no application to a case where neither of them had power to do it. Nor to the present case where the grant was entirely unauthorised and of no effect. The validity of the grant is here challenged on the ground that it was *ultra vires*: not on the ground that the other should have done it.

Conclusion

Underlying this case there is an important principle of planning law. It is that a grant of planning permission is made in the public interest—so as to ensure that the amenities of our countryside are preserved for the good of all. The protection of the public interest is entrusted to the represent- K
ative bodies and to the ministers. It would be quite wrong that it should be pre-empted by a mistaken issue by a clerk of a printed form—without any authority in that behalf. When such a mistake is brought to the attention of the court by the Attorney-General, or anyone else with a sufficient interest, the court can and should declare the grant to be void and of no effect. It does not follow that all the consequences that follow are void also. Thus the conveyance here by the owners to Tesco is not automatically avoided; that and other consequences must be left to be worked out by the parties themselves. But so far as the representative bodies themselves are concerned, they should not be bound by mistakes made by the L
clerk or people in the office—when the result would be to damage the interests of the public at large.

My conclusion is therefore that the grant of permission of the Tesco site was invalid and a nullity. The court should grant a declaration accordingly. It follows that the application for planning permission has never been determined. This opens the way to a just solution. The minister should call in both applications—both the Co-op and Tesco—and should determine them on their merits.

In these circumstances it is unnecessary to go on and consider all the nice points of planning law which were canvassed before us. I would only say that I agree with all that M
Ormrod LJ and Browne LJ say about these.

ORMROD LJ and BROWNE LJ delivered concurring judgments.

The appeal was allowed and declaration made that the borough council did not resolve to grant planning permission and that the notice given on November 4, but dated July 6 1976, purporting to grant planning permission, was of no effect. The appellants were awarded costs against both the borough council and Tesco Stores Ltd, the allocation of costs between them to be dealt with in contribution proceedings. The Department of the Environment was ordered to bear its own costs. Leave was given to appeal to the House of Lords.

VENDOR AND PURCHASER

A **CHANCERY DIVISION**

October 25 1978

(Before Mr Justice TEMPLEMAN)

TOPFELL LTD v GALLEY PROPERTIES LTD

Estates Gazette January 27 1979

(1978) 249 EG 341

Action for specific performance of contract of sale of house with an abatement of the purchase price—House advertised for sale with vacant possession of ground floor but subject to a tenancy of the first floor—Notice under section 19 of Housing Act 1961 that house was to be occupied by one B family only—Notice not disclosed to plaintiffs by defendants, but no allegation of concealment or fraud—Meaning of "vacant possession"—Not merely vacant physically, but available for occupation and enjoyment—Ground floor not available for occupation owing to Housing Act restriction —Specific performance ordered in purchasers' action with abatement of £1,000 in purchase price of £3,850

In this action the plaintiffs, Topfell Ltd, who had contracted to purchase a house at 9 Donald Road, West Croydon, sought a decree of specific performance subject to an abatement of the purchase price against the defendants, C Galley Properties Ltd, the vendors.

N B Primost (instructed by Rayner & Co) appeared on behalf of the plaintiffs; J R Reid (instructed by Armstrong & Co) represented the defendants.

Giving judgment, TEMPLEMAN J said: This case revolves round the meaning of the words "vacant possession."

In May 1976 Mr Firster, a director of the plaintiff company, received auction particulars relating to the house 9 Donald Road, West Croydon. On inspection, he found a tenant on the first floor and he found that the ground floor was unoccupied. He telephoned the local authority and was told that the house was situate in an improvement area but D he was told nothing else. He attended the auction on June 9 1976 but the house was withdrawn; apparently, it did not reach the reserve price. Mr Firster then attempted to negotiate by private treaty, but what he thought the place was worth did not accord with what the vendors thought and he lost interest.

After that, by a notice dated June 24 1976, the Croydon Council served on the owner of the house and on the house itself notice under section 19 of the Housing Act 1961 whereby they directed that the house should only be occupied by one household. I find that the defendants knew about this notice. Subsequently, they never disclosed it to the plaintiffs when E they became interested again, but it is right to say that there is no allegation of concealment or fraud. Exactly what happened I do not know and it is immaterial, but it probably was accidental.

In January 1977 Mr Firster noticed that, unusually, the house was up for auction again. He is a man who is experienced in property dealing and buying and renovating and exploiting generally. He said that it was unusual for a house to come up a second time for auction. He read the particulars and went to look at the house again, but he could not get in because the first-floor tenant was out. But, as far as F he could see, the situation and state and condition of the house were exactly the same as they had been when he had been there the previous year. He again took the precaution of telephoning the local authority, but was not told about the notice under the Housing Act or given any other information which might have deterred him from buying at auction. He says he did not make a personal search against the local land registry or make personal inquiries because, in his experience, it was a waste of time to do so. Either he would have been given very sparse information or he would have waited a long time to no good purpose. He did not make a requisition for written searches because there was no time between the G date he received the particulars and the date of the auction for him to get the replies back. He says he took the precautions which he usually takes and which usually stand him in good stead.

The second auction took place on January 27 1977, but the property was again withdrawn, the reserve price not being reached. After the auction Mr Firster again thought it worth his while to offer to negotiate by private treaty and he offered £3,500. The auctioneer, after taking instructions, asked for £3,850. Mr Firster demurred and thought that was on the high side, but the auctioneer said that it was good value; H where else could you get part possession? Eventually, Mr Firster agreed to pay £3,850 and signed a contract on behalf of the plaintiff company. As is usual in these matters after the auction, he signed the contract at the back of the auction brochure which incorporated the particulars which had been distributed for the purposes of the auction. At that stage, Mr Firster thought that he would have to spend about £1,400 on the property and then he could do one of two things. Either he could lease the ground flat for 99 years or, if this was not possible or desirable through shortage of money or something of that sort, then he would let the ground floor on a furnished tenancy to students or other persons who in J his judgment, and he was an expert in these matters, were not likely to be other than temporary occupants. Thus, when the tenant of the first floor, a lady whom he knew to be about 60 or 65 as he thought, living on her own, left or died, then or shortly thereafter he hoped to get vacant possession from his furnished tenants and sell the house with full vacant possession. In that event, there is no doubt he would have made a handsome profit.

Mr Firster knew what he was about, gave evidence very clearly and I accept his evidence. However, after Mr Firster had signed the contract on behalf of the plaintiff company, he went to see the house again and he found pinned to the K inside of the front door a copy of the notice under the Housing Act. His solicitors at once took up the matter with the vendors' solicitors and there the matter lies. The plaintiffs now sue for specific performance with an abatement of the purchase price and that action is defended on various grounds. In the alternative, the plaintiffs ask for damages for breach of contract.

The particulars of the auction which were incorporated in the contract and which were seen by Mr Firster and accorded with his knowledge of the interior set-out of the house by virtue of his visit, after setting out that it was a freehold terraced house, set out the accommodation which is two rooms, a kitchen and a separate wc on the first floor and three rooms, a scullery, outside wc and gardens front and rear. It continues:

> The first floor is let to Mrs E E Hiscock at a rental of £3.70 per week inclusive. . . .

and then, in bold black typing in capital letters, it says:

> WITH VACANT POSSESSION OF THE GROUND FLOOR.

Then it gives various other pieces of information which are not material.

In these circumstances it is rightly conceded that the vendors became contractually bound to give vacant possession of the ground floor. In the first place, the particulars say so and, in the second place, the special conditions of the contract say the property is sold subject to the tenancy mentioned in the particulars of sale. That, to my mind, reinforces the view that they were contracting to give vacant possession of the ground floor. One of the present conflicts is whether the defendants are contractually able to give vacant possession of the ground floor pursuant to their contract. The defendants say that all that vacant possession of the ground floor means is that it is empty and that, physically, there is nothing to stop the plaintiffs from entering it. The plaintiffs, on their part, say that vacant possession of the ground floor is a contractual requirement on the vendors which binds the vendors to deliver the property in a state in which it can be enjoyed. Vacant possession, they say, is the right to occupy and enjoy the property either by the purchaser himself or by his tenants or licensees.

In my judgment, the plaintiffs are right about this. The meaning of the words "vacant possession" can, I think, vary from context to context, but the background to this case is that, to all outward appearance, the house consisted of two separate occupations. To all outward appearance, the ground floor was capable of being occupied by the owner or his tenants and licensees. When the contract boasted that vacant possession of the ground floor would be given, it was saying that the great advantage of this property is that you can occupy the ground floor. In my judgment, when the vendors said they would give vacant possession in the context of these particulars in this contract and in the context of this property, the vendors cannot now say, "Oh, no; all we intended and all we contracted to give was the right to possession in the negative sense. There is no rubbish on the floor, no other tenants and nobody else was there. It was vacant."

I have come to the conclusion that the vendors were contractually bound, on completion, to hand over the ground floor in a condition which would allow the plaintiffs to occupy it. It is quite plain that at the date of the contract and at the date fixed for completion, the vendors cannot do that because, by reason of the Housing Act direction, in fact, nobody can occupy the ground floor. The vendors cannot occupy it themselves, they cannot sell it to somebody who wishes to purchase it in order to go and live there himself and they cannot let it.

To deal with the legal points first, Mr Reid submitted that even if the contract and the particulars did bind the vendors to give possession in the sense of handing over property which could legally and rightly be occupied, they are excused by the General Conditions of Sale and by the Special Conditions of Sale. The General Conditions of Sale on which he relies state in the usual form that the purchaser shall buy subject to any notices, whether registered or not. General Condition 9 says that the purchaser shall be deemed to have made local searches and inquiries and to have knowledge of matters that would be disclosed thereby and shall purchase subject to such matters. Mr Reid says that if the purchaser had in fact made local searches and inquiries there would have come to light the direction imposed under the Housing Act. Special Condition 10(15) provides that the purchaser shall be deemed to purchase subject to his acknowledgment that the contract is not entered into in reliance upon any representation whether written or oral made but not contained in its contract and he says that the statement in the particulars was a mere representation. Finally, he points to General Condition 10, which provides that the property is believed to be correctly described and no difference or discrepancy in the quantity or description shall entitle the purchaser to be discharged from his purchase nor shall the vendor or purchaser be allowed any compensation in respect thereof.

In my judgment, these Special and General Conditions cannot be allowed to contradict the contractual obligation into which the vendors entered by virtue of the particulars and the contract, the contractual obligation to give vacant possession, in the sense in which I have defined it, of the ground floor to the purchasers.

If there was a breach of contract, what is the appropriate remedy? The defendants called a valuer who gave evidence that the property was in 1977, in his view, worth £4,500. He gave evidence that, in his view, the property was worth about that sum whether it was subject to a direction under the Housing Act or not. He says that if he had been advising a purchaser, he would have allowed him to bid on the terms that he would have to spend a little money in drawing up plans but then he would have been able to get the Housing Act direction lifted and, at the same time, to get planning permission for the first floor and the ground floor to be used as two separate occupations. He says that the plaintiffs have no real grievance. First of all, they bought a bargain for £3,850 instead of £4,500 and, secondly, the imposition of the Housing Act direction did not make any difference.

The plaintiffs' valuer takes a different view. He says that the house was worth what it was sold for, namely £3,850. I find that it was. The defendants, who I gather dabble in property and have a manager, who gave evidence, of perhaps some 26 years' experience of property, chose to put the property up for auction not once but twice. They themselves now say through counsel that it would have been better to have sold by private treaty and they would have got the higher price which their valuer stated. I am not prepared to disregard the plain evidence that the defendants, who were not in this business philanthropically or ignorantly, did what they thought was right to secure the best price and the best price they could get was £3,850. The plaintiffs' valuer also says that there is a very great difference between a property subject to the Housing Act direction and one that is not, that if the vendors, consistently with their contract, had been able to deliver vacant possession so that the purchaser could go into occupation straight away, then the property would have been worth the sum of £3,850. But, he says, it stands to reason that with a Housing Act direction, the property was worth a good deal less. In the first place, the property could not be occupied immediately and, in the second place, the purchaser would have to spend on repairs and in providing facilities not the amount that he thought fit or proper or desirable, but the amount which was necessary to obtain the lifting of the Housing Act direction.

I accept the evidence of the defendants that it would have been, or it was, possible to get rid of the Housing Act direction almost certainly—one can never be certain about these things —and at the same time to get planning permission, but, as the defendants' own valuer said, it would require some expenditure and works, which would in effect be dictated by the local authority instead of being dictated by the purchaser. The defendants' valuer says that all that is by the way because a purchaser would buy and would assess a price on hope value; the hope that the first floor will fall vacant within the foreseeable future. To get that hope value, he will have to keep the ground floor vacant anyway.

A I do not accept that view. It seems to me that Mr Firster and other people interested in property of this sort would want to get some sort of income from the ground floor in the meantime and that any purchaser would either occupy it himself or do as Mr Firster planned to do, namely, to set out to tidy up the ground floor and then either sell it or wait for the upstairs tenant to leave.

It does not seem to me that it is consistent with commonsense to say that the price which this property fetched by boasting of vacant possession of the ground floor is no more than the property would have fetched if, instead, the vendors had disclosed the existence of the Housing Act direction. The
B difference between the two which is the proper measure of abatement is put by the plaintiffs' valuer at £1,600; it is put by the defendants' valuer at nil. That poses a little difficulty because both valuers gave credible evidence; both had some experience and yet they are poles apart. Because they are poles apart, I have to make up my mind with very little assistance except the maximum and minimum figures. £1,600 is a large proportion of the £3,850. In any event, the plaintiffs knew that they had to spend some money on this property with some delay while they brought it into reasonable condition and provide facilities. They might still have to get planning permission and even in what was said to be a
C depressed period of 1977 so far as house property is concerned, this is property in one of the suburbs of outer London.

Whether Mr Firster would have bought at all, I need not pause to consider. I have to consider what, he having bought and having wanted to go on with his purchase, is the sort of price that he might have been willing to pay, or the hypothetical reasonable purchaser might have been willing to pay, with full knowledge of the direction under the Housing Act. Having regard to the tenancy of the upper floor, the fact that it was occupied by one lady of middle age or more, the fact that the evidence of both sides is that the Housing Act direc-
D tion would be lifted or would stand a good chance of being lifted subject to the expenditure of money, I have come to the conclusion that the hypothetical purchaser would have paid £1,000 less than Mr Firster did. He would have been willing to offer £2,850, which is not a great deal of money by these standards, in the hope of getting vacant possession of the whole in the long run and in the hope and expectation of being able to make something of the ground floor even in the short term and even if he might have to spend more money than he would have otherwise have contemplated. If that is right, £1,000 is the measure of abatement and on the plaintiffs' alternative submission that they ought to have had
E damages for misrepresentation, £1,000 would represent the damages which would be attributable to any misrepresentation. But, in my judgment, the plaintiffs satisfy me that the defendants are in breach of contract, that this is a proper case in which specific performance ought to be ordered with abatement and that the proper sum for the abatement is £1,000.

The plaintiffs were granted an order for specific performance of the contract of sale subject to an abatement of £1,000 in the purchase price and were awarded the costs of the action.

F

COURT OF APPEAL

October 31 1978

(Before Lord Justice BUCKLEY and Lord Justice EVELEIGH)

UNIVERSAL CORPORATION v FIVE WAYS PROPERTIES LTD

Estates Gazette May 5 1979

(1978) 250 EG 447

Interlocutory appeal from order of judge striking out writ and statement of claim—Contract for the sale of property—Purchaser, a foreign company, unable to finance completion after time made essence—Vendor gave notice rescinding contract and forfeiting deposit—Purchaser claimed return of deposit in action which the judge considered was bound to fail—Construction of section 49(2) of Law of Property Act 1925 and Condition 22 of National Conditions of Sale —Jurisdiction under section 49(2) to be exercised when justice of the case requires it—Trial of action should not be prejudged—It might be just to order repayment of deposit, leaving vendor to remedy in damages—Judge's order set aside G

This was an appeal from an order of Walton J striking out the plaintiff's writ and statement of claim under Order 18, rule 19 or under the court's inherent jurisdiction in an action for the repayment of a deposit. Walton J held that the statement of claim disclosed no reasonable cause of action. The facts giving rise to the action are explained in the judgment of Buckley LJ. H

Terence Cullen QC and Anthony Thompson (instructed by Heald & Nickinson) appeared on behalf of the appellants; Gerald Godfrey QC and J C Harper (instructed by Linklaters & Paines) represented the respondents.

Giving judgment, BUCKLEY LJ said: This is an appeal from an order of Walton J, whereby he struck out the writ and statement of claim in this action under Order 18, rule 19, or under the inherent jurisdiction of the court, on the ground that the plaintiff was bound to fail—that the action was one J which could not possibly succeed.

The case arises on a contract for the sale of a piece of property in London to the plaintiff company at the price of £885,000, a deposit of £88,500 being paid on exchange of contracts. The facts, and statement of claim which has been served, are set out carefully by the learned judge in his judgment in the court below and I do not propose to recapitulate that material.

The plaintiff's case for the repayment of the deposit is founded upon two contentions: first, that the contract was frustrated, and secondly a claim under the Law of Property Act 1925, section 49(2). The plaintiff company is a Liberian K company; it proposed to finance the purchase out of funds in Nigeria. It has no place of business in this country and apparently it has no funds at its disposal in this country.

Unhappily for the plaintiff company, owing to complications arising out of a change in the Nigerian exchange control law, the plaintiff was unable to have the necessary moneys at its disposal in London by the date fixed for completion after time had been made of the essence of the contract. Consequently, the defendant vendor gave notice under Condition 22 of the National Conditions of Sale, 19th ed, which were incorporated in the contract, rescinding the contract and forfeiting the deposit; that was on February 22 1978. L

The money became available to the plaintiff company in London on March 6 1978, 12 days after the notice of rescission. On March 12 the plaintiff issued its writ, and on March 29 the defendants moved to strike out the writ and statement of claim. On April 28 the judge made the order appealed from.

The judge dealt with the topic of frustration quite shortly; he said: "But quite emphatically the doctrine of frustration cannot be brought into play merely because the purchaser finds, for whatever reason, he has not got the money to complete the contract." That seems to me to be an accurate and proper statement. Certainly the plaintiff was unable, by M reason of matters beyond its control, to complete the contract when it should have done so, but this is something quite different from the contract having become incapable of performance; nor, in my view, can it be suggested that anything had happened to make the performance of the contract, in the circumstances existing at the date for completion, significantly different from what was contracted for: see *Davis Contractors Ltd v Fareham Urban District Council* [1956] AC 696 at p 729. On the material before the court, in my judgment, no frustration is shown to have occurred.

The alternative claim is put forward under section 49(2) of

the Law of Property Act 1925, which is in these terms: "Where the court refuses to grant specific performance of a contract, or in any action for the return of a deposit, the court may, if it thinks fit, order the repayment of any deposit." The judge thought that it would be surprising if Parliament had, by that provision, conferred upon the courts an absolute unfettered jurisdiction to interfere in a bargain between vendor and purchaser whenever it thought fit to do so, without the remotest hint of any guidelines as to how what he described as "that quite extraordinary jurisdiction" should be exercised.

He reached a preferred view on the construction of the subsection, that the jurisdiction exists only where the vendor could, but has not, sued for specific performance, upon which basis the learned judge said that the subsection was simply not applicable to a case of the present nature, which is one in which the vendor, had he issued in time the writ for specific performance, must have been granted such a decree beyond all question. With deference to Walton J, this seems to me to confuse the question in what circumstances the jurisdiction exists with the question in what circumstances the court will exercise it in favour of the purchaser.

The judge, however, went on to say that if he were wrong in his preferred view as being too narrow, it was obvious that there must be severe limits on the operation of the subsection. It is not, he said, designed simply to do justice between vendor and purchaser. Looking at the subsection in its terms, having regard to the language used in it and without regard to extraneous considerations, this seems to be precisely what the section in fact is. By way of supporting, or establishing, his view, the learned judge asked: "When can it be just and equitable to deprive the vendor of the whole of the moneys to which he is at law entitled?" It seems to me that this is, or may be, precisely the problem which the section presents to the court, for it confers upon the judge a discretion which is unqualified by any language of the subsection, to order or refuse repayment of the deposit—a discretion which must, of course, be exercised judicially, and with regard to all relevant considerations, including the very important consideration of the terms of the contract into which the parties have chosen to enter.

With respect to the learned judge, it does not seem to me to follow, as he thought, that a purchaser can only succeed in a claim to repayment of the deposit if the vendor's conduct has been open to criticism in some way, which Walton J described as having some mark of equitable disfavour attached to it.

In the course of argument, we have been referred to the principle of construction which is stated in *Maxwell on the Interpretation of Statutes* at p 86, where it is said that "it is a canon of interpretation that all words, if they be general and not precise, are to be restricted to the fitness of the matter; that is, to be construed as particular if the intention be particular." That, as I understand it, is a reference to the well-known doctrine of having regard to the mischief which the enactment is intended to deal with; but that doctrine, as I understand it, does not entitle the court to disregard the plain and natural meaning of wide general terms in a statute. If the language is equivocal and requires construction, then the doctrine is a proper one to refer to; but if the language is quite plain, then the duty of the court is to give effect to what Parliament has said, and it seems to me that in the present case Parliament has conferred a wide and general discretion.

I prefer to the learned judge's approach to the construction of this subsection the approach of Megarry VC, who has expressed the view that the jurisdiction is one to be exercised where the justice of the case requires: see what he said in *Schindler v Pigault* (1975) 30 P&CR 328 at p 336. In this connection I take the word "justice" to be used in a wide sense, indicating that repayment must be ordered in any circumstances which make this the fairest course between the two parties. It is, I think, relevant in the present case that Condition 22 of the National Conditions does not confer on the vendor an unqualified right to forfeit a deposit. The words in paragraph 3 of the condition are: "... the purchaser's deposit may be forfeited (unless the court otherwise directs)." This formula may well have been adopted with the terms of section 42(2) in mind. However that may be, in my view the language makes clear that the vendor has not an absolute right to retain a deposit paid by a purchaser who is in default under the condition.

Walton J properly drew attention to the fact that the statement of claim contains no suggestion that the defendant knew that the plaintiff was going to obtain its finance for the purchase from Nigeria. That position has changed since the matter was before the learned judge, for counsel for the plaintiff has now stated that he is instructed that the plaintiff is in a position to allege that the defendant was at all relevant times aware that the plaintiff needed to bring funds from abroad in order to complete the purchase, and he has sought leave to amend the statement of claim accordingly. It is not for us, on this appeal, to attempt to prejudge the issues in the action, but we must remember that the statement of claim should be struck out only if it is clear that the claim cannot succeed, or that it is in some other way an abuse of the process of the court. It is not clear to me that when the circumstances are investigated at the trial, the trial judge might not justifiably reach the conclusion that, having regard to all the relevant circumstances, including those in which the notice of rescission and forfeiture were given, it would not be more just to order repayment of the deposit, leaving the defendant such remedy in damages as may be available to it, than to allow it to retain the very substantial deposit which was paid in this case.

I would allow this appeal.

Agreeing that the appeal should be allowed, EVELEIGH LJ said: I am not wholly confident of the extent of the jurisdiction of the judge under section 49(2) of the Law of Property Act 1925. But I am confident that the limit contended for by the respondent is not plain and obvious. That being so, in my judgment the procedure adopted was wrong. If there is an arguable point of law, which there clearly is in this case, then the matter should more appropriately be set down under Order 33, rule 3.

This case raises not only the question of construction of section 49, but also the effect of para 22 of the National Conditions of Sale, with the important words in that paragraph: "... the purchaser's deposit may be forfeited (unless the court otherwise directs)." The precise effect of those words, either standing alone in a contract, or perhaps read in conjunction with section 49(2), will give rise, I have little doubt, to a great deal of argument on either side.

It has been submitted in this case that section 49(2), although it has general words, has a specific intent, and consequently those words should be read in a narrow sense. That submission, as I see it, begs the whole question of what the intent of section 49(2) is. For those reasons, therefore, I would allow this appeal.

In so far as the question of frustration is concerned, I agree with all that has been said on that matter by my Lord.

The appeal was allowed and the order of the judge striking out the writ and statement of claim set aside. Certain consequential directions were given as to the statement of claim, and the costs in the Court of Appeal and below were reserved to the trial judge.

LANDS TRIBUNAL

COMPENSATION

A **SHARIF AND ANOTHER v BIRMINGHAM CITY COUNCIL**

(REF/164/1977)

(E C STRATHON FRICS)

July 21 1978

Estates Gazette January 13 1979

(1978) 249 EG 147

Underleasehold interest in Birmingham three-storey terraced house—Benefit of valid notice to enfranchise under Leasehold Reform Act—Acquiring authority valuer's deduction of contingency cost of enfranchisement and end-deduction B **for "time and trouble" of enfranchisement both rejected by tribunal—Award on basis of freehold value less enfranchisement price and surveyor's and legal fees**

C A Coode (instructed by Grazebrooks, of Birmingham) appeared for the claimants; J M Kilbey, assistant solicitor to Birmingham City Council, for the acquiring authority.

This is a reference to determine the compensation payable by the City of Birmingham District Council to M Sharif and A Rashid for the compulsory acquisition of their underleasehold interest in a house known as 44 Highgate Road, Sparkbrook, Birmingham.

The underleasehold interest is for a term of 99 years (less C six days) from September 29 1876 at a ground rent of £5 per annum, with full vacant possession available and with the benefit of a valid notice to enfranchise served on March 5 1975 under the provisions of the Leasehold Reform Act 1967, as amended. Therefore, at the date of the service of the notice, the underlease term had six months and three weeks unexpired. The City of Birmingham District Council own the freehold interest.

The property was comprised in an area declared as a clearance area on October 25 1973 and subsequently in the Birmingham (Highgate Road) Housing Compulsory Purchase Order (no 12) 1974 made by the council on November 22 1974 D pursuant to Part III of the Housing Act 1957. The order was confirmed with modification by the Secretary of State for the Environment on May 1 1975.

Notice to treat was served on September 10 1975, 13 days before the expiration of the term of the lease; the council entered the property on October 13 1975 (20 days after the expiry of the lease term) and this is agreed to be the date for the assessment of compensation; the house has been demolished.

The subject property is situated on the south side of Highgate Road, approximately one and three-quarters of a mile south east of the city centre. The house is described as E being approximately 100 years old and comprises a three-storey terraced dwelling-house of brick construction with a slate roof; it is in generally poor structural and decorative order; it has a rear yard and garden with a pedestrian right-of-way; there are two living rooms, two bedrooms, a bathroom, kitchen, two attics, a cellar and outside wc. The property is zoned for residential purposes in the approved development plan.

The valuations of D L Cariss FRICS for the claimants and P J A Taylor FRICS, principal valuer in the City of

D L Cariss FRICS (for the claimants) F

Freehold value	£2,500
Less Cost of enfranchisement	£625
Value at date of entry	£1,875

P J A Taylor FRICS (for the acquiring authority)

Value of freehold absolute with vacant possession		£2,500
Less Cost of enfranchisement:		
Reasonable minimum enfranchisement price	£625	
Appropriate surveyor's and legal fees ...	125	
Estimated total reasonable cost	750	
Allow for contingency that actual cost will exceed this figure, say 25% of £750	187	937
		1,563
Less deduction to reflect present imperfect title, say 10% of value of freehold absolute		250
Market value of claimants' interest ...		£1,313
		£1,325

G H

Birmingham estates department, are set out above.

(The deduction of £250 at the end of Mr Taylor's valuation was retitled by him during the hearing to "time and trouble of enfranchisement.")

Items which have been agreed by the two witnesses are the freehold value £2,500 and the enfranchisement price £625. J The witnesses also agreed that if compulsory purchase had not taken place the legal costs and surveyor's fees on an enfranchisement would have been £125.

The issues are thus identified as the following deductions which are made by Mr Taylor in his valuation:

£125 surveyor's-legal fees as a cost of enfranchisement;

£187 as a contingency cost of enfranchisement;

£250 end-deduction for "time and trouble of enfranchisement."

The parties agree that compensation is to be assessed in K accordance with the provisions of the Second Schedule to the Housing Act 1957 (as amended by the Fifth Schedule to the Housing Act 1969), section 5(6)(b) of the Leasehold Reform Act 1967 and the Land Compensation Act 1961. The claimants have established their right to a payment as owner-occupiers pursuant to the Fifth Schedule to the Housing Act 1969; it can be seen that the witnesses arrive at their respective opinions of the full compulsory purchase value of the claimants' interest, but for simplicity they do not proceed to identify the amount of the payment by apportioning out of the full value the amount of the site value which otherwise would have been payable.

The matter to be determined is the amount which the claimants' interest if sold in the open market by a willing seller might be expected to realise on October 13 1975. The interest is to be sold with the benefit of a valid notice of a desire to have the freehold served under the provisions of the Leasehold Reform Act 1967. I will deal with all the value items in the order in which they appear in the valuations.

	Mr Cariss £	*Mr Taylor* £
(1) *Freehold value with vacant possession*	2,500	2,500

This figure was not qualified by either valuer; I accept it.

(2) *Cost of enfranchisement*

	Mr Cariss £	*Mr Taylor* £
(a) *Enfranchisement price*	625	625

Neither valuer sought to invoke a deferment factor in respect of the six months' unexpired term of the lease, which would have produced a figure of £610. Mr Taylor describes this item as "reasonable minimum enfranchisement price." It is subsequently included in a total figure which is described as "total reasonable cost" and I am satisfied that both valuers regard £625 as the reasonable enfranchisement cost. I so accept it.

	Mr Cariss £	*Mr Taylor* £
(b) *Appropriate surveyor's and legal fees*	–	125

In a note to his valuation Mr Cariss had stated that if compulsory purchase had not taken place these costs and fees would not have been incurred. I think a purchaser of the interest would take into account this liability as part of the cost of enfranchisement. I therefore accept Mr Taylor's treatment of this item.

	Mr Cariss £	*Mr Taylor* £
(c) *Estimated total reasonable cost* ..	625	750

This item is so described by Mr Taylor, and as it is the sum of items (a) and (b), I accept Mr Taylor's figure of £750.

	Mr Cariss £	*Mr Taylor* £
(d) *Contingency allowance*	—	187

Mr Taylor advances two reasons for making a contingency allowance, namely (1) the £625 enfranchisement price itself might prove to be underestimated and (2) a price in excess of £625 might have to be paid in order to reach agreement without recourse to a reference to the Lands Tribunal. As to the first reason, the values agreed are reasonable estimates and Mr Taylor himself entitles the item £750 as "estimated total reasonable cost." Mr Cariss also considers the estimated price to be reasonable. In my view, values which are agreed to be reasonable by two surveyors of considerable experience in the specialist field of leasehold enfranchisement, should be accepted as their considered judgments and no such contingency adjustment is required to be made. As to the second reason, the owner of the freehold interest is the City of Birmingham District Council; I note that the council is so described in the schedule to the notice to treat. The hypothetical purchaser would be bidding for a known interest in land with the benefit of an actual valid notice of a desire to have the freehold from the council. Mr Taylor conceded that it is the policy of the council to negotiate reasonable enfranchisement prices with leaseholders. Mr Cariss said that a leaseholder of the "fag-end of a lease," if negotiating with "a difficult" freeholder, might pay £150 above a reasonable price in order to reach a settlement. In my view the prospective purchaser would expect to have a reasonable enfranchisement price negotiated with the council; he would not expect to negotiate in the climate which Mr Cariss described. I therefore decide that a contingency allowance should not be made.

	Mr Cariss £	*Mr Taylor* £
(3) *Time and trouble of enfranchisement*	—	250

This item is intended to take into account a purchaser's anxiety and risk. There may be cases where a bid should reflect these matters, but in circumstances of this case, a purchaser would know the reasonable enfranchisement price which his advisers expect to negotiate with the freeholder, the council, whose policy it is to negotiate reasonable prices with leaseholders. I have come to the conclusion that the transaction presents no special anxiety or risk which would cause a purchaser to reduce his bid. I therefore do not accept that the allowance of £250 should be made.

A summary of my findings is therefore:

Value of the freehold with vacant possession		£2,500
Less Cost of enfranchisement:		
Enfranchisement price	£625	
Surveyor's fees and legal costs	125	
Total reasonable cost	750	750
Market value of claimants' interest		£1,750

The claims for disturbance have been agreed at the following sums: M Sharif £94; A Rashid £75.

I determine that the acquiring authority will pay the claimants £1,750 compensation for the leasehold interest in 44 Highgate Road, Sparkbrook, Birmingham. The acquiring authority will also pay the claimants compensation for disturbance as follows: M Sharif £94; A Rashid £75.

The acquiring authority will also pay the claimants a surveyor's fee on scale 5(a) of the scales of charges of the Royal Institution of Chartered Surveyors and proper legal costs incurred from the date of the notice to treat to the date of the reference.

Having opened the sealed offer of compensation lodged by the acquiring authority, I find that the amount thereof is less than the amount of my award. The acquiring authority will pay the costs of the claimants of this reference such costs, if not agreed, to be taxed by the Registrar of the Lands Tribunal on Scale 4 of the County Court Scales of Costs.

A **THAMESIDE ESTATES LTD v GREATER LONDON COUNCIL**

(Ref/67/1976)

(V G WELLINGS QC AND E C STRATHON FRICS)

November 21 1977

Estates Gazette January 27 1979 and February 3 1979

(1977) 249 EG 346 & 448

Claim under Land Drainage Acts arising from flood prevention works on land fronting River Thames near Erith—Two-part claim: (1) £350,000 for diminution in value due to actual injury from executed works, together with interest; (2) over £1m for "entirely prospective" loss (namely cost of surcharging ad- B joining land with fill) in respect of certain works not yet begun—Existing planning permissions for warehousing and industry—Roads, sewers and four industrial buildings completed—Meaning of "injury"—A wrong which but for statute would be actionable—No exercise of any power giving rise to second claim—Held claim for cost of fill inadmissible—Alternatively, finding would have been that no compensation payable because purchaser would not embark on filling—Tribunal would have been "very sympathetic" to claim based on price difference—In valuation for first claim tribunal rejects GLC valuer's application of deferment factor and deduction of residual value—C Prospective purchaser would be discouraged from attaching any value—£293,000 awarded—Claim for interest prior to award rejected—No power in tribunal to award interest as compensation for detention of debt or delay in assessment of it

Michael Turner QC and Anthony Anderson (instructed by Gouldens) appeared for the claimants; Michael Mann QC and Michael Rich (instructed by the Solicitor to the GLC) for the compensating authority.

Giving their decision, Mr Wellings and Mr Strathon said: This reference concerns the amount of compensation to which the claimants, Thameside Estates Ltd, are entitled for such injury as they have sustained by reason of the exercise by the D Greater London Council (hereafter referred to as the "GLC") of powers under section 34 of the Land Drainage Act 1930 and section 40 of the Land Drainage Act 1961 relating to the maintenance, improvement and/or construction of drainage works on the claimants' land at Crayford Ness, in the London Borough of Bexley.

The claimants' land lies within what is known, for the purposes of statutes relating to water and land drainage, as the London excluded area. At all material times the GLC was, by virtue of the London Government Act 1963, the Greater London (Powers) Act 1968, the Water Act 1973, and other statutes, the drainage authority responsible within the London excluded area E for the execution of works for the protection of land from flooding by the River Thames and had and has within that area all the powers provided by section 34 of the Act of 1930 and section 40 of the Act of 1961. To the extent that the GLC has exercised or intends to exercise those powers, its general purpose has been, and will be, the maintenance, improvement and construction of works for the prevention of flooding by the tidal Thames. This much is common ground between the parties.

The claimants' land consists now of about 55 acres. On its northern side it adjoins the River Thames and its eastern flank is alongside the Dartford creek, which is part of the River F Darent. The land is about one-and-a-half miles due east of Erith.

The total amount of the claimants' claim is more than £1m. It is divided into two parts. The first part relates to the diminution in value of parts of the claimants' land attributable to drainage works which the GLC have carried out thereon. Those drainage works are part of works which we subsequently define as phase I works. This part of the claim, therefore, relates to actual injury suffered by the claimants by reason of the exercise by the GLC of the relevant statutory powers.

The second part of the claim is entirely prospective. It relates to loss which the claimants say they have suffered or will suffer by reason of works (which we subsequently define as phase II works) which the GLC have not yet carried out or begun but G which they intend to carry out. The claimants allege that the phase II works when executed will, upon the occurrence of a tide of the height which those works are designed to contain, cause part of the claimants' land otherwise unaffected by phase I and phase II works (namely that part which we subsequently describe as "the non-berm land") to become unstable and subject to catastrophic failure by reason of a condition known as uplift. The claimants allege that the appropriate manner in which to counter the risk of damage to the non-berm land from uplift is to surcharge the whole of it with one metre of fill. The claimants claim by way of compensation, under their second head of claim, the cost of that operation, limited, however, to a H sum not exceeding what would, apart from the effect of phase II works, be the value of the non-berm land.

The right of the claimants to make the first part of their claim is not disputed although the amount claimed thereunder is challenged. The GLC, however, dispute the claimants' entitlement in law to make the second part of their claim. The claimants claim interest on the amount of compensation to which they are entitled under the first part of their claim (only), but the GLC say that the Lands Tribunal has no power under the relevant provisions of the Acts of 1930 and 1961 to award interest. J

Prior to the hearing there were two preliminary hearings in the nature of pre-trial reviews, the purpose of all concerned being so far as possible to reduce the length and therefore the costs of the main hearing by defining the issues, promoting agreement wherever possible, both as to facts and documents. By consent of the parties we ordered full pleadings to be delivered and gave other appropriate directions. Our orders and directions were fully complied with and we are grateful to learned counsel, their instructing solicitors and the parties and their witnesses for their co-operation. The hearing lasted 18 days, but it would have lasted for a much longer period but for the co-operation which we K have mentioned.

We attach to this decision a glossary of terms provided by the parties which may be found to be useful:

Glossary of Terms

Aquifer	permeable layer of soil capable of rapidly transmitting pore water pressure.
Arisings	the product of excavation/demolition.
Bank Height	level of top of bank.
Berm	level or near-level steps in elevation in or adjacent to a slope.
Bursts/Blows/Blow-ups	local ground eruptions caused by uplift. L
Consolidation	load-induced squeezing out of pore water.
Cut-off	physical barrier inserted into ground to prevent/reduce under-seepage of ground water.
Design Water Level	water level with respect to ODN used for design purposes.
Heave	upward movement of soil.
Loading	increasing the influence of weight or pressure.
Ordnance Datum Newlyn (ODN)	mean sea level at Newlyn, Cornwall.
Overburden Pressure	vertical downwards pressure due to self weight of soil. M
Piezometers	devices for measuring pore water pressure.
Piping Failure	water erosion through the soil giving a hole or "pipe."
Pore Water Pressure	hydrostatic pressure of water in the pores of soil.
Settlement	downward movement of soil.
Sheer Strength	resistance of soil to slipping.
Slip Failure	mass sliding movement of soil.
Surcharge	vertically applied pressure.
Surge	rise in water level not due to astronomical influences, ie wind-induced and barometric pressure.

	step surge—water level rises instantaneously.
	peak surge—water level rises in the same way as for an astronomical tide.
Tide	astronomically predicted water level.
Uplift	upwards pressure exerted by pore water pressure against overlying stratum.
Tide-cum-Surge	combined level/extreme level/combined effect of tide and surge.

The following witnesses were called:

For the Claimants:

Professor J R D Francis MICE, Professor of Hydraulics, Imperial College of Science and Technology, London University. B K Menzies PhD MICE, Lecturer in Soil Mechanics, Department of Civil Engineering, Surrey University. J H R Haswell BSc FICE FI StructE, Chartered Engineer, principal of the firm of J H R Haswell & Partners, London. L P Hill BA FICE FIHE, chartered civil engineer, consulting engineer in private practice, E A Croucher FRICS, principal of the firm of Jennings & Barrett, Erith, Kent.

For the GLC:

N G Thurston B Eng FICE, chartered civil engineer, deputy project manager, Thames Flood Prevention, Tidal Rivers Division of the Department of Public Health Engineering of the GLC. A Marsland MSc, a principal scientific officer in the Geotechnics Division of the Building Research Station. J F Levy BSc FI StructE, chartered structural engineer, deputy structural engineer in the Department of Architecture and Civic Design of the GLC. E R Strouts BSc BA MICE, chartered civil engineer, sites manager in the Civil Engineering Division of the Department of Architecture and Civic Design of the GLC. D J Brackley ARICS, an area valuer in the Valuation and Estates Department of the GLC.

Agreed Facts

The following facts are taken from an agreed statement of facts. In setting them out we have for the most part followed the agreed statement verbatim, but we have taken the liberty in some instances of summarising the language there used, of omitting unnecessary words and of substituting for the sake of greater clarity, brevity and consistency, different words (but not, we think, in a manner which will be disputed). Fact no 26 (areas) is stated as agreed at the hearing.

Description of the site.

1. The site (that is to say, the totality of the claimants' land) comprises a large, mostly level and approximately rectangular, promontory on the south bank of the River Thames. It fronts on to the Thames at the western end of "Long Reach."

2. Owing to a bend in the River Thames, that river forms both the northern and western boundaries of the site. The easterly boundary is formed by the River Darent and the southern boundary is the only land boundary and is marked by a chain link fence running the whole length of the site, which is only broken at the access road known as Wallhouse Road, which links the site with Slade Green, approximately three-quarters of a mile distant to the south of the site.

3. The site is contained on its riverside frontages by an embankment. A public right-of-way exists on the crest of the embankment.

4. Between the embankment and the River Thames lie mud flats and saltings.

5. A main metropolitan watercourse runs through the site and in addition associated watercourses discharge into the main metropolitan watercourse. These associated watercourses are under the jurisdiction of the London Borough of Bexley and to some extent the GLC. The watercourses are shown on the agreed plan (exhibit no 1).

6. The Upper Petroleum Limit extends from the northern corner of the site, at Crayford Ness, to Purfleet Church. This imaginary line constitutes the upper limit for the daylight movement of petroleum-carrying vessels in the River Thames.

7. The nature of the subsoil underlying the site is as follows:

(a) Underlying the immediate surface layer from a depth at Ordnance Datum level to a depth varying from minus 6.26m to minus 8.00m is a layer of peat/clay.

(b) Below the peat/clay referred to in (a) is a stratum of sand and ballast which by its nature is pervious to water and which outcrops in the bed of the river.

The environs of the site.

8. *Access*

Access to the site is by way of Wallhouse Road, an unclassified road. It is maintained by the London Borough of Bexley as far as its junction with Jenningtree Road. From this point to the main entrance to the site, a distance of approximately 3,500 ft, the access road is a private roadway, although a public right-of-way exists over it.

The site is linked to the dual-carriageway North End Road (A206, which links the towns of Greenwich, Woolwich, Erith, Crayford and Dartford) by Bridge Road, Slade Green Road and Wallhouse Road, along which residential development extends for two-thirds of a mile.

9. *Surrounding development*

To the south of the site and to the east of the access road, there is an industrial area of approximately 18.20 acres. This area has been substantially developed over the past 10 years.

The site is separated from the residential and industrial development of Slade Green by a belt of open flat land, two-thirds of a mile in length, used principally for grazing purposes.

Existing development on the site.

10. A radar station, together with a navigation beacon, is situated on the site on a plot on the embankment leased to the PLA, who maintain it and have access rights for this purpose.

11. *Infrastructure.*

Sewers: Twin 5-in sewers, each 1,600 yds long, to connect with existing local authority sewers at Slade Green, together with pumphouse. A 6-in water main over 1 mile long has been constructed.

Roads: Roads to serve stage 1 of development constructed. Location shown on agreed plan.

12. *Stage 1 development* (defined in para 20, *infra*).

The following units have been developed:

Units 1 and 5	M Mullins Ltd
Unit 3	Lynton Waxes Ltd
Unit 4	Thames Conveyors Ltd

History of the site before 1968.

13. The business known as the Thames Ammunition Works was first established on the site by Sir W E Armstrong Whitworth & Co Ltd in 1902, and was used for the manufacture of military and commercial explosives.

14. In 1928, following the amalgamation of Sir Armstrong Whitworth and Vickers Ltd, the entire works were taken over by the newly-formed Vickers-Armstrongs Ltd and continued in use as an armaments factory.

15. This use was continued through the second world war, but in 1964 the site was acquired by Imperial Chemical Industries Ltd who continued to use it similarly until early 1968.

16. The site had been in continuous use as an explosives and ammunition works since 1902, but there exist no planning records prior to 1947. Six town planning consents were granted in respect of the site subsequent to the coming into force of the Town and Country Planning Act 1947, and prior to the purchase of the land by the claimants. The consents were in respect of a jolting hut, a blast wall, a block of lavatories, a building for X-ray plant and the reconstruction of a boiler house.

History of the site since 1968.

17. The claimants purchased the land in 1968. The total land originally purchased by the claimants amounted to approximately 65.8 acres of which approximately 50 acres were situated within the embankment.

18. On January 15 1969, the claimants submitted an outline planning application for the development of the site as an industrial and warehouse estate. The London Borough of Bexley did not determine the application within the statutory period of two months, and on July 2 1969 the claimants appealed to the Minister of Housing and Local Government.

19. Following meetings with officers of the London Borough of Bexley, the claimants submitted a further outline application on January 20 1970 for the layout of the site as a phased industrial and warehouse estate by the construction of roads and sewers. The then Minister of Housing and Local Government, in a letter dated February 19 1970, determined that he would decide the second application himself.

20. The public inquiry, concerning the two applications, was opened on March 3 1970, but it was immediately adjourned because the local planning authority and the claimants were prepared to enter into an agreement under section 37 of the Town and Country Planning Act 1962 to restrict the developable acreage per year, with the claimants given the right to develop 13 acres over a two-year period. The minister's inspector agreed that the matter could be dealt with by the two parties to the agreement and he adjourned the public inquiry. Subsequently the section 37 agreement was signed on November 26 1970, and planning permission was granted for the "layout of roads and sewers for warehousing, open storage and industry in respect of 13 acres as shown coloured pink on the deposited plan" (reference no 70/250). Those 13 acres which are in approximately the south-eastern corner of the site, constitute the area to which stage 1 (not to be confused with the GLC's phase I works) of the claimants' scheme of development for the non-berm land relates.

21. Since that planning permission the London Borough of Bexley have determined 16 planning applications, of which 15 were approved and one refused (on the ground that the applicant firm was outside the area of the GLC). Two applications are held in abeyance at the request of the applicants.

A 22. Many of the above planning permissions were not proceeded with but four industrial buildings have now been erected on the 13 acres described above (see fact no 12 *supra*) and the claimants have implemented the permission relating to roads and sewers.

The Southern Water Authority.

23. In January 1976, the claimants sold to the Southern Water Authority 6.7 acres of land, on the eastern flank of the land which they had purchased in 1968, in connection with the proposed construction of the Dartford Creek Barrier. The purchase price of £162,000 comprised 5.4 acres of land at £30,000 per acre and 1.3 acres of saltings for nothing.

Flood prevention works.

Works before 1974.

24. The level of the bank crest prior to the 1953 floods was 4.90 m B ODN. This bank level was raised and the bank itself strengthened by the Kent River Authority after the 1953 floods. The present level of the bank crest varies between 5.60 m ODN minimum and 6.00m ODN maximum.

Work since 1974.

25. In 1972 the GLC decided that the best solution for providing improved flood defences for the next 60 years was bank raising in conjunction with a flood barrier incorporating movable gates to be built across the river at Silvertown.

The Thames Barrier and Flood Prevention Act 1972 gave the necessary powers to the GLC for carrying out that solution. Bank raising is carried out under the Land Drainage Acts 1930 and 1961 (prior to the coming into force on January 17 1977 of the Land Drainage Act 1976).

The programme for Thames tidal flood prevention provides for C the construction of a tidal barrier at Silvertown and for raising the level of the flanking flood defences downstream of the barrier as far as Barking Creek on the north bank and Dartford Creek on the south bank. A small tidal barrier will be constructed across Dartford Creek as a joint venture by the Southern Water Authority and the GLC. This barrier will connect with the Southern Water Authority's flood defences extending downstream from Dartford Creek. The work of raising the level of flanking flood defences is being carried out under the Land Drainage Acts 1930 and 1961. The overall contract plan for the permanent bank raising downstream of the Thames Barrier in the London excluded area was contained in the Public Contract 17, ie, the first stage (ie phase I) of construction of the loading berm on the site which was approved by the Public Services Committee on June 25 1973 and by the Finance Board on D June 26 1973. The council's valuer and estates surveyor was authorised to make the necessary planning application in respect of this work.

Planning permission was applied for and obtained from the London Borough of Bexley. Notice of entry under the Land Drainage Acts was given to the claimants on July 23 1974 in respect of the construction work on the site. Approval for the award of a contract to Dredging and Construction Co Ltd for the construction of phase I of the embankment raising was given by the Public Services Committee and the Finance Board on June 14 1974. Construction work commenced in August 1974.

Areas.

26. During the hearing the following areas of parts of the claimants' land were agreed:

		Acres
E	(1) The berm (including 0.9 acre riverbank)	11.200
	(2) Land subject to slips	0.800
		12.000
	(3) Additional land required for regrading	0.341
	(4) Saltings	2.320
	(5) Riverbank not within the berms.	3.100

These areas are relevant to the first part of the claim.

As to part II of the claim, in respect of that part of the claimants' ownership landward of the berm (which we refer to as "the non-berm land"), the total area is agreed to be 35.72 acres.

Planning assumptions.

27. The site should be valued on the following planning assumptions:

F (1) Development of the 13 acres of stage 1 would be permitted by the local planning authority on application subject to phasing such that the first 6.5 acres were completed before the second 6.5 acres were begun, and could on this basis (subject to suitable occupant and IDCs being available) be fully developed by 1980 or thereabouts.

(2) Development of the rest of the site outside the 13 acres of stage 1 would not be permitted until the 13 acres were fully developed.

(3) It is the local planning authority's present policy to encourage and support development on a phased basis of the rest of the site (except the immediate environs of the path along the bank) but it is improbable that such development would be permitted to begin earlier than 1980 or proceed at a rate faster than 6¼ acres a year from January 1 1980.

Other Facts G

Other facts, which we conceive are not disputed but which in any event we find, are as follows.

In the year 1953 a flood in the River Thames which rose to a level of 5.25 m ODN at Crayford Ness overtopped the embankment defences which stood (see agreed fact no 24) at a minimum crest level of 4.9 m ODN. The claimants' land was inundated, but to a depth which is not recorded. After bank raising and reinforcement by the Kent River Authority the defences were capable of containing a 1953 flood. By these works ("the Kent works"), Mr Thurston told us, the risk of flooding by overtopping was reduced from a surge tide having a return period of 1 in 10 years to 1 in 60 years in 1972. H

The works executed by the GLC since August 1974 consist of the construction of a loading berm. It is formed by aggregate which was placed hydraulically on the landward side of the existing embankment. Its width is a distance varying between 40 and 60 m inland from the toe of the river bank; its height is 4.8 m ODN at the river bank and it is sloped at 1 in 60 to give a level at its landward edge of 4.0 m ODN before final settlement at 3.25 m ODN. The level of the claimants' adjacent land, between 1.5 and 2.0 m ODN, averages 1.7 m ODN. The berm has absorbed a smaller berm which formed part of "the Kent works."

The construction of the berm is described in the points of J defence as "the executed works," which at the date of the service of the notice of entry were "intended to constitute phase I of a scheme of work for which planning permission was granted by the London Borough of Bexley on August 12 1974, and that such permission permits the raising of an embankment upon the site of the 'executed works' to a height of 7.7 m above ODN, which said further works are hereafter called 'the further permitted works.'" At the hearing the further permitted works were referred to as phases II or II and III and for convenience of identification we shall refer to them as phase II works.

Arising from the construction of phase II works, the purpose of which would be to prevent flood inundation by overtopping, K the claimants' non-berm land would be subject to a risk of higher uplift pressures and instability. The possibility of uplift instability arises from the existence of the stratum of sand and gravel underlying the relatively impermeable alluvium layer, that is to say, the layer about eight metres thick of soft clay and peat. The sand and gravel outcrops in the River Thames and acts as an aquifer transmitting water pressure from the river to the underside of the alluvium layer. When this uplift pore water pressure equals the vertical overburden pressure of the alluvium, the factor of safety is unity and a critical situation is reached. Any further uplift pressure will cause the ground to heave. L

It was agreed that if it was necessary to remedy such a condition at Crayford Ness, the preferred engineering solution would be to surcharge the site or parts of it so as to increase the vertical overburden pressure, but the parties were in dispute as to how much of the non-berm land would need to be surcharged and as to the depth of fill attributable to the exercise by the GLC of the relevant powers.

Phase I works are not yet completed. It is the intention of the GLC to vacate the site in September 1978, but in the interim period, the berm is to be regraded to a slope of 1 in 10, top soil is to be spread and seeded, trees will be planted to meet a condition of the planning permission and piezometers will continue to M be monitored.

As to phase II works, we are satisfied, having regard to the points of defence and the further and better particulars thereof and the evidence, that it was and is the intention of the GLC at some stage in the future to carry out phase II in some form.

For the design of the complete flood defences, phases I and II, a tide level was adopted in the GLC's designs which at Crayford Ness is estimated to be 6.4 m ODN. Such a tide can be expected to have a return period of 1 in 1,000 years by the year AD 2030. It is a tide which at Southend would be one metre higher than the 1953 maximum recorded water-level. There is no certainty as to

A when such a tide will occur, but it was common ground at the hearing that tides in the south east of England are tending to increase in height.

The Main Issues

The claimants contend that what we will call "the berm land," consisting of 12 acres of berm, 2.32 acres of saltings and 3.10 acres of river bank outside the berm, have been sterilised and their value rendered nil by reason of the execution of the phase I works. Their valuer, Mr Croucher, in his valuations in evidence to which we shall refer, said that if the phase I works had not been executed, then, having regard to the relevant planning
B assumptions, the value of the berm land as at August 1974 would be £348,364. That, accordingly, is the sum which the claimants claim under the first part of their claim. If it were right to take into account under that part of the claim sterilisation attributable to the requirement of 0.341 acre of additional land for regrading the berm to a gradient of 1 in 10, notwithstanding that that work has not been carried out, then Mr Croucher would add to his figure of £348,364 a further sum of £9,548, making a grand total, under the first part of the claimants' claim of £357,912.

As will be seen, when we come to set out Mr Brackley's valuation, on behalf of the GLC in respect of the first part of the
C claim, he proceeds on the basis that the sterilisation of the berm land attributable to the phase I works is rather less than alleged by the claimants, and his valuation in respect of that part of the claim is £71,800. Mr Brackley makes no provision for the eventuality that it should be held that sterilisation attributable to regrading of the 0.341 acre should be included in the first part of the claim.

Under the second part of the claim, that is to say, in respect of the non-berm land, the claimants claim the cost of surcharging that land with fill one metre deep. The cost of that operation, according to the evidence of L P Hill, would be £1,118,100. The
D claimants, however, limit their claim under this head to what would be the value of the non-berm land if the phase II works were not intended and did not take place. That value, according to Mr Croucher, is £1,052,360, but, if it be right to include sterilisation of berm land arising from the requirement of 0.341 acre for regrading in the first part of the claim, then the value of the non-berm land, according to Mr Croucher, would be £1,042,812 (£1,052,360 minus £9,548) and the claimants therefore limit their claim under this head to £1,052,360 or, as the case may be, £1,042,812.

As we understand the claimants' case they are not, under the second part of their claim, in any sense claiming the diminution
E in value of the non-berm land attributable to the phase II works; they are claiming the cost of surcharging that land with fill, and the only relevance of the value of that land is that the claimants utilise it as an upper limit to the proportion of that cost which they claim.

The GLC's attitude to the second head of claim is that the claimants are entitled to nothing in respect of it, for the reasons:

(1) that it is, the GLC say, as a matter of law misconceived;

(2) that the amount of fill necessary and attributable to the phase II works, having regard to the danger of uplift, is no more than half a metre, but that that quantity of fill would be necessary
F in any event having regard to what the GLC describe as the poor quality of the non-berm land.

Further, the GLC allege that, having regard to the evidence, it should not be necessary to fill the entire non-berm land because, it was said, the buildings to be erected thereon and their surrounding hard standings would themselves constitute a sufficient counter-weight to uplift; and because, so it was said, uplift, if it occurred at all, was likely to be of local occurrence which could be dealt with satisfactorily as and when it occurred and would be less catastrophic to the non-berm land and the buildings thereon than inundation of the site, which might occur if the phase II works were not executed.

There are two other claims, ancillary to the claimants' second G
head, the first of which relates to 0.4 acre of riverbank in the south east corner of the claimants' land and the second to saltings of 0.96 acre lying between that part of the bank and the River Darent. That part of the bank and those saltings are not concerned in the first head of claim. In respect of the 0.4 acre of riverbank, Mr Croucher valued it at one-third of £28,000 per acre, producing a valuation of £3,733, and he valued the 0.96 acre of saltings at £100 per acre, namely, £96. These ancillary claims are dependent on the success of the second head of claim. If that head fails, the ancillary claims fail also. As we understand matters, the basis of the ancillary claims is that:

(1) the 0.4 acre of riverbank and 0.96 acre of saltings cannot be H
developed satisfactorily in isolation from adjoining non-berm land;

(2) that land will be sterilised in consequence of the bank-raising in the phase II works unless surcharged with one metre of fill;

(3) that surcharging will not in fact take place because the cost of it exceeds the value of the land;

(4) that part of the non-berm land will not be developed at all

with the result that the 0.4 acre of riverbank and 0.96 acre of saltings will be sterilised also. J

Are the second head of claim and the ancillary claims admissible in law?

Section 34 of the Land Drainage Act 1930, as amended, provides:

> (1) Every Drainage Board acting within its district shall have power:
> (a) to maintain existing works, that is to say, to cleanse, repair or otherwise maintain in a due state of efficiency any existing watercourse or drainage work;
> (b) to improve any existing works, that is to say, to deepen, widen, straighten or otherwise improve any existing watercourse, or remove or alter mill dams, weirs or other obstructions to watercourses, or raise, widen, or otherwise improve any existing drainage work; K
> (c) to construct new works, that is to say, to make any new watercourse or drainage work or erect any machinery or do any other act not hereinbefore referred to, required for the drainage of the area comprised within their district . . .
> (2) . . . .
> (3) Where injury is sustained by any person by reason of the exercise by a drainage board of any of its powers under this section the board shall be liable to make full compensation to the injured person, and in case of dispute the amount of the compensation shall be determined by the Lands Tribunal; . . .

Powers of entry for the purposes of exercising the powers L
contained in section 34 are contained in section 40 of the Act of 1961. Powers of entry for the purpose of surveying or taking levels of land or inspecting or examining the condition of any drainage work are contained in section 43 of the Act of 1930.

Have the claimants sustained any injury in relation to the non-berm land or the 0.4 acre of riverbank or the 0.96 acre of saltings by reason of the exercise by the GLC of any of its powers under section 34 of the Act of 1930?

Mr Mann, for the GLC, while accepting that the GLC had resolved and intended at some time in the future to carry out the phase II works and therefore to raise the height of the em- M
bankment, submitted that in relation to the phase II works there had been no exercise of any power in section 34 of the Act of 1930. The mere resolution to exercise one of the powers could not give rise to a claim for compensation because the powers were powers to do work and it was only the execution of work which gave rise to a claim under subsection (3).

Furthermore, said Mr Mann, there had been no "injury." That word had the same meaning as *injuria* in the civil law, that is to say, a wrong which apart from the statute would be actionable. There could be no injury in that sense in relation to the

A non-berm land unless and until the phase II works were executed. The kind of injury contemplated by the statute was in the nature of a tort. No claim for damages could in law be made for an anticipated tort: see *West Leigh Colliery Co v Tunnicliffe* [1908] AC 27. That case also showed that the fact that the market anticipated the commission of a tort was not relevant, although, of course, the claim had not been put on the basis of diminution in market value of the non-berm land.

Mr Turner, for the claimants, submitted that the GLC had exercised the relevant powers in relation to the phase II works, notwithstanding that none of them had as yet been carried out, because they had entered upon the land and executed the phase B I works and remained in control of 14.6 acres of embankment and berm (as admitted in the points of defence) for the purpose of investigation, design, planning and for the purpose of carrying out the future intended work. By reason of the fence surrounding that acreage of embankment and berm they effectively excluded the claimants therefrom. In any real sense they had exercised the relevant powers in relation to the phase II works.

Mr Turner said that the word "injury" was not a reference to a cause of action. It was not equivalent to "damage" but was a wider concept. It might include "damage" in a physical sense but it also included interference with legal rights. The Act of 1930 itself drew a distinction between injury and damage: see C section 44 (7) proviso. The Act also distinguished between injury to land and injury to a right: see section 35 (1) as originally enacted. "Injury" in section 34 (3) was not limited or tied to physical injury to land. The claimants had suffered injury because there had been an interference with their legal rights, that is to say, there had been what, apart from the Act would have been a trespass. They had suffered actual damage from that trespass, not only in relation to phase I but also in relation to phase II because it had become clear from the planning documents that the local planning authority as a consequence of the GLC's intentions with respect to phase II would as a condition of any planning permission granted for development of the non-D berm land require it to be surcharged with fill.

It was a principle of law, Mr Turner said, that the claimant must make his claim for compensation once and for all: he must include in his claim all damage which he can foresee will result from the exercise of the statutory power and he has no opportunity to come back for a second time. The *West Leigh Case* and other mining cases were *sui generis* and irrelevant. Decisions relating to compensation for compulsory purchase were, however, relevant: see *Chamberlain v West End of London Railway* (1863) 2 B&S 605; *Croft v London and North Western Railway* (1863) 3 B&S 436; re *Dudley Corporation* (1881) 8 E QBD 86; *Vincent v Thames Conservancy* (1953) 4 P&CR 66.

We agree with Mr Mann that no power given to the GLC by section 34 of the Act of 1930 in relation to the phase II works has been exercised. In relation to those works, in our opinion, the GLC has no more than expressed an intention to exercise the powers and that is not the equivalent of an actual exercise of them.

Meaning of "Injury"

If we were wrong in that view, it would be necessary for us to consider the meaning of the word "injury." We sympathise with F Mr Turner's interpretation, but having regard to a long line of authority relating to the use of the words "injury" and "damage" in other statutory contexts and having regard to the contrast which the Act of 1930 itself draws between "injury" and "damage" we think that the word "injury" refers to a wrong which but for the statute would be actionable. See the cases under section 68 of the Lands Clauses Consolidation Act 1845, which relates to compensation for injurious affection arising from works carried out on land not taken from the owner of the injuriously affected land: *Caledonian Railway v Ogilvy* (1856) 2 Macq 229; *New River Company v Johnson* (1860) E&E 435. See also *Hall v Borough of Bristol* (1867) LRCP 324 (under section 144 of the Public Health Act 1848); G *Rhodes v Airedale Drainage Commissioners* (1876) 1 CPD 390 and 412 (under a private Act incorporating the Lands Clauses Consolidation Act 1845); *Crofter Handwoven Harris Tweed Co v Veitch* [1942] AC 435, *per* Viscount Simon LC at 442. There is a line of cases relating to claims for compensation for injurious affection arising from land taken from the owner of the first mentioned land, which cases give support to Mr Turner's definition "injury." See *re Stockport, Timperley & Altrincham Railway Co* (1864) 33 LJQB 251, at 253 *per* Crompton J; *Cowper Essex v Acton Local Board* (1889) 14 App Cas 153; *Sisters of Charity of Rockingham v R* [1922] 2 AC 315 at 325 *per* Lord Parmoor. Unfortunately the feature, expressed in those cases, distinguishing them from H the cases under section 68 of the Act of 1845 is that land was taken and the injurious affection to land held with it was created on the land taken.

Finally, upon this part of the case, we are influenced by the interpretation placed upon the meaning of the word "injury" in this very subsection by Byrne J in *Marriage v East Norfolk Rivers Catchment Board* [1949] 2 KB 456, at 462, where he said:

> I take the word "injury" in that subsection to connote a legal right of action but for the powers which the Catchment Board have under this statute and in my view the subsection means that if that which is complained of would have been actionable at common law but for the statute, there is a right to compensation. J

The point was adverted to in the Court of Appeal in [1950] 1 KB 284 but not decided. Byrne J's dictum was applied by the tribunal (Erskine Simes QC) in *Vincent v Thames Conservancy* (1953) 4 P&CR 66. As we have said that is the view which we follow.

If the claimants had claimed the difference in the price which a purchaser in the open market would pay for the entirety of their land, having regard to the phase I works already carried out, the requirement of 0.341 acre of extra land for regrading of the berm and the view which that purchaser would take of the consequences of the intended phase II works, we would have K been very sympathetic. Our sympathy might, it is true, have been stultified by the views of the House of Lords in the *West Leigh Case*. The claim was not put on that basis but on a basis related entirely to the cost of surcharging the non-berm land with fill. The claimants have not surcharged the land and there is plainly no intention on their part to expend the sum claimed in surcharging the land. No evidence to that effect was given on behalf of the claimants and Mr Croucher's own evidence on their behalf assumed, in relation to the ancillary claims, that no fill would be placed on the non-berm land. We agree with him and so decide. In the circumstances the claim for the cost of the fill is inadmissible, if not for any of the reasons which we have so far advanced, L for the reason that the claimants have no intention of expending the money: see *Taylor (CR) (Wholesale) v Hepworths* [1977] 2 All ER 784; 244 EG 631.

It is no doubt a principle of law that all damages arising from a single cause of action should be assessed once and for all in one and the same proceedings, but that rule does not apply where there is more than one cause of action. The cases cited by Mr Turner on this topic were cases relating to compensation for compulsory purchase, which are distinguishable for the reason that all compensation of that nature, whether for disturbance or injurious affection or severance is treated in law as part of the M purchase price paid for the interest acquired: the purchase price cannot be ascertained unless all heads of claim are assessed at one time.

For the foregoing reasons we dismiss the claimants' claim relating to the non-berm land and their ancillary claims.

Valuations as to Berm Land

We will now deal with valuation evidence in respect of the berm land. For the purpose of comparison, the details of the valuations, as finally submitted, are rearranged in the following form:

A G

Claimants (E A Croucher FRICS)

Areas	*Acres*
A. Including part "Kent Works" old berm	11.1
Including part "Kent Works" old berm and part riverbank	0.9
	12.0
B. Saltings	2.32
C. Riverbank outside the berm	3.10
D. Additional land required for regrading berm to 1 in 10	0.341

Valuation Date August 1974

B

Area (acres)	*Value per Acre* £	*Compensation* £
11.1	28,000	310,800
0.9	9,333	8,400
		319,200
2.32	100	232
3.10	9,333	28,932
		348,364
		£348,364

GLC (D J Brackley ARICS)

Areas	*Acres*
Excluding old berm	9.7
Including old berm	2.3
	12.0
Saltings	2.32
—	—
—	—

Valuation Date May 1977

H

Area (acres)	*Value per acre* £	*Compensation* £
9.7	28,000	271,600
2.3	9,333	21,466
		293,066
Defer 5 years P.V. at 10%; tax at 52%	0.791	231,815
2.32	100	232
		232,047
Less Residual value		
9.7	18,670	181,099
2.3	9,333	21,466
12.0		202,565
Defer 5 years, P.V. at 10% tax at 52%	0.791	160,228
		£ 71,819
		say £ 71,800

C J

D K

The points of difference between the valuations may be identified as: (a) dates of valuation; (b) apportionment of the area of 12 acres for the purpose of the valuation; (c) application of a deferment factor by Mr Brackley; (d) the inclusion by Mr Croucher of 3.10 acres of riverbank; and (e) the deduction by Mr Brackley of a residual value of the land after the execution of the works.

Subject to deferment by Mr Brackley, £28,000 per acre is adopted by both valuers as what we shall call the basic value of

E the land. It is derived from sales of land in the locality. Mr Croucher's principal comparable is a site of 20.963 acres at Fishers Way, which lies about three-quarters of a mile from the subject site. He devalues the 1973 purchase price to produce £28,000 per acre and Mr Brackley, by a different method of analysis, arrives at £29,000 per acre.

Another sale cited by Mr Croucher is of 6.7 acres of land sold in December 1975 by the claimants to the Southern Water Authority. It abuts on the subject land, and has the similar physical features of land, riverbank and saltings; Mr Croucher, by analysis, ascribed £36,500 per acre sale price to the land element. Mr Brackley's principal comparable is 16.2 acres of

F land at Yarnton Way, Thamesmead, a purchase by the GLC from the British Gas Corporation. The sale price as analysed by Mr Brackley produces £28,000 per acre or, if a different area divisor is adopted, £29,500 per acre. Two further transactions cited by Mr Croucher do not carry support by comparables any further.

It is sufficient to say that on the material derived from the comparables, the valuers have ascribed to the subject land a basic value of £28,000 per acre. But in issue is to which part of the 12.0 acres the basic value should be applied. Mr Croucher values 11.1 acres at the basic value and 0.9 acre of riverbank at one-third. Mr Brackley values at one-third the 2.3 acres which comprise the old berm of the "Kent Works," which is at an average level of 1.1 m above the remainder of the 12 acres. We are satisfied that the land on which the old berm stands, although of a comparatively small area and of difference in level, should carry the less value because, before the GLC came on the scene, it had already been depreciated in usefulness and value by that old berm.

We therefore accept, at this stage of the valuations, Mr Brackley's valuation of the 12 acres at £293,066.

Deferment Factor L

Mr Brackley applies a deferment factor; Mr Croucher does not. The facts which we find bear on this issue are: (a) £28,000 per acre is an average value derived from the average sale prices per acre of the comparables, themselves, as yet, undeveloped; (b) while there was some discussion as to what date was relevant for the assessment of compensation, in the valuations as finally submitted, £28,000 per acre is taken to be the average basic value current in both August 1974 and May 1977; (c) development of each of the two principal comparables would be phased over years; this is reflected in the average sale price per acre; (d) the subject land would be developed and used in conjunction M with the remainder of the claimants' land; (e) the deferment factor applied by Mr Brackley relates to the phased development of the claimants' land.

On these findings we are in no doubt that Mr Brackley seeks to defer what is in truth a value already deferred. We therefore find that no further deferment is called for.

The saltings, 2.32 acres, stand at an agreed value of £232.

The next item in Mr Croucher's valuation is an area of 3.10 acres of riverbank, which he values at £9,333 per acre, being one-third of the basic £28,000 per acre. It is his opinion of its value,

A unsupported other than by an analysis of the Southern Water Authority's transaction in which he, himself, ascribed to the riverbank area a value of one-third of the value of the land element. The ground for inclusion of this item is that the sea wall, by reason of the construction of the berm, is not so readily available for development in relation to the claimants' remaining land. It is in essence a claim based on severance. We are not satisfied that the claimants will suffer loss either as to its use or the value which Mr Croucher puts on it.

Finally, we come to Mr Brackley's assessment of the residual value of the 12 acres of berm land in its physical state on May 2 1977, namely at the date of his evidence at the hearing. The
B basis of the valuation stems from the evidence of Mr Thurston in relation to two separate parcels of the 12 acres of berm land. First as to 6.4 acres which would later form the site of phase II works. Before these works were executed, there would be no objection to a use for light storage only, provided such use terminated in good time before the land was required for construction work. Proposals to build could be incorporated with the phase II works of embankment raising, provided no additional expenditure was incurred by the GLC. As to the other 5.6 acres, Mr Thurston thought that during the period of settlement of the berm, a storage use would be permitted and, thereafter, light building construction. The overriding criteria would
C be to safeguard the integrity of the defence works and to be satisfied that no appreciable increase in the cost of the works would be caused. Protection of the executed works is afforded by byelaws made under section 47 of the Land Drainage Act 1930 by the Kent Rivers Catchment Board. They remain in force and are now enforceable by the GLC. We were referred in particular to:

Byelaw 13. No person shall without the previous consent in writing of the board:
(a) erect or construct or cause or permit to be erected or constructed any building fence post pylon wall wharf jetty pier quay landing place
D piling groyne revetment steps or any other structure or erection whatsoever in the main river or in on or over any bank thereof or within fifteen feet of the foot of any bank on the landward side, or where there is no bank, within 20 ft of the brink of the main river.
Byelaw 29. No person shall without the previous consent in writing of the board erect or construct or cause or permit to be erected or constructed any building erection or structure whatsoever between the foot of the seaward side of any sea wall or tidal embankment and a line drawn on the landward side of any such sea wall or tidal embankment at a distance of 50 ft from the foot of the landward side thereof.

The claimants' engineering evidence was critical of the problems of access and provision of services, settlements of the berm,
E the requirement to terrace to create level sites, increase of construction costs, the possibility of having to resort to piling for foundations and the requirement to maintain the defence works profile; problems inherent in building upon the berm land should be examined after a waiting period of five to 10 years, that is to say, when the significant effects of settlement would have disappeared. Mr Brackley considers that the berm has a value for temporary uses, for future building development and for its inclusion in planning site cover calculations. Mr Croucher takes the view that a developer would be so discouraged that he would attach no value to the berm land as constructed and graded at 1 in 60 or, if and when regraded, at 1 in 10.

F In our judgment a prospective purchaser in May 1977 would be greatly concerned by the engineering evidence. Such a person would have had the benefit of inspecting the executed works. Contrary to the optimistic view taken by Mr Brackley, we accept Mr Croucher's opinion that such a prospective purchaser would be discouraged from attaching any value to those 12 acres of the claimants' land.

We find that the amount of compensation payable by the GLC to the claimants in respect of the executed works on the berm land, taking May 1977 as the valuation date and the land in its physical state at that date, is £293,298. A summary of the details is:

G

The berm land (12 acres)

Area (acres)	*Value per acre* £	*Compensation* £
9.7	28,000	271,600
2.3	9,333	21,466
		293,066
Saltings		
2.32	100	232
		£293,298

Compensation for Land Required for Regrading

It was conceded at the hearing by counsel for the GLC H that it is their intention by September 1978 to regrade the berm at its landward extremity to a gradient of 1 in 10 and that 0.341 acre of additional land would be required for this purpose. We would have liked to have included in the compensation awarded by us compensation for the depreciation in the value of this small piece of land attributable to that future operation. Unfortunately we feel that we are precluded from so doing on the same grounds as those we have given for rejecting the claimants' second head of claim (relating to the non-berm land), that is to say, because, in our opinion, there has been no exercise of the relevant powers in relation to the 0.341 acre and as yet the claimants have sustained no injury in respect of it. Had J we been entitled to include this item in our award we would have included a sum of £9,548 in respect of it, that is to say, we would have applied to the 0.341 acre Mr Croucher's basic value of £28,000 per acre.

Tribunal's Views on Facts of Second Head of Claim

We are conscious that the greater part of the hearing was devoted to engineering evidence on which such a claim was founded and the amount of compensation claimed greatly exceeds the claim in respect of the berm land. We feel, therefore, that we should proceed to come to our findings on the evidence K as we would have done had we come to different conclusions on the issues of law.

We do not propose to recount in detail the evidence which we heard from the highly qualified and experienced engineers whose names appear earlier in this decision. Of common interest, however, was a technique for estimating uplift pressures which was described in a paper by M F Randolph, of the Building Research Station, entitled "Modelling of tidal water flow in a gravel aquifer at Crayford Ness." The Randolph technique was described by Dr Menzies for the claimants as most promising as it makes use of actual measurements of uplift pressures during L a spring tidal cycle. It is based on a mathematical model or idealisation of the gravel aquifer underlying the site. For the GLC Mr Marsland described the studies as involving the measurements of the pore water pressures at typical positions throughout the marsh (Crayford Ness) and the formulation of theoretical models which reproduced the measured variations under given tides which enabled predictions to be made of the pore water pressures likely to develop under storm tides.

The Randolph model results could be tested against the known circumstance of the 1953 flood when the defences were overtopped and blow-ups by uplift pressure had not occurred. The parties were in agreement that to maintain the 1953 stability M of the claimants' land a depth of 0.5 m of "fill" would be required. But in the opinion of the claimants' engineers, surcharging the site should form part of the concept of the engineering design of the works; Dr Menzies would seek a factor of safety which would require 1.0 m of "fill."

The "Kent Works" in 1954 strengthened and raised the embankment crest from 4.9 m ODN to its present height of 5.6 m ODN. The claimants' expert witnesses say that, notwithstanding that fact, the 0.5 m of fill required to maintain the 1953 condition of stability, would still be required. A contrary view is held by the engineering witnesses of the GLC, who say that only

A 0.16 m would be required having regard to the 5.6 m ODN condition of stability.

We have come to the conclusion that it is the maintenance of the 5.6 m ODN stability conditions that should be taken as the basis of the requirement to surcharge and we are satisfied that the amount required, which, we accept, is of the order of 0.16 m, would not cause a purchaser or developer to embark on an operation of surcharging the site with "fill." Adequate precautions would be taken in the type of buildings to be erected and in the design and construction of foundations, not very different from those which would now be taken in this marshland area.

B There is a further matter which bears on this claim; it is about the consequences of flooding by overtopping. The present protection provided is that against a tide of 1953 proportions having a return period of once in 53 years. By the year 2030 a return period for such a tide of 1 in 22 years can be expected. By the phase II works the protection would provide against a surge tide having a return period of 1 in 1,000 years by the year AD 2030. It seems to us on the evidence that damage to buildings and plant by flood inundation could be as grave as damage which could be suffered by blow-ups due to uplift pressure.

On all the evidence, had we come to different conclusions
C on the legal issues and compensation fell to be assessed in respect of the phase II works, our finding would be that no compensation would be payable under the second head of claim and the ancillary claims relating to 0.4 acre of land and 0.96 acre of salting.

As already indicated we award to the claimants in respect of their first head of claim the sum of £293,298.

Interest

The claimants claim interest on the award from the date when the injury was sustained, that is to say August 1974 or
D thereabouts. There is, of course, in the Land Drainage Acts 1930 and 1961 no express power given to the tribunal to award interest in respect of any period prior to the date of the award. Mr Turner, however, puts his faith in the words "full compensation" in section 34(3) of the Act of 1930. The point was indeed expressed by the tribunal (J H Emlyn-Jones FRICS) in *Pattinson* v *Finningly Internal Drainage Board* (1971) 218 EG 1105, at 1108 as follows:

> I consider that interest should be allowed, if only because the payment of a sum of money in 1971 in respect of injury suffered in 1964—whatever may be the reason for the delay—cannot otherwise be said to be "full compensation" for that injury in the words of section
> E 34(3) of the 1930 Act.

We were also referred to *Hobbs (Quarries)* v *Somerset County Council* (1975) 234 EG 829; *Burlin* v *Manchester City Council* (1976) 32 P&CR 115; 238 EG 891, 974 (both decisions of the tribunal); Re: *Richard and Great Western Railway* [1905] 1 KB 68; *Swift & Co* v *Board of Trade* [1925] AC 520.

With respect to Mr Turner, his submission confuses compensation for the injury sustained by reason of the exercise of the relevant powers with compensation for a different matter altogether, namely, compensation for detention of a debt or delay in the assessment of it. Interest is compensation of the latter kind and the tribunal has no power to award it from any date
F prior to the date of the award. It appears to us that this is beyond argument: see *London, Chatham & Dover Railway Co* v *South Eastern Railway Co* [1893] AC 429; *Swift & Co* v *Board of Trade*, *supra*, both decisions of the House of Lords. See also the decision in *Burlin's* case, *supra*, where the question of interest is fully considered. Accordingly, we, with regret, reject the claim for interest from a date prior to the date of the award.

Under rule 38 of the Lands Tribunal Rules 1975 and section 20 of the Arbitration Act 1950, we order that interest on our award be paid at the rate of 10 per cent per annum from today until the full sum in our award be paid to the claimants.

The acquiring authority will pay to the claimants one third G of their total costs of this reference, such total costs, if not agreed, to be taxed by the Registrar of the Lands Tribunal on the High Court Scale.

RAKUSEN PROPERTIES LTD AND OTHERS v LEEDS CITY COUNCIL

(REF/289/1977)

(V G WELLINGS QC)

October 10 1978

Estates Gazette March 3 1979 H
(1978) 249 EG 859

Factory premises in Leeds—Deemed acquisition under purchase notice—Company owning factory different from company carrying on business, but both wholly-owned subsidiaries of a group company—Question whether "corporate veil" should be lifted—Interim decision on claim under rule (6) on basis of complete cessation of business—"Disastrous" fire after service of purchase notice but before public inquiry and confirmation—Possession not yet taken by acquiring authority nor notice of entry served—Suggestion that claimants would have resumed production J profitably after fire "unreal"—Companies held not entitled to rule (6) compensation other than fees incurred on freehold claim—Not necessary to determine applicability of "DHN" principle, but were it necessary tribunal member "very much doubts" whether he would apply it —DHN case distinguishable

William Glover QC and Charles Fay (instructed by Walker Morris & Coles, of Leeds) appeared for the claimants; Iain Glidewell QC and G Spafford (instructed by the Solicitor to Leeds City Council) for the acquiring authority.

Giving his decision, MR WELLINGS said: In this reference K the claimant companies, Rakusen Properties Ltd, Rakusen Group Ltd and Lloyd Rakusen & Son Ltd, or two or one of them, claim compensation for the acquisition of a factory and premises at Meanwood Road, Leeds 7, pursuant to a confirmed purchase notice. The reference raises for determination, *inter alia*, the question whether, in the circumstances of the present case, the "corporate veil" should be lifted in order to overcome the difficulty that the company which owned the factory (and served the purchase notice) and the company which carried on the business in the factory (and had no compensatable interest in it) were different companies, albeit that they were both wholly-owned subsidiaries of a L group of companies.

The first company to be incorporated was the Rakusen Group Ltd. It was formed in 1938, but it was then known as Lloyd Rakusen & Son Ltd, the change of name having taken place in 1974. I will call the Rakusen Group Ltd "the public company." The group owns two properties only, the factory in Meanwood Road, Leeds, and a factory at West Park, Leeds. The fee simple in both properties is vested in a wholly-owned subsidiary of the group, Rakusen Properties Ltd, which I will call "the properties company." Another subsidiary, Roland Minton Holdings Ltd, a building and development company, from time to time owns or has in its M possession, during the period of construction until sale, other properties. I will call that company "the building company."

There is a food manufacturing subsidiary, wholly owned, which carried on its business at the factory in Meanwood Road. Its name is Lloyd Rakusen & Son Ltd, namely, the name borne by the public company before its change of name. This subsidiary was incorporated in 1974. I will call it "the trading company."

Altogether there are eight subsidiaries in the group, the other five, so I was told, being mostly just names.

A The purchase notice was served under section 180 of the Town and Country Planning Act 1971 on the ground that the factory premises at Meanwood Road had become incapable of reasonably beneficial use in their existing state and was dated February 3 1976. On May 12 1977 the Secretary of State for the Environment confirmed the purchase notice without amendment and directed that the Leeds City Council were to be deemed to have served on that date a notice to treat in respect of the interest held by the properties company in the land in question.

Under section 180 of the Act of 1971 a purchase notice is to be served by the "owner," that is to say, by section B 290(1), a person who is entitled to receive the rack rent of the land or, where the land is not let at a rack rent, would be so entitled if it were so let. The company which fulfilled that definition was the company which served the purchase notice, namely, the properties company.

By the notice of reference (in its original form) in the present case, the properties company claimed compensation for the interest acquired. Originally the properties company made no claim for compensation for disturbance or any other matter referred to in rule (6) of section 5 of the Land Compensation Act 1961. However, a separate claim for compensation of that category was made by the trading com- C pany under reference number 290/1977. In the notice of reference under that number, the trading company alleged that it held "an irrevocable licence granted by the freeholders," in the Meanwood Road factory. On May 22 1978 I sat to determine the question whether the trading company had any right to claim compensation under rule (6). During the course of the argument, William Glover QC, on behalf of the trading company, said that he did not allege that the trading company had any licence in the Meanwood Road factory. On the contrary he sought to rely upon the principle of "lifting the corporate veil": see *D H N Food Distributors* v *London Borough of Tower Hamlets* [1976] 3 All ER 462. D To that end, the president having allocated the present reference (no 289/1977) to me, Mr Glover sought leave to amend the present reference by adding the public company and the properties company as claimants, jointly, severally and in the alternative, and by adding a claim for compensation under rule (6). With the agreement of Iain Glidewell QC, on behalf of the Leeds City Council, I permitted these amendments to be made, whereupon Mr Glover withdrew reference number 290/1977. I then proceeded to hear the present reference, as amended, and, at the request of the parties, I agreed to determine all questions, both of fact and law, other than the E question of valuation, which last-mentioned question would be deferred until a later date.

Heads of Claim

The heads of claim now made under rule (6) by one or two or all of the claimant companies (the claim being made on the basis of complete cessation of business) are as follows:

(1) Compensation for loss of goodwill;

(2) Compensation for loss on enforced sale of plant and machinery;

F (3) Compensation for loss on termination of

- (a) contract in respect of telephone;
- (b) contract with respect to metal closure machine;
- (c) contract with GPO;
- (d) gas contract;
- (e) electricity contract;

(4) Compensation for loss resulting from redundancy payments;

(5) Legal costs and surveyors' fees incurred prior to the commencement of proceedings.

I have no knowledge of the amounts claimed except that in a letter dated November 29 1977 the claim for compensation for loss on enforced sale of plant and machinery was G put in the sum of £329,500.

Some of the facts were agreed. I also heard evidence. From all the relevant sources, in addition to matters of fact which I have already mentioned, I find the following facts:

(1) The business of food manufacture and distribution was formerly owned by the public company. On June 27 1975 it sold the business to the trading company in consideration of the latter company agreeing to allot to the public company 100,000 fully paid-up shares of £1 each and agreeing further to discharge the public company's debts and liabilities in connection with the relevant business and to give to the public company appropriate indemnities. On August 1 1975 the trading H company issued to the public company 100,000 shares of £1 each.

(2) The Meanwood Road factory was formerly owned by the public company which, however, on June 27 1975 conveyed those premises to the properties company for an estate in fee simple, the consideration for the conveyance being £375,000, all of which sum was agreed to be left outstanding as an unsecured loan owing from the properties company to the public company.

(3) On November 28 1975 the properties company applied for planning permission for alterations to form three enlarged loading bay entrances to the Meanwood Road factory. That application was refused on the following grounds: "it is considered that the proposals to enlarge the three loading bays in order to accommodate large articulated vehicles would, as a result of the rever- J sing of such vehicles into and out of the premises, restrict the free flow of traffic on Meanwood Road and endanger the safety of pedestrians." Thereafter, on February 3 1976, as I have already mentioned, the properties company served the purchase notice.

(4) The number of directors on the boards of the respective companies were as follows:

- (a) the public company: 5
- (b) the trading company: 6
- (c) the properties company: 4

(5) The four directors of the properties company were also K directors of the public company and of the trading company, and they were indeed the active directors of the group. They had the same functions in each of the three companies.

(6) Very little activity other than statutory requirements as to share transfers and returns takes place on behalf of the public company. Apart from statutory returns, no activity takes place on behalf of the properties company.

(7) The public company and the properties company have no employees (other than the directors). The trading company has 80 employees (I do not know but I assume that the building company has some employees). Administration on behalf of the public company and the properties company is carried out by the trading company. L

Conducted as Single Unit

(8) The group is conducted very much as a single unit. Only such distinctions as between the individual companies are made as may be required by the companies' bankers, statutory provisions, hire purchase companies, insurers etc. Meetings of boards of directors tend to deal with problems rather than with individual companies, a single set of minutes being prepared for the group as a whole, the names of individual companies, wherever relevant, being underlined.

(9) Money for the payment of a debt incurred by one company is frequently taken from the funds of another company, recorded M in the accounts but not paid back (it being assumed that during the course of a year funds will flow back in the ordinary course of events). All profits from subsidiary companies are passed up to the public company which is the only company in the group which declares a dividend. If a guarantor is required for a transaction the public company acts as such.

(10) All the plant and equipment and materials used and made in the food production and distribution business belong to the trading company.

(11) The business of the building company is treated in practice as a separate division of one and the same business, namely the business carried on by the group as a whole.

A (12) On November 2 1976 a disastrous fire occurred at the Meanwood Road factory. It started in a large machine dryer and destroyed it. The fire and water used in extinguishing it brought production at the factory to a standstill. An insurance claim in respect of the fire was lodged by the insured, namely the trading company, which received, in respect of the premises, insurance money in the sum of £5,000 for the premises (had the premises been reinstated, which was not the case, a sum of £25,000 would have been received). A sum of £160,000 was received by the trading company on account of the fire damage to the plant, machinery and materials, finished goods and consequential loss.

(13) The trading company had a number of production lines
B or "works" known as, for example, "works 1," "works 2," etc. There was also a pasta plant, a bottling plant and a cereal room. Works 1 was concerned with the production of matzo meal, a material used in the manufacture of foodstuffs eaten at Passover. Works 1 was in use during part of the year only; that part varied with the date on which the next Passover should occur. Works 2 was a stand-by plant, not in general use, but capable of production of matzo meal. Works 3, 4 and 5 were moved from the Meanwood Road factory to the West **Park** factory before the date of the fire as were the machinery and equipment from the cereal room.

(14) At the date of the fire there were at the Meanwood Road factory: works 1, not in production but about to start pro-
C duction of matzo meal; works 2, incapable of use because of removal of parts; the pasta plant, which was in use for the production of rusk, a material used in the production of sausages; and the bottling plant, which was not in use because of the modern fashion for plastic bottles rather than glass bottles, which the plant was not capable of making.

(15) The Leeds City Council (the acquiring authority) has not taken possession of the Meanwood Road factory or any part thereof. The council has not served notice of entry and in law and in fact the trading company remains in possession of the factory premises.

(16) Until November 29 1977 (a date after the date of the con-
D firmation of the purchase notice) the council did not know of the interest in the matter of the trading company. Until then the council had been led to believe that the properties company was both the owner of the freehold and owner of the business carried on in the Meanwood Road factory. Indeed the case put forward on behalf of the trading company at a public inquiry into the question whether the purchase notice should be confirmed had been on the basis that the company owning the freehold and carrying on that business were one and the same.

The clearing-up process after the fire took some three months. Donald McFarlane, one of the claimants' directors, said in evidence that thereafter, but for the purchase notice and advice which the trading company had received with
E respect to the consequences arising from its service and confirmation, that company would have expended the insurance moneys on reinstating the Meanwood Road factory, would have installed there a replacement dryer and would have carried on production there of matzo meal and rusk for which, he said, there was great demand and from the production of which, he also said, that company could have reasonably expected substantial profits. The group in fact recommenced production of matzo meal but had to stop this production subsequently because of representations about conditions in the factory made by the hygiene department of
F the council.

Mr McFarlane said that the trading company purchased a dryer for the price of £100,000 and installed it in the West Park factory, a solution which was not satisfactory, because that factory is much smaller than the Meanwood Road factory and is less satisfactory for the installation of such a large machine (it is 60 ft long, 12 ft high and 10 ft wide).

Finally, Mr McFarlane said that he believed that the purchase notice involved not only the council buying the Meanwood Road factory but also their taking possession of it and that the trading company's business would have to be moved. Accordingly, he said, there would have been no way in which the company, if it had expended the insurance moneys on G
reinstatement of the Meanwood Road factory could have recovered them.

The Issue for Determination

The issue which I am asked to determine in advance of the valuation stage, as jointly settled by learned junior counsel, is as follows:

the respondent council having conceded that the Rakusen Group Properties Ltd are entitled in principle under rule (6), section 5 of the Land Compensation Act 1961, to compensation for surveyors' fees and other fees properly incurred by in and about the preparation of its claim in respect of the valuation of the H
freehold, whether or not in the events that have happened the claimants or any of them are entitled to compensation for disturbance or compensation in the nature of disturbance. If so, on what date should such compensation be assessed?

For the words "compensation for disturbance or compensation in the nature of disturbance," I would prefer to substitute the words of rule (6), namely, "compensation for disturbance or any other matter not directly based on the value of land."

In *D H N Food Distributors v London Borough of Tower Hamlets* [1976] 3 All ER 462, D H N was the parent company J
of a group of three companies. The fee simple in the premises in which D H N carried on business was vested in one of its wholly-owned subsidiaries. The directors in all three companies were the same. The premises were compulsorily acquired and D H N claimed compensation for disturbance to its business. The Court of Appeal held that it had a sufficient interest in the premises to qualify it to make the claim. It was held that D H N had an irrevocable contractual licence to carry on business in the premises, alternatively that D H N had an equitable interest in them. In the further alternative, it was held that in the circumstances it was right to pierce the corporate veil and treat the subsidiary which K
owned the premises and D H N as one company. Only the last-mentioned point is relevant to the present proceedings because Mr Glover has abjured the trading company's original claim to an irrevocable licence and because Mr Glover has not taken the point before me that the trading company is entitled to an equitable interest in the Meanwood Road factory.

In the *D H N* case, having said that the three companies could have put their house in order so as to make the claim impregnable by taking a very simple step, namely arranging for an assignment of the freehold from the subsidiary to
D H N, Lord Denning MR at p 467 said: L

This group is virtually the same as a partnership in which all the three companies are partners. They should not be treated separately so as to be defeated on a technical point. They should not be deprived of the compensation which should justly be payable for disturbance. The three companies should, for present purposes, be treated as one, and the parent company, D H N, should be treated as that one. So that D H N are entitled to claim compensation accordingly. It was not necessary for them to go through a conveyancing device to get it.

Goff LJ said at p 468:

... this is a case in which one is entitled to look at the realities of the situation and to pierce the corporate veil. I wish to M
safeguard myself by saying that so far as this ground is concerned, I am relying on the facts of this particular case. I would not at this juncture accept that in every case where one has a group of companies one is entitled to pierce the veil, but in this case, the two subsidiaries were both wholly owned; further, they had no separate business operations whatsoever; thirdly, in my judgment, the nature of the question involved is highly relevant, namely whether the owners of this business have been disturbed in their possession and enjoyment of it.

Shaw LJ agreed and pointed to the completeness of identity of the three companies which manifested itself in various

A ways: the directors were the same; the shareholders were the same and the three companies had a common interest in maintaining on the property concerned the business of the group.

I have very little knowledge as to how the claimants put any of their claims. I know absolutely nothing about their claims for compensation for the freehold interest. I have seen no calculations, assessments, valuations or indeed any figures whatsoever (other than the figure of £329,500, which may or may not now be apposite) relating to any of the claims. I have heard some evidence and have found some facts in relation to the claim for loss of goodwill or loss of profit
B (or however it might be put) and I have heard argument with respect to that claim. I have no idea as to how it might be quantified. I have heard no argument and no evidence in relation to any of the other claims.

With respect to the claim for compensation for loss of profits (or goodwill), Mr Glover submitted that, as a consequence of the compulsory acquisition brought about by the service and confirmation of the purchase notice, the group had lost the opportunity to return to the Meanwood Road factory and recommence production after the fire; it was the fact that the group required the factory for manufacture
C and that if there had been no purchase notice they would have gone back and recommenced production; they had been prevented from taking that course by the acquisition; expenditure of money would have been required; since the land was being acquired it was not unreasonable that such money should not be spent. It was possible to have "disturbance" without dispossession and the present case was an example thereof. Mr Glover invited me to treat the report of the inspector, who held a public inquiry into the question whether the purchase notice should be confirmed, as part of the agreed statement of facts.

On the question whether the corporate veil should be pierced
D Mr Glover saw no distinction on the facts between the present case and the *D H N* case: there was a group of companies conducted very much as a partnership; there was but one food business; the directors of the trading company had sufficient control to enable them, had they so decided, to vest the Meanwood Road factory and the business in one and the same company. Mr Glover and Mr Glidewell were in agreement that the test which determines whether the corporate veil should be pierced is: "look at the realities, not technicalities, and do what justice requires."

Mr Glidewell said that there was a difference between a
E case in which there was an actual compulsory purchase order and one in which there was a deemed compulsory purchase resulting from a purchase notice. In the case of an actual order, notice to treat had to be served on the owners of all estates and interests and land acquired (other than short interests of the kind to which section 20 of the Compulsory Purchase Act 1965 applies). In the case of a purchase notice, he said, however, all that was compulsorily acquired was the interest of the "owner" who served the notice, and notice to treat was to be deemed to have been served on that owner only: see section 186 of the Act of 1971 and the Secretary of State's decision letter in the present case. The council had
F therefore acquired the freehold interest of the properties company in the Meanwood Road factory subject to such rights therein as were possessed by the trading company. The council had taken no steps to dispossess that company and the council had not even been asked whether it intended to dispossess the company. The reason why the company had not resumed production at the factory was in no way attributable to any action on the part of the council but was attributable to a decision taken by the company and upon advice received by it. There was no requirement in the case of a purchase notice obliging the acquiring authority to take possession.

If it is true that the trading company has suffered a loss of G profit or loss of opportunity of earning profit at the Meanwood Road factory since the fire, then I am unwilling to hold that any such loss or lost opportunity was the consequence of the deemed acquisition resulting from the confirmation of the purchase notice, because:

(1) Upon the evidence, it is impossible to say whether, and if so, to what extent, any such loss or lost opportunity was attributable to the fire or to the requirements of the hygiene department of the council.

(2) Any such loss or lost opportunity was self-inflicted in the sense that it was attributable to a decision deliberately and freely taken by the trading company in the light of advice received by H it. Since the acquiring authority had not taken possession of the Meanwood Road factory the trading company was not precluded from continuing in possession or occupation of the factory and recommencing operations there. It is true, of course, that in the matter of security of tenure the trading company would have been at the mercy of the acquiring authority, but that no doubt is a matter which the trading company or the group could have taken into account before service of the purchase notice.

Dispossession a Necessary Ingredient J

Without the assistance of any application of the D H N principle, the trading company has no *locus standi* to make a claim under rule (6) at all. Moreover, a further point which destroys its own claim is that, in my view, dispossession by the acquiring authority is a necessary ingredient in such a claim and that there has been no dispossession: see *Harvey v Crawley Development Corporation* [1957] 1 QB 485; *Minister of Transport v Lee* (1965) 17 P & CR 181. If, however, the claim were treated under the D H N principle as having been made by the server of the purchase notice, namely, the properties company, it may be that the tribunal would not permit the claim to be defeated by the failure of the acquiring K authority to take possession but that it would assume that dispossession would ultimately occur and would endeavour, as best it could, to assess compensation in advance of that event: see Cripps' *Compulsory Acquisition of Land* 11th ed at p 922 (top).

I now come to the question whether the claimants or any of them have suffered the alleged loss of profit or opportunity of earning it. It is, in my judgment, unreal for the claimants to suggest that but for the deemed acquisition they would have resumed production at the Meanwood Road factory after the fire and would have done so profitably. The L purchase notice was served because of the refusal of planning permission and because the premises had become incapable of reasonably beneficial use in their existing state. If, after the fire, the claimants had taken the view that nevertheless the premises were capable of profitable production, they would surely have sought, if not to withdraw the purchase notice (a course which may not have been open to them), to persuade the Secretary of State not to confirm it. So far from doing that, as the report of the inspector shows, the claimants at the public inquiry (which took place on November 18 and 19 1976: that is to say after the fire) positively sought confirmation of the purchase notice and did so on the M ground that the business (which the inspector appears to have assumed to be that of the properties company or perhaps that of the group) was irretrievably and disastrously in a state of loss or would have been in such a state but for the profitability of the building company and that that condition was attributable to the refusal of planning permission, a view which found favour with the inspector: see paras 37 *et seq* and 121 *et seq* of the report. Accordingly, on this part of the case I do not accept the evidence of Mr McFarlane, although I do accept that the whole of his evidence was given in good faith.

A

Claim Dismissed

For these reasons I dismiss the claimants' claim for compensation for loss of goodwill (the first of their heads of claim). I have heard no argument or evidence upon their other claims, but it appears to me that they, other than the last of them. stand or fall with the main claim (the first of them). I propose to answer the question put to me by holding that the properties company is entitled to compensation for surveyors' fees and other fees properly incurred by it in and about the preparation of its claim in respect of the valuation of the freehold estate in the Meanwood Road
B
factory, but that, save as aforesaid, none of the claimant companies is entitled to compensation under rule (6) for disturbance or any other matter not directly based on the value of land.

In the circumstances it is not necessary for me to determine whether to apply the D H N principle in the present case. If it were necessary for me to make that decision, I very much doubt whether I would apply the principle. It appears to me that the present case is distinguishable from the *D H N* case for the reason that in that case the D H N company was the victim of the compulsory purchase, whereas in the present case the trading company was in effect ini-
C
tiator of it. It might be said with some force that it ought to have put its house in order before service of the purchase notice and that it could have done so by the simple expedient (" device " in this context would hardly seem to be apposite) of transferring the business to the properties company.

The existence of the building company may also be a distinguishing feature. If I were to apply the D H N principle I would do so in such a way as to treat the properties company as being the company representing the single entity of which Lord Denning MR spoke in the passage in his judgment in the *D H N* case which I have quoted.
D
I would think it right to choose that company because, by so doing, I would be restoring the situation which appeared to exist as at the public inquiry. namely the situation in which the properties company appeared to be both owner of the freehold and owner of the business. That course would also avoid, as it were, a variation of the definition of the " owner " in section 290(1) of the Act of 1971.

It is ordered that the costs in respect of the proceedings up until the date of this decision be paid by the claimants, such costs, if not agreed, to be taxed by the Registrar of the Lands Tribunal on the High Court Scale. No such
E
taxation to take place until further order.

HEARTS OF OAK BENEFIT SOCIETY v LEWISHAM BOROUGH COUNCIL

(REF/116/1977)

(W H REES FRICS)

September 1 1978

Estates Gazette March 10 1979
(1978) 249 EG 967

F
Freehold interest in eight terrace houses in Sydenham, S E London, subject to three leases—Question of extent to which special value of freehold interest to head lessees to be taken into account—Tribunal accepts that " there is a very large marriage value in this instance "—Some special allowance called for, but not established that head lessees would always be interested in purchasing—Council's figure of £4,350 too low and £5,500 awarded against £8,100 claimed

G N N Huskinson (instructed by Field Fisher & Martineau) appeared for the claimants; R A Campbell (instructed by the borough solicitor to the London Borough of Lewisham) for the acquiring authority.

Giving his decision, MR REES said: This is a reference to determine the compensation to be paid in respect of the free- G
hold interest in eight terrace houses in Sydenham, London SE26, 60 and 62 Holmshaw Road and 59, 61, 63, 65, 67 and 69 Miall Road. The freehold interest was subject to three leases which terminate in 1997 and the dispute between the parties concerns the extent to which the special value of the freehold interest to the head lessees should be taken into account. The claimants seek a figure of £8,100 and the acquiring authority consider that the proper sum to be paid is £4,350.

The acquisition was made under Part V of the Housing Act 1957, the relevant compulsory purchase order being the London Borough of Lewisham (Bell Green/Holmshaw Road H
Part V) Compulsory Purchase Order 1971, which was confirmed on July 7 1972. Three notices to treat were served. Two were dated September 7 1972, one covering the six houses in Miall Road and the other covering the two houses in Holmshaw Road. A third notice to treat, dated July 15 1974, was served in respect of a right of way close to the eight houses, but nothing was said concerning this right of way before me. In respect of the eight houses the agreed date of valuation was May 21 1973.

There is no dispute as to the compensation to be paid in respect of no 69: it was agreed to be £500. The parties further J
agree that the value of the freehold interest in the other seven houses, on the assumption that the lessees were not in the market, is £3,850 and that for these seven houses the value of the freehold interest if merged with the other two leasehold interests but subject to the existing tenancies is £20,100. Nos 60, 62, 59 and 61 were let by the freeholder for 99 years from June 24 1898 at a ground rent of £22 per annum. Nos 63, 65 and 67 were let for the same period at a ground rent of £16.50 per annum. The head leaseholders' interests in these seven houses were purchased by the acquiring authority for £4,750 and £4,000 respectively.

Nos 60, 62, 63 and 67 were sublet to controlled tenants K
(one tenant for each house). Nos 59, 61 and 65 had in each case been occupied by two tenants, one on the ground floor and one on the first floor. The tenancies of the ground and first floors of no 59 and the ground floor of no 61 were controlled. The tenant of the ground floor of no 65 was a regulated tenant and the first floors of nos 61 and 65 were vacant.

Counsel for the claimants called J S Moor FRICS, who practises in London EC2 [partner, Walker Son & Packman]; counsel for the acquiring authority called N A Hicks BSc ARICS, deputy valuer, London Borough of Lewisham. L

Mr Moor said that the subject houses were built in about 1900 and were in fair order. He had agreed a schedule of condition drawn up by the acquiring authority. The area was one allocated for residential development and there were shopping facilities close by. In his view, the freehold interest was not an attractive investment until 1997, when the leases fell in; until then there would only be a small income. On the other hand the head lessee was in receipt of income but the buildings would be expensive to maintain and there would be a liability for dilapidations at the end of the leases, which were too short to make it worthwhile to carry out the necessary work to turn the controlled tenancies into regulated M
tenancies although grants were available for this purpose. He did not consider that it would be easy to obtain a mortgage of a head lessee's interest.

In his experience leaseholders were interested in purchasing their freeholds in such circumstances as obtained here, where there was, in his view, a large marriage value. If a head lessee purchased the freehold interest he would have a much more satisfactory investment and the disadvantages of his present interest would no longer exist. The amount which would be arrived at in negotiation between a freeholder and

A a leaseholder would depend on the strengths of the parties, but in this instance he considered that an equal division of the marriage value was justified. He spoke to the following valuation (slightly amended):

Agreed value of freehold interest (after merger of freehold ground rents + head leaseholder's interests) in 59-67 Miall Road and 60-62 Holmshaw Road (excluding 69 Miall Road)		£20,100
Less		
(a) Agreed value of freehold ground rents (excluding 69 Miall Road) *Assuming lessees not in market	£3,850 (100 YP)	
(b) Compensation paid by London Borough of Lewisham for head leasehold interests: (excluding 69 Miall Road)	8,750	
		12,600
∴ Marriage value		7,500
	÷ 2	
∴ Half marriage value		3,750
Add (a) freehold ground rents		3,850
		7,600 (197 YP)
Add 69 Miall Road (agreed)		500
∴ Compensation for freehold interest		£8,100

B

C

* By analysis of the agreed value of 69 Miall Road, if 59-67 Miall Road and 60-62 Holmshaw Road were occupied by qualified lessees (under Leasehold Reform Act 1967) then the minimum value for the freehold interest would be 7 × £500) = £3,500 (90.91 YP).

D

In support, he submitted details of 26 transactions which took place in 1972 and 1973, where the claimants had sold their freehold interests to head lessees. He relied particularly on four of these where the numbers of years' purchase (of the ground rent) paid were 190, 200, 187.5 and 167 (these figures being on the same basis as his figure of 197 in his valuation). He did not consider that the acquiring authority's figure of £3,850 (£4,350 less £500) was realistic because of the large amount of marriage value in this instance.

Under cross-examination he said that the claimants owned "thousands of ground rents," parcels of which they put on E the market from time to time. It was their custom to offer them to the lessees, some of whom refused to purchase the freeholds at first, although, later, negotiations resulted in sales in some instances. Some lessees were not financially able to purchase the freehold interest which was offered to them. If head lessees did not purchase, his clients continued to hold the investment. No negotiations had ever taken place with the head lessees in the case of the eight houses, the subject of this reference. He gave at least one example of an equal division of marriage value from his comparables, which were in all parts of London.

F He accepted that the rate of interest used to arrive at the agreed value of £3,850 reflected possible capital appreciation, sales to sitting tenants and the possibility of getting vacant possession. He did not consider that the rates of interest used reflected the possibility of sales to the head lessees. However, £3,850 included the possibility that the head lessees might buy later. He did not consider that £3,850 reflected any element of marriage value. In his view, the £8,750 paid by the acquiring authority for the head lessees' interests excluded marriage value and he had used that figure because he considered that the acquiring authority should be deemed to have paid a proper price for them.

Mr Hicks analysed the agreed figure of £4,350 as follows: G

Total ground rents (excluding no 69)	£38.50	
YP for 24 yrs at 6%	12.55	£483
Reversion to estimated exclusive fair rents	2,379	
YP in perp def 24 yrs at 9%	1.404	3,340
		3,823
Agreed say		3,850
Add value of no 69 (agreed)		500
		£4,350

H

He had assumed that possible occupiers of the first floors of nos 61 and 65 would be able to purchase the freeholds of those houses under the Leasehold Reform Act 1967, as amended, after they had been in occupation for five years. He considered that £4,350 was the "investment value" in the open market, the lessee not then being interested in purchasing. The 6 per cent and 9 per cent reflected the possibility of capital appreciation, rental increases in the future, the possibilities of later sales to the head lessee and of obtaining vacant possession from time to time. The sales of the head lessees' interests in the subject properties had J been agreed in both cases with professionally qualified surveyors and neither figure included marriage value. Any possible bid by the freeholder was reflected in the rates of interest used. If the freeholds were to be offered in the open market for sale by private treaty, the lessees might or might not buy them: head lessees were not always interested in purchasing their freeholds and he considered that it was good policy for the claimants to put the ground rents back into their portfolio.

He relied particularly on the sale in 1974 at £5,100 of 27-65 (odd) Holmshaw Road to support his opinion that £4,350 was the correct figure. These were immediately oppo- K site some of the subject houses. He submitted an analysis of this price which showed the ground rent capitalised at 6 per cent for 27 years, with the reversion, on the basis of estimated fair rents, at 8½ per cent in perpetuity deferred 27 years.

Under cross-examination he agreed that the freehold interests in the eight subject houses were worth more than £4,350 to the head lessees: there could be an incentive for them to purchase the freehold, it depended on the views they took. They could pay more than £4,350 for the freehold interests and resell the merged interests, each at a profit. He did not L consider that a special allowance should be made where the marriage value was large; such value was nearly always present with this type of interest.

There is no dispute that the personalities of the freeholder and of the head lessees are not to be taken into account: "the parties are to be conceived as two passengers on the Clapham omnibus, devoid of prejudice or eccentricity, who, in any given circumstances, might have been expected to behave reasonably" (see *Mountview Estates Ltd v London Borough of Enfield* (1968) 210 EG 163 at p 165). I am in no doubt that the special interest of a freeholder in merging a leasehold with his own interest is an admissible market factor M (see *Lambe v Secretary of State for War* [1955] 2 QB 612).

I accept Mr Moor's evidence that if such freehold interests as I am concerned with here were offered for sale in the open market on the relevant date the head lessees might or might not be interested in purchasing. His experience of selling ground rents belonging to the claimants is a matter to which I attach considerable weight, but I reject his approach resulting in £8,100 because it has not been established that head lessees would always be interested in purchasing.

However, I accept the submission of counsel for the claimants that there is a very large marriage value in this

A instance and for this reason I think that the figure spoken to by the valuer for the acquiring authority (£4,350) is too low. I have no doubt that the rates of interest of 6 and 9 referred to by Mr Hicks do reflect to some extent the possibility of the head lessees buying later, possible capital appreciation and the possibility of obtaining vacant possession. If one was required to disregard these factors in arriving at the value of the freeholder's interest in this instance, I accept that 9 per cent would be too low. I also agree with Mr Hicks that that would be a very difficult basis to apply. Here there is a large marriage value and I come to the conclusion that some special allowance is called for.

B Weighing all the factors which were put in evidence before me, I come to a figure of £5,500 as the amount of compensation to be paid for the freehold interest in the eight houses. To this is to be added the claimants' surveyor's fee based on scale 5 of the scale of professional charges of the Royal Institution of Chartered Surveyors and any legal costs properly incurred between the date of service of the notices to treat and the date of the notice of reference.

Having opened the sealed offer of compensation lodged by the acquiring authority, I find that the amount thereof is less than the amount of my award; accordingly the acquiring authority will pay to the claimant his costs of this reference,
C such costs unless agreed to be taxed by the Registrar of the Lands Tribunal on the High Court Scale.

DICCONSON HOLDINGS LTD v ST HELENS METROPOLITAN BOROUGH

(REF/129/1977)

(V G WELLINGS QC)

December 18 1978

Estates Gazette March 17 1979 and March 24 1979

D (1978) 249 EG 1075 & 1178

Land at Billinge, Merseyside, of nearly half an acre acquired under purchase notice—Whether "golden strip" providing key access to rear "white" land—Planning permission refused several times for residential development of subject and rear land—Subject land held to be key which would unlock any development value rear land might have—Reasonable expectation that rear land would be developed within 5 to 7/8 years' time—Speculator would be willing to pay price for subject land above value for its intrinsic development—Award of £25,000, which was price paid by
E **claimants—£50,000 reasonably to be expected if planning consent received for rear land in 7 years—Obiter view that rule (3) inapplicable—Meaning of "particular purchaser" —Rule applies to actual not hypothetical bids**

Michael Fitzgerald (instructed by Compton Carr) appeared for the claimant company; Michael Rich (at the hearing) and Robert Carnwath (at the decision) (instructed by the solicitor to St Helens Metropolitan Borough) for the acquiring authority.

Giving his decision, Mr Wellings said: In this reference the claimants claim compensation for the acquisition of a piece of land (which I will call "the reference land"),
F formerly the site of the British Legion Club (now demolished), at Billinge, St Helens, Merseyside. The question for determination by the Lands Tribunal, broadly stated, is whether the reference land ought to be valued as a golden strip or key giving access to and unlocking the development value of land lying to the rear thereof and, in any event, what ought to be the amount of compensation payable?

The reference land has an area of 2,360 sq yd (0.488 acre). It lies on the north side of Ash Grove Crescent at its junction with Main Street, Billinge. It lies at the very centre of Billinge and most of the surrounding land has been developed. In particular, the land to the west and south-west of the reference land has been developed by a G
modern housing estate (I will call it "the Ash Grove Estate") and land to the south of that estate is now in the course of being developed as another housing estate (I will call it "the Sefton Fold Estate").

The reference land is considerably narrower at its northern end than it is at its frontage with Ash Grove Crescent. Its northern end leads to a narrow neck of land which in turn leads to what I will call "the rear land," an area of 14.14 acres of agricultural land owned and farmed by John Joseph Valentine, Mrs (I hope that that is her correct description) Cecelia Mary Valentine, Richard Valentine and Charles Roderick Valentine as part of their Ash Grove Farm, H
which consists of some 80 acres. I believe that the rear land includes both the farmbuildings of Ash Grove Farm and the neck of land which I have described.

Immediately to the west of the rear land and directly adjoining the Ash Grove Estate on its northern side is an area of land, approximately rectangular in shape, of 7 acres (I will call it "the 7-acre land"). To the west of that land is a piece of land of irregular shape consisting of 5.06 acres (I will call it "the 5-acre land"). To the south of that land and on the western side of the Ash Grove Estate and of the Sefton Fold Estate is a narrow tongue of land of 10.06 acres, the broadest part of which is in the north and the narrowest part in the south (I will call this piece of land J
"the tongue of land"). I believe that the rear land, the 7-acre land, the 5-acre land and the tongue of land are all part of Ash Grove Farm; it is my understanding that all these parts, other than the 5-acre land (which is woodland) are in agricultural use by the Valentine Family. I believe (although I have only counsel's unchallenged statement for this) that Ash Grove Farm is farmed by the Valentine Family in conjunction with two other farms in the vicinity: Houghwood Farm and Sefton Fold Farm: a total (not including the rear land) of about 200 acres.

At the present time an access to the rear land exists on K
its northern side from Beacon Road (which leads into Main Street), but it is common ground that it would not be a suitable access for residential development on the rear land. Accordingly, if residential development of the rear land is to take place it will be necessary to provide a suitable access for that land. The questions arise whether it is possible to provide a suitable access to the rear land at all and, if so, whether the reference land or some other means of access would be the more suitable.

The claimants purchased the reference land on July 24 1975 for the price of £25,000. They purchased it because
Thomas Nicholson (whom, if I may be permitted moment- L
arily to pierce the corporate veil, I regard as in truth the claimant), one of their directors, believed that the rear land was ripe for development and that the reference land was the key access to it and because Mr Nicholson, or Mr Nicholson and his wife, or one of the companies (the claimant company or Tomnik Ltd) controlled by Mr Nicholson had the benefit of an option granted by the Valentine Family to purchase the rear land. I have seen no documentary evidence of the option, but I believe that the purchase price contemplated in the event of exercise of it was £50,000 and that it expired in April 1976 without having been exercised. M
As will be seen, when I come to deal with Mr Nicholson's evidence, there was subsequently a new option.

Prior to the expiration of the first option a number of applications for planning permission for residential development of the rear land and the reference land had all failed and on June 26 1976 or thereabouts the claimants served a purchase notice. The notice was served on the ground that the reference land had become incapable of reasonably beneficial use in its existing state and that it could not be rendered capable of reasonably beneficial use by the carrying out of any other development for which permission has

A been granted or is deemed to be granted or for which the local planning authority or the Secretary of State for the Environment has undertaken to grant permission. The acquiring authority has not yet taken possession of the reference land.

Michael Fitzgerald, for the claimants, called Thomas Nicholson, whom I have already described; James Cannon FRTPI; Robert William Arthur Foreman FRICS MICE; and Thomas Roy Aspinwall FRICS, senior partner in the firm of Thomas Aspinwall & Partners, of Skelmersdale, Lancs. Michael Rich, for the acquiring authority, called John Woods MICE, principal assistant engineer in the B Department of Technical Services of the acquiring authority; Vernon Arthur Lewis BSc FIME, deputy director of technical services of the acquiring authority; John Parker Wesencraft BA MRTPI, principal planning officer of the Merseyside County Council; and S W Monham ARICS, district valuer, St Helens.

Planning History

1. The approved Lancashire County Development Plan (approved in July 1956) contained no detailed land use C proposals relevant to the present case, but the Billinge, Orrell and Up Holland development proposals plan, which has been approved by the acquiring authority and the Merseyside County Council as a basis for development control from April 1 1974, showed the reference land as zoned for primary residential use, the rear land as white land and the 7-acre land and remainder of the land of Ash Grove Farm to the west as provisional green belt. The Ultimate Village Plan for Billinge (which plan has not received any official recognition) identified Ash Grove Farm (apparently the whole of it) as a further possible area for housing.

2. On January 20 1964, the then Lancashire County Council granted to the Valentine Family planning permission D for residential development of what is now the Ash Grove Estate. The same document also refused permission for residential development of the piece of land which I have called the neck of land lying to the north of the reference land and did so on the ground that it was not suitable for development by reason of its physical characteristics, and suitable access to it was not practicable either from Main Street or from what is now the Ash Grove Estate. The document also refused permission for residential development for what is now the Sefton Fold Estate. I need not go into the grounds of that refusal because, as I have indicated, that development is in the course of construction.

E 3. In 1973 Robinson Construction Ltd made two applications for planning permission in respect of the reference land, the one for the erection of four shops and 15 flats, the other for 20 flats. On the first application no decision was given and the second was withdrawn.

4. On July 17 1973, Mr Nicholson, acting as agent for the Valentine Family, made an application in respect of the whole of Ash Grove Farm and the reference land for residential development (192 detached and semi-detached houses and bungalows at 12 units per acre), siting of access roads and public open-space areas. In a letter dated August 7 1973 to the surveyor of Billinge Urban District Council, the divisional planning officer said that there might be F arguments in favour of releasing the rear land, but there was the problem of access to overcome and that this would seem to be via the reference land. In a further letter dated November 8 1973 to the same surveyor the divisional planning officer said that the present indications were that the application was likely to be viewed favourably. Nevertheless on December 5 1973 the Billinge Urban District Council recommended that there should be a refusal on the grounds of inadequate access to the proposed development and that the land was of high agricultural class and was agreed should remain in agricultural use. The Ministry of Agriculture, however, made no such objection. Subsequently the application was withdrawn.

5. On July 1 1974 Mr Nicholson, on behalf of the Valentine Family, made an application for development similar G to that last mentioned, but in this case, in a plan accompanying the application, the reference land was shown as an access to the residential development and what I will call "the gap" was shown as another. The gap lies in Ash Grove Crescent at a point between numbers 22 and 24 in that road and consists of the adjoining sites of two proposed houses which were omitted to be built specifically with the object of affording access to the rear land. At the time of this application the gap was owned by Moore Brothers (Builders) Ltd, a company which I believe to be unconnected with Mr Nicholson. The Ministry of Agriculture, Fisheries and Food, having been consulted, stated H that it did not propose to enter an objection. Nevertheless, on October 16 1974, the application was refused on the grounds that: "(1) The proposed development is contrary to the provisions of the Billinge, Orrell and Up Holland Development Proposals Map which allocates this site as white land. (2) Access to the site cannot be provided to allow for the satisfactory development of the land."

6. On July 5 1974 Robinson Construction Ltd made an application for development of the reference land by the erection of 11 two-storey town houses. The application was refused on September 9 1974 on the grounds that the proposed density was excessive and that the area of the site was insufficient for the type of development proposed. J

7. On February 6 1975 Robinson Construction Ltd applied for permission to develop the reference land by the erection of "dwelling-houses." This application was refused on the ground that "the development of the site as proposed would prejudice the future development of the land forming part of Ash Grove Farm." This is no doubt a reference to the rear land.

8. On April 28 1975 Tomnik Ltd made an application in respect of Ash Grove Farm less the tongue of land for residential development and use as public open space. The plan accompanying the application showed access as being across the reference land. No access through the gap was K shown. In respect of this application, the county planning officer, on August 19 1975, wrote to the director of technical services of the acquiring authority as follows:

> "I note that although the site is shown "white" land on the approved development plan for Billinge and Winstanley, it is shown residential on the draft village plan for Billinge and that past investment in infrastructure has been made in anticipation of future residential development. The application will also provide a planning gain in that a considerable area of open space, required as part of the Billinge Hill to Carr Mill linear park will be designated to the local authority.
>
> Whilst I am prepared to recommend to my council that the L application be approved, I must express my reservations about the release of "green field" sites in the St Helens area. As you are well aware, a considerable amount of derelict land or despoiled sites exist in the St Helens Metropolitan District, some of which will be suitable for residential purposes. It is likely that the County Council will adopt Structure Plan policies which will encourage the development of such sites, whilst the development of productive agricultural land would be discouraged."

9. With respect to the last-mentioned application, Mr Nicholson, on September 10 1975, wrote to the director of technical services of the acquiring authority that he had agreed terms with the Valentine Family for the purchase of 4.11 acres of land (part of the 5-acre land I think) com- M prised in the application which would be dedicated to the local authority if and when planning consent is granted for the remaining part of the application site, and that the Valentine Family would be prepared, if planning consent were granted, to enter into negotiations for the sale to the local authority of the tongue of land at a price to be agreed with the district valuer. Discussions then took place between Mr Nicholson and the county planning officer and other persons with respect to improving the reference land as a main access to the rear land and the use of the gap as a secondary access thereto but limited to use by the emergency services only.

10. With respect to the application dated April 28 1975, a report dated October 14 1975 to the planning and development committee of the acquiring authority by a subcommittee of that committee stated that the question of access points had now been resolved and that the Ministry of Agriculture had no objections. However, in a letter, also dated October 14 1975, the Ministry of Agriculture, Fisheries and Food informed the county planning officer that planning permission should not be granted for the reason that "the land is of high agricultural value, being Grade 2 on the Ministry's Agricultural Land Classification Maps." Meanwhile, on October 2 1975, a public meeting took place in Billinge with respect to the application dated April 28 1975. One hundred and fourteen people attended and, according to a written report of the meeting, the "unanimous feeling" was "that the application be turned down."

11. In a document, exhibit 41, which purports to be a report dated October 16 1975 of the Development Control Subcommittee of the acquiring authority, there is what purports to be a recommendation that the application dated April 28 1975 be approved subject to conditions which include a restriction of access through the reference land and the use of the gap for access by pedestrians and emergency services only. In another document, exhibit 41A, also purporting to be a report of the same subcommittee and also purporting to be dated October 16 1975, there is a recommendation that the application dated April 28 1975 be refused. On October 20 1975 that application was in fact refused and on the following grounds:

"(1) The site is allocated as "white land" on the Lancashire County Development Plan, where existing uses for the time being are intended to remain substantially undisturbed.

(2) The proposal is premature in that the site lies within an area not allocated for residential purposes whilst land in the area allocated for residential purposes remains undeveloped.

(3) The proposal would result in the loss of high quality agricultural land, being Grade 2 Land in the Ministry of Agriculture Classification."

The first and third such grounds were recommended in exhibit 41A. The second ground was not included in that exhibit in so many words although it was stated there that the application would conflict with county strategic policy of concentrating resources in the older urban areas of the county and away from the peripheral areas not allocated for development. It appears to me that exhibit 41 was a draft by a junior officer whose views were not accepted and that exhibit 41A represents the views of the relevant subcommittee.

12. On January 8 1976 the county council resolved that in the event of an appeal being lodged against the refusal of permission dated October 20 1975, the county council would support the acquiring authority at a public inquiry on its decision to refuse the application for three stated reasons. The first two reasons were identical to the first and third reasons in the refusal dated October 20 1975. The third reason was as follows:

"The approved Stage 1 Report of the County Structure Plan proposes the concentration of investment and development within the urban county and particularly in those areas with the most acute problems, enhancing the environment and encouraging housing and economic expansion on derelict and disused sites, and it would restrict development on the edge of the built-up areas to a minimum; there would be a reciprocal effort to enhance and conserve the natural features of the county's open land and its agriculture while ensuring its capacity to meet the county's needs for leisure, recreation and informal education."

13. Meanwhile on December 17 1975 the Valentine Family and Tomnik Ltd made a joint application in respect of Ash Grove Farm less the tongue of land. The proposal was for development of the rear land for private housing with part reserved for sheltered housing, use of the 7-acre land for recreational purposes and use of the 5-acre land for public open space. The farm buildings were shown on the plan accompanying the application as excluded from the rear land and access was shown as being over the reference land and through the gap. The application was refused on April 8 1976 on precisely the same grounds as those contained in the refusal dated October 20 1975.

14. On December 17 1975 Tomnik Ltd applied for permission to carry out residential development on the reference land. On June 3 1976 the application was refused on the ground that "the development of this site for residential purposes is premature in that it would prejudice the future development of land at the rear, forming part of Ash Grove Farm, which is at the present time zoned as 'white land'."

15. All the above-mentioned planning applications were outline applications.

Current Planning Policy

1. The Merseyside Structure Plan—Stage 1 Report was approved by the county council on September 30 1975 as an interim policy statement of the county council, the relevant resolution stating that pending consultation with other bodies as mentioned in the report, the strategy there set out be adopted by the county council as an interim measure for the purpose of controlling development. On January 5 1976 the acquiring authority accepted the Stage 1 strategy in principle and resolved that it be borne in mind when considering development decisions but that it should not be used as a rigid development control policy until questions of cost, resources and other matters mentioned in the resolution had been discussed with the county council.

2. The Stage 1 strategy, as stated in the Stage 1 Report is as follows:

"The proposed strategy would concentrate investment and development within the urban county and particularly in those areas with the most acute problems, enhancing the environment and encouraging housing and economic expansion on derelict and disused sites. It would restrict development on the edge of the built-up areas to a minimum.

There would be a reciprocal effort to enhance and conserve the natural features of the county's open land and its agriculture while ensuring that its capacity to meet the county's needs for leisure, recreation and informal education is exploited."

The two paragraphs quoted above were underlined in the report.

3. The Stage 1 strategy has not yet received official Government approval but, so far as can be judged, from certain documents to which I shall refer, the strategy has received at least unofficial Government approval. The documents in question are: "Targets for Merseyside: a Response by Government Departments" (exhibit 82), a letter from the Secretary of State for the Environment dated May 3 1977 (exhibit 82) addressed to Councillor W H Sefton and accompanying the last-mentioned document, and a document (exhibit 113) which purports to be the text of a speech intended to be delivered by the Secretary of State on September 17 1976.

4. Exhibit 109 is an important document upon which the claimants much rely. I refer to it under the heading of Planning Policy, but it is also material to the question of the need or lack of need in St Helens for land for housing, to which question I shall refer. The document is a joint report by the Town Clerk and the Technical Services Department of the acquiring authority to the Policy and Resources Committee of that authority, which report was approved by the authority on April 7 1978. The report expresses reservations about the Stage 1 strategy. It states *inter alia:*

"Housing conditions and Land Supply—unlike some of the other problems referred to, this may be one unique to St Helens in the Merseyside district. The objective of improving housing conditions, and the question of land supply are inextricably bound together. There is a direct relationship between the clearance of older, high-density housing and their replacement to modern standards of amenity and density and the retention of population (which is the essence of the draft structure plan) on the one hand and the supply of sufficient land on which to erect houses for people moving within the district, as well as those wishing to live here, on the other hand. Current indications are of an acute shortage of housing land arising on and after 1981. Land implications of the housing population estimates have not been examined in detail, although a survey by the district has shown that the land required

A to maintain the population at around 190,000 (see section 3 below) is about 200 acres over and above the current sites identified in the land availability resources survey. There is considerable uncertainty over the exact acreage that would have to be released but two aspects in particular are worth noting:

(i) The land would need to be released after 1981;

(ii) The sites would almost certainly have to be 'green field.' This last factor goes against the aims of 'Targets'."

In section 3 the report referred to the relationship of housing, population and land supply and said:

"There is some difference between the figures adopted by the county council and those which the Borough's officers regard as B realistic. On the basis of the housing investment programme, the Borough's estimated housing stock in 1981 will be 71,110 reflecting a population of some 189,000. By 1986 the Borough Council's officers estimate that at the lowest the housing stock will be 70,310 with a population of between 178,500 and 179,000."

The report contained the following recommendations:

(a) The need to revise the strategic approach towards St Helens and to define the problems referred to in section 2 (quoted above) and specifically state the policies aimed towards their solution;

(b) To accept that 189,000 is the most realistic minimum population for 1986 and that the structure plan will C aim at securing that figure within the framework of the housing investment programme and other county and district policies;

(c) The structure plan housing strategy be based on housing investment programmes;

(d) The structure plan accept in relation to St Helens the importance and necessity of a policy for the release of "green field" sites for housing so far as sufficient capacity does not exist or cannot be made available satisfactorily elsewhere in the Borough.

Highway Considerations

D The acquiring authority do not accept that the reference land would be a suitable means of access to residential development on the rear land. Mr Woods, on their behalf, said that if the rear land were to be developed residentially, for the size of housing estate which might be expected thereon, two means of access, a primary access and a secondary access, would be necessary in order to reduce the possibility that the estate might be isolated by ice, snow, traffic accident, burst water main or other cause. Mr Woods said that the reference land was not suitable for either a primary access or a secondary access because:

E (1) Standard sight lines (agreed between the parties to be 15 ft by 200 ft) could not be achieved on the frontage of the reference land;

(2) The traffic flows in Main Street would be increased undesirably (within a five-year period, by some 2,750 vehicles over a 16-hour day, being approximately 160 vehicles more in both the morning and the evening peak hour than at present).

(3) Traffic at the junction of Ash Grove Crescent and Main Street ought to be reduced rather than increased because of an undesirably high accident record at that point.

F (4) If there were but one access to the housing estate, the length of road across the reference land thereto would be 700 ft, which was too long for a cul-de-sac, the maximum desirable length of which was 600 ft.

(5) The exit from the housing estate would be at the foot of a 10 per cent gradient (1 in 10) contrary to design desiderata.

Mr Woods thought that, if the tongue of land were to be sold to the acquiring authority as suggested in Tomnik Ltd's letter dated September 10 1975 (to which I have already referred), a suitable primary access could be provided over that land, the 5-acre land and the 7-acre land to the rear G land. I will call this proposed access "the linear park road" because the acquiring authority have a proposal to use the tongue of land as park land and that proposal is known as "the linear park." Mr Woods was of the opinion that the linear park road would be a suitable access notwithstanding that it would run across park land.

Having formed that view and being of the opinion that the reference land was not suitable either as a primary or as a secondary access, Mr Woods said that there was no suitable secondary access to the proposed housing estate and for that reason, he believed, planning permission for residential development of the rear land ought to be resisted. H He expressed that opinion notwithstanding that none of the refusals of planning permission which I have mentioned above was on highway or traffic grounds. Indeed, of course, the last two refusals in respect of the reference land tacitly accepted its suitability as a means of access to a housing estate on the rear land.

The witnesses on both sides of course considered the question of the suitability of the gap as a means of access to the rear land, but they appeared to be in agreement, for reasons which I need not specify, that it was unsuitable and that, if used at all, it should be limited to use by pedestrians and by the emergency services. J

Mr Foreman thought that the proposed linear park road was quite unsuitable. It was unthinkable and the very negation of highway and town and country planning to put a road to a housing estate across park land. It was also an expensive operation (the cost of construction of such a road was agreed to be £60,000). He ruled out the proposed linear park road altogether. He rejected the contention that a secondary access to the proposed housing estate on the rear land would be necessary. One suitable access would be sufficient. He believed that the reference land would constitute a satisfactory access. It could achieve sight lines, K curve radii and width which were suitable. The highway design standards were not absolute and could be varied to suit individual cases. There would be no adverse effect in environmental terms and the position at the junction with Main Street would be acceptable. The cost of construction (agreed to be £35,000) was reasonable. Some part (8/13ths) of that sum could, he said, be saved if, in addition to the road, part of the reference land was used for development of two or three houses, which development was feasible. That would reduce the cost of the road to £13,461.

Need for Land for Housing L

According to figures produced by Mr Lewis and Mr Wesencraft there is a supply of land (that is to say, sites with planning permission not yet implemented, sites part covered by planning permission not yet implemented, sites zoned for residential development in a statutory or non-statutory plan and sites with a residential option) as follows:

(1) In St Helens district, 7.1 years' supply if the rate of completion of houses in 1976 be applied;

(2) In that district, 4.4 years' supply if the average rate of completion over the 5-year period between 1971 and 1976 be applied; M

(3) In Merseyside, there is an 8.6 years' supply based on 1976 completions or nearly 6 years' supply based upon the average rate of completions between 1971 and 1976.

In the light of this supply situation, Mr Lewis and Mr Wesencraft said that, having regard to the Stage 1 strategy of the Merseyside Structure Plan (which I have already described) the release of further green field land, namely the rear land, is not justified and the grant of planning permission for development of that land for residential purposes would be strongly resisted by both the acquiring authority

A and the county council. Mr Lewis, however, conceded that if one accepts the premise contained, or apparently contained, in exhibit 109 that the population of the St Helens district ought to be maintained, then there is in that district a shortage of land available for housing.

Valuations

Mr Aspinwall's primary valuation was as follows:

Area of developable land (the rear land) forming part of Ash Grove Farm which would be available for residential development on the acquisition by the acquiring authority of the reference land—14.14 acres	
B 14.14 acres at £27,500 per acre (on the basis of a notional residential planning permission)	£388,850
Deferred 5 years at 6% YP	0.747
	£290,470
Estimated cost of purchasing access for residential development	
Deferred value of land with access referred to above	£290,470
Less value as accommodation land 14.14 acres at £1,250 per acre	£ 17,675
C Increase in value due to access	£272,795
For the purpose of this valuation it is considered that a fair and equitable approach in assessing the amount a prospective purchaser would expect to pay for the reference land as providing access to realise the increase in value in the adjoining rear land may be taken to be	£ 95,000

Add claimants' surveyor's fees in accordance with Scale 5(a) of the scale of charges of the Royal Institution of Chartered Surveyors.

Mr Aspinwall said that £95,000 represents the price which a reasonably minded purchaser or indeed speculator in the D open market would expect to pay for the reference land on the assumptions that the reference land is both a necessary and a sufficient access to the rear land, that the rear land is ripe for development for residential purposes and may be expected to be developed within three to five years and that planning permission for such purpose will be forthcoming. Mr Aspinwall would have no difficulty in finding a purchaser who would be prepared to pay that sum for the reference land in the expectation that he would share in the profit realised by the owner for the time being or the developer of the rear land, that price and that profit being calculated in the course of negotiations between them conducted in a reason- E able manner.

Mr Aspinwall's price of £27,500 per acre represented his own opinion supported by transactions (to which he referred) which he regarded as comparable. The figure represented the value of vi-gin land with the benefit of planning permission with sewers and services available but needing to be brought to the site. He deferred at 6 per cent rather than at 10 per cent because he regarded the rate at which money could be borrowed as being irrelevant for this purpose.

Mr Aspinwall produced an alternative valuation intended to be utilised in the event that it should be held that rule F (3) of section 5 of the Land Compensation Act 1961 applies in the present case. It was as follows:

Site area:	2,360 sq yd or thereabouts	
Basis:	Assumed that planning permission would be granted for housing development by 9 linked detached houses with garages. The valuation is in accordance with Part II of the Land Compensation Act 1961, provided rule (3) of the rules for assessing compensation under section 5 of the Land Compensation Act 1961 would apply.	
	2,360 sq yd:	
	9 house plots at £2,500 per plot =	£22,500

Add claimants' surveyor's fees in accordance with G Scale 5(a) of the scale of charges of the Royal Institution of Chartered Surveyors.

He explained that he had based his valuation of £2,500 on a residual valuation resulting in a figure of £2,875 per plot which had been prepared by Mr Nicholson and to which I shall refer.

Mr Monham, the district valuer, did not have a valuation corresponding to Mr Aspinwall's primary valuation. Nevertheless he expressed the view that the value of the rear land for housing purposes was £15,000 per acre, but only where planning permission actually exists or there is a guarantee of H it within a period of five years. If planning permission were not guaranteed but there was merely a reasonable hope of it being obtained within a period of five years, he did not think that a purchaser would take the rear land at a price in excess of its agricultural value. He did not think that anyone would pay a hope value for the rear land.

If he were to prepare a valuation on the basis that the reference land is the key to the unlocking of the development value of the rear land, he would multiply 14 acres (rather than 14.14 acres) by £15,000 (£210,000) and deduct from the total the cost of construction (£35,000), leaving a balance of £175,000. He would then defer that figure over 5 years J at a rate of 10 per cent, producing a multiplier of 0.621 YP. The product is £108,675. After deducting £17,500 (the value of 14 acres as accommodation land at £1,250 per acre) the increase in value due to access would be £91,175. Like Mr Aspinwall, he would divide that sum by three, giving a balance of £30,392. He would defer at 10 per cent because a developer would have regard to loss of interest on his capital, to the cost of finance and would expect a remunerative rate equivalent to an unsecured ground rent.

Mr Monham, having formed the view that the reference land has no key or merger value, produced a valuation which had no regard to the intrinsic development potential K of the reference land itself only. That valuation was as follows:

Statutory basis of compensation:	Section 180, Town and Country Planning Act 1971. Part II, Land Compensation Act 1961.
Valuation date:	Possession has not been taken by the acquiring authority. Valuation date is as set at the date of agreement or the date of the tribunal's decision.
Interest to be valued:	Freehold with vacant possession. L
Planning assumptions:	Land available for redevelopment for residential purposes or for social club.
Area of site:	2,360 sq yd (0.488 acre).
2,360 sq yd @ £7.50 per sq yd	£17,700
Say	£18,000

In this valuation Mr Monham gives the reference land its full development value, notwithstanding that no actual planning permission in respect of that land exists. Mr Monham, however, takes account of the zoning in the rele- M vant development plan of the reference land for residential development and, for that reason, assumes that there is a "guarantee" of the grant of planning permission. He had assumed that three houses could be built on the reference land and, although he had not stated as much in his valuation, he preferred to regard them as being worth £6,000 per plot. He supported that figure by reference to transactions to which he referred me and which he regarded as comparable. If the reference land were developed by the erection of nine linked houses on the land (as assumed by Mr Aspinwall) his valuation would not have differed.

A
Other Evidence

Mr Nicholson said that in his personal capacity, he had acted as agent for the Valentine Family for some 20 years and had acted for them and advised them in and about the sale by them of the land which now constitutes the Ash Grove Estate and the Sefton Fold Estate. He had had the benefit of an option granted in December 1975 by the Valentine Family to purchase the rear land at the price of £50,000. It was for that reason and because he took the view that the reference land was the key to the residential development of the rear land that he purchased the reference land. The option had expired in April 1976 and after that
B
date the Valentine Family were not willing to sell the rear land because they were afraid of the effect upon their tax position of the Development Land Tax Act 1976 which shortly was to come into force. It was for that reason that he had not appealed against the refusals of planning permission which his company had received in respect of the rear land and the reference land in 1976.

He said that the rear land would certainly be developed residentially; it was the last piece of suitable land for residential development in Billinge. There was no land available for building in the area of Billinge now. The rear land represented one of the finest sites in south-west Lancashire.
C
The Valentine Family continued to farm the rear land; he did not know the agricultural quality of that land. He had received two offers for the rear land, one of them subject to contract, and both of them subject, *inter alia,* to satisfactory planning permission for residential development being granted. The one was for the price of £26,000 per acre and the other for the price of £27,500 per acre.

Mr Nicholson said that, in addition to being the key to unlock the development of the rear land, the reference land was suitable for residential development on its own: it was capable of being developed by the erection of nine housing units in semi-detached or stepped form. He valued each
D
housing plot at £2,875. He had done this by a residual valuation [see next column].

Through his company, Tomnik Ltd, on September 28 1977, he had obtained from Moore Brothers (Builders) Ltd an option to purchase the freehold estate in the gap for the price of £4,000. He had not exercised the option.

After Mr Nicholson had given evidence, there was an adjournment of about two months, and when the hearing resumed Mr Nicholson was recalled. He then produced an agreement made between the Valentine Family of the one part and himself and his wife of the other part giving to
E
them the option to purchase the freehold in an area of 18 acres, including the rear land, for the price of £100,000. The agreement required that the option should be exercisable within one year from its date. Mr Nicholson said that he did not intend to exercise this option because, by reason of the notice and its acceptance, he was committed to part with the reference land.

James Cannon had been until February 1974 divisional planning officer at Wigan, responsible for an area which included Billinge. He said that there was no land in Billinge available for housing except the rear land. Outside Billinge there was little such land available. There was no land
F
available to the private sector other than the rear land. In the adjoining district there appeared to be large areas available, but in fact none were available because the owners would not sell. He said that it was the adoption of Stage 1 of the structure plan which changed the county council's mind with respect to the grant of planning permission. His personal view was that one could not overnight change a situation by passing a policy. He thought that if there had been an appeal to the Secretary of State it would have succeeded. He could not say that success would have been certain. There were features which justified the grant of permission and other features which justified refusal.

G

Selling price of each plot and one house erected thereon		**£12,500**
Deduct:		
(i) Building costs 900 sq ft @ £8 per sq ft	**£7,200**	
(ii) *add* cost of roadworks and roadsurfacing per plot	**£ 500**	
	£7,700	
(iii) *add* 25% for overheads, estate agent's commission, legal fees and profit	**£1,925**	
	£9,625	
Residual value		**£ 2,875**

H

Argument

Michael Rich, for the acquiring authority, said that this was not a planning appeal. The question was not whether the planning refusals were right or whether planning permission could now reasonably be expected. The issue was: would the purchaser in the open market attribute to the reference land an access value enabling him to realise the development of the rear land? That was an issue of fact on which the onus of proof lay on the claimants. The answer
J
to the question was in the negative because, as a matter of fact, it was clear that no one existed in the market who would purchase the reference land in the belief that enabled the development value of the rear land to be realised. The reasons for this were:

(1) No purchaser would assume that planning permission for the rear land would be granted;

(2) The reference land was not necessary in order to develop the rear land;

(3) No speculator existed who would assume otherwise.

Mr Rich said that, while the officers of the county council
K
at one time on balance were prepared to recommend a grant of planning permission for the rear land, that was a view which had been resiled from. The officers of the district council likewise had been prepared to recommend planning permission for the rear land. But the position was that planning permission had always been refused. It followed that any prospective purchaser of the rear land would have to face the history of the planning refusals, the hardening of the view of the Ministry of Agriculture, Fisheries and Food, the advent of the Stage 1 strategy and the certainty of refusal at the level of the local planning authority. The view that
L
no speculator would be interested in the reference land was, he said, supported by the fact that Mr Nicholson had not exercised the 1975 option to purchase the rear land notwithstanding that, according to the evidence of Mr Aspinwall, he could have made a profit of £135,000.

Mr Rich had not objected to the production of the new option at the price of £100,000 because the Valentine Family were apparently willing to sell for that price notwithstanding Mr Aspinwall's valuation of £295,000. There could be no better evidence that no one else was in the market for the rear land. On what other basis could they have agreed to sell at such an undervalue? Other pointers were that Mr
M
Nicholson was not prepared to exercise the new option and his service of the purchase notice. The proper basis on which to value the reference land was the basis on which the district valuer had valued it, namely on the assumption of planning permission for the erection of three houses on that land alone. The acquiring authority conceded that planning permission would be granted because it was, as it were, "guaranteed" by reason of the notation in the development plan.

Michael Fitzgerald submitted that if planning permission were refused at the present day for development of the rear land the refusal would not be on the same grounds as those

A on which the earlier refusals were made. On the question of the need for land for housing he emphasised the importance of exhibit 109. He said that the district valuer's valuation of the reference land was, in effect, £36,000 per acre, which figure supported Mr Aspinwall's valuation of £27,500 per acre for the rear land. That was an extraordinary contrast with the district valuer's valuation of £15,000 per acre for the rear land. Of course, Mr Aspinwall's valuation of £27,500 per acre for the rear land was open to the same criticism because his valuation (£22,500) for the reference land was the equivalent of £45,000 per acre. Nevertheless, his figure of £27,500 per acre was nearer to £36,000 B and £45,000 than was the district valuer's figure of £15,000 per acre.

Tribunal's Decision

I have inspected the reference land, the town of Billinge, the rear land and the tongue of land (in those two cases, as best I could without trespassing thereon). I have also inspected all the sites which were put forward by the parties as comparables.

I accept the evidence of Mr Foreman that the reference C land would constitute a satisfactory access to residential development on the rear land. I agree with him that access through the reference land can achieve adequate sight lines, curve radii and width, that the cost of construction would be reasonable, that there would be no adverse environmental effects and that the traffic and highway objections are not of major proportions. I agree also that the linear park road would, for the reasons given by him, be unthinkable as a means of access to the rear land. I do not see the need for a secondary access and I take the view that if the gap were used as an access road at all, its use should be limited to use by pedestrians and the emergency services.

D For these reasons I hold that the reference land is the key which will unlock any development value which the rear land may have. It follows that planning permission for residential development of the rear land would not be refused on highway or traffic grounds. I also accept from Mr Foreman that because of the opportunity to develop parts of the reference land in the vicinity of the access road, thereby reducing the length of road which would be unproductive, the burden of cost attributable to the access would be reduced to 5/13 of the agreed figure of £35,000, that is to say, to £13,461.

It is a difficult matter to determine the quantity of land E in St Helens district and Merseyside which is available for housing. This is because the figures tend to be inflated by the bad years for housing completions (more particularly 1975 and 1976), thus producing inaccuracies. I leave out of account for the moment any need for an extra supply of land having regard to considerations of the kind mentioned in exhibit 109. Apart from those considerations, I think that a reasonably-minded purchaser would take a cautious view and would assume that there is in St Helens district a five years' supply of land available for housing and that in Merseyside the supply is about the same or slightly greater but that in Billinge itself the supply of land for housing F is nil.

In my opinion I can accept that speculators would be interested in the reference land with a view to purchase of it. That opinion receives support not only from the evidence of Mr Aspinwall but also, by implication, from the case which the acquiring authority itself put forward at the hearing. They say that a purchaser in the open market would pay £18,000 for the reference land and would do so on the assumption that planning permission would be granted for residential development of the reference land. They say this notwithstanding that planning permission for such development has been refused on a number of occasions in the past. Notwithstanding that fact, they say that planning permission G is "guaranteed" because the reference land is zoned residential in the relevant development plan. There is no such thing as a guarantee that planning permission will be granted because, despite the zoning, the local planning authority has a discretion in the matter. No doubt there would be a reasonable expectation that planning permission for development of the reference land for the erection of three dwelling-houses, or indeed for the erection of nine semi-detached units, would be granted.

If the reference land has no key or merger value or if rule (3) of section 5 of the Act of 1961 applies, then I accept Mr Monham's valuation of £18,000. Having regard to comparable sites, it is well supported, whereas Mr Aspinwall's H alternative valuation of £22,500 is related to Mr Nicholson's residual valuation. I agree with Mr Monham that it is a calculation rather than a valuation and I reject it.

Returning to the main issue, the valuation of £18,000 which I have conditionally accepted, is, for the reasons which I have given, a hope value and the purchaser in the open market willing to pay it is necessarily a speculator. Being a speculator and a reasonably-minded speculator, he will investigate the question whether the reference land has a key or merger value and with some care. He will be influenced by the history of planning refusals and by the J Stage 1 strategy. If he is unimpressed by exhibit 109, he will, having regard to those refusals, that strategy and the available supply of land for housing, conclude that planning permission will not be forthcoming for residential development of the rear land. I think that the effect which exhibit 109 has on his mind must depend on the view which he takes on the question whether it is desirable that the existing population of the district should be maintained. It may be that exhibit 109 itself tacitly gives an affirmative answer to that question. Whether that be so or not, I think that a speculator, having regard to the pleasant nature of much of K the district and the function of Billinge and other parts of the district as dormitory areas for Liverpool, would on balance conclude that an affirmative answer is appropriate. He would not regard the matter as entirely free from doubt but he would expect that ultimately the rear land would be needed for housing development. He would not be over-impressed by the objections of the Ministry of Agriculture, Fisheries and Food, which have not been supported by evidence before me. He would not regard the ministry's classification of land, which includes the rear land, as being Grade II as proving that the rear land has any particular agricultural quality.

L Having regard to all the considerations which I have mentioned, in my judgment a speculator would not expect planning permission for residential development of the rear land to be granted within the period of five years from now. Nevertheless, he would expect planning permission to be granted within seven or eight years from now. That is not to be regarded as a guarantee or certainty of the grant of planning permission but a reasonable expectation or strong hope of it. It appears to me that a speculator, having formed such an opinion, would be willing to pay a price for the reference land in excess of its value (£18,000) for intrinsic develop-M ment. The question is: how much extra would he be willing to pay? In order to determine that question it is necessary to adjust Mr Aspinwall's primary valuation. His figure of £27,500 per acre represents the maximum price per acre which the rear land might realise if an actual planning permission existed. I think it is too high, even on that basis, and I would reduce the figure to £25,000 per acre. If there were an actual planning permission I would defer at 6 per cent as does Mr Aspinwall, but having regard to the facts and to the probabilities, I would defer over seven years at 10 per cent. In addition to deducting the value of the rear land as accommodation land, I would deduct £13,461 on

A account of the cost of construction of the access road. The result would be as follows:

14.14 acres @ £25,000 per acre (on the basis of a notional residential planning permission)	£353,500
Deferred 7 years at 10% YP	0.513
	£181,346
Less value as accommodation land 14.14 acres at £1,250 per acre	£ 17,675
	£163,671
Less cost of construction of road 5/13ths of £35,000	£ 13,461
B	£150,210

Take one-third of £150,210 = £50,070
Say £50,000

Accordingly, a speculator would know that if planning permission is granted within seven years from now he can expect to receive from the owner for the time being of the rear land, the price of £50,000 for the reference land. He will not necessarily expect the Valentine Family to be the owners in seven years' time, but he will expect the owners
C for the time being to act reasonably. Having regard to all the circumstances and doing the best I can I hold that a speculator would be willing to pay £7,000 for the reference land in addition to its intrinsic value of £18,000, bringing the price up to £25,000, the same price as that paid by the claimants. That accordingly is the value of the reference land in the open market under rule (2) of section 5 of the Act of 1961.

Rule (3)

Mr Rich made no submission that rule (3) applies. No doubt this was because of his submission that no speculator would be interested in the reference land and, if the contrary
D view were taken, because Mr Rich would be of the opinion that rule (3) is inapplicable. The decision of the tribunal (to that effect) in *Rathgar Property Co v London Borough of Haringey* (1978) 248 EG 693 (W H Rees FRICS) was referred to at the hearing and its correctness was not challenged. Nevertheless, the parties in their alternative valuations proceeded on the basis, first that the rule did apply and, secondly, that it did not, and it is at least desirable that I shall indicate my views on the matter.

The relevant words are "The special suitability or adaptability of the land for any purpose shall not be taken into
E account if that purpose is a purpose . . . for which there is no market apart from the special needs of a particular purchaser. . . ." It appears to me that it is not correct to read these words as though they say: "for which there is no market apart from what may be assumed to be the special needs of some hypothetical prospective purchaser." The word "requirements" which follows the relevant words means actual requirements and the "special needs" means actual needs of a particular actual purchaser.

It may well be that the word "purchaser" means somebody who has actually purchased and not a hypothetical purchaser. That view is by no means impossible: see
F *Inland Revenue Commissioners v Clay* [1914] 3 KB 466 (the case which the relevant words when originally enacted were no doubt intended to overrule), where there was an actual purchaser. Moreover, of course, to be a purchaser a person need not have completed the contract; it is sufficient if he has entered into a binding contract.

Nevertheless, if the relevant words do not refer to an actual purchaser, they mean an actual identifiable living person who for his own special purposes (such as a pressing need to expand his factory or nursing home on adjoining land) has made an actual inflated bid, thereby, because of the nature of the market, forcing up the market price.

The view that rule (3) applies only to actual bids and G
not to hypothetical bids is shown by the words of the transitional proviso which followed the rule when it was originally enacted in section 2 of the Acquisition of Land (Assessment of Compensation) Act 1919. That proviso was: "Provided that any bona fide offer for the purchase of the land made before the passing of this Act which may be brought to the notice of the arbitrator shall be taken into consideration."

It appears to me that having regard to my findings the reference land has a natural suitability, arising from its position for marrying-up or conjoint development with the rear land. The value which the market would attribute to the reference land on that account is a natural value, based H
on the assumption that all persons concerned will act reasonably, and not on any artificially inflated bid. In the circumstances the application of rule (3) would mean that the claimants would receive in compensation less than their true loss and that would be wrong: see *Horn v Sunderland Corporation* [1941] 2 KB 26. In my view the rule does not apply in the present case.

For these reasons I award to the claimants the sum of £25,000 together with their surveyor's fees on Scale 5(a) of the scales of fees of the Royal Institution of Chartered Surveyors.

J

LAING HOMES LTD V EASTLEIGH BOROUGH COUNCIL

(REF/176/1977)

(E C STRATHON FRICS)

December 18 1978

Estates Gazette April 28 1979 and May 5 1979

(1978) 250 EG 350 & 459

Land at Eastleigh, Hampshire, acquired for spine road through claimants' housing development site from adjoining larger development site in other ownership—Both sites former "white land" released for development—Spine road constructed on adjoining site under section 52 agreement as varied by two later supplemental agreements—Claimants' valuation of £1m on basis of enhancement in value of adjoining site—Council's before-and-after valuation with betterment accruing—Offer of nominal £1 compensation—Tribunal rejects both valuation approaches—Potential development value of adjoining land held already released at valuation date—Reference land not an access key to unlock enhancement in value—Rule (3) excludes special value of reference land as key to completion of spine road—No diminution in value in terms of section 9—Betterment pursuant to section 7 and "Pointe Gourde" principle does not arise—No evidence that claimants' land could not be developed in absence of spine road—Release of land for private development does not equate with scheme within "Pointe Gourde" principle—No scheme, therefore, whereby value to be disregarded—Estoppel point on alteration of section 52 agreement, but tribunal decides it does not have adequate material to find facts on which to found argument in law—Award of £12,750, being proportion of site development value with 25% allowance for cost of acquisition of access K L M

M Fitzgerald (instructed by George Carter & Co) appeared for the claimant; J Harman QC and O Albery (instructed by the Solicitor to the Borough of Eastleigh) for the acquiring authority.

The decision is published by consent of the parties.

Giving his decision, MR STRATHON said: This is a reference by consent to the Lands Tribunal to determine the price to be paid to Laing Homes Ltd (hereinafter called "Laing") by Eastleigh Borough Council (hereinafter called "Eastleigh") for the freehold of 0.68 acre of land at Fair Oak Road,

Bishopstoke, Hampshire. By an agreement dated March 4 1977 the price shall be the value of the land ascertained in accordance with the provisions of the Land Compensation Acts 1961 and 1973 and the Compulsory Purchase Act 1965 as if the corporation, pursuant to the obligation on its behalf contained in an agreement dated May 18 1973 under section 52 of the Town and Country Planning Act 1971 between Hampshire County Council, Eastleigh and Mill Lodge Properties Ltd (hereinafter called "Mill Lodge") and two supplemental agreements between Eastleigh and Mill Lodge has lawfully exercised powers of compulsory acquisition under the provisions of the Town and Country Planning Act 1971.

The value of the reference land is to be assessed at March 7 1977, on which date vacant possession was granted to Eastleigh.

The subject land lies on the north side of Fair Oak Road (B3037), which runs between the town of Eastleigh and the village of Fair Oak, two miles to the east.

For the claimants, evidence was called from I H D Jones, a director of Laing; D J Pearson, managing director of S Regent Ltd, a firm of builders; R W A Foreman C Eng FRICS MICE, as to engineering evidence; and J G Harkus ARICS as to valuation evidence.

For the acquiring authority, evidence was called from E S Benefield ARICS, chief planning officer, Eastleigh; R J Welch CE MICE MIStrucE, chief engineer in the Department of Technical Services, Eastleigh; and J W Gregory ARICS, estates surveyor and valuer, Eastleigh, as to valuation evidence.

Agreed facts

The following facts agreed between the parties have been extracted from a statement:

1. The planning history goes back to May 1972 when the South Hampshire Interim Policy Plan Stage Two (SHIPP 2) was published and which released certain areas for development in advance of the South Hampshire Structure Plan including approximately 85 acres to the north of Fair Oak Road (Area 12 page 23 of SHIPP 2 and Plan no 2 thereof). Although SHIPP 2 states that the land in Area 12 totals 85 acres, the actual acreage of the whole site has been measured by planimeter as being 90 acres (the school site extension being approximately 4.7 acres). A major consideration in this proposed development was a diversion of the existing Fair Oak Road through the proposed development along a spine or distributal road so as to render unnecessary works to widen the existing unsatisfactory Fair Oak Road. The existing Fair Oak Road carries a high volume of traffic, particularly at peak periods, and the section to be relieved by the spine road is tortuous and has many individual accesses.

2. On May 2 1972, the then borough engineer, surveyor and town planning officer submitted a "proposed developers' brief" for the development of Area 12. This brief indicated the access points to serve the new development (limited to three points on the north side of the spine road and two points from the relieved section of the existing Fair Oak Road). The land was then in five separate ownerships and outline planning applications had already been submitted by four owners.

3. The events which took place between August 1972 and May 1973 are as follows:

(a) by letter dated August 14 1972 the council invited the owners to a meeting to discuss the development of the area, the letter setting out the general considerations and matters proposed to be included in the proposed section 52 agreement. The meeting took place on September 26 1972; but did not produce any concluded agreement.

(b) On November 3 1972 approximately 76 acres (mentioned in paragraph below) was offered for sale on behalf of three owners.

(c) On November 14 1972 Laing received acceptance of their offer to purchase 3.4 acres of land within which the eastern junction of the proposed spine road was to be located.

(d) On November 15 1972 Laing submitted an offer to purchase the land referred to in (b) above (the 76 acres).

(e) On December 8 1972 Laing contracted to purchase Laing's land (3.4 acres) and completed the purchase thereof on December 19 1972.

(f) On November 10 1972 the town clerk of Eastleigh sent a letter to interested developers setting out the general requirements of the council in accordance with the draft developers' brief.

(g) Towards the end of 1972 Mill Lodge Properties Ltd acquired the land referred to in (b) above (Mill Lodge land) (76 acres).

(h) On January 2 1973 the town planning committee of Eastleigh approved a developers' brief modified to require the provision of 5 acres for local authority housing.

(i) By letter dated February 9 1973 the town clerk of Eastleigh informed Laing Homes that Mill Lodge Properties had purchased Mill Lodge land and requested Laing Homes to get in touch with Mill Lodge who were preparing a detailed planning application.

(j) By a letter dated February 20 1973 Mill Lodge requested Laing Homes to submit planning application in respect of Laing's land (for the spine road).

(k) Between February 1973 and the end of November 1973 there was correspondence and inconclusive discussions between Mill Lodge Properties, Eastleigh Council and Laing Homes concerning a possible exchange of land between Mill Lodge and Laing Homes and the acquisition from Laing of land required for the extension to the spine road.

Section 52 agreement

4. (a) On May 18 1973 an agreement was entered into under section 52 of the 1971 Planning Act and section 7 of the Hampshire County Council Act. The agreement was made between the former Hampshire County Council, the former Eastleigh Borough Council and Mill Lodge. Hampshire was a party to the agreement as the local planning authority; Eastleigh was authorised by Hampshire under section 251 of the Highways Act 1959 to exercise highway authority functions for the construction of that part of the spine road within the area of the former Winchester Rural District Council, and Eastleigh was made a party to the section 52 agreement in view of clause 14 thereof.

(b) On May 18 1973 Mill Lodge were granted outline permission in respect of Mill Lodge land.

(c) On June 19 1973 planning permission was given to Mill Lodge for the construction of the spine road on Mill Lodge land.

5. (a) By clause 6 of the section 52 agreement the spine road between A (western end of Mill Lodge) and B (western boundary of Laing's land) was to be constructed by Mill Lodge to Eastleigh Council's specifications and alignment. By clause 11 the construction of houses on the area south of the spine road was restricted to 150 units and the construction of any more required a certificate by Eastleigh that the spine road (including the section on Laing Homes' land) and subway had been completed to Eastleigh's satisfaction. By clause 12 no dwelling should be occupied until an approved foul and surface water system to serve that dwelling had been completed to the council's satisfaction. By clause 13 the development was to be carried out in accordance with the provisions and principles of the planning brief, copy of which was annexed to the agreement. By clause 14 it was agreed that if Mill Lodge were impeded or prevented from carrying out works on sections B-D (which included part of Laing Homes' land) then Eastleigh would exercise its statutory powers to enable such works to be carried out, with a provision whereby Mill Lodge indemnified Eastleigh against costs.

(b) On December 4 1973 Mill Lodge applied for planning permission for the construction of the section of spine road crossing Laing Homes' land and full planning permission for the spine road was granted on January 11 1974.

Supplemental agreements

6. On local government reorganisation on April 1 1974 the boundaries were changed and the area of the present Eastleigh Borough Council includes the whole of the SHIPP 2 site. On May 6 1974 a first supplemental agreement was entered into between Eastleigh and Mill Lodge which varied the section 52 agreement. The county council were not a party to the first supplemental agreement in view of Article 10 of the Town and Country Planning (Transitional Provisions) Order 1974, and in

A addition Eastleigh now acted under highway agency powers. By the first supplemental agreement Mill Lodge were allowed to construct more than 150 houses (though not on the area coloured yellow (18 acres at the north-east of Mill Lodge land)) once Eastleigh certified *inter alia* that the spine road between points A-B had been completed and that a temporary roadway had been constructed (on Mill Lodge land) to link the eastern end of the spine road with the existing Fair Oak Road, and subject to the completion of foul and surface water drainage systems. Development of the area coloured yellow was to be allowed only when Eastleigh certified that that part of the spine road B-D which included the section across Laing Homes' land had been completed.

7. (a) On October 13 1975 a second supplemental agreement was
B entered into between Eastleigh and Mill Lodge which replaced the amended clause 6 of the section 52 agreement and required the construction of the spine road between A-B within 12 months of the grant of planning permission for the yellow land to the north-east of the spine road, with provision for a temporary access at the eastern end of the spine road across Mill Lodge's land (to Fair Oak Road). In place of the restriction on the number of houses to be constructed before completion of the spine road, Mill Lodge agreed to deposit with Eastleigh a sum to represent the cost of constructing the spine road across Laing Homes' land and a further sum to cover the market price of the land to be purchased to enable the section of the spine road on Laing's land to be constructed. By clause 14 Mill Lodge were required to grant rights
C of access and drainage for the benefit of Laing's land.

(b) No planning application has been made by Mill Lodge in connection with the construction of the temporary roadway referred to in the first and second supplemental agreements and such temporary roadway has not been constructed.

(c) On May 10 1974 there was a meeting at Town Hall, Eastleigh, to discuss the possibility of agreement being reached between Laing Homes and Mill Lodge for the acquisition of Laing's land in order to permit the development of the SHIPP 2 release in accordance with the terms of the planning brief.

(d) On September 25 1974 the Borough of Eastleigh wrote to Laing Homes confirming the above meeting and emphasising the council's concern at the lack of progress in implementing the road
D proposals which are essential to the development of the SHIPP 2 land.

(e) Between June 1974 and March 1977 there were further discussions, negotiations and correspondence between Mill Lodge, Laing Homes and Borough of Eastleigh concerning the provision by Laing Homes of land required for the construction of section B-D of the spine road.

(f) In April 1976 construction of section A-B of the spine road was commenced by Mill Lodge.

8. On March 4 1977 Laing Homes were granted outline planning permission for residential development of their land. This was subject to a section 52 agreement of the same date between Laing Homes and Eastleigh. The planning consent allowed permanent
E access on to the existing Fair Oak Road to the west of the proposed eastern junction (of the spine road and Fair Oak Road) as indicated on the plans to the section 52 agreement and the agreement allowed temporary access from the spine road to be constructed to the land to the north of the spine road.

9. Also on March 4 1977 the purchase agreement was entered into between Laing Homes and Eastleigh. Eastleigh agreed to purchase the land required for section B-D of the spine road and possession was given on March 7 1977. The land includes not only the amount required for the initial single carriageway of the spine road; but also sufficient to provide for the proposed future dualling of the spine road.

F **Facts found on evidence**

Further facts which I have found on the evidence include:

1. The planning brief annexed to Mill Lodge's section 52 agreement provides: "Before any dwellings are occupied, the new Fair Oak Road diversion and subways should be completed together with the construction of the surface water culvert, all to the satisfaction of the borough council." This provision was not amended by the two supplemental section 52 agreements.

2. Laing purchased the area of 3.4 acres in November 1972 at a price of £140,000.

3. The offer by Laing to purchase the 76 acres from three owners on November 15 1972 was expressed in terms of willing-

ness to negotiate a purchase at a price in the region of £2.5m. G
The purchase price paid by Mill Lodge for the land was £3.1m.

4. From the eastern boundary of Laing's land the spine road is aligned to cross a corner of land which was owned by Brown Brothers, a firm of builders, one of the four owners who had applied for planning permission to develop within area 12 before SHIPP 2. By March 7 1977 planning permission had been granted for development with access through an existing road system to the east; the owners had dedicated the site of the spine road to Eastleigh free of charge.

5. By the acquisition of 0.68 acre by Eastleigh the land remaining to Laing is severed into two parts, namely 2.30 acres and 0.40 acre, respectively, north and south of the spine road.

6. By the date of valuation, March 7 1977: H

(a) the spine road had been constructed to the western boundary of the reference land at B.

(b) Mill Lodge land south of the spine road was being developed by the construction of houses and flats, some of which were completed, sold and occupied.

(c) the relief sewer to serve SHIPP 2 land had been provided.

(d) 5 acres of Mill Lodge land had been sold to Eastleigh for housing purposes.

I take note that SHIPP 2 is a non-statutory document. It supplements the local planning authority's existing policies for the various parts of the area. Before SHIPP 2, area J
12 was unallocated in the approved statutory town map—it was "white land"—and it was also in an area proposed for green belt. The proposals in SHIPP 2 relating to area 12 are expressed as follows:

Area 12 34.0 ha (85.0 acres) north of Fair Oak Road, Bishopstoke. (Category)

Planning Requirements

Fair Oak Road should be diverted through the development. Provision should be made for dual carriageways; one carriageway should be provided as part of the development. Provision must be made in the development to extend the site of the Bishopstoke K
County Junior School to 4 hectares and it is possible that a site of 4 hectares for a new primary school at the eastern end of the site will be required. About 3.7 hectares (9.0 acres) of open space, including the appropriate number of small children's play spaces, should be provided within the development.

Laing claim £1m compensation, which is the amount of the valuation produced by Mr Harkus. Mr Gregory, for Eastleigh, considers that the amount of compensation should be nominal, £1.

Each valuation is prefaced by a statement of the basis, or of the assumptions, on which the valuation has been made; L
the valuations within the respective statements are reproduced as follows:

J G Harkus ARICS (for Laing)

Valuation Basis

1. All ownerships are valued at the relevant date, which is March 7th 1977.
2. The land was ripe for residential development at the relevant date subject to the securing of access through the frontage land owned by Laing Homes Ltd.
3. The valuation of Laing Homes Ltd ownership is based on the enhancement value attributable to the adjoining development area created by the release of the subject land for the purposes of completing the spine road. M
4. All prices per acre are net and agreed with Eastleigh borough valuer.

Valuation

Areas A and B net of road and reservation totalling 18 acres at £25,000 per acre.	£450,000
Area C net of road and reservation totalling 18.4 acres at £23,000 per acre	423,200
Areas D, E and F net of road and reservation totalling 33.6 acres at £24,500 per acre	823,200
Total land values	1,696,400

A *Add*

value of work in progress as agreed with Eastleigh borough valuer:			
Area D	38 dwellings at £6,500 each	£247,000	
	30 dwellings at £5,500 each	165,000	
Area E	124 houses at £6,500 each	806,000	
	12 flats at £5,500 each	66,000	
	12 houses at £3,000 each	36,000	
	12 houses at £1,000 each	12,000	
	12 flats at £2,000 each	24,000	1,356,000
Value of land and work in progress at 7/3/77			3,052,400
Marriage value at 50 per cent available to purchase access requirements			1,526,200
But negotiated reasonable settlement			£1,000,000

B

J W Gregory ARICS (for Eastleigh)

Valuation Assumptions

The following assumptions have been adopted for the purpose of valuing the land to be acquired from Laings:

(a) That the special purchase element should be ignored in assessing compensation.

(b) That without the new single carriageway road no development could take place but that the second carriageway is a future highway requirement which is not directly related to the development.

C (c) That by the council's involvement Laing have been saved the cost of constructing their section of the spine road resulting in betterment to Laing's site.

Valuation

Before Valuation

Value of claimant's interest without borough council involvement.			
2.863 acres at £25,000 per acre	£71,575		
Less cost of spine road attributable to claimant	34,715		
Value of land:	36,860	say	£37,000

After Valuation

Value of claimant's land with spine road constructed and land needed for second carriageway excluded.	
2.72 acres at £25,000 per acre	68,000
Benefit to claimant	£31,000
Compensation—nominal figure only.	£1

D

There are issues of law and of fact which bear on these valuations. Having shown the extent in terms of value of the difference between the valuers I propose to leave the valuations for later consideration.

A volume of correspondence and another of discovery documents were canvassed throughout the hearing. I do not E propose to record the details, but, when appropriate, reference will be made.

Claimants' evidence

I turn first to the evidence of Mr Jones, who was responsible for negotiating the purchase of Laing's land in November 1972 following the publication of SHIPP 2 in May 1972. It was his view that the purchase of the 3.4 acres at the eastern end of the 90 acres represented the purchase of control of SHIPP 2 land. The purchase at a price of £140,000 was in excess of its development value, but it had been made to secure a foothold to acquire the whole or part of the 75.5 acres of F adjoining land. Mr Jones was aware of the alignment proposed for the spine road through the proposed development and also of factors which should be covered by a section 52 planning agreement, including, in particular, a requirement that no buildings on the site were to be occupied until the spine road was completed within the site.

Having failed to purchase the adjoining 75.5 acres, Laing informed Eastleigh in February 1973 that they were prepared to discuss proposals for the overall development of the land identified in SHIPP 2, and in April 1973 Mill Lodge was told of Laing's desire to acquire a further 15 or 20 acres of land for development on reasonable terms. In return Laing would be willing to participate in making the necessary land available G for the eastern end of the spine road and also to be a party to a section 52 agreement.

Inconclusive negotiations with Mill Lodge followed: a parcel of 18½ acres to the north-east of the spine road which Mill Lodge intended to offer for sale by tender was not, in Mr Jones's view, a viable proposition for development. It was his contention that throughout the whole of the negotiations Laing had offered to exchange or purchase on terms of market value, having regard to the special value of Laing's 3.4 acres.

In correspondence Eastleigh had informed Laing about the adoption of compulsory powers to acquire land for the spine road—the following is an extract from a letter of October 17 H 1973 from the town clerk:

> As you probably know, the borough council has already resolved to exercise compulsory powers for the acquisition of any land required for the construction of the new road crossing the SHIPP 2 site, part of which affects your own land and I have been asked by Mill Lodge Properties Ltd to proceed with the service of notices in respect of the compulsory purchase order as you are unable to reach any agreement and as they are particularly interested in constructing the new road as one operation rather than in separate parts. I have informed Mill Lodge that before serving any notice on you I would wish to discuss the matter fully with you so that you are aware of the council's position and so that neither side is under any misapprehension from a bargaining point J of view as to the situation as the council sees it.

It seemed to Laing that the only practical course that was open to them was to offer their land to Eastleigh. But as Mill Lodge remained in the field of negotiation and the first supplemental section 52 agreement had been executed, Laing, who in Mr Jones's view were under pressure, made an offer to sell their holding, 3.4 acres, May 28 1974 as follows:

> On Friday May 10 last at a meeting held at the Town Hall, Eastleigh, to discuss the above land, [land at Bishopstoke] we defined to the meeting our estimation that the value of our land holding at Bishopstoke, having regard to all the circumstances, was £430,000. K
>
> We would like to place this valuation on record and to confirm that we are willing to dispose of the land at this point in time.
>
> Although we have been advised that our holding attracts a percentage of the value of the site with planning permission, we are prepared to compromise on this to the extent that we will take into account that the land south of the diversionary spine road may be developed with access from Fair Oak Road. We have also allowed that some 5.0 acres of land is non-productive financially due to the spine road and the reserved area. This leaves approximately 41.0 acres which we maintain benefits for the access across our land.
>
> Accordingly we are prepared to dispose of our total land holding for the sum of £300,000, subject to contract. L

After a further six months, Laing, in Mr Jones's view still under pressure, offered to sell the 3.4 acres to Mill Lodge at a price of £200,000 or, for the land required for the spine road only, a price of £110,000. Finally, 12 months later, the 3.4 acres were offered for sale at £170,000, which Laing regarded as rock-bottom, representing their purchase price plus interest.

As has been said, between the eastern boundary of Laing's land and the proposed eastern junction of the spine road with Fair Oak Road, there is land owned by Brown Brothers. Mr Jones said he had recognised the importance of this land as an access from the east via the terminal stretch of the spine road to Laing's land; he had seen Brown Brothers before and M after Laing's purchase, and he was of the opinion that there would be no difficulty about access to the 3.4 acres from there.

By his appointments as divisional planning officer for South Hampshire and, from 1973, chief planning officer of Eastleigh Borough, Mr Benefield was aware of the ownerships within the 90 acres at the time of the publication of SHIPP 2. The alignment of the spine road was known in the autumn of 1972; in his view the line could have been taken further south, which would have made the eastern junction with Fair Oak Road further to the west; the alignment as then designed and adopted was not the only acceptable solution but probably the best. Mr Benefield had been concerned with the contents of the section 52 agreement with Mill Lodge and it was clear to him that the exercise of compulsory powers by Eastleigh provided in clause 14 was aimed at those with whom Mill Lodge were to negotiate in order to acquire land for the construction of the road.

Pressure for houses

At that time, there was much pressure on local authorities to develop land for housing and since Eastleigh wanted the SHIPP 2 land to be developed, support was given to Mill Lodge so that the spine road would be built. This motive, to achieve development, had led to the first supplemental agreement, which permitted Mill Lodge to build houses on the 75.5 acres (except in the north-east corner), with the requirement for completion of the spine road only to the west boundary of Laing's land with a temporary road link to Fair Oak Road. He was aware of the negotiations in which Laing were maintaining that they held the key to development of SHIPP 2 land, the sort of situation which, as a planner, he had always endeavoured to avoid and it was envisaged that the supplemental agreement would affect the terms on which Laing would negotiate. The intention was not to damage Laing but to put Mill Lodge in a more favourable position.

The second supplemental agreement, which released the land to the north-east within 12 months of the completion of the spine road on Mill Lodge land, was also designed to encourage development, and thereafter houses were built and occupied. It had not occurred to Mr Benefield before the hearing that occupation of completed houses and flats was in breach of clause 13 of the section 52 agreement; in his view the clause was not altered because of a drafting error; in his view Eastleigh would not take any proceedings against Mill Lodge or against householders; any such action would not be contemplated by Eastleigh and would be contrary to the whole objective to complete the development.

As to the development of Laing's land itself, Mr Benefield referred to the refusal of outline planning permission to build houses in October 1974 for the reason that "(1) the site should be developed only in accordance with a comprehensive plan which incorporates adjoining land to the north and west and (2) no access will be permitted from the site on to the existing Fair Oak Road or on to the proposed spine road and the applicants have not satisfied the planning authority that an adequate means of access is available." On the submission, by agreement, of a revised application in July 1976, outline permission for residential development was granted on December 2 and again on March 4 1977.

Mr Benefield was of the opinion that without the spine road Laing's land could not be developed; previous to SHIPP 2 there were no prospects of planning permission being granted for residential development and after SHIPP 2 in 1972 he would have expected a minimum of between two and four years before the spine road would be completed to enable development to proceed.

The spine road was the subject of evidence by engineer witnesses. Mr Foreman, for Laing, had concluded from a very detailed examination of traffic conditions in 1972 and of estimates of future requirements that the construction of the spine road was a vital requirement of the development of 800 houses which otherwise would be served from Fair Oak Road. It was his view that it was essential to relieve Fair Oak Road of through traffic before 150 houses, the subject of the section 52 agreement, were built so as to provide capacity for traffic generated by them. The spine road should have been built by March 7 1977, since, by that date, the construction of the houses was in progress. Further evidence by Mr Foreman, which was accepted by Mr Welch, was directed to the feasibility of so aligning the eastern end of the spine road that it would not encroach on the land owned by Brown Brothers.

Mr Welch, chief engineer in the Department of Technical Services of Eastleigh, considered that development of the SHIPP 2 land could have been served by a major improvement and widening of Fair Oak Road at a cost of about £500,000, but the spine road was to be preferred; similarly, an alternative line for the spine road through the 90 acres, but further to the south and to emerge into Fair Oak Road to the west of Laing's land, would have been acceptable but less desirable than the line of the road as constructed.

I return to the valuation evidence.

As can be seen in the preface to his valuation, Mr Harkus has valued the reference land on the basis that its release for the purposes of completing the spine road creates enhancement in the value of the adjoining development area (Mill Lodge land), and he ascribes to the reference land a proportion of the enhancement in value. Mr Harkus envisaged that a purchaser at March 7 1977 would value at an amount of £3,052,400 the 75.5 acres of Mill Lodge land together with the buildings that had been built upon it, and in order to protect that value an offer of £2m would be made to Mill Lodge and £1m to Laing. The value of houses and flats sold by Mill Lodge was included because occupation of them was in breach of the requirement in the brief (which was incorporated in the section 52 agreement) that the spine road should first be constructed; Mr Harkus considered that Mill Lodge would be at risk of action by householders who would wish "to legitimise their position." Mr Harkus said in cross-examination that in so far as Mill Lodge had no title to the properties, the householders themselves would be potential vendors.

"Marriage value"

As to "marriage value" (also referred to by the parties as "ransom value"), Mr Harkus said that a purchaser would take into account that access to the 75.5 acres had not been secured at the eastern end. An allowance of 50 per cent which had been taken as a starting point was considered by Mr Harkus to be supported by recent decisions of the Lands Tribunal and by recent market transactions; his estimate of £1m as the reasonable settlement represented a ransom proportion of one-third which he considered would result from bargaining between the parties.

Mr Gregory thought that development of the Mill Lodge land could proceed without access through the reference land; by reason of the two supplemental agreements no value remained to be unlocked, and on March 7 1977, a ransom strip did not exist.

Mr Gregory's valuation assumption that the special purchase element should be ignored in the assessment of compensation refers to rule (3) in section 5 of the Land Compensation Act 1961. There were submissions on the law which I will come to later, but Mr Gregory considered that as the construction of the spine road was a requirement of Eastleigh, an authority which possesses compulsory powers, and as Mill Lodge, a particular purchaser, was the only other party who would have gained, the increased value, if any, either to Eastleigh or Mill Lodge should be ignored.

Mr Gregory had adopted the "before and after" method of valuation as a means of computing, as a single assessment, the value of the reference land (0.68 acre) and the amount of betterment which, in his opinion, has accrued to the remaining 2.72 acres of Laing's 3.4 acres. Mr Gregory proceeds on four assumptions: (1) before the acquisition, planning permission to develop Laing's land would have been subject to a condition that provision should be made for a road of single carriageway width (0.54 acre) on the line of the spine road; (2) the cost of the road would fall on the landowner; (3) the number of dwelling units permitted on the 3.4 acres would be no more than the 31 units specified by the detailed permission of April 20 1978 for the two severed areas after the acquisition; and (4) after the acquisition a further 0.14 acre would have been acquired for the provision of a dual carriageway and the costs of the construction of the spine road would fall on Eastleigh (in fact, it would fall on Mill

A Lodge). The assessment of betterment is therefore derived solely from the cost of the road on part of the reference land.

There are several points of law in issue.

Special suitability or adaptability

I will first take Mr Harman's submissions as to the operation in this case of rules (2) and (3) in section 5 of the Land Compensation Act 1961. These rules provide:

> Rule (2) The value of land shall, subject as hereinafter provided, be taken to be the amount which the land if sold in the open market by a willing seller might be expected to realise.
>
> Rule (3) The special suitability or adaptability of the land for B any purpose shall not be taken into account if that purpose is a purpose to which it could be applied only in pursuance of statutory powers or for which there is no market apart from the needs of a particular purchaser or the requirements of any authority possessing compulsory purchase powers.

Mr Harman argued that if it is found that Laing's land did not hold the key that enabled the vendor to squeeze the purchaser, then in law under rule (2) there is no special value in the hands of the vendor. If special or ransom value is found in the hands of the vendor, Mr Harman submitted, the only reason for the existence of any key is the special suitability of the reference land for the building of the spine C road. He relied on *Lambe v Secretary of State for War* [1955] 2 QB 612 to support his argument that geographical location of the spine road met the criterion of the "quality of land" which was identified in that case with "special suitability" in rule (3). If the alignment of the spine road was deemed to be the best line for the construction of a spine road, the special suitability for the purpose of a spine road brought rule (3) into operation as a purpose for which there is no market apart from the needs of a particular purchaser (Mill Lodge) or the requirements of an authority (Eastleigh) possessing compulsory purchase powers. The special needs of Mill Lodge are not as an owner of adjoining land but as an D owner to carry out works on behalf of Eastleigh to enable Eastleigh to adopt the road.

Mr Fitzgerald submitted that there is no quality of special suitability or adaptability inherent in the reference land as is referred to in *Lambe*. The reference land is not subject to a special geographical location because, until the construction of the spine road to the western boundary of Laing's land, access could have been provided from the east by different routes. Mill Lodge by their action have put themselves in the position of having to pass through the reference land if they wish to secure access from the east. This points to the needs of a particular purchaser of which Parker LJ in *Lambe* E said:

> The expression "special suitability or adaptability of the land" in rule (3) is, I think, clearly referring to the quality of the land as opposed to the needs of a particular purchaser. The fact that the sitting tenant may be prepared to pay more than an investor in order not to be turned out does not clothe the land with a special suitability within that rule, the value of which is to be ignored.

The purpose of the acquisition is concerned with legislation relating to planning and not to highways; the purpose of the spine road was not to provide a highway but to serve the new development and as a benefit to relieve Fair Oak Road. The implementation of the planning requirement was attained by F the construction of the road by the developers, Mill Lodge, through land in their ownership, and where not in their ownership, Eastleigh, as agents, acquired the land for and at the cost of Mill Lodge. Mr Fitzgerald also submitted that on the evidence the existence of a market causes rule (3) not to be brought into operation.

Betterment, a valuation factor of which account is taken by Mr Gregory, arises, Mr Harman said, by the operation of section 7 of the Land Compensation Act 1961 and by the application of the principles in *Pointe Gourde Quarrying & Transport Co Ltd v Sub-Intendent of Crown Lands* [1947] AC 565, PC. In the instant case, Mr Harman said, the section directs that from the compensation payable for the reference G land of 0.68 acre, there shall be deducted such part of the increase in value of the remaining land of 2.7 acres as is attributable to the prospect of the development of the 0.68 acre as a highway, that being the development for the purpose for which the reference land is being acquired and is development not likely to be carried out if the acquiring authority does not acquire. Increase in value of the remaining land was attributable to the SHIPP 2 release, and as the development of the 0.68 acre of reference land as a highway is an essential part of SHIPP 2, without the highway there would be no SHIPP 2 release.

Was there a "scheme"? H

If that analysis does not satisfy the conditions of section 7, the alternative is the assessment of compensation in the light of the gain to Laing as the owner of other land pursuant to the scheme of SHIPP 2 (Area 12) as a whole, for which submission Mr Harman sought support from *Pointe Gourde* and from *Camrose (Viscount) v Basingstoke Corporation* [1966] 1 WLR 1100 and *Wilson v Liverpool Corporation* [1971] 1 WLR 302.

Mr Fitzgerald submitted that the existence of a scheme is a matter of fact, as it was said by Lord Buckmaster in *Fraser v City of Fraserville* [1917] AC 187, PC: J

> The principles of those cases are carefully and correctly considered in the judgments the subject of appeal, and the substance of them is this; that the value to be ascertained is the value to the seller of the property in its actual condition at the time of expropriation with all its existing advantages and with all its possibilities, excluding any advantage due to the carrying out of the scheme for which the property is compulsorily acquired, the question of what is the scheme being a question of fact for the arbitrator in each case.

Further, it is the effect of fact on value that has to be identified, as Lord Denning MR said in *Wilson* at p 309:

> A scheme is a progressive thing. It starts vague and known to K few. It becomes more precise and better known as time goes on. Eventually it becomes precise and definite, and known to all. Correspondingly, its impact has a progressive effect on values. At first it has little effect because it is so vague and uncertain. As it becomes more precise and better known, so its impact increases until it has an important effect. It is this increase, whether big or small, which is to be disregarded at the time when the value is to be assessed.

A scheme in the concept of *Pointe Gourde*, Mr Fitzgerald said, is related not to planning but to compulsory purchase, where there has to be disregarded value attributable to the purpose of acquisition. SHIPP 2 does not involve the acquisition of land; the only indication of the potential exercise of L compulsory powers was to be found in the minute of the Highways and Works Committee of Eastleigh on July 10 1973:

> 133. SHIPP 2—Land north of Fair Oak Road.— The town clerk reported on the progress being made towards the development of this land and in particular the construction of the proposed spine road. It appeared that there was a possibility of some delay in construction of the spine road and discussions were to take place with the developers concerned to see if the problems which had arisen could be solved. It might, however, be necessary for the council to acquire the land to secure early progress. Recommended that a compulsory purchase order be made under the provisions of M section 214 of the Highways Act 1959 in respect of the land required for completion of the spine road as shown edged and hatched in red on the plan produced by the borough engineer and surveyor.

Mr Fitzgerald submitted that there is no authority which equates the release of land for private housing, or the allocation in a development plan, or the grant of planning permission for private housing development, with a scheme within the concept of *Pointe Gourde*. Mr Fitzgerald also referred me to *Sprinz v Kingston upon Hull City Council* (1975) 235 EG 225. As to the alternative betterment issue pursuant

A to section 7, Mr Fitzgerald submitted there was no evidence that Laing's land could not be developed in the absence of the spine road. A finding under section 7 became a matter of fact.

I was referred by Mr Fitzgerald to the other side of the coin, the requirement by section 9 of the Land Compensation Act 1961 to disregard any diminution in value attributable to the prospect of compulsory acquisition:

> 9: No account shall be taken of any depreciation of the value of the relevant interests which is attributable to the fact that (whether by way of designation, allocation or other particulars contained in the current development plan, or by any other means) an indication has been given that the relevant land is, or is likely, B to be acquired by an authority possessing compulsory purchase powers.

It was unnecessary to rely on any "indication" earlier than the minute of July 10 1973 which has been quoted. There was an earlier indication in November 1972, but it was given to Mill Lodge—it was not an indication to Laing. Submitting that the wording of the section "which is attributable" is very wide, Mr Fitzgerald referred to *Jelson Ltd v Blaby District Council* (1977) 243 EG 47. The section provided that there must be determined what would have happened in the absence of the exercise of compulsory powers. In the instant case no C houses would have been permitted to be constructed. Mr Harman agreed that as to "indication" there was no evidence that Laing were aware of the recommendation of July 10 1973 to make a compulsory purchase order. Mr Harman also said that the land would not have possessed value for development without the proposal to have a spine road. As the provision of a spine road was a basis for the SHIPP 2 release of land, there could be no diminution in value in the terms of section 9, since the reference land would have possessed only "white land" value.

Finally, Mr Fitzgerald submitted that the alteration of the principal section 52 agreement by the two supplemental agree- D ments raises the question of estoppel. I was referred to Spencer Bower's book on *Estoppel by Representation* by Sir Alexander Turner and to the case in the Court of Appeal of *Evenden v Guildford City Association Football Club Ltd* [1975] 3 All ER 269. I was not asked to come to a decision on a point of law, but I was urged by Mr Fitzgerald to find the facts on which an argument in law could be founded. Mr Harman contended that facts on this issue should not and could not be found; no mention of this issue was to be found in the statement of facts or other exchanged documents, so Eastleigh had not pursued further enquiry by discovery; nor were representations produced in the cross-examination of E witnesses; it was also an issue which was only taken by Mr Fitzgerald in his reply.

Having considered the evidence, I am in no doubt that I have not adequate material on which I could find with confidence the facts as I have been asked to find by Mr Fitzgerald. Estoppel was mentioned in the opening of the claimants' case, but as a matter that was not for my decision, and there was but brief reference in examination and cross-examination of witnesses to the matter of representation which would go to the issue of estoppel. I have come to the conclusion that I am unable to make the finding of facts which is sought by Mr Fitzgerald.

F

Tribunal's findings

I come to my findings.

In the beginning there was a block of 90 acres of unallocated land in the statutory town map of the development plan. It was released for development by SHIPP 2. There were five owners. Mill Lodge purchased from a consortium of three owners an area of 75.5 acres at the western end. Laing purchased from one owner, a Mr Sole, 3.4 acres at the eastern end. The fifth owner, Brown Brothers, owned the remaining area adjoining Laing's eastern boundary. The spine road was aligned through and bisected the 90 acres. Mill Lodge has access to the spine road at its western end where it joins Fair G Oak Road and the land has frontage to Fair Oak Road. Laing's 3.4 acres lies astride the line of the spine road and has frontage to Fair Oak Road; the line of the spine road continues eastwards from Laing through land which was owned by Brown Brothers before it was released to Eastleigh by dedication.

As an alternative to the spine road, Fair Oak Road could have been widened and improved. The cost would have been greater and it would not have satisfied the planning requirement of SHIPP 2 that "Fair Oak Road should be diverted through the development." It would have been practicable to align the spine road through the 90 acres without passing H through Laing's land although part of the Fair Oak Road frontage of that land would have been required. It would have been practicable to realign the eastern end of the spine road to pass through Laing's land and to avoid the land of Brown Brothers. It is not in issue that at March 7 1977 none of these alternatives was available. I accept Mr Welch's evidence that when the line of the spine road was first produced no difficulties were envisaged and no alternatives were suggested.

Valuation conclusions

I now turn to the valuations.

By the valuation date, the spine road had been constructed J through Mill Lodge land as far as the western boundary of the reference land; at that date there was no realistic permanent access to Mill Lodge land available from the east other than through the reference land; outline planning permission for residential development had been granted in respect of the whole of Mill Lodge land; detailed planning permission had been granted in respect of Mill Lodge land for the building of about 700 dwellings of which 174 had been completed and 66 partly completed; an unspecified number of dwellings were occupied; Mill Lodge could be required by Eastleigh to construct a temporary access on Mill Lodge land from the eastern end of the spine road K southwards to Fair Oak Road. The development of areas D and E on the south side of the spine road (28.8 acres net) could continue with access to the Fair Oak Road frontage; similarly, development of area F (4.8 acres net), which Mill Lodge had sold to Eastleigh, could continue. Development of areas A, B and C (36.4 acres net) north of the spine road could proceed.

I have come to the conclusion that on the facts which I have found, at March 7 1977 potential development value of Mill Lodge land had already been released; the reference land at that date was not an access key which was required by the adjoining owner to unlock enhancement in value. If, L therefore, value has been, and may continue to be, realised without the reference land as an access, a purchaser of Mill Lodge land would not have to retain a proportion of its value for bargaining for the purchase of the reference land. Mr Fitzgerald submitted that access to Fair Oak Road at the eastern end of the spine road at the valuation date could be provided only by a temporary road which Mill Lodge could be called upon to construct and that the necessity for permanent access was only deferred for two or three years.

In so far as the reference land attracts a special value as the key to the completion of the spine road, I am persuaded by Mr Harman's submission that the effect of rule (3) in sec- M tion 5 of the 1961 Act would be to exclude such special value. On the evidence, the line of the spine road was considered to be the acceptable line for the purpose and I am satisfied that geographically it is the best line for its purpose. Within the terms of rule (3), following the judgments in *Lambe*, the reference land possesses the quality of special suitability for the purpose of building the spine road. The market for such a purpose would be limited; it is clear that Eastleigh would be in the market and Mill Lodge, as the owner of land, to carry out the construction works on behalf of Eastleigh to enable Eastleigh to adopt the road.

The evidence of Mr Pearson concerned his purchase of the north and south parts of Laing's land in October 1977 after the date of valuation; if the whole of Laing's land had been on offer Mr Pearson would have included the site of the spine road in the purchase so as to improve cash flow by reducing the purchase price and to make a profit by building the road at a lower cost than Eastleigh's estimate. I am not satisfied that on all the facts at March 7 1977 Mr Pearson would have been in the market for the purpose of building a section of the spine road.

It follows that on this issue I accept the submission of Mr Harman that the special suitability of the reference land for the purpose of building the spine road shall not be taken into account because for that purpose there is no market apart from the requirements of Eastleigh or the special needs of Mill Lodge on behalf of Eastleigh.

I have come to the conclusion in the light of these findings that no assistance is to be derived from the valuation of Mr Harkus to determine the value of the reference land at March 7 1977.

I should say at this point of my decision that, on the evidence, the Mill Lodge section 52 agreement and the two supplemental agreements enabled the development of Mill Lodge land to proceed and planning permissions to be granted. At the outset, it was recognised that development of the SHIPP 2 land could be affected by multiple ownership, and in order to ensure that the construction of the spine road could be completed, the section 52 agreement contained clause 14 as to the exercise of statutory powers by Eastleigh.

Mr Benefield, in cross-examination, said that if compulsory powers had not been available, the situation in 1973 would have been very different; it was his view that development would not have been permitted in advance of securing the spine road access. Other solutions (if there were any) to the multiple-ownership problem were not canvassed; so in the absence of a reservation to the council as to the exercise of compulsory powers, the evidence is that the parties would not have entered into the section 52 agreement, and no conclusion can be reached as to what course would have been taken to implement SHIPP 2.

I should consider here the submissions of counsel as to the requirement by section 9 of the 1961 Act to disregard any diminution in value attributable to the prospect of compulsory acquisition. It follows from my conclusions that I find Mr Harman's argument on section 9 the more acceptable, namely, that the land would not have possessed value for development without the proposal to have a spine road, and without the provision of a spine road development would not be permitted; thus within the terms of section 9 the reference land would suffer no diminution in value.

Before I turn to the detail of Mr Gregory's valuation evidence I will consider two of the issues of law to which I have referred. The first is about betterment, which Mr Harman submitted arises pursuant to section 7 of the 1961 Act and by the application of the principle expressed in *Pointe Gourde*. I accept Mr Fitzgerald's submission that there was no evidence that Laing's land could not be developed in the absence of the spine road through Laing's land. I so find as a matter of fact and I conclude that betterment under section 7 does not arise.

I accept also Mr Fitzgerald's submission that SHIPP 2 release for private development, which is akin to the allocation in a development plan, does not equate with a scheme within the concept of *Pointe Gourde*, where there has to be disregarded value attributable to the purpose of acquisition. It would seem that the release of land created the increase in value, as is shown in Mr Gregory's valuation, where the value of the remaining land before and after acquisition is put at the same figure of £25,000 per acre. I therefore find that there is no scheme whereby any value of the relevant land is to be disregarded.

I have come to the conclusion that Mr Gregory's valuation approach is not supported by the evidence. First, there is no evidence that a planning condition to a permission in respect of Laing's land would require Laing to bear the cost of constructing the spine road; secondly, there was no evidence that if the reference land was not to be acquired for the purpose of the spine road, permission would not be granted on March 7 1977 for residential development; on the contrary, there was evidence that development of Laing's land and the adjoining Brown Brothers' land would be a proposition to which there would be no highway objections, provided that an emergency exit was made available at the southern boundary to Fair Oak Road. Laing would require access through Brown Brothers' land, which Mr Fitzgerald conceded would require Laing themselves to experience and bear ransom treatment.

Ransom proportions

During the hearing a great deal was said about ransom proportions of values released; I was referred to several decisions of the Lands Tribunal and I was told by witnesses of their own experience in the land market where ransom or key value was a feature of importance; in my judgment an allowance of 25 per cent would be appropriate in the instant case. I accept Mr Gregory's development value of Laing's land at £25,000 per acre and I determine the value of the reference land at £12,750. The details are:

Development value of Laing's land: 3.4 acres with access at £25,000 per acre	£85,000
Deduct cost of acquisition of access 25%	£21,250
Net value (equivalent to £18,750 per acre)	£63,750
Value of reference land 0.68 acre at £18,750 per acre	£12,750

I therefore determine that £12,750 is the price to be paid by Eastleigh Borough Council to Laing Homes Ltd for the purchase of 0.68 acre at Fair Oak Road, Bishopstoke, pursuant to the agreement dated March 4 1977.

Having opened the sealed offer of compensation lodged by the acquiring authority, I find that the amount thereof is less than the amount of my award. Having heard submissions by counsel as to costs I determine that the acquiring authority will pay the costs of the claimant of this reference, such costs, if not agreed, to be taxed by the Registrar of the Lands Tribunal on the High Court Scale.

MARCHANT v SECRETARY OF STATE FOR TRANSPORT

(REF/294/1978)

(V G WELLINGS QC)

February 28 1979

Estates Gazette May 12 1979

(1979) 250 EG 559

Depreciation claim under Part I of Land Compensation Act 1973—Bungalow in small rural village near Maidstone affected by noise from M20 597 m away—No settlements of compensation in locality at greater distance than 244 m—Evidence of acoustics expert accepted—Main effect of motorway was to raise background level of noise considerably—On tribunal's inspection, noise "impressive" and caused some discomfort—Conversation difficult in garden—Level of sound certain to have affected prospective purchaser's bid—Award of £1,000

D A Marchant, the claimant, appeared in person; Michael Rich (instructed by the Treasury Solicitor) for the compensating authority.

Giving his decision, MR WELLINGS said: By this reference Donald Alec Marchant, the owner and occupier of Timberley,

A a detached dwelling-house in Masters Lane, Birling, Maidstone, Kent, claims from the Secretary of State for Transport compensation under Part I of the Land Compensation Act 1973 for depreciation in the value of his interest in that house and land by physical factors caused by the use of public works. More specifically he claims compensation for depreciation in the value of that interest by reason of noise arising from use of the M20 motorway, which lies to the south west, south and south east of the property.

The motorway (namely, the Ditton by-pass) at this point runs, broadly speaking, from west to east. At no relevant point is the motorway in a cutting or on an embankment. Indeed

B it is on, broadly speaking, the same level as that of the claimant's property, as is the land in between.

Timberley is a Colt bungalow consisting of three bedrooms, living-room, kitchen, bathroom, wc and hall and outbuildings. The house itself faces due south towards the motorway. Birling is a small rural village; Masters Lane runs southward from approximately the centre of the village. The situation of Timberley, its garden and the aspect southwards to and beyond the motorway are pastoral and pleasant, which is a word apt also to describe the village itself. There are a few trees between the house, its garden and the motorway but the intervening land is otherwise uninterrupted.

C

Agreed facts

Certain facts and matters were agreed between the parties. They were:

(i) The claimant's interest in the property is an owner's interest qualifying for compensation;

(ii) the motorway was opened for traffic on December 8 1971;

(iii) the relevant date for the purposes of valuation in the present case is December 8 1972;

(iv) the period in which a claim for compensation could lawfully be made was from June 23 1973 (the date of coming into force of Part I of the Act) until June 23 1975;

D (v) the claimant's claim for compensation was made on March 6 1974 and was duly made;

(vi) the area of the house itself (not including an extension thereof made after December 8 1972) is 112.13 m²;

(vii) the distance from the house to the edge of the hard shoulder of the motorway is 582 m (636 yds);

(viii) the distance from the house to the centre line of the motorway is 597 m (653 yds);

(ix) the length of the frontage of the land to Masters Lane is 64 m (70 yds).

It was further agreed that the issue to be determined by the Lands Tribunal is whether there has or has not been a diminu-

E tion in the market value of the subject property by reason of noise attributable to use of the motorway (M20 Ditton by-pass) as at December 8 1972 and use of the motorway expected as at that date.

The issue is put in that form because of the provision in section 4(2) of the Act of 1973 that, in assessing depreciation due to the physical factors caused by the use of any public works, account shall be taken not only of the use of those works as that use exists on the valuation date, but also of any intensification that may then be reasonably expected of the use of those works in the state that they are on that date.

F Finally it was agreed between the parties that this is not a case in which the value of any benefit derived from the motorway is required to be deducted from any compensation which would otherwise be payable.

The claimant conducted his own case and gave evidence. He called no valuation or other expert evidence. Michael Rich, counsel for the Secretary of State for Transport, called Jeremy Sargent, a chartered engineer BSc (Eng) MICE MIMunE MIOA. He is an acoustics expert employed by the Kent County Council for the purpose of the environmental assessment of its highway schemes. Mr Rich also called the district valuer, Cecil Percy Cowell ARICS.

Mr Marchant said that he had owned the freehold of the site since 1959 and had occupied the property since November 11 1960, the date of completion of the house. In 1960 the property was tranquil. He said that for 75 per cent of the time the wind is in the quadrant from the south west to the south east. On some days there was no wind at all and on such days the noise emanating from the motorway was at its loudest. The bulk of his garden was on the motorway side of the property. The garden was mostly lawn. He and his family enjoyed sitting and working and playing in the garden. But when the noise was at its worst such activities were rendered uncomfortable; conversation was rendered difficult and impossible at 10 yds. Such occasions occurred twice a H
day on 90 days in a year. In a further 180 days of a year the noise was substantial but not so as to cause great discomfort or make conversation difficult. In the remaining 90 days or so the garden would be quiet with only negligible noise from the motorway. On the occasions on which noise from the motorway was at its worst (that is to say, on 90 days in the year) it could be heard inside or outside the house; it did not keep him or his wife awake because they slept at the back of the house, but nevertheless it kept his visitors awake. He claimed as compensation a sum equal to 10 per cent of what would have been the capital value of the property as at December 8 1972 if noise from the motorway were disregarded. Originally J
his claim on that basis was £1,500 (10 per cent of £15,000) but during the course of the hearing the sum which he claimed was, for reasons I shall mention, increased.

Acoustics expert's evidence

Mr Sargent referred me to the *Final Report of the Committee on the Problem of Noise* (Cmnd 2056), of which the chairman was Sir Alan Wilson FRS. I will call that report: "The Wilson Report." Para 117 of that report is concerned with dwellings and states the conclusion of the committee as follows: "a very tentative estimate is that the following levels inside living-rooms and bedrooms should not be exceeded for K
more than 10 per cent of the time" and there follow various figures. In respect of country areas the relevant daytime level is stated to be 40 dB(A) and the nighttime level is stated to be 30 dB(A).

The relevant levels for suburban areas, away from main traffic routes, are respectively 45 dB(A) and 35 dB(A).

No doubt the above-mentioned levels are expressed in terms of L10 over a period of 18 hours. I should explain that L10 is the sound level in dB(A) which is exceeded for one-tenth of a period of one hour and L10 (18-hour) is the arithmetic mean of all the levels of L10 during a period from L
0600 to 2400 hours on a normal working day (see Regulation 2 of the Noise Insulation Regulations 1975). At this stage it is convenient to mention also L90 which, as defined by Mr Sargent, is the sound level in dB(A) exceeded for 90 per cent of a period.

In addition to the Wilson Report and the Noise Insulation Regulations 1975, Mr Sargent also referred me to Department of Environment Circular 10/73 *(Planning and Noise)*, British Standard 4142: 1967 *(Method of Rating Industrial Noise Affecting Mixed Residential and Industrial Areas)* and Department of the Environment and Welsh Office document: *Calculation of Road Traffic Noise.* M

Mr Sargent said that in the absence of any other standard he would accept the standard mentioned in the Wilson Report as being an absolute standard. If exceeded, some intrusion would be caused. If not exceeded, there would be no intrusion. That standard represented what most reasonable people would regard as the limit.

Mr Sargent had carried out three investigations into the noise levels experienced and likely to be experienced at and in the vicinity of Timberley:

(i) He had employed the "prediction" method to calculate the levels of noise at and in the vicinity of Timberley on

December 8 1972, in 1978 and 1986. In so doing, he had followed the prediction method described in section I of *Calculation of Road Traffic Noise, supra.* For 1972 and 1978 he had applied figures of traffic flows taken from traffic census records of the M20 for those years. For 1986 he had made an estimate of the traffic flows. In each case he applied the figures and traffic flow in the manner recommended in that document. The prediction method assumed, he said, that a slight wind was blowing from the motorway towards the receiver: he had made that assumption. The predicted noise levels in terms of L10 (18-hour) were as follows: (a) as at December 8 1972: 51.5 dB(A); (b) as in 1978: 52.1 dB(A); (c) as in 1968: 55.2 dB(A).

(ii) In February and March 1978 he had used a sound-level meter to take sample measurements of noise levels in terms of L10 and L90 at "Point C" (outside the south-western boundary of the claimant's property) and at "Point F" (as quiet and rural a situation as could be found to represent the pre-existing background noise and just over a mile to the north of Timberley). The results were as follows:

(a) on February 3 1978:

Instruments employed B and K 4426

Weather Dry/cool; *Windspeed* Light. Direction north

Microphone station	*Time* From	To	*Noise levels* (State peak, background, etc) L10	L90
C	1120	1130	49·5	44·5

(b) on March 6 1978:

Instruments employed B and K 4426

Weather Sunny; *Windspeed* Light. Direction south west

Microphone station	*Time* From	To	*Noise levels* (State peak, background, etc) L10	L90
C	1310	1320	52	46·5
C	1545	1555	51	44
F	1415	1425	42	36·5

(iii) on August 3 1978 he took measurements at Point A (outside the southern wall of the bungalow), Point B (at the southern edge of the lawn: that is south of Point A), Point C (referred to above), Point D (a point half-way between Timberley and the motorway) and Point E (inside the living-room of the bungalow with windows open and closed). The results were [see opposite]:

The difference between the prediction for 1978 (52.1 dB(A)) and the measurement for August 3 1978 (53.6 dB(A)) at Point C is only 1.5 dB(A). Mr Sargent said that this showed that there was a fair correspondence between the results reached by prediction and those reached by measurement and that the predictions were reliable. He said that in order to compare the predictions and measurements on the one hand with the standards in the Wilson Report on the other, it was necessary to deduct from the predictions and measurements 20 in order to ascertain the level in a living-room or bedroom with windows closed and 10 to ascertain the level therein with windows open. On that basis, the level in such a room in Timberley with windows closed was satisfactory; with windows open, the level was marginally in excess of the appropriate standard.

Mr Sargent said that the effect of the northerly wind at Point C on February 3 1978 was to reduce the traffic noise component of the overall noise. That component was responsible for raising the L10 (18-hour) figure to between 50 and 55 db(A) whenever the wind was in the west, south west, east or south east. He agreed that on a majority of days, not necessarily on three out of every four days, the more adverse conditions applied. During something less than half of the year the Wilson Report standards were complied with. He would, however, not apply those standards to the garden of the claimant's property. When noise levels reach 58-60 dB(A) conversation at 1.3 m. becomes difficult. Such conditions would not apply in the present case even in adverse wind conditions. Nevertheless, L90 (the level of background noise) was the most significant factor, he said, in the present case. The main effect of the motorway was to raise the background level, including noise from the brick works at Ryarsh (½-mile to the south west) considerably: by 13 dB(A). He was not surprised that that level gave rise to complaint. The outlook of the property was rural and the background level was unexpected. If he had experienced the two situations, before and after the coming of the motorway, he would not be very happy and he would complain. He, however, expressed the view that people who come to an area without having experienced the pre-existing situation are likely to notice an increase in noise less than long-standing residents would do.

I have inspected the claimant's property, the village and the vicinity and various properties. Upon my inspection, in order to enable me to correlate the sounds experienced by me with those referred to in his evidence, Mr Sargent kindly set up at Point C a meter of the kind used at that point on August 3 1978. The wind was blowing from the south east and was moderate in force and the weather was fine. The noise, to my ear, to say the least, was impressive and, in my view, caused some discomfort. The readings taken by Mr Sargent on this occasion over a 10-minute period were the equivalent of the following: L10 (18 hour): 63.5 dB(A), and L90: 52 dB(A).

(a)

Instruments employed B and K 4426

Weather Dull; *Windspeed* 0900 6-8, 1200 5-6, 1500 7-9, 1900 0-3; Direction westerly

Microphone station	*Time* From	To	*Noise levels* (State peak, background, etc) L10	L90
A	0655	0705	54·5	49·5
	0733	0743	56·0	51·0
	0830	0840	56·0	52·0
	0940	0950	53·0	47·0
	1045	1055	53·0	48·0
	1145	1155	52·0	47·5
	1230	1240	50·0	44·5
	1325	1335	52·0	46·5
	1430	1440	52·5	48·0
	1535	1545	56.0	50·5
	1620	1630	59·5	53·0
	1742	1752	54·0	51·0
	1817	1827	55·0	50·0
	1915	1925	55·0	50·5
	2010	2020	53·0	46·5
	2105	2115	52·5	47·5
	2205	2215	50·5	44·0
	2300	2330	49·0	41·3
	24		18 HR Ave 53·6 dB(A)	18 HR Ave 48·2 dB(A)

(b)

Instruments employed 4426

Weather Dull; *Windspeed* 0900 6-8, 1200 5-6, 1500 7-9, 1900 0-3; Direction westerly

M crophone station	*Time* From	To	*Noise levels* (State peak, background, etc) L10	L90	Remarks
E	1010	1020	34	29	windows closed
E	1025	1035	44	39·5	windows open
C	1130	1140	52	47·5	
B	1210	1220	54	46·5	
D	1255	1305	35·7	51·5	
E	1310	1320	33	27	windows closed
C	1510	1520	56	50·5	
E	1605	1615	48	42	windows open
B	1808	1818	55·5	50·5	
D	1855	1705x	58	52·5	

xNote by tribunal: presumably this is an error for 1905

A Mr Sargent said that the noise was significantly louder than he had previously heard it. There is no doubt that it made conversation difficult in the garden. I may add that the L10 (18 hour) level was but 5 dB(A) short of the level at which under the Noise Insulation Regulations grants for insulation of residential buildings become payable.

District valuer's evidence

Cecil Percy Cowell ARICS has been district valuer and valuation officer in the Tunbridge Wells area for the last six years. He has had 45 years' experience in valuation. He produced a written valuation showing that as at December 8
B 1972 the before-value (that is without regard to the effect of motorway noise) of Timberley was £18,000 and that as at the same date the after-value (with regard to motorway noise) was also £18,000. In the course of cross-examination, he said that he was prepared to accept that the before-value as at December 8 1972 was £20,000. He none the less expressed the view that the effect of motorway noise on the value was nil. There was, however, he said, no direct comparable transaction on which reliance could be placed.

Mr Cowell produced a schedule of 116 properties in the locality in respect of which compensation under Part I of the Act of 1973 on account of depreciation attributable to motorway noise had been agreed and paid. In 105 cases estate agents
C had acted for the claimants. No settlement (and no claim other than in respect of Timberley) had been made in respect of any property at a greater distance than 244 m from the motorway.

Mr Cowell also produced a schedule of 89 properties in the locality in respect of which an agreement had been reached that no compensation on account of depreciation attributable to motorway noise was payable; in 76 of the cases estate agents had acted on behalf of the claimants.

Upon my inspection, in the company of Mr Cowell, I inspected (externally only) a selection of properties referred to in both schedules.

D Mr Rich submitted that the claimant had failed to prove his case: the onus of proof was on him and he had not discharged it. I ought, said Mr Rich, to accept the evidence of Mr Cowell.

Tribunal's findings

I accept Mr Sargent's evidence. It may be that a prospective purchaser would not employ an expert, whether or not of the calibre of Mr Sargent, in order to ascertain the level of noise attributable to the motorway. Nevertheless Mr Sargent's evidence provides useful precision in the assessment of the
E level of noise which a purchaser could reasonably be expected to encounter. It is a most valuable addition to the inexpert evidence of the claimant, and of my own ears, and it was indeed gracious of the Secretary of State to provide that evidence. I recognise that a purchaser might not be so impressed by the level of noise found by Mr Sargent as would someone who had experienced the situation existing before the coming of the motorway. On the other hand, the purchaser can use his eyes and would, in my view, expect to find at Timberley a peaceful and rural, if not idyllic, situation. Like Mr Sargent, and indeed myself, he would be surprised by the levels of sound experienced there. It is indeed surprising to
F find such levels so far from the motorway. No explanation for the cause of this has been provided but the situation is proved to exist.

I am aware of the force of the district valuer's evidence. Nevertheless, it appears to me that a level of sound of the order which the evidence reveals is certain to have affected a prospective purchaser's bid for the property at the relevant date and to have depreciated its value. Some idea of the view which a potential purchaser would take is revealed by the schedules of settlements. I bear in mind that they are settlements and not open market transactions and I regard them with a mildly benevolent eye. Looked at in that way, it appears to me that they do not belie the view that this very G pleasant property in its rather special rural setting has been depreciated in value; on the contrary they could be said to give some support to that view. It is not an easy matter to determine the amount of the depreciation. Doing the best I can, I determine that amount to be £1,000. I award that sum to the claimant.

The compensating authority will pay £200 towards the costs of the claimant.

FREDERICK POWELL & SON LTD v DEVON COUNTY COUNCIL H

(REF/293/1977)

(E C STRATHON FRICS)

February 16 1979

Estates Gazette May 19 1979

(1979) 250 EG 659

Residential building land at Cullompton, Devon, acquired for school—Part of site for which detailed planning consent granted—Claim for injurious affection to retained land by severance rejected by tribunal as not substantiated—Inability of claimant builders to continue to average costs over whole site does not depreciate value of retained land—Value of J **land taken derived from comparable transactions—Weight of evidence towards claimant's valuer's opinion of value**

Sir Frederick Corfield QC and E R E Caws (instructed by Riley & Co, of Blackburn) appeared for the claimant; Viscount Colville of Culross QC (instructed by the solicitor to Devon County Council) for the acquiring authority.

Giving his decision, MR STRATHON said: This is a reference to determine the amount of compensation payable to Frederick Powell & Son Ltd by the Devon County Council upon the compulsory acquisition of the claimant's freehold interest in 1.6 acres of land at Crow Bridge, Cullompton, Devon.

The following facts are included in an agreed statement K by the parties:

(A) General facts:

(1) The claimant is a subsidiary of the Broseley Group of Companies, one of the major housebuilding companies in the country. The claimant is principally concerned with housing development in south-west England.

(2) The acquiring authority is a local education authority and requires the subject land for the purposes of a proposed infants school.

(3) The subject land has an area of 1.6 acres and lies on the western side of Cullompton, being approximately 750 L yds from the town centre.

(4) The site consists of part of an enclosure of pasture land with a moderate slope from the western corner to the north-east and south-east. There is an 8ft depression in the eastern corner of the site which has an area of 0.13-acre. To the east and south of the site there are council housing estates.

(5) Formerly a market town, Cullompton is 15 miles from Exeter by the M5 motorway and 21 miles by motorway from Taunton.

(6) The population of the town of Cullompton according M to a population census carried out by the Central Information Service in 1976 was 3,750.

(7) Possession of the subject land was delivered up to the county council by agreement on April 10 1978.

(8) The appropriate date for the valuation of the subject land is April 10 1978.

(B) Chronological facts:

(1) On May 9 1972 outline planning permission for residential development was granted by Tiverton District Council in respect of, *inter alia*, the subject land.

A (2) The claimant purchased the 9-acre Brook Close Estate, including the subject land, in 1973 with the benefit of the above outline consent, contracts having been exchanged on August 23 1973.

(3) The Brook Close Estate adjoins the North Farm Estate which is being developed by the claimant for private housing; approximately 17.8 acres have been completed and sold. Of the Brook Close Estate 3 acres are reserved for public open space and approximately 6 acres, including the subject land, have the benefit of detailed planning consent for the erection of 105 dwellings together with garages, roads and sewers.

B (4) On October 23 1973 the county education officer lodged an application for planning permission under the Town and Country Planning (General) Regulations 1969 for deemed planning consent for development of the subject land as an infants school. On the same day the county education officer wrote to the claimant inviting it to sell the subject land to the county council for education purposes. This was the first indication to the claimant that the county council required the subject land.

(5) The claimant replied on October 26 1973 indicating that it was not prepared to sell to the county council by agreement.

C (6) A resolution to make a compulsory purchase order was made by the county council's Estates Sub-Committee on September 13 1974, and at its meeting on October 17 1974 the county council approved the minutes of the Estates Sub-Committee.

(7) Planning permission for the school site was deemed to have been granted by the Secretary of State for the Environment by virtue of a resolution of the Education Committee of the Devon County Council on September 19 1974.

(8) The compulsory purchase order was sealed by the county council on March 13 1975 and the claimant was served
D with notice accordingly on March 17 1975.

(9) The claimant lodged an objection to the compulsory purchase order on March 27 1975.

(10) At the time that the compulsory purchase order was made, the claimant had not formally submitted, for the approval of the local planning authority, its detailed proposals for the development of the Brook Close Estate of which the subject land is part.

(11) On April 29 1975 the claimant applied to the local planning authority for approval of details reserved by the outline planning permission dated May 9 1972 relating to
E the layout of 105 houses, garages, roads and sewers. Following a public inquiry on February 26 and 27 1976 the compulsory purchase order was confirmed by the Secretary of State on March 10 1977 with modifications. The inspector had recommended adjustment of the order boundaries to coincide with the claimant's detailed proposals for development to achieve a more logical result and to reduce the compensation arising from severance. The Secretary of State accepted the recommendation, but was only able in law to amend the boundaries by reduction and not addition. This left the question of the addition of any extra land to be agreed between the parties.

F (12) The details for the development of the Brook Close Estate submitted to the Tiverton District Council as local planning authority on April 29 1975, were approved by the Secretary of State on November 29 1976 following an appeal.

Blight notice served

(13) On May 5 1977 the claimant served a blight notice under section 193(1) of the Town and Country Planning Act 1971. By virtue of section 196(1) of the said Act of 1971 the acquiring authority were deemed to have served a notice to treat in respect of the subject land on July 5 1977.

(14) The claimant lodged a notice of reference at the Lands Tribunal on December 9 1977. G

(15) It is agreed that the compensation shall be assessed on the basis of the detailed planning consent dated November 29 1976 together with the claimant's plan number 20040 dated April 1975 (ie DCC1) which shows the detailed layout of the 105 dwellings out of which the county council requires the sites of 29 dwellings.

(16) At the date of being granted possession of the subject land by the claimant, ie April 10 1978, excavation work had been carried out on roads nos one and three, together with services therein, but neither of these two roads had H

Primary valuation

29 house plots at £3,600 per plot*	£104,400
Less	
cost of roads nos 4 and 5	20,700
	£83,700
Compensation, say	£83,000

*Note by Lands Tribunal. Derived from Schedule 1 which follows.

Schedule 1

*House sale prices (as at April 10 1978).		
105 houses (6 types)		£1,431,550
**Deduct* cost of house-building (as at April 10 1978)		
105 houses (the above 6 types)	£808,276	
Interest on building costs 3 months at 10%	20,207	
Building operation profit at 5%	40,414	868,897
		562,653
Interest and developers' profit 12½% on £1,431,550		178,944
Value of land with roads, sewers, services etc.		£383,709

J

£383,709 ÷ 105 plots = £3,654 per plot
say = £3,600 per plot

*Note by Lands Tribunal: Details of sale prices and costs of all types of houses were included in the schedule. K

Supporting valuation I (Residual)

Predicted sale prices		
17 inner-terrace at £12,700 each	£215,900	
12 end-terrace at £13,350 each	160,200	
	376,100	
Less		
developers' profit 10%	37,610	£338,490
Building costs		
17 inner-terrace at £7,146 each	121,482	
12 end-terrace at £7,922 each	95,064	
Cost of roads 4 and 5	20,700	
	237,246	
Building profit 5%	11,862	
Interest 3 months at 10%	5,931	255,039
		83,451
Interest on site for 6 months at 10%		4,173
		£79,278
Site value, say		£80,000

L

Supporting valuation II (Before and after) M

Value of the estate including school site		
105 house plots at £3,600 each	£378,000	
Less		
cost of roads and sewers	212,220	£165,780
Value of the estate excluding school site		
76 house plots at £3,600 each	273,600	
Less		
cost of roads and sewers	191,520	82,080
		£83,700
Say		£83,000

A *Supporting valuation III*

(Value of the land to be acquired if separately held and excluding any sum which would be paid by the owner of the adjoining estate as a special purchaser.)

1.6 acres at £25,000 per acre			£40,000
Injurious affection to remainder of land held therewith:			
Costs to be spread over houses to be erected:			
Erection and dismantling of site compounds accommodation, electricity supply, telephones, etc		£4,000	
Salaries during road construction (site agent, clerk, engineer)		4,500	
Cost of non-productive plant required for associated works (pumps, rollers, fork-lift)		8,400	
B Roads, services etc		212,220	
		229,120	
Interest at 10%—9 months		17,184	
		£246,304	
£246,304 ÷ 105 units = cost per unit		£2,346	
Cost on 29 houses		29	
		68,034	
Less			
Savings on roads 4 and 5	£20,700		
Plus			
Interest at 10%—9 months	1,552		
C Saving of overheads	1,290	23,542	44,492
		Compensation	£84,492

For the acquiring authority, C B Keene ARICS, assistant county estates surveyor, submitted the following valuations:

Primary valuation (Ripe acreage basis)	
1.6 acres at £20,000 per acre	£32,000
Check valuation (Serviced plot basis)	
29 plots at £1,936 per plot	£56,144
Less	
Estimated cost of constructing roads 4 and 5 with sewers. Abnormal site costs.	24,130
D	£32,014

been completed. No building works had begun within the boundaries of the subject land. House-building operations had started on land held therewith.

Mr Hughes Brackett FRICS gave evidence on behalf of the claimant; he submitted a primary and three supporting valuations, the primary valuation of £83,000 being the amount of compensation claimed. The amount is the sum of two E heads of claim (a) the value of the land taken and (b) injurious affection to the land held therewith. The valuations, which are rearranged for ease of reference, are set out above.

As has been said, the claim includes compensation for injurious affection to the retained 4.40 acres of building land by severance from the 1.60 acres of land taken. It was submitted by Sir Frederick Corfield for the claimant that it is fundamental in this case that injurious affection is suffered. The reason for such a claim is described by Mr Brackett to be that a smaller number of 76 houses are left to cover the overheads, which virtually remain the same as for 105 houses.

F

Evidence about injurious affection

Before dealing with the submitted valuations, I will consider the evidence about injurious affection for which the amount of compensation is not separately identified in Mr Brackett's primary valuation or in supporting valuations I and II; Mr Brackett regards supporting valuation III as a check and in his view it is not the most reliable method of assessing the loss. In his opinion injurious affection is there, it is severe and it is difficult to assess. Each house has to bear its share of development costs, being the actual cost of the house itself and its share of common costs over the whole estate of six acres.

Mr Brackett considers that it is misleading to suggest that the G reference land, which on the approved layout would be developed at higher density and at less cost per unit, would have subsidised the remainder of the estate on which the infrastructure costs are heavier. The claim is made only for the reduction in the value of the 4.40 acres severed; it is not a claim, Mr Brackett says, for loss of profits on development of a building site, a proposition which was not approved in *Wimpey & Co v Middlesex CC* [1938] 3 All ER 781. The retained land has to be developed in accordance with the approved layout. The planning process took two years and any attempt to revise the pattern would result in loss of time and money. H

The value approach to this item is to be found in paragraph 5 of the preface to Mr Brackett's valuation. It is as follows:

> In practical terms the most reliable method of calculating the compensation is to value the land taken as part of the estate from which it is being severed, or to value the estate as a whole and as a diminished part, the difference representing the loss to the landowner. The Primary Valuation is on the basis of the land being taken as part of the whole estate and supports the total amount claimed, namely £83,000 both in respect of land taken and injurious affection due to severance. Any division of this figure is really arbitrary, but in order to meet the statutory requirement an attempt has been made in Supporting Valuation III in which the value of J the land taken and the injurious affection are separately calculated. The result shows a higher figure than that claimed, but this does not invalidate the principal claim.

Before turning to Mr Keene's evidence I should refer to Sir Frederick Corfield's submission that *Wimpey* is to be distinguished, since it concerned acquisition of the whole ownership. However, Mr Brackett firmly expresses his valuation of injurious affection as depreciation in the value of the land retained. I was also referred by a submission by Lord Colville to *D McEwing & Sons Ltd v Renfrew County Council* (1959) 11 P & CR 306.

Mr Keene's approach to the claim is to endeavour to identify the actual injurious affection suffered. His first conclusion K is that cost of the roads on the retained land is high, much higher per unit than the cost of roads on the reference land. In his experience injury suffered is usually clearly identifiable as a direct consequence of the acquisition of land. But in the instant case it is his view that every metre of road and services on the land retained is put to useful purpose and service; no money is thrown away. The acquisition has not created the need for additional lengths of roads; Road no 1 is required for its full length through the estate for development and the value of the land acquired takes into account its access to that road. High costs of the retained land are inherent in that land and L no higher costs are imposed by acquisition of the reference land.

In criticism of Mr Brackett's averaging of costs, Mr Keene's view is that averaging is an accounting transfer of profits from the more profitable development of the reference land to the high infrastructure cost of the retained land. Mr Keene is unable to identify any injurious affection suffered by severance of the land retained.

"Costs to be spread over houses"

The only valuation in which injurious affection is the subject of separate assessment is Mr Brackett's supporting valuation III, where items feature under the heading "injurious affection M to remainder of land held therewith." They are described as "costs to be spread over houses to be erected" and I will consider them seriatim:

1. Erection and dismantling of site compounds accommodation, electricity supply, telephones etc. £4,000

C D Morgan, a director of the claimant's company, said that the compound came into use in December 1977/January 1978 as road construction was started. The cost of erection and dismantling would be the same for development of 105 units as of 76 units; the loss of the land taken rendered the retained land less profitable.

A 2. Salaries during road construction (site agent, clerk, engineer) £4,500

This item covers the part of the salaries not included in the bill of quantities for road construction, but charged to house building. If 105 houses were built the cost per unit for this item would be reduced.

3. Cost of non-productive plant required for associated works (pumps, rollers, forklift) £8,400

Much of this plant, required for housebuilding, is hired at a weekly rate of approximately £120. The saving of hire for plant not required for construction on 29 units on the reference land is estimated by Mr Brackett to be £1,290, an amount B which can be seen in his valuation as "saving of overheads." The balance of the cost, £7,110, is spread over 105 units and thus a proportion is borne by the reference land.

4. Roads, services etc £212,220

This amount comprises the actual and estimated cost of roads and services for the 105 units on the whole estate of 6.0 acres. The equivalent cost per unit is £2,021; the estimated saving on roads nos 4 and 5, which would have been built to serve 29 units on the reference land, appears in Mr Brackett's valuation at £20,700 at £714 per unit.

I accept Mr Keene's view that, subject to the allowances C made in respect of savings, the development of the retained 4.4 acres requires the expenditure of the four items of costs. The acquisition of the reference land does not render the expenditure abortive. The consequence of the acquisition is that the claimants are denied the development of 1.6 acres at costs which are lower than those which prevail on the retained land. Mr Brackett does not regard supporting valuation III as the most reliable method of assessing the loss which is claimed, but it is only this valuation which defines the substance of the claim for injurious affection separately from the valuation of the land taken.

The inability of the claimant, by reason of the acquisition, D to continue a practice of averaging costs over the 6.0 acres of ownership does not in my judgment depreciate the value of the 4.4 acres of retained land, as is the contention of Mr Brackett. It is not in issue that the value of the land taken is to be assessed on the basis that roads, sewers and all services are available at its boundary, so that it reflects the benefit of actual and anticipated expenditure which contributes to its ripeness for development.

I have come to the conclusion that the evidence does not support the claim that the value of the 4.4 acres of retained land is depreciated by severance by the acquisition of the reference land. My finding is not affected by the agreement of E the parties during the hearing that the costs of roads and services which appear in the submitted valuations, and which are reproduced in this decision, required amendment. The agreed amendments are: for the whole of the 6.0 acres, £212,220 to £210,456; for the 1.6 acres of land taken £20,700 to £32,300; and for the 4.4 acres retained £191,520 to £178,156. The acquiring authority says that the effect of the amendments would be to reduce the amount of Mr Brackett's supporting valuation III from £85,782 to £72,780. The primary and supporting valuations I and II would be similarly affected.

I turn to the value of the land taken. I again refer to para F 5 of the "Basis of Valuation" which prefaces Mr Brackett's valuations and in particular the statement that "The primary valuation is on the basis of the land being taken as part of the whole estate and supports the total amount claimed, namely £83,000 both in respect of land taken and injurious affection due to severance." The primary valuation, at £3,600 per plot, is derived from Schedule 1, which itself is a valuation, on the residual method, of the whole estate of 6.0 acres with roads, sewers and services. Mr Brackett himself says that any division of the primary valuation is really arbitrary and Mr Keene is unable to suggest an apportionment.

Supporting valuation I is a residual valuation using the data in Schedule 1 as to the costs of construction and sale prices of 29 houses which, but for the compulsory purchase, it was G proposed to build on the reference land. Supporting valuation II, which is on the "before-and-after" method, also takes the plot value of £3,600 from Schedule 1 as the basis for both "before" and "after" values. Supporting valuations I and II both attract the similar criticism that they are said by Mr Brackett to include injurious affection in respect of depreciation in the value of the land retained whereas I have found that such a claim has not been substantiated.

I find the primary valuation and supporting valuations I and II do not assist in determining the value of the land taken. Much has been said in evidence about the residual method of valuation, but in the light of my finding further examination H of its use in the instant case is unnecessary.

Comparable transactions

There is, however, other evidence on which the parties rely for support, namely transactions of the sales of parcels of land which I inspected accompanied by Mr Brackett and Mr Keene. The 100-mile tour of inspection of eight sites extended from Cullompton in the north to Silverlands (the south-western outskirts of Exeter) and from Crediton in the west to Colyton in the east. Three of the comparables were cited by both surveyors, another by Mr Brackett and four others by Mr Keene. I will now examine this evidence. J

In his supporting valuation III, Mr Brackett values the reference land of 1.6 acres at £25,000 per acre = £40,000. He says it is the value of the land if separately held and excluding any sum which would be paid by the owner of the adjoining estate as a special purchaser. This exclusion is on account of Mr Brackett's understanding that section 5 rule (3) of the Land Compensation Act 1961 would be brought into play; but both Sir Frederick Corfield and Lord Colville considered that the rule is not applicable on the facts of the instant case. Mr Brackett says that his valuation of £25,000 per acre is derived exclusively from the comparable transactions. In his primary valuation Mr Keene values the reference land of 1.6 acres at K £20,000 per acre = £32,000; he relies for support of this value on the comparables.

The eight transactions comprise one sale by public auction, one sale by public tender, two sales by private treaty and five purchases by local authorities. They span a period from March 1976 to July 1978 (three months after the date of valuation). Sale prices per acre adjusted by Mr Keene by addition for abnormal on-site and off-site costs vary from £34,580 to £15,084. The smallest site is 1.04 acres and the largest 14.10 acres. Planning densities of dwellings per acre vary from 15 to 7.23 to the acre by comparison with 18.2 for the reference land. Mr Brackett made further adjustments to L the sale prices for the value difference between the dates of sale and the date of valuation. The percentage increases vary between 50% and 23% and a decrease of 12½% in respect of the post-valuation date sale. For these adjustments Mr Brackett relied on two sales by the claimant by compulsory purchase procedure at Sidmouth which demonstrated an increase in price of 50% between August 1976 and April 1978, a difference which accorded with his general experience.

Award

On a review of all the evidence, which was confirmed by my M inspections, I accept that the Sidmouth difference of levels of values should not be applied to all the transactions, but I do not accept Mr Keene's contention that prices remained stable throughout the relevant period. I also consider that Cullompton is not entirely beyond the influence of the higher land values that prevail in the area around Exeter. I have come to the conclusion that the weight of evidence is towards Mr Brackett's opinion of value and I determine the value of the reference land at April 10 1978 to be £24,000 per acre. The amount of compensation is therefore £38,400, being 1.6 acres at £24,000 per acre.

A The acquiring authority will pay the claimant £38,400 compensation for the compulsory purchase of the freehold interest of 1.6 acres of land at Cullompton. The authority will also pay to the claimant a surveyor's fee on Scale 5 of the Scales of Charges of the Royal Institution of Chartered Surveyors and legal costs properly incurred from the date of the notice to treat to the date of the notice of reference.

Having opened the sealed offer of compensation lodged by the acquiring authority, I find the amount thereof exceeds the amount of my award. The acquiring authority will pay the costs of the claimant of this reference up to the date of the sealed offer and the claimant will pay the costs of the
B acquiring authority as from that date, such costs, if not agreed, to be taxed by the Registrar of the Lands Tribunal on the High Court Scale.

WILKINSON AND OTHERS v MIDDLESBROUGH BOROUGH COUNCIL

(REF/63/1978)

(E C STRATHON FRICS)

April 23 1979

C Estates Gazette June 2 1979
(1979) 250 EG 867

Veterinary surgeons' multi-principal practice in Middlesbrough —Whether compensation assessable on equivalent reinstatement basis under rule (5)—Four essentials of "Sparks" case examined—Partner described premises as "hopelessly inadequate," but tribunal holds they would have continued to be devoted to purposes of claimants' practice—Vet's practices not normally bought or sold in open market— Customary to take in partners—Claimants fail to satisfy tribunal that there is no general demand or market—
D Normal method of selling single-principal practices to incoming partner points to existence of market—Although multi-principal practices never reach market, member sees no reason to doubt that there "would be" a market— Compensation to be paid under rule (2)

K Bagnall QC and S Lightwing (instructed by Newby, Robson & Cadle, of Middlesbrough) appeared for the claimants; A Anderson (instructed by the solicitor to Middlesbrough Borough Council) for the acquiring authority.

Giving his decision, MR STRATHON said: This is a decision on a preliminary point of law concerning the basis of compensation payable upon the compulsory purchase of premises
E at 126 Newport Road, Middlesbrough. The point in issue is whether the claimants are entitled to compensation assessed under rule (5) of section 5 of the Land Compensation Act 1961 or under rule (2).

The claimants own the freehold and carry on the practice of veterinary surgeons in 126 Newport Road. The subject property, about 90 years old, was formerly an end-terrace dwelling-house; there have been alterations and additions over the years. It now comprises on the ground floor: a waiting-room, inquiry office, small accounts office, consulting room, X-ray/consulting room, laboratory/operating room, preparation room, a range of four kennels and food preparation room, an
F exercising yard and additional kennels and stores; on the first floor: a caretaker's flat of three rooms and bathroom.

The subject property lies about 550 m to the south-west of the town centre in a locality which has been the subject of compulsory purchase orders made under Part III of the Housing Act 1957. A blight notice dated July 8 1977 in respect of the subject property, was served under the provisions of the Town and Country Planning Act 1971 and of the Land Compensation Act 1973. It was accepted by the Middlesbrough Borough Council, and consequently the council shall be deemed to be authorised to acquire compulsorily the claimants' interest in the subject property and to have served a notice to treat on September 10 1977, being two months from the date of the service of the blight notice. G

K Bagnall QC and S Lightwing for the claimants called evidence from J B Gale BVSc MRCVS, L Wilkinson BSc MRCVS, two of the three claimant partners practising as Wilkinson, Gale & Hall; from W B Elgee ARICS, a building surveyor and a partner in the firm of Darnton Elgee Wrightson Jackman & O'Connor; from J M Catterall FRICS, a senior partner in the firm of Sanderson Townend & Gilbert, and from Mrs I P Houston, the honorary secretary of the RSPCA clinic at Halifax. No evidence was called by A Anderson for the acquiring authority.

The claimants contend that compensation should be assessed under rule (5), which is set out hereunder: H

> (5) Where land is, and but for the compulsory acquisition would continue to be, devoted to a purpose of such a nature that there is no general demand or market for land for that purpose, the compensation may, if the Lands Tribunal is satisfied that reinstatement in some place is bona fide intended, be assessed on the basis of the reasonable cost of equivalent reinstatement.

A veterinary practice was established at 126 Newport Road, Middlesbrough, in 1900. Mr Wilkinson joined the practice in 1948, Mr Gale in 1963 and Mr Hall in 1968; no 126 is the headquarters, with branches at Hartlepool, Stockton and Billingham. The whole of the ground floor of no 126 is used J
for the practice, and the caretaker, who occupies the first-floor flat, deals with telephone calls at night and also helps in emergencies. In addition to the three partners there are five veterinary surgeons in the practice; at no 126 the total workforce comprises: a partner, four veterinary surgeons and 10 lay staff.

Work at no 126 and at the branches is closely related, with communication by telephone, and each surgeon has radiotelephone equipment in his car. The practice recognises the requirement of the Royal College of Veterinary Surgeons that a 24-hour service must be provided. Headquarters is responsible for the purchase and storage of drugs and equipment for the K
practice and has expertise on call for the branches; operations are mainly carried out at no 126 but some at Halifax.

Until March 1977 the partnership carried on two types of practice, namely, a small animal practice, which is concerned with "pets" such as cats, dogs, birds, snakes and fish, and a large animal practice, which is concerned with the treatment of horses and farm animals. Mr Gale said that the large animal practice demanded a large area for the storage of drugs and equipment and a great deal of clerical time spent in connection with the government department work that is involved. By an agreement dated March 2 1977 the large animal practice was sold to R G Russ, a veterinary surgeon L
in the practice, at a price of £14,500 with, in addition, certain stock-in-trade at valuation; there were reciprocal undertakings as to the carrying out of certain types of veterinary work during a period of five years and within a radius of 25 miles from the respective addresses of the parties.

Mr Gale considered that the large animal practice would have accounted for about one-fifth of the total turnover of the practice. In 1978 the practice carried out about 51,000 separate treatments, and another indication of workload is shown by his estimate that of the total amount of veterinary work in the area, 50 per cent is undertaken by the partnership practice. M

The claimants say that the goodwill of the practice is attached to the subject site, where there has been a veterinary practice since 1900; they would not voluntarily leave the site. Mr Gale said that in the Cleveland Area veterinary practices are not bought or sold in the open market. The normal course is to take in partners. A single-principal practice would advertise for a partner; the newcomer would work with the principal for a period and then would take over the practice on paying out the principal. In the claimants' type of practice, if a partner wished to retire, the remaining partners would, if they so desired, take in another partner. In his 20 years in

A practice Mr Gale had not seen a practice advertised for sale; no 126 could be sold only with the practice, on which he placed a value of £100,000, an amount which in his view would not be forthcoming in the market.

W B Elgee ARICS produced a series of costs of the construction of a building for the claimants' practice on a site to be made available by the Middlesbrough Borough Council in Victoria Road, not far removed from no 126. Of the various schemes I refer only to the one which is considered by Mr Elgee to be in the nature of equivalent reinstatement; at levels of prices at December 1978 and at June 1979 (estimated) the respective costs are said by Mr Elgee to be £117,600 and B £134,600. If the cost of equivalent reinstatement is not provided as compensation, it is Mr Gale's view that the practice would be unable to finance the setting up of a practice elsewhere; in such circumstances the Stockton branch could not survive and although Hartlepool and Billingham could exist, only inferior service could be given.

J M Catterall said that neither he nor his partners in a practice that covers about 1,000 square miles have had experience of, or have seen advertisements about, the sale and purchase of veterinary surgeons' premises in the area; he had seen only one advertisement about a sale elsewhere—it was of a practice in Luton, Bedfordshire. In his experience there C was no general demand or market for the purpose of a veterinary surgeon's practice.

Mr Catterall said that if rule (5) is held not to apply, his valuation of compensation under rules (2) and (6) would amount to £194,450 plus costs and fees. The sum is made up as follows:

Rule (2)			
The value of no 126 for purposes other than that of a veterinary clinic. (An agreed amount)			£12,000
Rule (6)			
Various items			£43,070
D Loss of goodwill			
Adjusted net profits	£46,460		
Capitalised at 3 YP		3	£139,380
			£194,450

Mr Catterall considered that three years' purchase would be the multiplier to apply where the practice is so well established. It is the figure which Mr Catterall would consider appropriate to adopt in his own practice when selling to incoming partners. It was supported, Mr Catterall said, by the three years' purchase of profits paid by Mr Hall on entering the practice in E 1968.

Mrs I P Houston's evidence concerned compensation assessed under rule (5) in respect of the acquisition in 1978 of the RSPCA clinic at Halifax, which had been built in 1961. The following extracts from a letter dated May 3 1978 from the chief administrative officer of the Metropolitan Borough of Calderdale to the society are relevant:

> The resolution of the appointed members was as follows, namely:
> 1. That the RSPCA be informed that after detailed consideration of the alternative suggestion relating to the provision of the new clinic the committee is of the view that the normal rule (5) procedure should be followed;
> F 2. This committee appreciates the co-operation of the society in this matter and looks forward to continued assistance in the future.
>
> Accordingly I have, therefore, instructed the district valuer today to proceed to negotiate with your agents on the basis of the normal rule (5) procedure. . . .

Mrs Houston said that in terms of service to animals the RSPCA would not differ from a private veterinary practice. The society is a charity, it does not charge fees, it asks foı donations. The Halifax branch carries out first aid on the premises but other work is sent to a private veterinary surgeon. Mrs Houston has not heard of a sale of a veterinary practice in Halifax.

Mr Bagnall and Mr Anderson referring to the decision of G the Lands Tribunal (J D Russell-Davis FRICS) in *Sparks and Others (Trustees of East Hunslet Liberal Club) v Leeds City Council* (1977) 244 EG 56 adopt the analysis of rule (5) by the member at p 57:

> There are four essentials in rule (5) to be satisfied by the claimants, on whom is the burden of proof:
> 1. That the subject land is devoted to the purpose, and but for the compulsory acquisition would continue to be so devoted;
> 2. that the purpose is one for which there is no general demand or market for the land;
> 3. the bona fide intention to reinstate on another site; and
> 4. these conditions being satisfied, that the tribunal's reasonable discretion should be exercised in their favour. H

Mr Anderson concedes that the claimants satisfy essential (3), that it is their bona fide intention to reinstate on another site. I turn to the other essentials.

Essential (1). Mr Bagnall referred to *Aston Charities Trust Ltd v Stepney Borough Council* [1952] 3 P & CR 82 where it was held that the words "devoted to a purpose" imported the concept of intention and were not confined to the *de facto* use of land at a particular date and that on the facts the premises at the date of the notice to treat had not ceased to be devoted to the original purpose. Mr Bagnall submitted that in the instant case the *de facto* use has remained constant since 1900 and that at September 1970 (when the local auth- J ority indicated that the premises would be acquired) the claimants had no intention of moving from no 126, the whole of which is used and devoted to the practice of veterinary surgeons. Mr Anderson accepts that the purpose is identified as offices and premises of the practice of veterinary surgeons and that the subject property is clearly devoted to that purpose. But, Mr Anderson said, there is nothing to show that at the date of the notice to treat, September 8 1977, no 126 would continue to be devoted to that purpose. Mr Anderson relies on a letter of June 10 1976 to the chief planning officer, Middlesbrough Borough Council from L Wilkinson, one of the claimants whose evidence was called at the hearing. The K letter is reproduced hereunder:

> I would like to point out the difficulties we have had over the last five years.
>
> In 1970 we were told that our present premises at 126 Newport Road would be required for redevelopment probably within 12 months. A request for relocation by the council was placed and since then we have tried to find a suitable site privately without success.
>
> As we were recently informed that the council had no plans for compulsory purchase in the foreseeable future, we approached the owners of the adjacent premises with a view to purchasing some or all of their site to allow redevelopment laterally as vertical development of a veterinary hospital is not feasible. We applied for L outline planning permission, and this has been turned down.
>
> Since coming to the premises, which were largely custom built in the 1920s, my partners and I have spent a lot of time, effort, and money in further developing them. Though the premises were adequate in the 1960s when four veterinary surgeons were running the practice, it has now grown to require nine veterinary surgeons, and this makes the present premises hopelessly inadequate.
>
> Had we been given a site and compulsorily purchased in 1971 or 1972, we could have built an effective hospital for approximately £40,000. Due entirely to change in planning, we will now need to find about £100,000 to build a similar building, and as this is urgently needed to enable us to provide an efficient service for the animals of the area, and better working conditions for our staff, M I would ask for a CPO on the grounds of planning blight, to enable us to develop the site in Victoria Street, which you have recently made available.

Mr Anderson submitted that the admission that the present premises are "hopelessly inadequate" and that a new building is urgently needed to enable the practice to provide an efficient service, was supported by the evidence of Mr Gale that the premises had become unsuitable, in view of the requirement for space and modern facilities. Mr Anderson referred me to *Zoar Independent Church Trustees v Rochester Corporation* [1975] QB 246 where Russell LJ said at p 259:

A Suppose this had been an ordinary compulsory acquisition with an actual notice to treat, and the acquiring authority had moved in with bulldozers in the following week, it could not have been contended that this requirement of rule (5) had not been fulfilled. Nor do I consider, in such a case, without the authority so taking possession, that it would be right to conclude otherwise, simply because in the ensuing two years the congregation had dwindled to three or four; for the process of acquisition might well itself have had an adverse effect upon the numbers. And I do not consider that any different approach is justified in a case of a deemed notice to treat. The requirement of continuity of devotion is not directed to perpetuity; I regard it as sufficiently complied with if it appears that at the time of the notice to treat it is then intended by the owners of the site to continue to devote it to the relevant purpose
B with no future time limit.

Mr Anderson suggested that Mr Wilkinson's description of the premises as being "hopelessly inadequate" was supported by an application, some 10 days after the letter of June 10, for planning permission to redevelop no 126 with the adjoining house no 128 for use by the practice. The application was refused on July 30 1976 for reasons that "the redevelopment of this site in this property block could prejudice the council's declared policy of redeveloping the site for residential purposes."

I do not accept on the evidence that a time limit to the
C continuance of the practice could be envisaged. I had the advantage of hearing evidence from Mr Wilkinson, whose purpose was to endeavour to stir the acquiring authority into action, having been told in 1970 that the subject property would be acquired and that it might be acquired within 12 months. It was of great importance to the practice that it should continue at no 126 to which its goodwill is attached. The "large animal" part of the practice was sold in 1977 and, although it resulted in a very modest saving in terms of space and staff, it provided some easement of the conditions which prevailed at the time of Mr Wilkinson's letter.

D I am satisfied that but for the acquisition by the local authority the subject land would continue to be devoted to the purpose of the claimants' practice and essential (1) is therefore satisfied by the claimants.

Essential (2). Mr Bagnall urges me to accept on the evidence that there is no general demand or market for no 126 for the purpose of a veterinary surgeon's practice. Mr Catterall has not heard of a sale in the extensive area covered by his practice; Mr Gale says multi-practices evolve and are not sold and, in single-principal practices, the newcomer takes over by paying out the old partner. I was referred to the decision of the Lands Tribunal (J Stuart Daniel QC) in *Trustees of the*
E *Manchester Homeopathic Clinic v Manchester Corporation* (1970) 215 EG 1033; Mr Bagnall suggested that it was close in character to the instant case. The claimants had compared the homeopathic clinic with other medical establishments and the acquiring authority had relied on 24 transactions to support the proposition that by reason of design and layout, the clinic was suitable for use as a clinic in a general sense. Mr Bagnall said that the tribunal had determined that there was no general demand or market for land for the purpose of a clinic and contrasted the instant case where no transactions had been cited to support the existence of a market or demand.

F Mr Anderson submitted that the tribunal should be slow to widen the purposes which have been accepted as coming within rule (5), such as churches and social clubs. In *Festiniog Railway Co v Central Electricity Board* (1962) 181 EG 787, Harman LJ at p 791 said:

> Reinstatement is usually resorted to in cases where the displaced undertaking was some non-productive enterprise such as a church or a hospital which was not intended to make a profit but to perform some public service to the community which could not equally well be performed in another situation. There are, however, cases in the books which show that reinstatement may be applied to a commercial concern.

Mr Anderson said it may also be considered inherently G unlikely that veterinary practices, which are found countrywide, are not subject to a general demand or market and the referred to a decision of the Lands Tribunal (Erskine Simes QC, R C Walmsley FRICS) in *Vaughan (Viscount) v Cardiganshire Water Board* (1963) 185 EG 949 where the members said:

> We agree with Mr Bowen that the "purpose" must be a general and not a localised purpose, and that in the present case it is not enough to say that there would be no general demand for premises devoted to the management of *this* estate. What is necessary is to prove that there is no general demand or market for premises devoted to estate management, and there was no evidence to that effect. H

I come first to Mr Anderson's submission that I should be slow to widen the purposes which have been accepted as coming within rule (5). I do not think I should allow my judgment to be affected by this proposition; my task is to determine the issues on the facts found and on the evidence adduced in the instant case. Similarly, I find I derive no asistance from the evidence that compensation based on rule (5) was paid in respect of the acquisition of the RSPCA clinic at Halifax which was built for the purpose in 1961. The purpose to which the clinic was devoted was not the same as a veterinary surgeons' practice.

In the instant case the evidence about general demand or J market comes from Mr Gale and Mr Catterall. Mr Gale says that veterinary practices in the Middlesbrough area are not bought or sold in the open market; it is customary to take partners into the firm. First the single-principal practice, which is sold by the owner to the incoming principal after a period of working in dual harness. I think that as such a transaction is the normal method of disposal, it points to the existence of a market. I had no evidence as to the extent of such a market; on the one hand the witnesses said they had no knowledge of sales of practices and, on the other hand, said that practices are normally disposed of in the manner described.

Secondly, incoming partners to the multi-principal practice K purchase shares in the practice, which continues to operate as a partnership. The instant case provides an example, by the purchase of a partnership by Mr Hall in 1968 on terms which Mr Catterall said he would expect from an incoming partner to his own multi-partnership practice of surveyors. There is also the example of a disposal of the "large animal" part of the claimants' practice to Mr Russ to carry on that part of the business elsewhere. Thus in a multi-partner practice, so long as partners are taken into the business to add to the number of working partners or to take the places of retiring partners, the practice as a whole is not offered for sale.

In the instant case Mr Gale puts the capital value of the L whole practice at about £100,000 and Mr Catterall's valuation of the goodwill is of the order of £139,000. Mr Gale accepts that a number of veterinary surgeons might band together into one firm and that the subject premises might prove an attraction to such a partnership, but he did not know of any source from which finance could be secured for such a purchase. In the instant case, the practice is flourishing, it is commercially viable with every prospect of improving in performance, and Mr Catterall considers that having regard to the practice being so well established with income secure, monopoly value attaches to the subject property.

It seems to me that there is a demand and market for M premises for the purposes of the single-principal practice of a veterinary surgeon. In respect, however, of multi-principal veterinary practices, these never do reach the market, for the simple reason that the aim of such a partnership is to continue to carry on the practice as veterinary surgeons rather than to dispose of the partnership assets. I see no reason to doubt, nevertheless, that for such a practice there *would be* a market, just as there is for the single-principal practice.

On consideration of the whole of the evidence I find the claimants have not satisfied me that there is no general demand or market for the subject property for the purpose of carrying on the practice of veterinary surgeons and the claimants therefore fail to satisfy essential (2) of the rule.

As has been said, Mr Anderson concedes that essential (3) is satisfied, namely that reinstatement in some other place is bona fide intended.

As the claimants have failed to satisfy me in respect of essential (2), the question to answer in essential (4) as to the exercise of the tribunal's reasonable discretion, does not arise. But as I was addressed on this essential (4) and heard evidence, I have come to the conclusion that I should record that Mr Bagnall said the evidence as to cost and value was put before me to indicate the witnesses' opinions of the order of magnitude of the claimants' case. He added that the reason why the figures were not challenged by Mr Anderson was because they were not the subject of dispute before me. The evidence, Mr Bagnall said, had been adduced having regard to the case of *Festiniog Railway Co (supra),* where Ormerod LJ said at p 789:

> Mr Blundell [for the claimants] has argued that the reasons for his decision given by the learned President show that he took into account circumstances which were not relevant to this particular inquiry, and therefore that the discretion could not have been exercised reasonably. He argued that the tribunal accepted the submission of Mr MacKenna, as he then was, that this was a commercial undertaking and in the circumstances the only principle to be applied in considering the question of reasonableness was whether the cost of reinstatement on other land would exceed the value of the business. I may say here that although there may be many circumstances when this consideration is a factor, and in some cases even a conclusive factor, I am far from saying or agreeing that it is a principle which is in itself conclusive. However, it follows, I think, beyond any doubt that if the undertaking in question is a business undertaking then the question of the relation between the cost of reinstatement and the value of the undertaking is relevant and may be paramount in considering the question of reasonableness.

In this context, Mr Anderson said that it was no part of his case to show there was a disproportion in balance between reinstatement cost and value of the business.

My decision is that compensation shall be paid on the basis of rule (2) of section 5 of the Land Compensation Act 1961.

The claimants will pay the costs of the acquiring authority of this reference, such costs if not agreed to be taxed by the Registrar of the Lands Tribunal on the High Court Scale.

DAWSON v NORWICH CITY COUNCIL
(REF/125/1977)
(V G WELLINGS QC)
July 31 1978

Estates Gazette June 30 1979
(1978) 250 EG 1297

Tenant's claim in respect of two parts of agricultural holding at Bowthorpe, Norfolk—Council possessing compulsory purchase powers bought land from landlords by voluntary agreement, thereafter giving tenant notice to quit on grounds that land was required for non-agricultural use—Planning permissions for development obtained by landlords prior to sale—Section 59 of 1973 Act held to apply equally to notice to quit served by council as by private landlord—Section 48 of 1973 Act therefore applies but does not overrule ordinary basic principles for assessing compensation under section 20 of 1965 Act and Horn v Sunderland—Meaning of "required" in section 24(2)(b) of Agricultural Holdings Act 1948—On question of landlord's right to serve effective notice to quit, position in real world and no-scheme world distinguished—Tribunal holds that valuer not bound to assume that claimant tenant has indefinite security of tenure—Valuer entitled to give effect to potential in no-scheme world for private development if it is such that notice to quit without acquiring authority's intervention could have been expected

George Dobry QC and Roger Cooke (instructed by Ellis & Fairbairn) appeared for the claimant; Alan Fletcher (instructed by the solicitor to Norwich City Council) for the acquiring **authority.**

Giving his decision, Mr Wellings said: This is an application for the determination of preliminary points of law arising in respect of a notice of reference claiming compensation for compulsory purchase of two parts of an agricultural holding served by the claimant, Captain R T Dawson, the tenant of the holding.

Questions for determination

The points of law in question were stated in the claimant's application to be as follows:

(a) whether a claimant who is a tenant who has made an election and who otherwise satisfies the requirements of section 59 of the Land Compensation Act 1973 ought to have compensation assessed according to the provisions of section 59(2) (thereof) when

(i) the compensating authority being an authority possessing compulsory purchase powers have acquired the landlord's interest by agreement, and

(ii) the compensating authority have thereafter given the claimant notice to quit;

(b) whether compensation ought to be assessed and paid on the footing that there ought to be disregarded any right of the landlord to serve a notice to quit under section 24(2)(b) and section 25(1)(e) of the Agricultural Holdings Act 1948 as provided in section 48(3) of the said Act of 1973 when:

(i) the compensating authority have taken possession of the holding within the meaning of subsection (1) of the said section;

(ii) notice to quit has been given by the compensating authority only;

(iii) the landlord has shortly prior to purchase of his interest by the compensating authority received a grant of outline planning permission for development of the holding;

(iii) (a) the landlord has decided not to develop in accordance with the said permission and has not at the date of purchase of his interest found an identified purchaser other than the compensating authority;

(iv) such outline planning permission was for development substantially similar to the proposals of the compensating authority;

(c) further or alternatively to (b), whether in the events set out in (b)(i) to (iii)(a) inclusive the landlord would have had any right to serve an effective notice to quit under the above-mentioned sections of the Agricultural Holdings Act 1948.

The above questions may be stated more shortly as follows:

(1) was the claimant's election under section 59 of the Land Compensation Act 1973 effective having regard to the circumstance that the notices to quit served on him in respect of his agricultural holding were served, not by a private landlord, but by an authority possessing compulsory purchase powers?

(2) in assessing compensation in circumstances where section 48 of the Land Compensation Act applies, is it permissible to take into account such power as the tenant's landlord for the time being might have had as at the date relevant for the assessment of compensation, if there had been no compulsory acquisition of the tenant's interest, to determine his tenancy, on the ground of a requirement of the land by that landlord or some other person for a non-agricultural use in accordance with section 24(2) of the Agricultural Holdings Act 1948?

Relevant statutory provisions

The relevant subsections of section 59 of the Land Compensation Act 1973 are:

(1) This section has effect where the person in occupation of an agricultural holding, being a person having no greater interest therein than as tenant for a year or from year to year, is served with a notice to quit the holding; and

(a) the notice is served after an acquiring authority have served notice to treat on the landlord of the holding or, being an authority possessing compulsory purchase powers, have agreed to acquire his interest in the holding; and

(b) either—

(i) subsection(1) of section 24 of the Agricultural Holdings Act 1948 does not apply to the notice by virtue of subsection (2)(b) of that section (land required for non-agricultural use for which planning permission has been granted etc); or

(ii) . . .

(2) If the person served with the notice to quit elects that this subsection shall apply to the notice and gives up possession of the holding to the acquiring authority on or before the date on which his tenancy terminates in accordance with the notice,

(a) section 20 of the Compulsory Purchase Act 1965 (compensation for tenants from year to year etc) and section 12 of the Agriculture (Miscellaneous Provisions) Act 1968 shall have effect as if the notice to quit had not been served and the acquiring authority had taken possession of the holding in pursuance of a notice of entry under section 11(1) of the said Act of 1965 on the day before that on which the tenancy terminates in accordance with the notice to quit; and

(b) the provisions of the Agricultural Holdings Act 1948 relating to compensation to a tenant on the termination of his tenancy and sections 9 and 15(2) of the Agriculture (Miscellaneous Provisions) Act 1968 (additional payment and compensation in cases of notice to quit) shall not have effect in relation to the termination of the tenancy by reason of the notice to quit.

(3) . . .

(4) Any election under subsection (2) above shall be made by notice in writing served on the acquiring authority not later than the date on which possession of the holding is given up.

(5) This section shall have effect in relation to a notice to quit part of an agricultural holding as it has effect in relation to a notice to quit an entire holding and references to a holding and the termination of the tenancy shall be construed accordingly.

The relevant subsections of section 48 of the Act of 1973 are:

(1) This section has effect where in pursuance of any enactment providing for the acquisition or taking of possession of land compulsorily an acquiring authority,

(a) acquire the interest of the landlord in an agricultural holding or any part of it; or

(b) acquire the interest of the tenant in, or take possession of, an agricultural holding or any part of it.

(2) In assessing the compensation payable by the acquiring authority to the landlord in connection with any such acquisition of an interest as is mentioned in subsection (1)(a) above,

(a) there shall be disregarded any right of the landlord to serve a notice to quit, and any notice to quit already served by the landlord, which would not be or would not have been effective if—

(i) in section 24(2)(b) of the Agricultural Holdings Act 1948 (land required for non-agricultural use for which planning permission has been granted etc) the reference to the land being required did not include a reference to its being required by an acquiring authority; and (ii) in section 25(1)(e) of that Act (proposed termination of tenancy for purpose of land's being used for non-agricultural use not falling within section 24(2)(b)) the reference to the land's being used did not include a reference to its being used by an acquiring authority; and

(b) if the tenant has quitted the holding or any part of it by reason of a notice to quit which is to be so disregarded, it shall be assumed that he has not done so.

(3) In assessing the compensation payable by the acquiring authority to the tenant in connection with any such acquisition of an interest or taking of possession of land as is mentioned in subsection (1)(b) above (hereafter referred to as "the tenant's compensation"), there shall be disregarded any right of the landlord to serve a notice to quit, and any notice to quit already served by the landlord, which would not be or would not have been effective if the said sections 24(2)(b) and 25(1)(e) were construed in accordance with subsection (2)(a)(i) and (ii) above.

(4) Section 42 of the Agriculture (Miscellaneous Provisions) Act 1968 (tenant's compensation to be assessed without regard to his prospects of remaining in possession after contractual date) and section 15(1) of that Act (effect on tenant's compensation of provision enabling landlord to resume possession on non-agricultural use) shall cease to have effect.

Having regard to the facts of the present case (to which I shall refer) I need not quote the provisions of section 25(1)(e) of the Agricultural Holdings Act 1948, but the relevant subsections of section 24 of that Act are as follows:

(1) Where notice to quit an agricultural holding or part of an agricultural holding is given to the tenant thereof, and not later than one month from the giving of the notice to quit the tenant serves on the landlord a counternotice in writing requiring that this subsection shall apply to the notice to quit, then, subject to the provisions of the next following subsection, the notice to quit shall not have effect, unless the [Agricultural Land Tribunal] consent to the operation thereof.

(2) The foregoing subsection shall not apply where

(a) . . .

(b) the notice to quit is given on the ground that the land is required for a use, other than for agriculture, for which permission has been granted on an application made under the enactments relating to town and country planning, or for which (otherwise than by virtue of any provision of those enactments) such permission is not required, and that fact is stated in the notice.

Hereafter in this decision any reference to the "trustees" is a reference to the persons constituting at the date and time referred to the trustees of the Overland Estate situate *inter alia* at Bowthorpe in the County of Norfolk.

Facts of the case

The following facts appear from agreed documents, an agreed statement of facts and from agreements made between counsel at the hearing:

(1) Under a tenancy agreement dated October 10 1964, as subsequently amended, the claimant was tenant from year to year to the trustees of an agricultural holding consisting of 676.344 acres or thereabouts and known as Bowthorpe Hall Farm.

(2) Between 1954 and 1966 the trustees had made a number of unsuccessful planning applications to develop land comprised in the tenancy. On June 30 1970 they made a further such application and, no decision having been given within the statutory period, an appeal to the Secretary of State was made and he on March 30 1972 granted to the trustees planning permission for the development of 562 acres of land forming part of the trustees' estate (not all of it within the agricultural holding) for residential and ancillary uses, an area of warehousing and the layout of an industrial estate.

(3) On April 12 1972 the trustees and their beneficiaries decided to prepare a master plan, not to take any steps themselves to develop the land benefited by the permission but rather to dispose of it to the best advantage and in parcels.

(4) Between April 1972 and January 1973 there were discussions between the trustees and the former council. At a meeting between representatives of both of them on July 20 1972 the former council expressed an intention to acquire the whole of the land benefited by the permission. The trustees expressed their opposition both to this and to a proposal that there should be a joint venture between the two of them.

(5) In November 1972 the trustees decided to offer the land benefited by the permission for sale by tender and so informed the former council on December 13 1972. On January 12 1973 the Bowthorpe Development Committee of the former

A council recommended the making of a compulsory purchase order.

(6) On February 28 1973 the former council purchased, by voluntary agreement, from the trustees 504.23 acres of the land benefited by the permission, 486.82 acres being land comprised in the agricultural holding (that land was shown edged in green on a plan produced at the hearing and I will call it " the green land ").

(7) On November 7 1973 the former council granted to the trustees a planning permission for development of a further 19.1 acres, all of them part of the agricultural holding, for housing purposes.

B (8) On April 17 1974 the respondents, as successors to the former council, purchased, by voluntary agreement, the 19.1 acres (that land was shown coloured red on the plan and I will refer to it as " the red land ").

(9) At all material times both the former council and the respondents were authorities possessing compulsory purchase powers within the meaning of section 59 of the Land Compensation Act 1973, although the purchases mentioned above were not made under such powers.

(10) By a notice in writing dated August 23 1973 the former council gave to the claimant notice to quit the green land on C October 11 1974. It was stated in the notice that it was given for the reasons specified in section 24(2)(b) of the Agricultural Holdings Act 1948, that the green land was required for a use other than agriculture, namely, development for housing purposes, for which permission had been granted on an application made under the enactments relating to town and country planning.

(11) On October 4 1974 the claimant gave notice in writing to the respondents that he elected that section 59(2) of the Act of 1973 and any other appropriate statutory provisions should apply to the above-mentioned notice to quit in respect of all the green land.

D (12) By a notice in writing dated October 4 1974 the respondents gave to the claimant notice to quit the red land on October 11 1975. The said notice contained a statement that it was given for the reasons specified in section 24(2)(b) of the Act of 1948, that the red land was required for a use other than agriculture, namely, development for housing purposes for which permission had been granted on an application made under the enactments relating to town and country planning.

(13) On October 7 1975 the claimant gave to the respondents notice in writing that he elected that section 59(2) of E the Act of 1973 and any other appropriate statutory provisions should apply to the notice to quit dated October 4 1974 in respect of the red land.

It was common ground between the parties at the hearing that the notices to quit, being notices to quit parts of the holding, were validated by section 31 of the Act of 1948. It was equally common ground between them that section 48 of the Act of 1973 applies to the circumstances of the present case if, and only if, the claimant's elections under section 59 of that Act were effective.

The question as to section 59

F Alan Fletcher, for the respondents, submitted that section 59 had no application in any case such as the present where notice to quit was served by an acquiring authority or authority possessing compulsory purchase powers; it applied, he said, only in the case of a notice to quit served by the landlord from whom the reversionary interest in the tenancy is being acquired. His reasons for this submission were:

(1) That section 30 of the Act of 1948 (which invalidates a notice to quit served by a landlord if during its currency the landlord enters into a contract for the sale of his interest in the land or any part thereof and does not give notice to the tenant of the contract or the tenant does not agree that the notice shall continue in effect or does not give up possession G in consequence of the notice) contains no reference to a notice to treat;

(2) Although section 59 does not apply unless possession is given up to an authority, if the reference to service to a notice to quit is to or includes a notice to quit by an authority, there is no limit of time in the section in which notice to quit is to be served; thus in the present case the authority might have, for example, waited until 1980 before serving notice to quit.

To avoid that situation, Mr Fletcher said, the right solution was to read the provision as not applying to notice to quit by an authority. H

The mischief at which the section aims is indicated by subsection (2)(a) (which, where a valid election is made, deems the acquiring authority to have taken possession under a notice of entry under section 11 of the Act of 1965 on the day before that on which the tenancy terminates in accordance with the notice to quit) and is the practice of local authorities, who, having purchased by compulsion or threat (expressed or implied) of compulsion the reversion on a tenancy of an agricultural holding, then seek possession from the tenant not as acquiring authority, by notice of entry, but *as landlord*, by notice to quit, thereby avoiding payment of compensation otherwise than under the Agricultural Holdings Act 1948. J

The references in section 59 to service of a notice to quit are in the passive tense and there is no mention of the server of the notice. Those references are equally capable of applying to a notice to quit served by a local authority as they are to one served by a private landlord.

In my opinion, the section applies to the notices to quit served upon the claimant by the former council and the respondents. Accordingly the claimant's elections were properly made. In the result both section 48 of the Act of 1973 and section 20 of the Compulsory Purchase Act 1965 apply in the circumstances of the present case.

Therefore, the answer to question (a) in the claimant's K application, set out at the beginning of this decision, is: Yes.

The question as to section 48

George Dobry QC, for the claimant, said that the purpose of section 48 was to reverse (1) *Minister of Transport v Pettitt* (1969) 20 P&CR 344; (2) *Rugby Joint Water Board v Shaw-Fox* [1973] AC 202 (except as between landlord and tenant); (3) Section 42 of the Agriculture (Miscellaneous Provisions) Act 1968 (tenant's compensation for compulsory purchase to be assessed without regard to his statutory L security of tenure).

From this, Mr Dobry drew the inference that section 48 is concerned with the real world situation; that at the stage of ascertaining the nature and extent of the tenant's interest, the valuer cannot take into account any right to serve a notice to quit under section 24(2)(b) of the Act of 1948, unless it would be effective in the real world on the facts as they exist at the date of valuation (agreed to be the date of deemed possession) and the possibility of service of any other notice is irrelevant; the fact that at the relevant date the reversion is vested in the acquiring authority and the fact that no other person could at that date serve a notice to quit are not required M by the section to be disregarded. Accordingly, said Mr Dobry, the tenant is to be treated as having the benefit of indefinite and indefeasible security of tenure, in effect, equivalent to that of a tenancy for life.

Moreover, Mr Dobry said that the valuer could not take into account the possibility of service of a notice to quit on the ground of a requirement for a non-agricultural use unless there is an actual requirement for such use by an identified person other than an acquiring authority on whom the landlord for the time being could rely; see *Jones v Gates* [1954] 1 WLR 222. The requirement referred to in section 24(2)(b) of

A the Act of 1948 is equivalent to an intention of the kind referred to in section 30(1)(f) of the Landlord and Tenant Act 1954; see *Woodfall* (27th edition) vol 2, para 2520; alternatively the word meant bona fide wanted and genuinely intended to be occupied for the purpose of carrying out the permitted development; see *Kennealy v Dunne* [1977] QB 837.

Mr Fletcher said that if one looked at the real world only, there was no date for which any other landlord could have given an effective notice to quit because the acquiring authority had already acquired the reversion; they had given a notice to quit and were going to get possession on the expiration of it; it was, therefore, impossible to contend that in the real world
B anyone other than the acquiring authority could serve a notice to quit relying on his own requirement or that of anyone other than the acquiring authority.

Mr Fletcher submitted, however, that the no-scheme world must be looked at. In that event, the question was whether, on the agreed facts and documents, the only possible inference was that at the relevant date the landlord for the time being (who might or might not have been the trustees) could have given notices to quit expiring on October 11 1975 (in the case of the green land) and October 11 1976 (in the case of the red land) or indeed for later dates. Section 48 was part of the compensation code, adding new assumptions to it. The appli-
C cation of that code always involved a comparison between the claimant's financial position in the real world at the time and after the acquisition with what it would have been in the no-scheme world, that is to say, a world in which there was no proposal by an acquiring authority to acquire the land for the purposes of the scheme. Notwithstanding section 48, it was necessary to ascertain the claimant's true loss in accordance with the settled principles stated in *Horn v Sunderland Corporation* [1941] 2 KB 26. Section 20 of the Compulsory Purchase Act 1965 was to be applied in the ordinary way.

In reply, Mr Dobry said that if Mr Fletcher were right and if the valuer was not precluded from investigating the question
D whether the tenant's security of tenure was defeasible under section 24(2)(b) of the Act of 1948, he could have regard nevertheless only to the prospect of a requirement for development capable on the facts of being carried out by a private landlord; not to a requirement for a development by or involving an acquiring authority. Section 20 of the Act of 1965 was to be applied in the light of the assumptions required by section 48 of the 1973 Act.

During the argument, in addition to the authorities which I have mentioned, I was referred to *Myers v Milton Keynes Development Corporation* [1974] 1 WLR 696; *Worlock v Sodbury Rural District Council* (1961) 12 P&CR 315 (John
E Watson FRICS); *Wakerley v St Edmondsbury Borough Council* (1977) 241 EG 921; 242 EG 49 (R C Walmsley FRICS).

I do not think that section 48 of the Act of 1973 should be treated as requiring to be disregarded anything other than the specific matters mentioned in this section. There is nothing in the section which overrules the ordinary basic principles relating to the assessment of compensation under section 20 of the Act of 1965 and in particular nothing which overrules the principle of *Horn v Sunderland Corporation (supra)*, namely, that the claimant is entitled to compensation representing neither less than nor more than his loss. The purpose of the inquiry as to compensation is to ascertain his true loss.
F It is just as much wrong for the claimant to receive more than his true loss as it is for him to receive less than his true loss (to vary the words of Lord Hodson in the *Rugby Water Board* case *(supra)*, at 219). G

In the present case the inquiry as to what is the claimant's true loss arises under section 20 of the Act of 1965, which section is applied by section 59 of the Act of 1973. Provided that the particular matters required by section 48 to be disregarded are in fact disregarded, there is no provision, express or implied, in that section which requires an assumption to be made that invariably in all circumstances the tenant whose interest is acquired has the benefit of indefinite security of tenure under the Act of 1948. It is, notwithstanding that section, proper and indeed necessary to compare the value of the claimant's interest in the circumstances actually existing at the relevant date with what would have been its value in the light of circumstances which could reasonably have been expected to exist at that date if there had been no acquisition. It is proper for the valuer to consider whether at the relevant date, if there had been no acquisition, there would have been any risk (and, if so, the degree of risk) that the tenant's statutory security of tenure would have become defeasible by notice to quit served under the Act of 1948 and, if so, on what grounds. H

An important fact in the present case is that the planning permissions were obtained by the trustees. They are planning permissions which would enure for the benefit of the landlord for the time being. Whether or not the word "required" in section 24(2)(b) of the 1948 Act means needed or wanted or intended to be occupied, it is not possible to infer from the agreed facts that in the no-scheme world at the relevant date the trustees or the landlord for the time being would or might have served notice to quit the green land or the red land or any part or parts thereof, relying on a requirement for a relevant non-agricultural use without the intervention of an acquiring authority. It seems to me that questions of that kind are better left for the valuation stage, when the potential of the land for private development in accordance with the planning permissions without the assistance of an authority possessing compulsory powers can be investigated. J K

It is sufficient for the purposes of the present application that I hold, as I do, that as a matter of law, the valuer is not bound to assume that the claimant has indefinite security of tenure under the Agricultural Holdings Act. If at the valuation stage it is found that at the relevant date in the no-scheme world the potential of the green land or the red land or some part or parts thereof for private development is such that a notice or notices to quit relying upon a relevant requirement without the intervention of an acquiring authority could have been expected, the valuer will be entitled to give effect thereto. L

My answers to the questions relating to section 48 of the 1973 Act raised in the claimant's application and mentioned at the beginning of his decision are as follows:

Question (b): No; not as a matter of law;

Question (c): in the real world the answer would be: No. In the no-scheme world, the answer would depend on the results of the kind of investigation at the valuation stage as to the potential of the land for development without the aid of an acquiring authority, which I have mentioned.

There will be no order as to costs. M

LANDS TRIBUNAL

RATING

A CRESSWELL (VO) v BOC LTD
(LVC/923/1976)
(Sir Douglas FRANK QC, President)
February 26 1979

Estates Gazette June 23 1979
(1979) 250 EG 1195

Fish farm at Castlerigg, Cumbria—Whether exempt from rating—President follows his decision in Jones v Bateman and finds for ratepayers—View not changed by decision to contrary in subsequent Scottish case—Meaning of "livestock"—Open to doubt whether a matter of law—Ordinary meaning in agricultural context is "something which is B alive and is stocked for the purposes of providing food"

Alan Fletcher (instructed by the Solicitor of Inland Revenue) appeared for the appellant valuation officer, G O Cresswell; Anthony Anderson (instructed by the solicitor to BOC Ltd) for the respondent ratepayers.

Giving his decision in this appeal from the Cumbria Local Valuation Court, THE PRESIDENT said: The question in this case is whether buildings associated with and used for the production and rearing of trout are exempt from rating as agricultural buildings by reason of the provisions of the General Rate Act 1967 and the Rating Act 1971. The agreed facts are set out in the statement which is appended to and forms part of C the decision.

It is common ground that the question to be decided is whether trout are "livestock" within the meaning of section 1 of the Rating Act 1971.

Alan Fletcher for the appellant valuation officer pointed out that the relief is primarily given to land and no buildings are exempt unless they are occupied with agricultural land. Thus, relief would not be given to carp kept in a pond and so there was no logical reason for giving it to fish farms. He referred to the definition of livestock in section 109 of the Agriculture Act 1947, section 109 of the Town and Country Planning Act D 1947, and section 94 of the Agriculture Act 1948 for the purpose of construing the word "livestock" in the 1971 Act. He submitted that the meaning of the word in section 2(1) of the 1971 Act must be ascertained by reading it with section 1(3) and that section 1(3) is both limiting and extending in that it limits the class to be exempt to mammals and birds and the class of mammals and birds to those kept for the production of food and wool. He conceded that the word "livestock" can include fish but does not in the context of the Act. He referred to *Belmont Farm Ltd v Minister of Housing and Local Government* (1962) 13 P & CR 417, *Minister of Agriculture, Fisheries and Food v Appleton* [1969] 3 All ER 1051 E and *Wallace v Assessor for Perth and Kinross* [1975] SLT 118.

It is common ground that the issue in this case is the same as that in *R E Jones (VO) v Bateman* (1974) 232 EG 1392, which I decided in favour of the ratepayers. The following is the material part of that decision:

Subsection (3) of section 1 of the Rating Act 1971 provides: "In this Part of this Act 'livestock' includes any mammal or bird kept for the production of food or wool or for the purpose of its use in the farming of land." Clearly that provision does not cover the breeding and rearing of trout; so that the question is whether that activity falls within section 2(1)(a), namely "any building used for the keeping or breeding of livestock." It was F argued that the operation carried on was all to do with food production and that the keeping of fishponds is an ancient agricultural operation. It was further said that the definition in section 109(3) of the Agricultural Act 1947 serves as a useful guide in deciding what is an agricultural operation. Certainly if section 2(1) of the Act of 1971 stood, as it were, unencumbered by subsection (3) of section 1, I would have little difficulty in concluding that the respondent's buildings are agricultural buildings, but on the other hand, if I must read the two subsections together, then I think the position is very difficult.

I am, of course, bound by the two Divisional Court cases if they apply. They do not apply precisely, for the definitions are different, but they raise somewhat similar considerations. In both G those cases the activities in question were, in essence, found to be outside the definitions because they were not carried on for the right purposes; the test was really a functional one. In the present case there is no difficulty at that stage, for the trout are quite clearly kept for the production of food.

But then comes the question whether the words "any mammal or bird" must be read as limiting the classes of animals intended, or rather whether they can be regarded as illustrations only or as extending the class of creatures intended. The word "include" is generally used in interpretation clauses in order to enlarge the meaning of words or phrases occurring elsewhere and when it is so used those words or phrases must be construed as comprehending not only such things as they signify, according to their natural import, but also those other things which the interpretation H clause declares that they shall include. There are also cases where the word "include" is employed to give an exhaustive or limiting definition of the phrase or word. Such an intention should not, I think, be imputed to the draftsman unless it is imperative, for it involves an abuse of language, and in this case using the word in a sense inconsistent with subsection (1) of section 1. But in the present case it is very difficult to suppose that the "inclusion" of mammals and birds was meant to extend or enlarge the natural import of the word "livestock." For mammals and birds are the ordinary typical examples of "livestock."

Thus the argument for an exhaustive definition is very strong; yet I find it repugnant, because if everything else fits it seems highly absurd that fish should be singled out from all other J creatures. I think the true answer may be that the definition was indeed intended to be exhaustive but was meant essentially to limit the purposes for which animals were kept rather than the class of animals or creatures who live in the buildings. I am bound to say that I feel some doubt about this, but on balance I do not find it imperative to read the word "livestock" as being limited to mammals and birds. I think that what the draftsman was doing in section 1(3) was, in effect, to say to himself "if the ordinary meaning of 'livestock' should at any particular time not include all mammals and birds kept for . . . etc, then my section 1(3) will ensure that it does include these categories." On that basis section 1(3) has a meaning.

K Further, I think that the method of construction favouring the appellant is repugnant in that the definition of "agricultural building" in section 2 being unambiguous in its application to this case, it would be wrong to look elsewhere to arouse doubt. If the words used in section 2(1)(a) of the Act of 1971 in their natural sense include a building used for the keeping and breeding of trout, it would be wrong to write in something which is not there. I think that the *Belmont* and *Appleton* cases are distinguishable in that there was an agricultural concept from the beginning. The *rationes* of those decisions appear to be that because the word "agriculture" governed the whole definition, only creatures used in connection with the farming of the land could be said to be livestock. It seems to be logical and consistent with all the four cases cited that in an agricultural context "livestock" means all live creatures reared for the production of food.

Since I gave that decision the Lands Valuation Appeal Court have reached the opposite view in *Wallace v Assessor for Perth and Kinross*. Their *ratio decidendi* seems to have been that the breeding or keeping of trout would not amount to an agricultural use of land in the widest sense. Lord Robertson said that trout may become animals in the future and, "if so, it may be necessary for some ratepayer to prove the fact to a committee by leading evidence, but that situation is not before us here." It seems, therefore, that the decision was based on one of fact, and for that reason Mr Anderson called a farming journalist to prove that both in government and the National Farmers' Union fish farming is regarded as an agricultural activity.

It is perhaps important to bear in mind that in the *Wallace* case the ratepayer appeared in person and the main arguments which were advanced in the instant case and in *Jones v Bateman* do not seem to have been fully explored. Be that as it may, I do not feel compelled by the Scottish case to change the view I formed in *Jones v Bateman*. I think I should mention that in all three cases it seems to have been assumed that the meaning of the word "livestock" is a matter of law but, having regard to the speech of Lord Reid in *Brutus v Cozens* [1973] AC 854 at 861, that is open to doubt. There Lord Reid said:

> The meaning of an ordinary word of the English language is not a question of law. The proper construction of a statute is a question of law. If the context shows that a word is used in an unusual sense the court will determine in other words what that unusual sense is. But here there is in my opinion no question of the word "insulting" being used in any unusual sense. It appears to me, for reasons which I shall give later, to be intended to have its ordinary meaning. It is for the tribunal which decides the case to consider, not as law but as fact, whether in the whole circumstances the words of the statute do or do not as a matter of ordinary usage of the English language cover or apply to the facts which have been proved. If it is alleged that the tribunal has reached a wrong decision then there can be a question of law but only of a limited character. The question would normally be whether their decision was unreasonable in the sense that no tribunal acquainted with the ordinary use of language could reasonably reach that decision.
>
> Were it otherwise we should reach an impossible position. When considering the meaning of a word one often goes to a dictionary. There one finds other words set out. And if one wants to pursue the matter and find the meaning of those other words the dictionary will give the meaning of those other words in still further words which often include the word for whose meaning one is searching.

If, then, the word "livestock" can reasonably include fish, and Mr Fletcher conceded that it can, then no question of law is involved. In my judgment the ordinary meaning of the word in an agricultural context is something which is alive and is stocked for the purposes of providing food.

It follows from the foregoing that I confirm the decision of the local valuation court and dismiss the appeal.

The appellant will pay to the respondent its costs of this appeal, such costs if not agreed to be taxed by the Registrar of the Lands Tribunal on the High Court Scale.

Appendix

Statement of facts

1. *Assessment No:* ED 14265020.
2. *Hereditament:* Fish Farm and Premises, Low Plains, Castlerigg, Cumbria.
3. *Occupier:* Shearwater Fish Farming Ltd, Hammersmith House, London W6 9DX. Shearwater Fish Farming Ltd is a subsidiary of BOC Ltd.
4. *Owner:* BOC Ltd, Hammersmith House, London W6 9DX.
5. *Current assessment at date of originating proposal:* Weighbridge and Premises, Low Plains, Castlerigg. Net Annual Value, Rateable Value £100.
6. *Proposed Assessment:* Dated March 20 1975: Fish Farm and Premises, Low Plains, Castlerigg: gross value £4,600, rateable value £3.805.
7. *Objection:* made in letter dated April 7 1975. Grounds of objection ". . . the assessment appearing is illegal, incorrect, excessive and unfair."
8. *Local valuation court decision:* "Valuation officer's appeal dismissed. Entry to be deleted." *Reasons:* "After very careful consideration of all the evidence put before us we find this fish farm and premises to be an agricultural hereditament and exempt from rating. It is our opinion that under section 1(3) of the Rating Act 1971 the word 'livestock' can be interpreted to include fish. The appeal of the valuation officer is, therefore, dismissed. The entry now appearing in the valuation list will be deleted."
9. *Assessment:* agreed if it is held whole hereditament is rateable: gross value £2,850, rateable value £2,347.
10. *The hereditament:* The fish farm is situated 3 miles due south of the village of Armathwait and 7 miles due north of Penrith. It has a plentiful supply of unpolluted water drawn by way of boreholes from sandstone strata of rock at relatively shallow depths. The site has an area of 12½ acres or thereabouts and its height varies from about 400 ft to 450 ft above sea level, the higher part being along its eastern boundary. The layout of the site is shown on . . . drawing No 6007.

An internal road with roughly-metalled surface gives access to the interior of the site from a minor county road. At the relevant date there were on the site the buildings and structure etc, set out below; on the plan 6007 these are numbered to correspond with the item numbers.

(1) Wood and felt administration office, 20 ft by 12 ft.

(2) "Vic-Hallam Compact" office prefabricated mineral-finished timber with flat roof and timber floor, lined internally, 23 ft 9 in by 8 ft.

Both items (1) and (2) were near the entrance to the site.

(3) Pump house. Wood, felt-covered pitched roof, 8 ft by 6 ft.

(4) Pump house. Rendered breeze-block walls. CA roof. 6 ft by 5 ft.

(8) Willerby caravan occupied for night-manning security purposes, 29 ft 4 in by 9 ft 4 in.

(8a) "Vic-Hallam Compact" used as a laboratory, 23 ft 9 in by 8 ft 11 in.

(9) Sprite caravan used as a portable laboratory, 12 ft 7 in by 6 ft 4 in.

(10) WC, brick with CI roof.

(11) Mess room and store built of brick with CI roof, 11 ft 8 in by 19 ft.

(12) General-purpose building. Steel framed, brick construction with corrugated metal roof and internal concrete block partitions. It measures 29 ft 6 in by 23 ft 1 in and contains electrical installations.

(13) Manager's office, lean-to with solid walls rendered externally and corrugated metal covered roof, 12 ft by 9 ft.

(15) Generator shed built of brick and corrugated asbestos sheeting, 13 ft 4 in by 7 ft 6 in.

(16) Atcost building built of concrete and corrugated asbestos, 75 ft by 27 ft.

(16a) Preparation room built of concrete blocks, corrugated asbestos roofing and fitted with two stainless-steel sinks and worktop bench, 24 ft by 8 ft.

(18) Atcost building of similar construction and size to (16).

(19) Food store built of concrete blocks with felt-covered roof, 18 ft by 27 ft.

(21) Small switch house. Wood with felt-covered pitched roof.

(25) Boiler house built of brick and corrugated asbestos. Measures 11 ft by 8 ft and is fitted with a "Hoval" electric boiler.

Other items not in the nature of buildings comprise:

One large lake and three small lakes with a total capacity of 18,278 cu yds or thereabouts.

Three bore holes: 150 ft lined 12 in-50 ft; 220 ft lined 18 in-60 ft; 240 ft lined to 60 ft.

Three bore hole chambers.

Small tank-support walls.

Propane tank base and three concrete support walls.

A General concrete area 3,480 sq yds or thereabouts.
Header tank base 177 sq yds.
Settling pit—breeze-block lined and rendered, average depth 6 ft 6 in, capacity about 4,609 cu ft.
Settling pit with brick walls and outlet channel, average depth 4 ft 6 in, capacity approx 2,066 cu ft.
Standby " Dale " generator with a Rolls-Royce engine (147 kVA).

Items not rateable:
Two header tanks with intake and output piping.
20 growout tanks standing on concrete area. These tanks measure 25 ft in diameter and are constructed by bonding together 16 sections of expanded polystyrene. The joints between the sections are filled with mastic and the interiors of the tanks are covered with Pallidux,
B a waterproof paint. The tanks are kept in place on concrete bases by lengths of angle iron bolted to the concrete. Each tank has a capacity of 6,240 gals of water, the weight of which is sufficient to keep the tanks in place without other means of support. Each tank is protected from the weather by a polystyrene roof and marine plywood sides. A small amount of water in the tanks is recycled and each tank is equipped with electric light, meters to indicate the oxygen content of the water and a system to give alarm when the oxygen content or water level vary beyond certain tolerances. Each tank has attached to it a 1½ hp pump which powers the oxygenation equipment.
Propane gas tank of 12 tons capacity.
Vacuum-insulated evaporator tank containing liquid oxygen.

C *Processes carried out at the site*

The purpose of the operations on the site is the production of edible fish, mainly rainbow trout. Fish eggs are imported from California and other parts of the world in rotation according to the season, and when they are received at Low Plains they are placed in small circular tanks on racks in the first Atcost building known as the hatchery.

Infertile eggs are removed and the remainder are hatched out after 3-4 weeks. At this stage the young fish are known as alevins. After about two weeks or so the alevins are transferred to other circular D tanks about 5 ft in diameter, where they remain for about three months and develop into fry.

At the relevant date there were in the hatchery 38 fry tanks and racks of alevin tanks, three small tanks for hatching out salmon and six other fibre-glass tanks (5 ft in diameter, used for experiments with various types of sea fish). During the time the fry are in the tanks in the Atcost building they are fed manually on granulated fish protein. When large enough, the fry are transferred to the outside growout tanks described above, where they remain for about 12 months. At regular intervals they are graded for size and fish of similar size are kept in the same tank. Fish in the growout tanks are fed manually and when they weigh about 8 oz they are removed and prepared for sale. Sea-water fish on the site consist of sole and turbot. E

Water supply and oxygenation

Water used on the site is pumped from underground by three electric pumps to two header tanks, from which it is conveyed by PVC piping, channels etc, to various tanks. About 1,500 gals per hour of water pass through each large growout tank.

Effluent passes to settling pits. After the extraction of effluent the water passes through three polishing lakes and the large polishing lake from which it travels to a nearby stream. This process removes by natural biological action the high ammonia content which is found to be present in the water after having passed through the growout tanks.

The oxygenation of the water in all tanks ensures the most economic F use of the water.

Marketing

In March 1975 trout were being disposed of locally to hotels and shops in the area, but it was then anticipated that freezing plant would be installed on the site to enable fish to be sold in other parts of the country. The aim was to produce 100 tons of edible fish per annum.